Ocean Modeling and Parameterization

NATO Science Series

A Series presenting the results of activities sponsored by the NATO Science Committee. The Series is published by IOS Press and Kluwer Academic Publishers, in conjunction with the NATO Scientific Affairs Division.

General Sub-Series

A. Life Sciences	IOS Press
B. Physics	Kluwer Academic Publishers
C. Mathematical and Physical Sciences	Kluwer Academic Publishers
D. Behavioural and Social Sciences	Kluwer Academic Publishers
E. Applied Sciences	Kluwer Academic Publishers
F. Computer and Systems Sciences	IOS Press

Partnership Sub-Series

1. Disarmament Technologies	Kluwer Academic Publishers
2. Environmental Security	Kluwer Academic Publishers
3. High Technology	Kluwer Academic Publishers
4. Science and Technology Policy	IOS Press
5. Computer Networking	IOS Press

The Partnership Sub-Series incorporates activities undertaken in collaboration with NATO's Partners in the Euro-Atlantic Partnership Council – countries of the CIS and Central and Eastern Europe – in Priority Areas of concern to those countries.

NATO-PCO-DATA BASE

The NATO Science Series continues the series of books published formerly in the NATO ASI Series. An electronic index to the NATO ASI Series provides full bibliographical references (with keywords and/or abstracts) to more than 50000 contributions from international scientists published in all sections of the NATO ASI Series.
Access to the NATO-PCO-DATA BASE is possible via CD-ROM "NATO-PCO-DATA BASE" with user-friendly retrieval software in English, French and German (© WTV GmbH and DATAWARE Technologies Inc. 1989).

The CD-ROM of the NATO ASI Series can be ordered from: PCO, Overijse, Belgium.

Series C: Mathematical and Physical Sciences – Vol. 516

Ocean Modeling and Parameterization

edited by

Eric P. Chassignet

Department of Meteorology and Physical Oceanography,
Rosenstiel School of Marine and Atmospheric Science,
University of Miami,
Miami, FL, U.S.A.

and

Jacques Verron

Centre National de la Recherche Scientifique,
Laboratoire des Écoulements Géophysiques et Industriels,
Grenoble, France

Kluwer Academic Publishers

Dordrecht / Boston / London

Published in cooperation with NATO Scientific Affairs Division

Proceedings of the NATO Advanced Study Institute on
Ocean Modeling and Parameterization
Les Houches, France
January 20–30, 1998

A C.I.P. Catalogue record for this book is available from the Library of Congress.

ISBN 0-7923-5228-9

Published by Kluwer Academic Publishers,
P.O. Box 17, 3300 AA Dordrecht, The Netherlands.

Sold and distributed in North, Central and South America
by Kluwer Academic Publishers,
101 Philip Drive, Norwell, MA 02061, U.S.A.

In all other countries, sold and distributed
by Kluwer Academic Publishers,
P.O. Box 322, 3300 AH Dordrecht, The Netherlands.

Printed on acid-free paper

Printed in the Netherlands

TABLE OF CONTENTS

PREFACE

The realism of large scale numerical ocean models has improved dramatically in recent years, in part because modern computers permit a more faithful representation of the differential equations by their algebraic analogs. Equally significant, if not more so, has been the improved understanding of physical processes on space and time scales smaller than those that can be represented in such models. Today, some of the most challenging issues remaining in ocean modeling are associated with parameterizing the effects of these high-frequency, small-space scale processes. Accurate parameterizations are especially needed in long term integrations of coarse resolution ocean models that are designed to understand the ocean variability within the climate system on seasonal to decadal time scales.

Traditionally, parameterizations of subgrid-scale, high-frequency motions in ocean modeling have been based on simple formulations, such as the Reynolds decomposition with constant diffusivity values. Until recently, modelers were concerned with first order issues such as a correct representation of the basic features of the ocean circulation. As the numerical simulations become better and less dependent on the discretization choices, the focus is turning to the physics of the needed parameterizations and their numerical implementation. At the present time, the success of any large scale numerical simulation is directly dependent upon the choices that are made for the parameterization of various subgrid processes. These choices are often constrained by the overall model architecture and may be more or less natural to the particular model design. The question of this dependence has led to several comparison exercises between various models such as DYNAMO in Europe and DAMÉE in the U.S. A good understanding of parameterizations and their impact on the modeled oceanic circulations is crucial to the large international projects presently seeking to achieve global simulations.

In parallel, several large observational programs are underway to advance our understanding of the short and long time scales, such as observations from space (Topex/Poseidon, Geosat-Follow-On, ERS1, ERS2, ADEOS, ENVISAT, JASON, ...) or in situ (WOCE, CLIVAR, GOOS, GCOS, ...). These observations in turn are leading to improvements in available parameterizations. We therefore considered it timely to bring together leading scientists to summarize our present knowledege of those processes requiring parameterization in ocean models, to fully consider their optimal applications, and to introduce young scientists to the current state of the field.

We would like to thank all the speakers for providing a stimulating series of lectures, collected in this volume, and especially Jim McWilliams for his active support before, during, and after the school. We also express our appreciation to the members of the scientific committee (in particular David Anderson and Jürgen Willebrand), who contributed in numerous ways to the success of this Advanced Study Institute. We thank all the attendees for participating actively in the lecture review process and for creating a most cordial atmosphere. Finally, our thanks also go to Linda Smith, who spent a considerable amount of time proofreading the manuscripts.

Primary support for this Advanced Study Institute was provided by the North Atlantic Treaty Organization (NATO) and the European Community (EC/TMR). Additional funding was provided by the National Science Foundation (NSF), the Centre National d'Etudes Spatiales (CNES), and the Service Hydrographique et Océanographique de la Marine (SHOM). This support is gratefully acknowledged.

Eric P. Chassignet
Jacques Verron

April 1998

OCEANIC GENERAL CIRCULATION MODELS

JAMES C. MCWILLIAMS
Department of Atmospheric Sciences and
Institute of Geophysics and Planetary Physics,
UCLA, Los Angeles, CA, U.S.A,
and National Center for Atmospheric Research,
Boulder, CO, U.S.A.

1. Introduction

The practice of oceanic numerical modeling is growing rapidly. Among the reasons for this are the following: a widespread realization that model solutions can, either now or at least in the near future, be skillful in mimicking observed oceanic features; an understanding of the limitations of the alternative and more traditional scientific methodologies of making measurements in the oceans and developing analytic theories for highly nonlinear dynamical systems; an appreciation of the importance of the oceans in the socially compelling problems of anthropogenic changes in climate and the environment; and an exploitation of the steady increases in computing power that make meaningfully comprehensive oceanic calculations ever more feasible.

The *oceanic general circulation* is defined as the currents on horizontal space and time scales larger than the mesoscale (of order 100 km and 3 months), the associated pressure, density, temperature, and salinity fields, plus all other elements involved in establishing the dynamical balances for these fields. The latter includes the forcing fields, the domain geometry, and the transport contributions by currents on meso- and micro-scales. In some contexts the general circulation also includes the biogeochemical processes associated with other material property fields such as nitrate or freon. We will use the term *tracers* to denote the scalar variables moving with fluid parcels, including T and S.

These lectures give an overview of the formulation and solution behavior of *oceanic general circulation models* (OGCMs). The purpose is primarily pedagogic, so there will be less of the history of the subject than is custom-

E. P. Chassignet and J. Verron (eds.), Ocean Modeling and Parameterization, 1–44.

ary in review articles (*e.g.*, Haidvogel & Bryan, 1992; McWilliams, 1996). OGCMs are defined as numerical models that include all of the major influences on the general circulation, constrained by limitations in both our knowledge of how to formulate the model and the available computing power. Of course, this is a somewhat loose definition, but I take it to imply that OGCMs necessarily have wind, heat, and water forcing, geographically correct domain geometry, and the equation of state for seawater—all with their inescapable approximations.

Why are OGCMs useful? It is obvious that they would be so for the study of ocean currents, but they have many other important applications as well: dynamical coupling with the atmosphere, sea ice, and land run-off that in reality jointly determine the oceanic boundary fluxes (*e.g.*, El Niño); transport of biogeochemical materials; interpretation of the paleoclimate record; climate prediction for both natural variability and anthropogenic changes; assimilation of observations to provide dynamically consistent syntheses; pollution dispersal; and fisheries and other biospheric management.

The scientific practices of using OGCMs include a continuing development process to improve their accuracy, a testing process against observations to assess their accuracy, and conceptual inquiries about how different physical influences combine and compete to produce the general circulation. A particularly instructive way to use an OGCM, once it has captured some particular behavior of interest, is to systematically reduce its component processes to distill the essential cause. Particularly interesting uses of OGCMs are to explore hypothetical realities, such as the ocean circulation of ancient earth epochs, the consequences of deliberate biological fertilization of the oceans, or future climates under alternative planetary management strategies.

2. Dynamics

Broadly summarized, the historical path of OGCM usage is firstly to model only the largest scales of motion with an excessively coarse numerical grid and a concomitant excessively linear and diffusive dynamics, where the "diffusion" represents the effects of, or *parameterizes*, the unresolved, or *sub-grid-scale* (SGS), currents. Subsequently, more ambitious problems are posed that refine the grid; reduce the SGS diffusivities and thereby increase the nonlinearity of the dynamics; and examine both atmospherically forced fluctuations and intrinsic variability arising from instabilities of large-scale currents. The most energetic forms of the intrinsic variability are *mesoscale eddies* (and even larger-scale, lower-frequency fluctuations), generated by instability of the *wind-driven gyres*, and transient, three-dimensional *overturning cells*, generated by instability of the *thermohaline circulations*. An

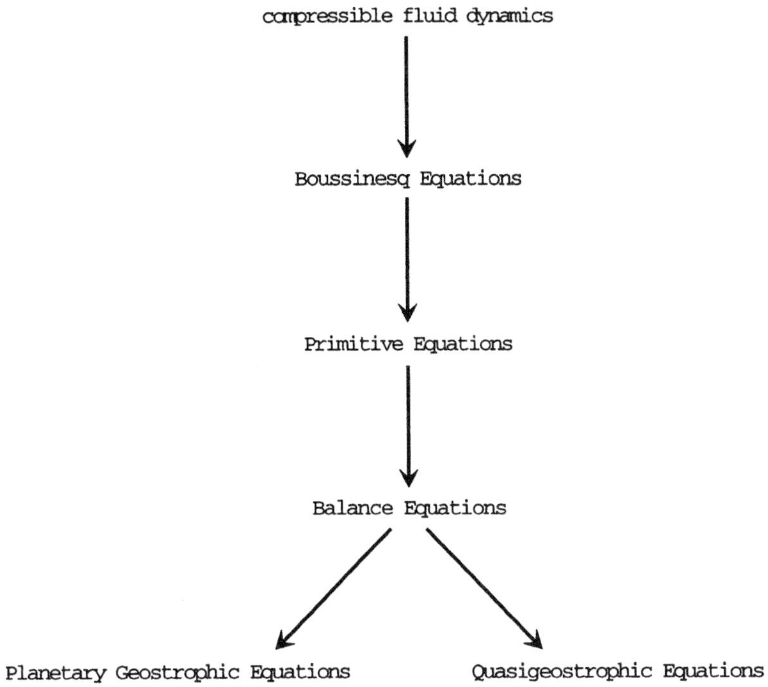

Figure 1. A hierarchy of dynamical approximations.

increasingly broad range of scales of motion is thereby incorporated into the OGCM solutions, thus decreasing the scope of the requisite SGS parameterizations. Nevertheless, it is inconceivable that a numerical model could incorporate all excited degrees of freedom (of $\mathcal{O}(10^{40})$, if 1 mm and 1 s are taken as the minimum scales, and the whole ocean and 10^4 yr are taken as the maximum scales), so SGS parameterizations will always be essential elements of an OGCM. The present frontiers for OGCM solutions lie either in fluctuations of decadal or longer periods and their relation to the atmosphere or in the turbulent behavior of mesoscale currents and eddies.

The fundamental fluid dynamics of oceanic circulation is the Navier-Stokes equations for the rotating earth and a compressible liquid, seawater, comprised of water plus a suite of dissolved salts that occur in nearly constant ratio but variable amount (the salinity S), with an empirically determined equation of state. However, since the target of OGCMs is large-scale, low-frequency currents, we can choose among several levels of dynamical approximation (see Fig. 1):

* The *Boussinesq Equations* neglect variations of density ρ everywhere in the momentum and mass balances except for the gravitational force, while retaining the full effects of compressibility in the equation of state. This excludes acoustic modes. This is a safe approximation for ocean currents since $\delta\rho/\rho \ll 1$.

* The *Primitive Equations* further make the *hydrostatic approximation* in the vertical momentum balance:

$$\frac{\partial P}{\partial z} = -g\rho, \tag{1}$$

where P is the pressure, ρ is the density, z is the local vertical coordinate (parallel to the gravitational force), and g is the gravitational acceleration. This excludes convection, which requires vertical accelerations, and distorts high-frequency gravity waves. This is a safe approximation as long as $H/L \ll 1$, where H and L are vertical and horizontal length scales characteristic of the solutions; this is true for all OGCM solutions to date (but just barely so). The same condition of small aspect ratio justifies the neglect of the earth's rotation vector $\mathbf{\Omega}_e$ except for its local vertical projection, *i.e.*, the *Coriolis frequency*, $f = 2\mathbf{\Omega}_e \cdot \hat{\mathbf{z}}$, where $\hat{\mathbf{z}}$ denotes the local unit vertical vector.

* The *Balance Equations* further make approximations to the horizontal momentum equations that are consistent with *gradient-wind balance*,

$$\frac{1}{\rho_o}\nabla_h^2 P = \hat{\mathbf{z}} \cdot \mathbf{\nabla}_h \times f\mathbf{u}_h + 2\frac{\partial(u,v)}{\partial(x,y)}, \tag{2}$$

where ρ_o is the mean density, boldface denotes a vector, the subscript h denotes horizontal component, $\mathbf{u}_h = (u,v)$, and $\mathbf{x}_h = (x,y)$. This is an approximate balance of force divergences among the pressure gradient, Coriolis, and centrifugal forces that is usually a quite accurate approximation for general circulation currents (Gent & McWilliams, 1983). This model excludes gravity and inertial waves. There are a variety of particular formulations of balanced models, but among those whose balance relation asymptotically coincides with (2) for small *Ro*, the differences may not be quantitatively significant.

* The *Quasigeostrophic* and *Planetary Geostrophic Equations* are leading-order asymptotic approximations for small Rossby number, $Ro = V/fL \ll 1$ (where V is a characteristic horizontal

velocity). They satisfy *geostrophic balance*,

$$\frac{1}{\rho_o}\nabla_h P = -\hat{\mathbf{z}} \times f\mathbf{u}_h. \tag{3}$$

The two approximations differ by whether L/R_d is assumed to be order one or large, respectively, where $R_d \sim NH/f$ is the *baroclinic deformation radius* and N is the *buoyancy frequency* for the vertical density stratification; the size of R_d is typically in the range 10-100 km, with smaller values at higher latitudes. Each of these models is a subset of the Balance Equations. They suffer from having non-uniform validity in L/R_d and Ro; from inaccuracy of eq. (3) near the equator where $f = 0$; and, for the Quasi-geostrophic model, from the neglect of horizontal and temporal variations in $N(z)$. Nevertheless, these two maximally simplified models are accurate for many particular phenomena in the general circulation and are thus useful simple models for the theory of ocean currents, hence for the interpretation of OGCM solutions.

When faced with this range of approximations, the early general circulation modelers chose the Primitive Equations because they were confident of the underlying assumptions and the necessary numerical solution methods were simpler than those required for the Boussinesq or Balance Equations. From a modern dynamical perspective, the alternative choices of either the Boussinesq or Balance Equations could be defended, but their respective advantages—allowing solutions with H/L not small or having both an intrinsically smoother evolutionary rate (hence permitting a large step size, Δt) and a reduced dynamical phase space within which to interpret the solution—are unlikely to be enormous in most OGCM applications. Another basis for distinguishing among these three models has been their permissible time step size under the constraint of computational stability for explicit time integration methods: the permissible step size is larger the greater the degree of physical approximation, because of the implied exclusion of faster modes of behavior. However, this distinction would disappear if unconditionally stable implicit time integration methods were used.

Thus, the common basis for an OGCM is the Primitive Equations for a thin fluid layer in spherical coordinates (*i.e.*, replacing the radial coordinate $r = z + a$ by the mean radius of the earth a wherever it appears as an algebraic factor):

$$\frac{Du}{Dt} - \left(f + \frac{u\tan\phi}{a}\right)v = -\frac{1}{a\cos\phi}\frac{\partial P}{\partial\lambda} + SGS$$
$$\frac{Dv}{Dt} + \left(f + \frac{u\tan\phi}{a}\right)u = -\frac{1}{a}\frac{\partial P}{\partial\phi} + SGS$$

$$
\begin{aligned}
\frac{\partial P}{\partial z} + \frac{g\rho}{\rho_0} &= 0 \\
\frac{1}{a\cos\phi}\left(\frac{\partial u}{\partial \lambda} + \frac{\partial v\cos\phi}{\partial \phi}\right) + \frac{\partial w}{\partial z} &= 0 \\
\frac{D(T,S)}{Dt} &= SGS \\
\rho &= \rho[T,S,P]. \qquad (4)
\end{aligned}
$$

Here (λ, ϕ, z) are longitude, latitude, and height; (u, v, w) are the associated velocities; T is the potential temperature (*i.e.*, invariant under adiabatic compression); S is the salinity; and D/Dt is the substantial time derivative,

$$
\frac{D}{Dt} = \frac{\partial}{\partial t} + \frac{u}{a\cos\phi}\frac{\partial}{\partial \lambda} + \frac{v}{a}\frac{\partial}{\partial \phi} + w\frac{\partial}{\partial z}. \qquad (5)
$$

Where SGS appears in eq. (4), there are important, small-scale, non-conservative processes; these provide spatial transports and pathways to the dissipation that actually occurs by molecular kinetic processes on scales of $\mathcal{O}(1)$ mm. Implicitly these equations are the result of a low-pass filter at the space-time resolution of the numerical model, with SGS denoting the contributions from the unresolved scales. These equations can be augmented by additional tracer equations similar to those for (T, S), adding whichever chemical reactions are required.

The boundary conditions for an OGCM are comprised of the appropriate kinematic conditions on the normal component of velocity and flux rules for the momentum and tracers. These conditions are particularly simple at solid boundaries, where the normal velocity is zero and tracer fluxes are usually neglected. A bottom stress law is also required there, but this is part of the SGS parameterization (Section 7.5). Since OGCM domains usually have sides of finite depth, unlike the ocean shoreline in most places, as part of their spatial discretization (Section 5); the usual practice is to set the horizontal tangent velocity to zero there (*i.e.*, *no-slip*), although alternative choices of *free-slip* (no stress) and *partial-slip* (stress proportional to slip) are sometimes used. These artificial side-wall conditions exert significant influence on OGCM solutions with fine resolution and strong boundary currents, and thus are a quite problematic aspect of the model formulation.

At the upper free surface, $z = \eta(\lambda, \phi, t)$, the kinematic condition is $w = D\eta/Dt$. Current practice is divided between the full use of this condition and the *rigid-lid* approximation, where $w = 0$ at $z = 0$.[1] This approximation is based on $\eta/H \ll 1$ and an assumption of slow evolution

[1]There is an associated diagnostic relation in this approximation, where the implied sea level is calculated hydrostatically from the model's surface pressure, $\eta = P(\lambda, \phi, 0, t)/g\rho_o$.

compared with *surface gravity waves* (with a long-wave speed of $\sqrt{gH} \sim$ 200 m s^{-1} with $H/L \ll 1$) and long *barotropic Rossby waves* (with speed $\beta gH/f^2 \sim 100$ m s^{-1}, where $\beta = 1/a \; \partial f/\partial\phi$). This is an accurate approximation for most types of currents, excluding the tides and surface gravity waves and distorting modestly the response to synoptic weather events, and thus this choice is often made for computational convenience or efficiency. The surface momentum and tracer flux conditions are discussed below.

When the domain is less than global or full depth, *open boundary conditions* are required at an internal fluid boundary. There is no fundamentally correct basis for specifying such conditions. However, various rules have been devised that sometimes suffice for the target phenomena of the calculation (*e.g.*, Barnier *et al.*, 1998). These often include a specified inflow (if any), outward wave radiation, restoring tendencies towards climatological values of the tracers, and enhanced damping in the neighborhood of the internal boundary (a.k.a., a sponge layer). This is another problematic aspect of OGCM formulation.

3. Forcing

The primary forcing of a full-depth, global OGCM is by surface fluxes of momentum (stress), T (heat), S (water), and other material properties, while side (*e.g.*, rivers) and bottom tracer fluxes may provide secondary forcing but usually are neglected.

The surface stress is due to the drag by the overlying wind. It is calculated from an empirical wind climatology using bulk regression formulas for stress. This climatology is readily available in several forms, and it is now being systematically improved through satellite wind observations and climatological *reanalysis* projects at operational weather forecast centers (*e.g.*, Kalnay *et al.*, 1996). In polar regions the stress transmission may be mediated by sea ice. Sea ice changes the drag coefficient in the bulk regression formula for surface stress, and it can inhibit stress transmission to the ocean if ice jams develop. An ice model may be needed to adequately incorporate these effects.

The heat and water fluxes are more problematic, since there is no comparably good climatology for them. Bulk regression formulas can be used together with atmospheric surface climatologies for some locally determined components (*e.g.*, sensible and latent heat, evaporation), but other components are non-locally determined (*e.g.*, precipitation and radiation). Again the presence of sea ice modifies the fluxes, both by blocking air-sea material exchanges and through freezing and melting. The historically most common practice has been to replace the uncertain flux boundary conditions with *restoring* terms of the form $1/\tau \cdot (T_{clim} - T)$ in the temperature tendency equation for the uppermost model grid level (ditto for S), where T_{clim} is the observed sea surface temperature and τ is a specified relaxation time (usually on the order of many days); this term can then be diagnostically interpreted as a surface heat flux divided by the grid thickness Δz. This has the seeming virtue of giving model solutions whose T and S fields are close to the observations (but note that the agreement cannot be exact or the implied surface flux would be zero). However, OGCM experience shows that the implied fluxes obtained by this method are not physically plausible, because of too much small-scale variation and probably even some large-scale bias; these defects are especially severe in S. The physical error in using restoring conditions is that they imply excessively strong local *atmospheric feedbacks*, so that any tendency of the ocean to depart from T_{clim} is met with an atmospheric response that supplies compensating fluxes. On the other hand, choosing the opposite extreme, *specified flux*, has the dual problems of uncertainty in what to specify and the implied absence of any feedback. The lack of feedback allows the ocean solution to drift far away from climatology due to errors in the model and the fluxes.

The reality of atmospheric feedbacks is in between these extremes. The flux components that are locally determined do have a greater degree of negative feedback than do the non-locally determined ones. Thus the approach that is preferable to either restoring or specified flux is a mixture of specifying certain components and calculating others with negative feedback forms, using an atmospheric climatology from reanalyses as the underlying data base (*e.g.*, Barnier *et al.*, 1995; Large *et al.*, 1997). In most aspects the accuracy in the climatology of atmospheric state variables (*e.g.*, surface air temperature and cloudiness) is better than for the air-sea fluxes; thus, the former may provide a better empirical basis for specifying surface fluxes than the latter. This approach is a complicated one with many arguable steps, but its resulting OGCM solutions have been found to be better than with the alternatives. Ultimately, of course, the fluxes should be self-consistently determined with sea-ice and atmospheric general circulation models.

4. Initial Conditions and Equilibrium

The state of the oceanic general circulation is not observed in anywhere near the detail that is required for a complete initialization of an OGCM, nor is it likely to be anytime soon. A better global observing system would, however, greatly improve the quality of approximate initial conditions. I believe there is also some utility in retrospective spin-ups using the atmospheric climatology for forcing and data-assimilation methods to provide oceanic constraints; however, this has not yet been done in a substantial fashion, and there are interesting open questions about in what ways the ocean is sufficiently *predictable* for this approach to yield a unique answer. Some aspects of OGCM solutions, such as wind-driven Rossby waves and upper-ocean thermal fluctuations, do seem to be largely predictable from surface fluxes, but intrinsic variability modes are likely to be much less so.

Typical OGCM initial conditions are climatological T and S fields (*e.g.*, as in Levitus *et al.*, 1994) and zero motion. From such a state there is a *geostrophic adjustment* to the tracer fields within days, followed by *boundary*, *Kelvin*, and *Rossby wave* adjustments to the wind forcing within a few years, that leave behind currents that at least grossly resemble the currents present in the long-time equilibrium state. The true equilibrium state occurs only after thousands of years when advection and SGS transport have redistributed the T and S fields consistent with the OGCM problem as posed (see Suginohara & Fukasawa, 1988, and Danabasoglu *et al.*, 1996, for analyses of spin-up). Unless approximately correct tracer fields are given as initial conditions, an OGCM solution will differ greatly from its equilibrium state after an integration of only several years. Scientists who wish to avoid the costly computations to reach full equilibrium nevertheless bear the burden of demonstrating that shorter integration times do not excessively bias their solutions. At present it is computationally infeasible to integrate a global OGCM to equilibrium with mesoscale resolution.

Another obvious approach to initialization is by *bootstrapping* (*i.e.*, using one equilibrium OGCM solution as an initial condition for a differently posed problem). We might expect this method would be helpful in shortening the approach to equilibrium if the family of solutions were all sufficiently close to each other; however, the common experience to date is that this method rarely is cheaper than restarting from a stratified state of rest when achieving full equilibrium is the goal.

5. Numerical Methods

The computational algorithms that have been used for OGCMs have mostly been rather simple ones, as presented in Bryan (1969) and by others with minor variations. They are a finite-difference discretization of eqs. (4)-(5), using centered, nearest neighbor differences that are second-order accurate in the grid spacing for uniform grids (the usual choice in (λ, ϕ)) and formally first-order for the (usually weakly) non-uniform vertical grids with finer resolution in the more stably stratified upper ocean. The grids are staggered in their distribution of the dependent variables, using one of several alternative schemes, with the B- and C-schemes most commonly used. The spatial difference operators are integrally conservative for tracer content and variance and for kinetic energy. Near the poles extra smoothing is required if the grid spacing becomes very small, as it does on a uniform longitude-latitude grid.

The time-stepping is by a mixture of first- and second-order accurate procedures, with time-splitting often used to solve separately for the SGS vertical mixing by an implicit algorithm that is computationally stable for arbitrarily large SGS vertical diffusivity and for everything else by an explicit algorithm. If the full free surface kinematic condition is used, then it too is handled by time splitting. Thus, the time step size Δt is limited by CFL stability conditions for advection, internal gravity and barotropic Rossby wave propagation, and SGS lateral transport. Over a rather wide range of spatial grid size, $\Delta t = \mathcal{O}(1)$ hr; thus, integrating for $\mathcal{O}(10^3)$ yr to approach equilibrium requires $\mathcal{O}(10^7)$ time steps or more.

The coarsest spatial grids used in global OGCMs have $\mathcal{O}(10^5)$ grid points, corresponding to a horizontal spacing of hundreds of kilometers and a vertical spacing of hundreds of meters. To refine the grid resolution to be adequate for mesoscale eddies, with horizontal spacing finer than the internal deformation radius, R_d, *i.e.*, of $\mathcal{O}(10)$ km (see Section 10), increases the size of a global grid to at least $\mathcal{O}(10^8)$ points. Thus, an OGCM calculation is indeed a very large one. At present it is infeasible to combine full mesoscale resolution with full equilibrium in a single global calculation, although it likely will become so within the next decade with faster computers. Since this computer power probably will be achieved only with substantial parallelism, there is a considerable incentive to creating suitably structured OGCM codes (*e.g.*, Dukowicz *et al.*, 1993; Bleck *et al.*, 1995; Marshall *et al.*, 1997).

A shortcut for reaching equilibrium can often be taken for coarse-resolution OGCMs by *acceleration*. This is a time-stepping technique in eq. (4) with a larger Δt for (T, S) than for (u, v), and a further increase in Δt for the deep tracers (Bryan, 1984). The rationale is that the CFL constraint is usually most severe on (u, v) through the SGS horizontal momentum diffusion and the fastest waves, whereas the constraint for tracers is set by their slower SGS lateral transport and advection (which decrease with depth). Since deep tracers are the slowest solution components to approach their equilibrium values, this acceleration technique can reduce the integration time by a factor of $\mathcal{O}(100)$. If the OGCM approaches a steady solution under steady forcing, then this distortion of the transient dynamics causes only a transient error. Danabasoglu *et al.* (1996) shows that, even with periodic seasonal forcing, acceleration leads to a valid equilibrium solution if it is followed by a *synchronous* integration, with uniform Δt, for $\mathcal{O}(10)$ yr to allow the seasonal cycle to equilibrate. No doubt this particular technique has a limited range of validity in model resolution and solution transience, but more robust acceleration methods are well worth seeking because of the high computational cost of calculating OGCM equilibria.

Since far more sophisticated numerical methods than those described above are now well established in other computational contexts, a spate of numerical developments for OGCMs is likely to occur soon. Some of the currently interesting areas of development are higher order spatial and temporal discretizations (*e.g.*, Haidvogel *et al.*, 1995); advection operators not so prone to creating erroneous extrema (*e.g.*, Leonard, 1979; Shchepetkin & McWilliams, 1997); fuller temporal implicitness to escape physically unimportant computational stability limits (*e.g.*, Yavneh & McWilliams, 1995); more efficient procedures for solving for a reference-level $P(\lambda, \psi, t)$ in the vertical integration of the hydrostatic relation, via the barotropic streamfunction, rigid-lid surface pressure (Pinardi *et al.*, 1995), or free-surface elevation (*e.g.*, Dukowicz & Smith, 1994); and local grid refinement or nesting to allow a regional focus with less physical uncertainty and numerical error in the open-ocean side boundary conditions.

The governing equations (4)-(5) are physically equivalent under any coordinate transformation; however, uniform grids in different coordinate systems imply different discretizations of the equations. One useful application is to achieve a more uniform tiling of the sphere in the horizontal coordinates than does a regular latitude-longitude grid, most importantly near the north pole where $\Delta\lambda \to 0$ (*e.g.*, Ebby & Holloway, 1994a; Smith *et al.*, 1995). Another potentially useful application is through transformation of the vertical height coordinate z to either a *potential-density* (*e.g.*, Bleck & Chassignet, 1994; Lunkeit *et al.*, 1996) or a *topography-following* (*i.e.*, σ) coordinate (*e.g.*, Ezer & Mellor, 1994; Song & Haidvogel, 1994; Barnier *et al.*, 1998). Advantages of a potential-density coordinate are a natural concentration of resolution in the pycnocline where vertical gradients are often largest; the simplification of the conservative nonlinear terms in eqs. (4)-(5) to involve only the horizontal velocity; and a relative ease in assuring integral conservation properties for the discretized, isopycnally oriented SGS operator forms (although this has also been achieved satisfactorily with other coordinates; see Section 7.2). Disadvantages are an uncontrollable sparseness of resolution in weakly stratified regions, such as in boundary layers and convective sites; errors in implementing the equation of state associated with its nonlinear form of compressibility, such that globally defined density or potential density surfaces are not uniformly accurate in representing the static stability (McDougall, 1987); and a greater complexity in specifying the discrete diapycnal SGS operator forms. Advantages of a topography-following coordinate are a more accurate representation of the bottom kinematic boundary condition and bottom boundary layer SGS mixing. A well known disadvantage is the error in the horizontal pressure gradient force where the topography is especially steep (though there may be similar, albeit less obvious, errors in other coordinate systems), and another potential disadvantage is the vanishing grid spacing as the depth becomes very shallow at the ocean margins, with a possibly severe constraint on the time step size for computational stability. In my opinion no vertical coordinate choice is clearly the superior one; probably, the more important choices are for the discrete operators that arise within a given coordinate system. The recent DYNAMO comparison among OGCM solutions with different vertical coordinates finds mixed advantages (DYNAMO, 1997). Arakawa (1988) gives an interesting perspective on these different vertical coordinates and discretizations in atmospheric GCMs.

6. Domain Geometry

The shape of an ocean basin is defined by its bottom topography, including its intersection with the top surface, the coastline. Topography has a profound influence on the direction of currents, especially near the bottom. The reason for this is the approximate material conservation of potential vorticity, whereby a change in the thickness h between a deep interior isopycnal surface and the bottom implies a compensating change in the fluid vorticity that acts to turn the fluid trajectory towards a path of constant f/h.

It is far less certain, however, when and where topography fundamentally controls the existence and strength of the large-scale currents. The latter are largely accelerated by internal pressure gradient forces, which are closely related to (T, S) distributions through hydrostatic balance and the equation of state. Momentum balance against this acceleration can be provided either internally by large-scale advection and SGS transport, or by topographic effects such as *topographic form stress* (*i.e.*, integrated horizontal pressure force on the bottom), *boundary-layer drag*, and *hydraulic control* (*i.e.*, throughflow limited by a critical Froude number, where the velocity equals the gravity-wave speed)—mediated, of course, by Coriolis force for approximately geostrophic dynamics, as in eq. (3). This competition can be posed conceptually as a comparison of model solutions with simpler, smoother topography and ones with rougher, more complex topography. While the desire for geographical realism favors the latter, achievement of computational accuracy favors the former (*i.e.*, accurate numerical solutions require smoothness near the grid scale), and the choice of numerical algorithms for topography may be quite important in determining the balance point between these competing goals. There is no satisfactory resolution of these issues at present, and the common practice is one of trial and error. Increasing the grid resolution allows the incorporation of finer topographic features, hence diminishes the scope of the ambiguity in specifying topographic smoothness; however, the true topography in the oceans remains rough on all conceivable OGCM scales.

Some aspects of topographic influence must be represented in the SGS parameterizations: the bottom boundary layer, coastal shoaling, small-scale roughness, and, probably, hydraulic control (I am not aware of any OGCM solutions that achieve this state in their resolved dynamics, although the governing equations admit the possibility). However, form stress and impedance by constricting straits are at least partly among the resolved dynamical processes.

Topographic form stress is of central importance in maintaining the depth-integrated transport of the Antarctic Circumpolar Current against the surface wind stress. This was first suggested by Munk & Palmen (1951), and it has since been demonstrated in many OGCM solutions (*e.g.*, Ivchenko & Stevens, 1995). Treguier & McWilliams (1990) found that most of the form stress in this current is associated with topographic features of large spatial scale, and this, of course, is favorable to its modelability. The role form stress plays in other large-scale currents is as yet less certain, but it probably is important near mid-latitude western boundary currents (*e.g.*, Hurlburt *et al.*, 1995; Sakamoto & Yamagata, 1996; Kagimoto & Yamagata, 1997).

Constricting straits can limit throughflow, either by boundary stress or hydraulic control (*e.g.*, Pratt, 1990). In OGCM solutions, a current through a strait whose width is too close to the grid scale will have a primarily viscous dynamics; a small width probably exaggerates the boundary stress and precludes the inertially controlled dynamics of hydraulic control. This presents a modeler with the temptation to artificially widen or deepen the channel to compensate for these biases, but this is an essentially arbitrary choice. The two most important straits in the global ocean are the Drake Passage and the Indonesian Archepelago, both of which have large throughflows; demonstrations of the extreme consequences of blocking them entirely are in Ishikawa *et al.* (1994) and Hirst & Godfrey (1993), respectively. Coarsely resolved OGCM solutions have about the same transport as is observed when these straits are artificially widened (*e.g.*, Danabasoglu & McWilliams, 1995). Straits occur in all sizes, of course, and some will remain part of the SGS specification at all foreseeable resolutions. Probably the most important of these is the Strait of Gibraltar, whose outflow of salty Mediterranean seawater has significance in the S budget of at least the North Atlantic Ocean (see Fig. 17); Armi & Farmer (1988) argue for its hydraulic control. There are analogous difficulties with flow over sills and ridges whose size is too close to the grid scale; a notorious example is the flow of deep seawater from the Greenland Sea into the Atlantic across the Icelandic Ridge (*e.g.*, Roberts *et al.*, 1995).

7. Parameterizations

The art of parameterization is to augment the resolved dynamics in eq. (4) with mathematical operators that accomplish the necessary physical effects by unresolved SGS processes, all of which involve turbulence and thus do not have complete analytic theories to make use of. I believe it is impossible to make a parameterization rule that is entirely correct, in the sense that it will yield the same answer as in a model that fully resolves the process. Therefore, a hypothesis must be declared about what the minimum necessary effects are. Then, after finding a suitable operator that satisfies this hypothesis, any free parameters are chosen to either achieve some quantitative effect in an OGCM solution or match independent data about the process—both, if possible. The most common parameterization hypothesis about turbulent processes is that they mix material properties, hence the most common operator form is *eddy diffusion* (*e.g.*, by spatial Laplacians) with an *eddy diffusivity* as the free parameter.

We should be humble about of our ability to represent turbulent processes. So, a guideline for choosing a good parameterization is to keep it as simple as possible, with as few free parameters as possible, consistent with achieving both the hypothesized effect and a significant impact on the solution. Nevertheless, for an OGCM there are many needed effects associated with many different SGS processes, hence many different parameterization forms are involved.

Material transport in the oceans is strongly constrained by the iso-surfaces of entropy, which are locally tangent to iso-density surfaces (Fig. 2). (Momentum transport is less constrained because of the pressure-gradient force.) In a stably stratified region, gravitational work is required to move matter across surfaces of neutral buoyancy (*i.e.*, *isopycnal surfaces*); thus transport along these surfaces is much more efficient than across them, with an observed ratio of cross-isopycnal (*i.e.*, *diapycnal*) to isopycnal eddy diffusivities of about 10^{-8} in the oceanic thermocline. The principal agents of turbulent isopycnal tracer flux are mesoscale eddies, which crinkle the basin-scale isopycnal surfaces without disrupting their stable vertical ordering. The agents of diapycnal tracer flux are micro-scale motions that can cause overturning and fragment the surfaces. Within and near the surface and bottom boundary layers, where the density stratification is weak, this constraint is lifted and micro-scale transports are much more efficient than in the interior.

7.1. LATERAL MOMENTUM TRANSPORT

For the general circulation, the dominant mechanism for lateral momentum transport is by mesoscale eddies acting through the horizontal Reynolds

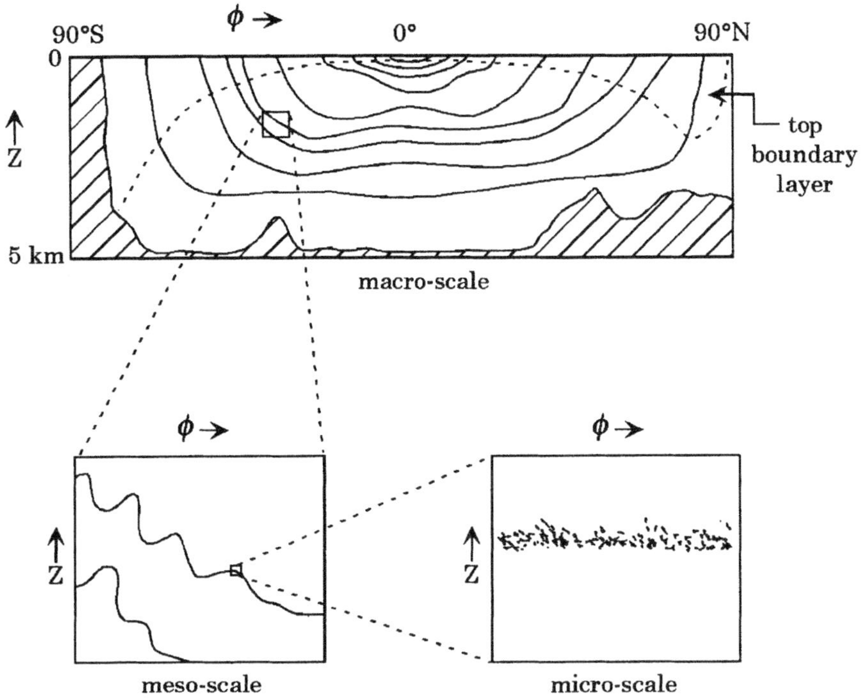

Figure 2. Macro-, meso-, and micro-scale views of oceanic isopycnal surfaces in a meridional plane. Material fluxes across these surfaces are much weaker than along them, except in the surface and bottom boundary layers and perhaps above regions of rough topography as well.

stress, $\overline{\mathbf{u}'_h \mathbf{u}'_h}$. This process can either be wholly parameterized or be partially resolved in an OGCM. The common parameterization is horizontal eddy diffusion, with an eddy viscosity $\nu_h \geq 0$, although hyperdiffusion (*i.e.*, an iterated Laplacian operator) is sometimes used to strengthen a marginally resolved instability process such as Gulf Stream meandering. The value of ν_h depends upon the grid size: it must be large enough to resolve the viscous boundary layers of width $\sim (\nu_h/\beta)^{1/3}$ and to suppress nonlinear computational instability on the grid scale (*i.e.*, a grid Reynolds number, $Re = V\Delta s/\nu_h$ cannot be large). Fundamental symmetry principles about the nature of a stress divergence imply that the operator must have additional terms beyond a simple Laplacian operator in a spherical domain (*e.g.*, Wajsowicz, 1993). A similar consideration applies to spatially non-uniform or anisotropic eddy viscosities that are sometimes used when the grid spacing has the same attributes.

There is a well-known counter-example for this parameterization—this prompted Victor Starr to talk about "negative eddy viscosity"—viz., counter-gradient eddy momentum flux in the core of a broad (*i.e.*, with $L/R_d \gg 1$), baroclinicly unstable current, such as the Antarctic Circumpolar Current (*e.g.*, McWilliams & Chow, 1981). This has led some to suggest replacing lateral eddy viscosity with potential vorticity diffusion (see Section 7.2), although this has not yet been implemented in an OGCM as far as I know. Furthermore, since the momentum flux need not be related to the local large-scale flow, as with Rossby wave propagation through a region, an eddy diffusion form may not always be valid. Analyses of eddy-resolving OGCMs indicate the resolved horizontal Reynolds stress divergence patterns are locally much different from the eddy viscosity parameterization, especially near mid-latitude western boundary currents where they are largest. With fine enough grid resolution and small enough ν_h in an OGCM, however, these confounding behaviors can become part of the resolved dynamics, and an eddy viscosity parameterization may suffice.

7.2. ISOPYCNAL MATERIAL TRANSPORT

The traditional parameterization for lateral tracer transport is horizontal eddy diffusion, but this violates the constraint of tracers mostly staying on isopycnal surfaces, with $\overline{\mathbf{u}c'}$ nearly perpendicular to $\nabla\overline{\rho}$ (here c is any tracer concentration). A parameterization by Gent & McWilliams (1990) that is isopycnally oriented and *integrally adiabatic* (*i.e.*, without interior sources or sinks of any material property that alters its inventory on isopycnal surfaces) has had quite beneficial effects on OGCM tracer distributions and fluxes (Danabasoglu *et al.*, 1994; Boening *et al.*, 1995). In addition to tracer diffusion along isopycnals (Redi, 1982), it has an incompressible *eddy-induced transport velocity*, whose horizontal component is

$$\mathbf{u}_h^* \equiv \left(\frac{\partial \overline{z}}{\partial \rho}\right)^{-1} \overline{\frac{\partial z'}{\partial \rho}\mathbf{u}_h'}, \tag{6}$$

where z is height of an isopycnal surface and the overbar is an average over the mesoscale on that surface. $\mathbf{u}^*$ combines with the large-scale $\mathbf{u}$ as the large-scale Lagrangian velocity that advects the large-scale tracers (Gent *et al.*, 1995), including the isopycnal surface itself, thereby causing vertical transport of momentum, as in *isopycnal form stress*, and depletion of available potential energy, as in *baroclinic instability*. In present OGCM implementations, both $\mathbf{u}^*$ and the isopycnal mixing are represented as eddy diffusion of isopycnal layer thickness and tracers. The form of the eddy-induced velocity is

$$\mathbf{u}_h^* = -\frac{\partial}{\partial z}\left[\kappa_i \mathbf{L}_h\right], \quad w^* = \nabla_h \cdot \left[\kappa_i \mathbf{L}_h\right] \tag{7}$$

where $\mathbf{L}_h = -\nabla_h \rho \big/ (\partial \rho / \partial z)$ is the slope of the isopycnal surface, and the eddy diffusivity, $\kappa_i \sim 10^3 \text{ m}^2 \text{ s}^{-1}$, can be identified with the rate of dispersion of floats in the ocean. Thus far, κ_i has been used in coarse resolution OGCMs with essentially no grid-size dependence in its value; however, even in fine resolution calculations that partially resolve the mesoscale eddies, there is benefit to using this parameterization with a smaller κ_i, if only to avoid false diapycnal flux (Roberts & Marshall, 1997). Visbeck *et al.* (1997) and Killworth (1998) suggest flow-dependent rules for κ_i based on linear baroclinic instability theory.

If one assumes a simplified $\rho[T, S, P]$ and neglects SGS terms, then the hydrostatic form of *Ertel potential vorticity, i.e.*,

$$q = \left(f(\phi) + \frac{1}{a \cos\phi} \left[\frac{\partial v}{\partial \lambda} - \frac{\partial (u \cos\phi)}{\partial \phi} \right] \right) \frac{\partial \rho}{\partial z} - \frac{1}{a \cos\phi} \frac{\partial u}{\partial z} \frac{\partial \rho}{\partial \lambda} + \frac{1}{a} \frac{\partial v}{\partial z} \frac{\partial \rho}{\partial \phi}, \tag{8}$$

is conserved along parcel trajectories, which lie on isopycnal surfaces, in eq. (4). (This is also true in the other models in Fig. 1 with alternative definitions for q.) If they are strong enough, mesoscale eddies can mix, or *homogenize*, large-scale q (Marshall, 1981; McWilliams & Chow, 1981; Rhines & Young, 1982), although they do not always do so: a large-scale approximation to q is $f \rho_z$, and there are some regions in the ocean where this latter quantity is observed to be nearly uniform along some isopycnal surfaces. Also, at least in zonally symmetric channel flows, eddy potential vorticity fluxes may be more uniformly down-gradient than eddy thickness fluxes (Treguier *et al.*, 1996). Thus, *isopycnal potential vorticity mixing* is a possible parameterization hypothesis. It would be a composite of the SGS terms in the momentum, T, and S equations in (4), and as such it must be reconciled with them in a way that does not involve double counting of processes. An eddy diffusion of potential vorticity should avoid the fallacy of generating a large-scale flow from a resting stratified state[2], and this suggests that the diffusivity has to be flow dependent (or be replaced by a different operator form; see Sommeria, 1998). Finally, the usual methods for solving eqs. (4)-(5) do not involve an equation for q explicitly, so a different method is required. There are thus significant challenges to be overcome in implementing a q-mixing parameterization in an OGCM, and potential vorticity mixing may prove to be more useful as a theoretical concept than as a practical parameterization form—but see Killworth (1998) for a more favorable view.

[2]Some have argued that a resting state is physically unrealizable, given all the sources of current excitation in nature, and therefore this consideration can be ignored (*e.g.*, Holloway, 1992).

7.3. SURFACE BOUNDARY LAYER AND SURFACE GRAVITY WAVES

In the surface *planetary boundary layer* (PBL), small-scale turbulence is relatively strong because of shear and buoyancy instabilities resulting from the surface fluxes, hence $\overline{w'\mathbf{u}'_h}$ and $\overline{w'c'}$ are much larger than in the interior. In the default SGS form, the PBL is implicitly contained within the top grid cell of an OGCM, since there are no elevated diffusivities in the grid interior. An essential extension beyond the default form is to increase the vertical SGS transport where the density stratification is gravitationally unstable (*i.e.*, with a positive vertical gradient in potential density), mimicking the convection that cannot be dynamically resolved in (4)). This is often done either by *convective adjustment* (a minimal vertical rearrangement of tracers until stable stratification is achieved) or by a substantial enhancement of the vertical diffusivities in the unstable region. Insofar as the convection arises from an upward surface buoyancy flux—as it usually does—then the PBL extends from the surface throughout the depth interval of adjustment or enhanced mixing. In this situation an explicit boundary layer model for the PBL transport is preferable to a local mixing rule. The PBL model types most commonly used in OGCMs are either bulk *mixed-layer models* (*e.g.*, Kraus & Turner, 1967) or single-point, moment-closure turbulence models (*e.g.*, Mellor & Yamada, 1982). Large *et al.* (1994) reviews the usage of oceanic PBL models and proposes a *K-profile parameterization* (KPP). KPP does not impose such a strong constraint on the boundary layer profiles of large-scale variables as does a mixed-layer model; KPP escapes the dubious spatial locality assumption of moment-closure models; and KPP also includes counter-gradient buoyancy and fluxes in convective situations. The KPP hypotheses are that the boundary layer thickness h keeps a bulk Richardson number, $Ri_b = -gh\Delta\rho\big/(\Delta\mathbf{u})^2$, at a critical $\mathcal{O}(1)$ value, and that the diffusivity profile smoothly joins with the Monin-Obukhov form near the surface and the interior form at the outer edge of the boundary layer. In the few assessments of its performance in OGCMs to date, KPP seems to do well (Large *et al.*, 1997; Li *et al.*, 1998; Large & Gent, 1998). It is important that PBL vertical mixing not occur simultaneously with isopycnal tracer transport (which presumes a strongly stable stratification) in an OGCM, or else the exchange between the boundary layer and the interior will be too rapid; accordingly, a rule is used that turns off κ_i where $z \geq -h$ or $\nabla_h\rho|\big/\rho_z$ is too large.

As yet no OGCM includes several SGS effects of surface gravity waves that may be important under some conditions. These include enhanced mixing by breaking waves near the surface and by wave-driven Langmuir cells throughout the PBL; extra wave-averaged terms, somewhat analogous to the mesoscale-eddy $\mathbf{u}^*$ term above, in both the horizontal momentum and tracer equations in (4) due to the mean Lagrangian velocity of the waves, the Stokes drift; and extra wave-averaged terms in the surface kinematic and pressure boundary conditions (McWilliams & Restrepo, 1998).

7.4. INTERIOR VERTICAL OR DIAPYCNAL MIXING

The default SGS parameterizations for diapycnal momentum and tracer transports in the oceanic interior—*i.e.*, $\overline{\mathbf{u}'\mathbf{u}'_h} \cdot \nabla\rho$ and $\overline{\mathbf{u}'c'} \cdot \nabla\rho$—are eddy diffusion, with uniform $\nu_v \sim 10^{-4}$ m^2 s^{-1} and $\kappa_v \sim 10^{-5}$ m^2 s^{-1}; not uncommonly, these values are made to increase with depth in the deep ocean, which has some observational support (Toole, 1998). Such a tracer diffusivity is inherently non-adiabatic, in the sense defined above, and it implies a breaking of the isopycnal surfaces (Fig. 2). These are usually represented as a vertical flux, since this direction is nearly perpendicular to isopycnal surfaces in most places. (Isopycnal slopes are typically $\mathcal{O}(10^{-4})$.) For a given, non-uniform vertical grid, a sufficiently large vertical eddy diffusivity is needed to dominate the implicit numerical diffusion arising from discretization errors (Yin & Fung, 1991; but see a response by Treguier *et al.*, 1996, challenging this interpretation); therefore, larger (ν_v, κ_v) values are often chosen for coarser vertical grids. There is a continuing controversy over whether the global oceanic circulation is consistent with the small diffusivity values above, or whether "missing mixing" processes must be found; my opinion is that the case for missing diapycnal mixing has not yet been conclusively made since OGCM solutions can approach these small values without developing hard discrepancies with observations (McWilliams *et al.*, 1996).

However, there are various small-scale processes, in addition to convection, that make the interior diabatic mixing spatially and temporally non-uniform and sometimes elevated above the default values. Most notable among these processes are ***stratified shear instability***, ***breaking internal waves***, and ***double diffusion*** caused by the larger molecular diffusivity of T than of S (see Large *et al.*, 1994, for a review of the SGS parameterization forms). Pacanowski & Philander (1981) shows that shear instability influences the equatorial undercurrent and thermocline profiles in OGCM solutions using a parameterization where (ν_v, κ_v) vary inversely with the gradient Richardson number, $Ri_g = N^2 \big/ (\partial \mathbf{u}/\partial z)^2$. Demonstrations of potentially substantial consequences on the abyssal tracer distributions are given in Cummins *et al.* (1990) and Cummins (1991) for a κ_v varying inversely with N, and in Gargett & Holloway (1992) for double diffusion with unequal κ_v values for T and S. Zhang *et al.* (1998) shows moderate sensitivities of the meridional overturning circulation and heat flux to the double diffusion parameterization. The nonlinearity in $\rho[T, S, P]$ also allows for locally enhanced vertical transport through thermobaric and cabbeling instabilities. The effects of $\rho[T, S, P]$ nonlinearity are present in principle in OGCM solutions that have enhanced vertical diffusivity in the presence of convective instability, but their significance has not been broadly assessed.

7.5. BOTTOM BOUNDARY LAYER AND GRAVITY CURRENTS

There is an analogous PBL at the ocean bottom, driven principally by boundary stress without any significant buoyancy flux. A primary effect of the bottom PBL is to provide a drag force on the adjacent flow, as in a simple Ekman layer. Usually in OGCMs this is represented as a boundary stress proportional to $\mathbf{u}$ or $|\mathbf{u}|\mathbf{u}$, acting only in the deepest grid cell. Obviously, an explicit PBL model, of the types discussed above, could be used at the bottom. Armi (1978) and Garrett (1991) argue for the possible significance of tracer transport parallel to the boundary by PBL turbulence; this implies a diapycnal flux for any isopycnal surfaces intersecting the bottom and thus locally diminishes the impedance to diapycnal exchanges (Fig. 2). This effect has not yet been examined in an OGCM, although Marotzke (1997) shows plausible solutions in an simple domain geometry while limiting κ_v to be non-zero only near the sides.

Gravity currents are down-slope, bottom flows with a density greater than in their surroundings at the same level. They are thus doubly turbulent, due both to the stress against the bottom and to gravitational instability compared to their downstream environment; thus, they efficiently entrain and mix with the ambient fluid as they flow. Gravity currents occur where convection over a slope penetrates to the bottom or dense water overflows a continental shelf or a sill, and they will continue either until entrainment quenches the density anomaly, geostrophic adjustment turns the flow to be across the slope (though bottom stress keeps the flow turbulent and allows a continuing, albeit slower, downslope progression), or the bottom of the slope is reached. Several sites in the global ocean have significant water mass flows in gravity currents; these include the Mediterranean outflow, the Denmark Straits overflow, and outflows from the Antarctic coastal shelves. The usual practice in OGCMs is to ignore this process, and this is probably unwise. Beckman & Doescher (1997) proposes a particular OGCM parameterization that combines the kinematic condition with a stress-driven boundary layer but neglects entrainment, and Gnanadesikan (1998) makes a somewhat more general proposal. Price & Beringer (1994) presents a gravity current model that includes entrainment that perhaps could be adapted as an OGCM parameterization.

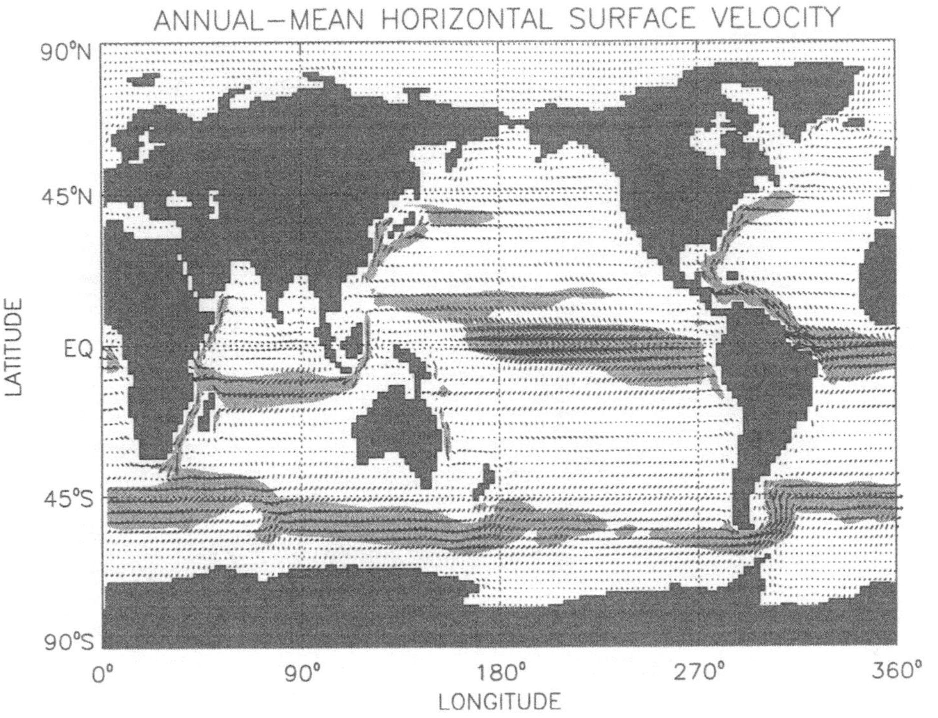

Figure 3. Time-mean surface currents from a coarse-resolution OGCM. Speeds greater than 0.1 m s^{-1} are shaded, and the maximum vector length is 0.55 m s^{-1}.

7.6. TOPOGRAPHIC EFFECTS

The common practice with OGCMs is to include only the effects of topography that are resolved on the model grid, and these aspects are discussed in Section 6 above. Here we consider some additional SGS effects that might be parameterized.

Small-scale topography influences large-scale circulation. The most direct possibility is through a turbulent rectification process wherein mesoscale eddies above topography cause a form stress that drives large-scale currents along large-scale contours of f/h (Bretherton & Karweit, 1975; Bretherton & Haidvogel, 1976; Salmon *et al.*, 1976). Holloway (1992) argues that this is a significant effect on continental slopes by driving barotropic, cyclonic rim currents (*i.e.*, having the same rotational sense around the basin as f), and Merryfield & Holloway (1998) examines the modifications due to stable stratification in a quasigeostrophic model. Ebby & Holloway (1994b), using

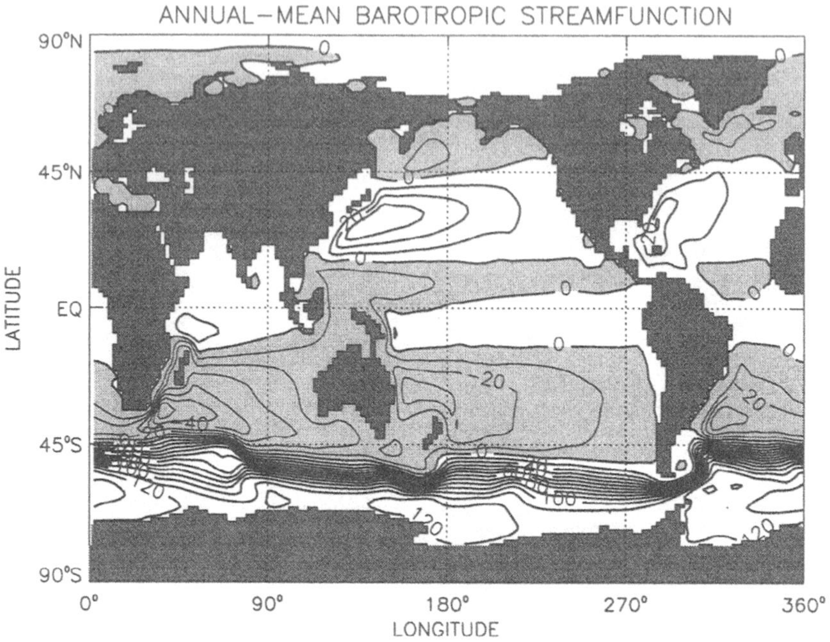

Figure 4. Time-mean horizontal transport streamfunction from a coarse-resolution OGCM. Counterclockwise circulation is shaded. The contour interval is 10 Sv (1 Sv $= 10^6\ m^3\ s^{-1}$).

a parameterization for this rectification, and Hurlburt *et al.* (1995), with very fine mesoscale resolution, show examples of this possibility in OGCM solutions. I believe that SGS form stress is a qualitatively important effect to include in OGCMs, although its strength, its vertical structure, and its relationship to the resolved form stresses all need further clarification. A less direct effect of topographic roughness is its tendency to make the vertical structure of mesoscale eddies more surface intensified, with implications for their transport properties in the general circulation (Rhines, 1977; Owens & Bretherton, 1978; Boening, 1989). Large-scale currents over rough topography can generate gravity waves that can propagate vertically into the interior, depositing their momentum and enhancing diapycnal mixing where they break and dissipate. This is an important mechanism for the atmospheric general circulation, and recent observations indicate the occurrence of the process in the oceans as well (Polzin *et al.*, 1997). Finally, topography on the scale of the bottom boundary layer generates local secondary circulations that may substantially alter the vertical momentum flux and diapycnal mixing rate, and no SGS proposal has yet been made to account for this effect.

7.7. RIVERS AND MARGINAL SEAS

The common practice for OGCMs is to neglect river inflows and SGS exchanges with marginal seas. Restoring terms, if used, may indirectly contribute some of these otherwise missing effects. If restoring terms are not used for the surface fluxes, then lateral tracer restoring conditions could be used in places with rivers or significant marginal sea exchanges, in the form of an open boundary condition.

It is better physics to have flux boundary conditions for river inflow, with the inflow specified either from data or a land hydrology model. Then the river inflow appears as just another component of the surface water flux for that location. The exchanges with SGS marginal seas or straits (*e.g.*, Bering and Gibraltar) could be specified as lateral flux conditions, determined either from data or from coupling with a local model of their circulation.

8. Global Solutions

In this and the following section, several common diagnostic measures are presented from two examples of modern OGCM solutions: a global equilibrium solution with relatively coarse horizontal resolution and an Atlantic basin solution with relatively fine resolution.

The global solution is one reported in Gent *et al.* (1998), which was partly motivated by the development of the Climate System Model by the U.S. National Science Foundation and National Center for Atmospheric Research (NSF/NCAR). The model resolution is 2.4^o in longitude, 1.2-2.4^o in latitude (finer in tropical and polar regions), and 45 levels in the vertical with a vertical spacing expanding with depth from 12.5 to 250 m. Its surface forcing is a mean annual cycle with a combination of specified fluxes and feedback/restoring relations as described in Large *et al.* (1997). The model uses the isopycnal tracer transport parameterization of Gent & McWilliams (1990) and the KPP boundary layer and interior vertical diffusivity parameterizations of Large *et al.* (1994). The calculation is carried to equilibrium using the acceleration technique, with a final 17 years of synchronous integration.

The solution behavior is presented here primarily through figures, with limited discussion of their interpretation and accuracy.

The time-mean surface currents in Fig. 3 show strong westward currents near the equator, eastwards currents around Antarctica, and western boundary currents in the sub-tropical and sub-polar gyres. The interior,

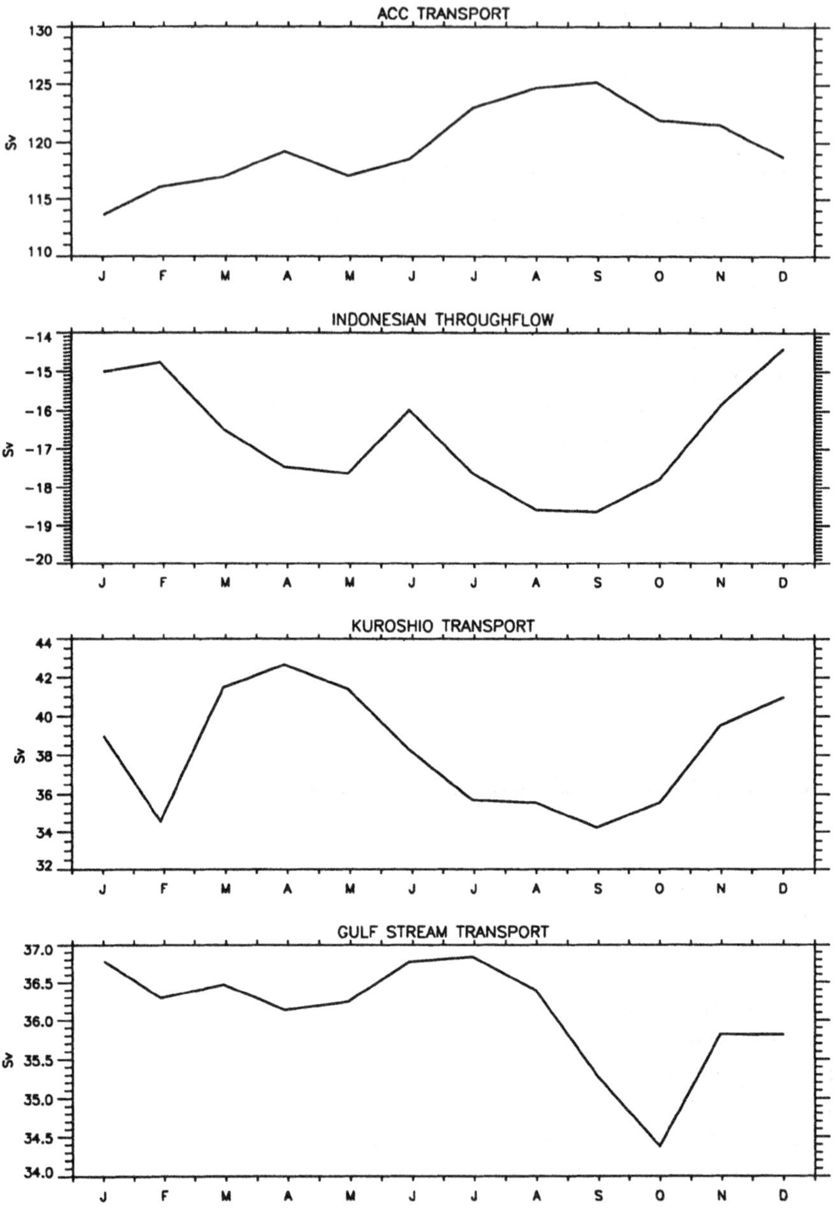

Figure 5. Annual cycle of horizontal transport across selected currents from a coarse-resolution OGCM: (a) Antarctic Circumpolar Current (Drake Passage); (b) Indonesian Throughflow; (c) Kuroshio (26° N); (d) Gulf Stream (29° N).

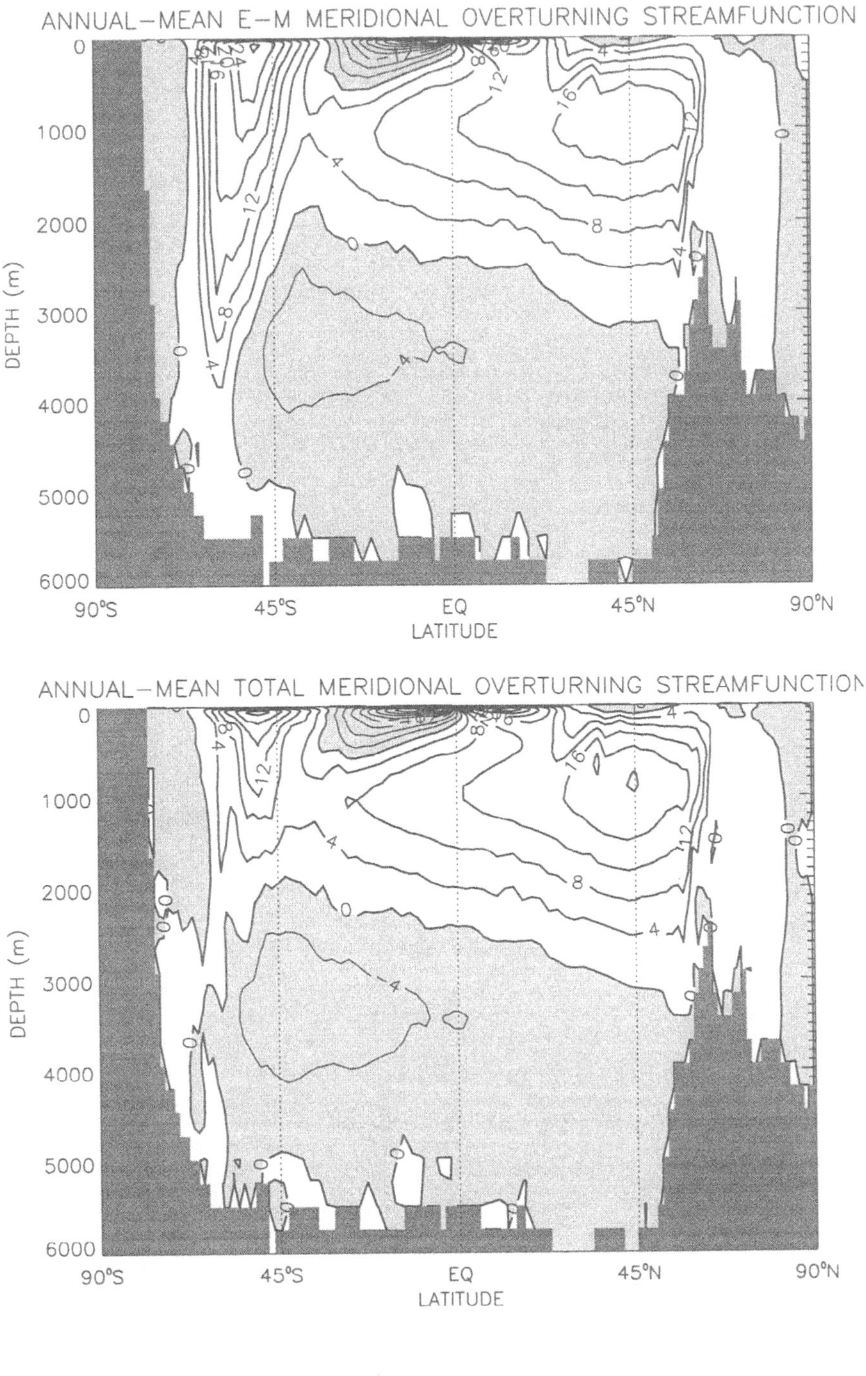

Figure 6. Time-mean meridional transport streamfunction from a coarse-resolution OGCM: (a) Eulerian flow; (b) total Lagrangian flow including the eddy-induced transport velocity. Counterclockwise circulation is shaded. The contour interval is 4 Sv.

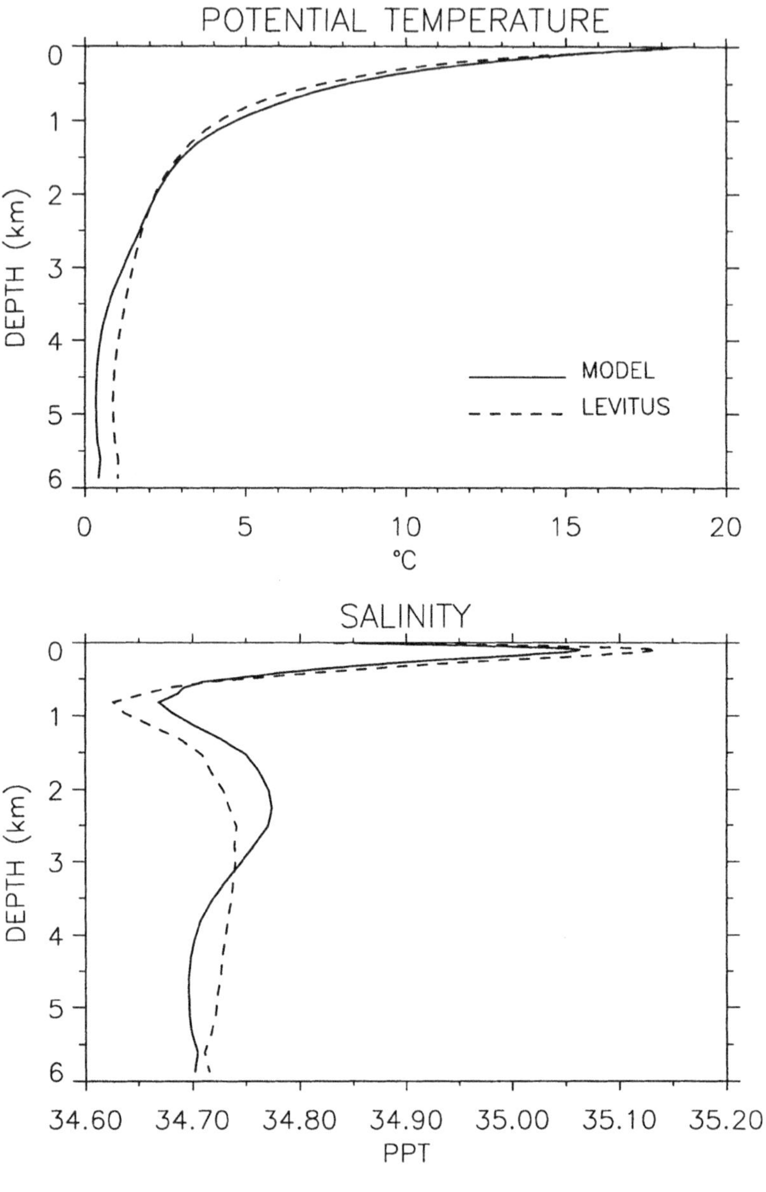

Figure 7. Time- and horizontal-mean profiles of T and S from a coarse-resolution OGCM and from Levitus (1982) and Levitus *et al.* (1994).

extra-tropical currents exhibit Ekman-layer turning from the surface wind stress, to the right (left) in the northern (southern) hemisphere.

The time-mean horizontal transport streamfunction in Fig. 4 shows only

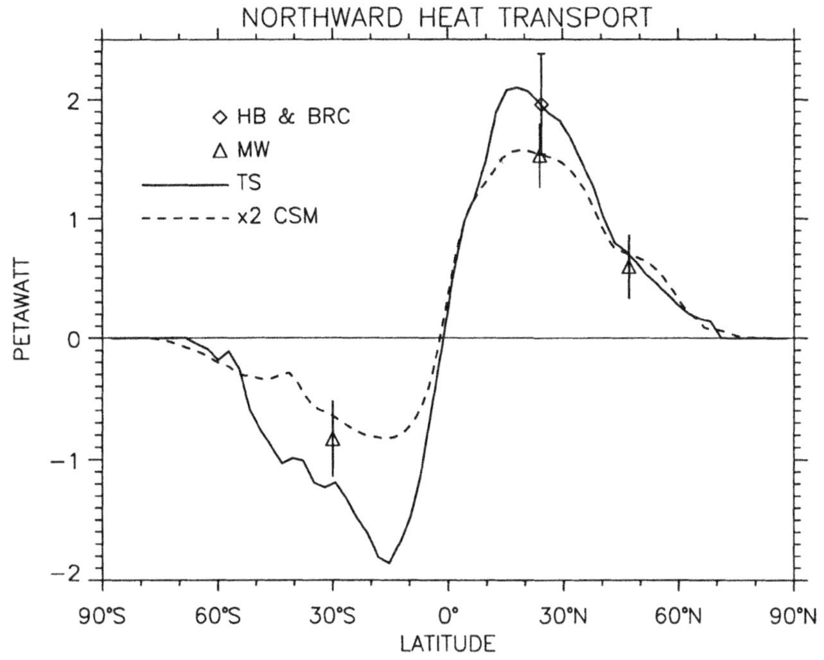

Figure 8. Time-mean northward heat transport from a coarse-resolution OGCM and from empirical estimates (and their uncertainty) by Hall and Bryden (1982), Bryden *et al.* (1991), Trenberth & Solomon (1994), and MacDonald and Wunsch (1996).

weak barotropic flow near the equator, anti-cyclonic sub-tropical gyres in all basins, cyclonic sub-polar gyres in the northern basins, and the Antarctic Circumpolar Current. Time series of the annual cycle of transport in the Antarctic Circumpolar Current, Indonesian Throughflow, Kuroshio, and Gulf Stream are shown in Fig. 5.

The time-mean, global meridional transport streamfunction is in Fig. 6a. It shows shallow, wind-driven (Ekman) cells near the equator and in the northern sub-polar region, a deep wind-driven Deacon Cell in the Antarctic Circumpolar region, a strong northern thermohaline cell with sinking mostly in the North Atlantic and extension into the southern hemisphere at middle depths, and a weaker Antarctic thermohaline cell that extends along the bottom into the northern hemisphere. Fig. 6b is the sum of this meridional transport plus the time-mean, zonally integrated eddy-induced transport from (7). Its largest effect is a near cancellation of the Deacon Cell except in the diabatically forced region in the upper ocean.

Time- and horizontal-mean vertical profiles of $T(z)$ and $S(z)$ are in Fig. 7, both from the model solution and from a hydrographic climatology.

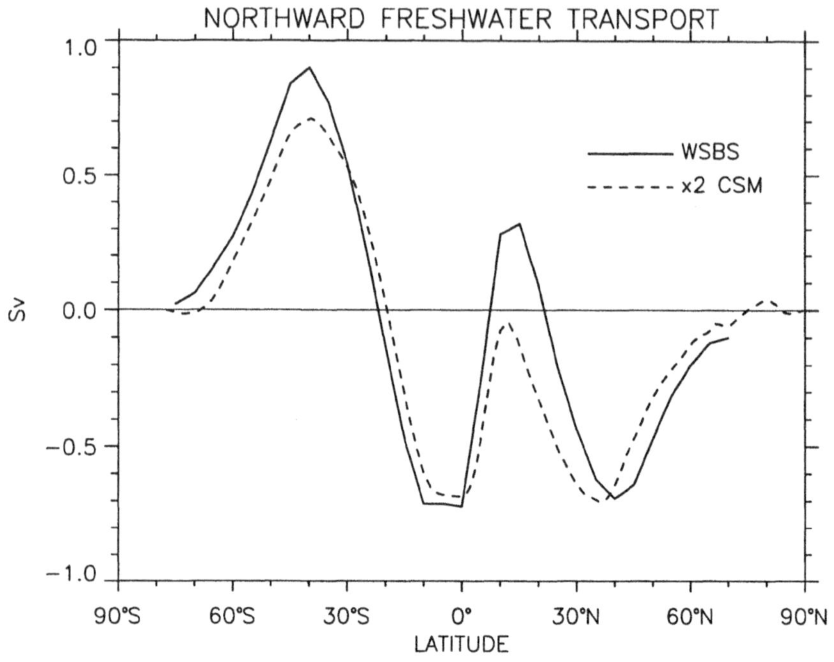

Figure 9. Time-mean northward water transport from a coarse-resolution OGCM and from empirical estimates by Wijffels *et al.* (1992).

These profiles show a sharp thermocline and halocline in the upper ocean and more uniform water properties in the deep ocean. $S(z)$ shows a shallow maximum and an intermediate-depth minimum.

Time-mean, longitude-integrated meridional transports of heat and freshwater are shown in Figs. 8-9, both from the model solution and from several empirical estimates. The heat transport is poleward in both hemispheres, with its peaks near $\pm 20^o$. In the southern hemisphere, the heat transport becomes quite small across the Antarctic Circumpolar Current, due to the canceling effect there of the meridional overturning by the eddy-induced circulation. The model heat transport is within the uncertainties of the empirical estimates. Although the general shapes of the model and empirical water transport curves agree fairly well, estimating the uncertainties is more problematic than for heat.

9. Basin Solutions

The basin solution is one reported in Smith *et al.* (1998). It was partly motivated by the World Ocean Circulation Experiment Community Modeling

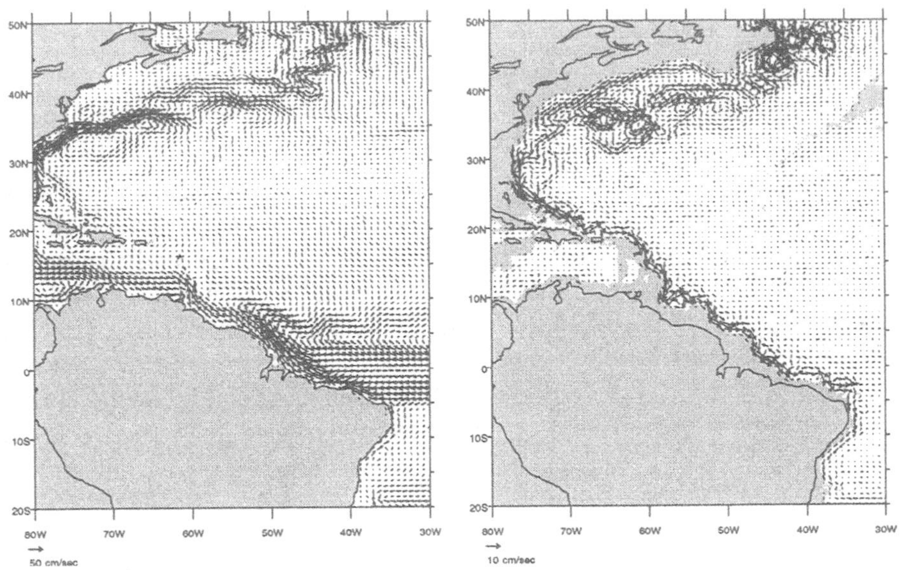

Figure 10. Time-mean currents from a fine-resolution OGCM for the Atlantic basin: (a) 15 m depth and (b) 2500 m depth. Only a portion of the domain is plotted here.

Effort for comparing high-resolution OGCM solutions. The model domain is the tropical and North Atlantic basin, between 82° W and 17.2^o E (Hudson Bay excluded) and 20° S and 72.6^o N; it also includes the eastern part of the Gulf of Mexico mapped into Africa and part of the Mediterranean Sea. Its horizontal grid is in Mercator coordinates with a resolution of 0.1^o at the equator, and it has 40 vertical levels, varying smoothly from 10 m at the top to 250 m at depth. The forcing is by daily winds (1985-1995), heat flux as in Barnier *et al.* (1995), and surface restoring to mean monthly S from Levitus (1982). Lateral buffer zones with T, S restoring are used along all open boundaries. The solution is spun up from climatological T, S and $\mathbf{u} = 0$, and the solution averages below are for years 6-10.

The time-mean currents are shown in Fig. 10 at two levels, one near the surface and the other at a middle depth. Near the surface, the western boundary currents have a similar pattern to those in the coarse-resolution solution (Fig. 3), except they are faster and narrower here, especially in the off-shore Gulf Stream. At the middle depth, the western boundary current is southward and is part of the thermohaline circulation seen in Figs. 6 and 12. It shows complex standing eddies in the off-shore Gulf Stream region and near the Grand Banks.

The time-mean horizontal transport streamfunction is in Fig. 11. Com-

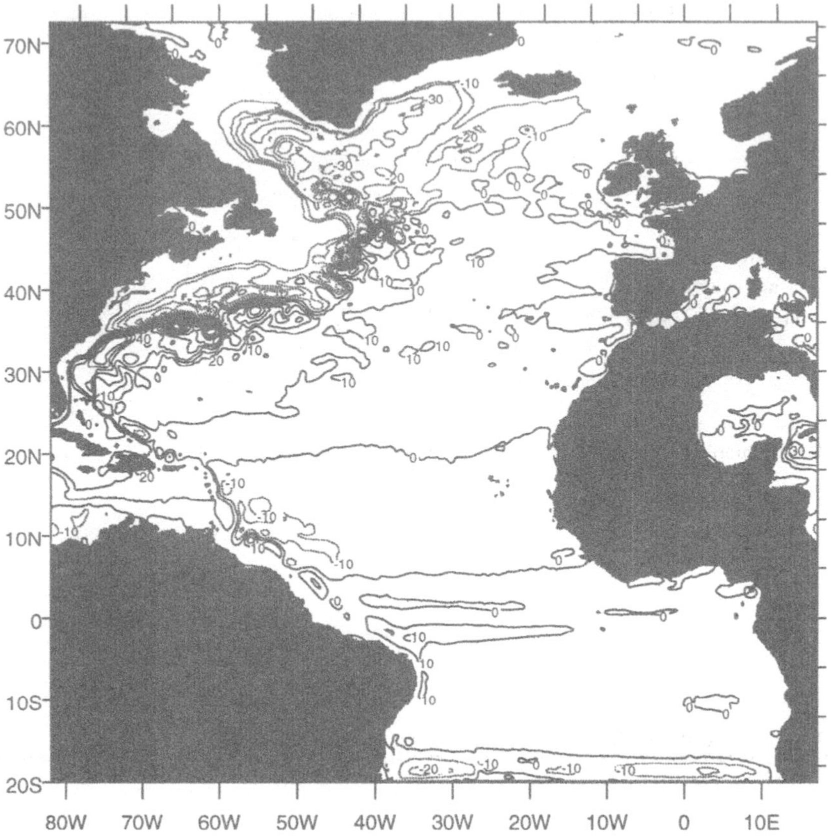

Figure 11. Time-mean horizontal transport streamfunction from a fine-resolution OGCM for the Atlantic basin. The contour interval is 10 Sv.

pared to Fig. 4, it shows stronger recirculation cells near the western boundary and a greater tendency for following the topographic slopes of the basin. The time-mean meridional transport streamfunction is shown in Fig. 12. Its overall amplitude is similar to the global thermohaline circulations in Fig. 5, although the patterns are somewhat different, particularly with respect to the sub-tropical maximum in the basin solution. This circulation component is substantially influenced by the lateral buffer zones near the northern and southern boundaries.

The time-mean sea surface elevation is in Fig. 13. It shows the broad, geostrophic, sub-tropical and sub-polar gyre circulations, and the enhanced recirculation cells near the western boundary. A zonal pressure gradient is evident along the equator, which drives the eastward Equatorial Undercurrent.

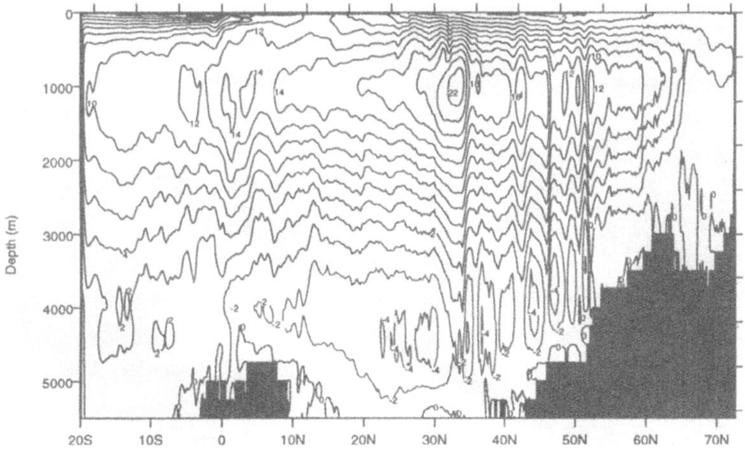

Figure 12. Time-mean meridional transport streamfunction from a fine-resolution OGCM for the Atlantic basin. This is the Eulerian flow only. The contour interval is 2 Sv.

A plot of instantaneous sea surface temperature is in Fig. 14. In addition to the broad-scale pattern imposed by the climatological surface heating, it shows meanders and filaments associated with advection by the mesoscale eddies and boundary currents.

The time-mean eddy kinetic energy density near the surface is in Fig. 15. It shows higher values where the mean currents are stronger, but the spatial extent of such regions is broader (*cf.*, Figs. 10a and 13). Note in particular the elevated eddy energy along the Azores Current and the North Atlantic Drift. Wavenumber-frequency spectra of sea-surface elevation in the Gulf Stream region are in Fig. 16, both from the model solution and from altimeter measurements. In the mesoscale bands, the OGCM variability matches quite well the observations. On the longest scales the model variability seems slightly higher, and at fine scales the model variability is clearly less.

The horizontal pattern of S at a middle depth near the Mediterranean is in Fig. 17. The outflow water, diluted until it achieves neutral buoyancy with its surrounding, is being spread and mixed by mesoscale eddies that are qualitatively similar to the observed Mediterranean Eddies (Meddies), although they are only marginally resolved in this solution.

From the perspective of a computational mathematician, there is nothing mysterious about the influence of the spatial grid spacing in a numerical

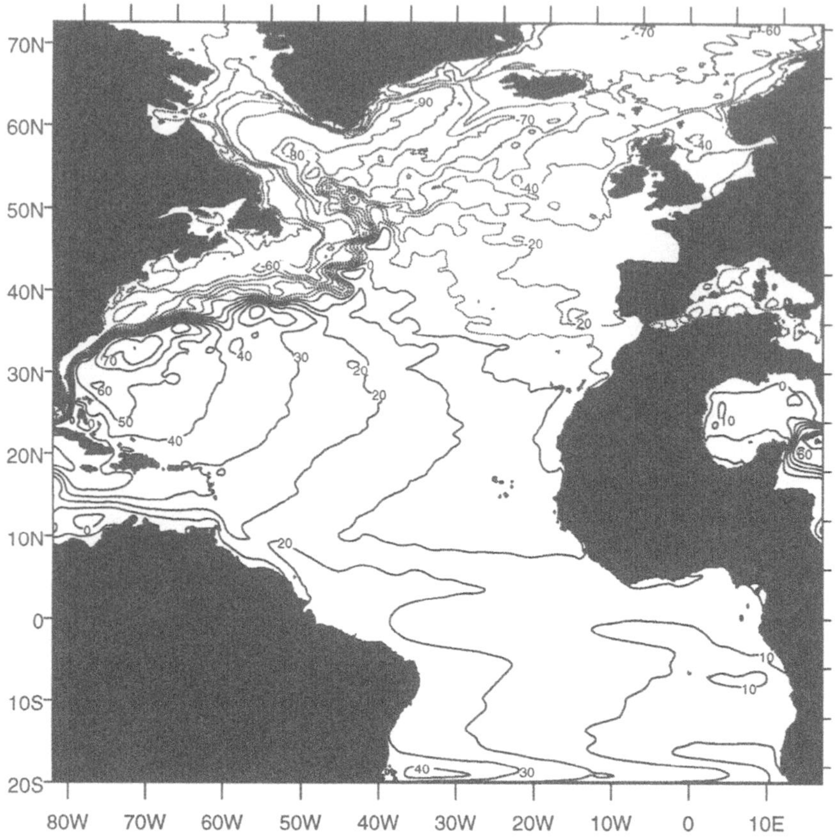

Figure 13. Time-mean sea surface elevation from a fine-resolution OGCM for the Atlantic basin. The contour interval is 10 Sv.

calculation: in a given problem, once the resolution is fine enough for the solution to be smooth on the grid scale, the solution converges with increasing resolution according to the order of the discretization method. For oceanographers the influence of resolution in OGCMs often has been a matter of confusion, and public discussions have often generated more heat than light. There are at least three reasons why the oceanographic issues go beyond the computational mathematical dictum:

> * It is common practice to change the problem with the resolution, both by decreasing at least some of the eddy diffusivities (the lateral viscosity, in particular) to its computational stability limit and by adding finer scales to the forcing and domain geometry.

* Qualitative changes in the solution behavior occur with resolution, due to changing the eddy diffusivity, as instability thresholds are passed; the most prominent change is the emergence of mesoscale eddies, but this actually involves a suite of thresholds for the many different currents in the general circulation.

* The general circulation has many measures, and no complete understanding yet exists for how different features differently depend on resolution and each other.

10. Spatial Resolution

There are remarkably few published studies of the influence of resolution in OGCMs, and more certainly would be worthwhile. The evidence in hand suggests that it is relatively unimportant for those solution properties, such as tracer fluxes and water mass distributions, that are simulated reasonably well in OGCMs without resolved mesoscale instabilities (*i.e.*, with horizontal grid spacings $dx > 50$ km); see Covey (1995). In such models, the boundary currents typically are too weak and broad but are not as incorrect in their transport, the eddy-driven western recirculation gyres are missing, and the equatorial currents are too weak and broad except when the meridional resolution is locally refined to a scale smaller than 50 km.

Once the grid is fine enough for eddies to arise, the resolution seems to be extremely important, up to an as yet poorly determined threshold at least as small as 10 km, in order to calculate mesoscale eddies and intense, narrow currents credibly (*i.e.*, with qualitative similarity to observations in eddy energy level and current location). For eddy-resolving models, resolution convergence has not yet been demonstrated and remains an important open issue. However, several recent OGCM solutions for the Atlantic with especially high vertical or horizontal resolution do appear to have achieved certain correspondences with the observations lacking at lower resolution (*e.g.*, eddy kinetic energy density, Gulf Stream separation site and off-shore path, and meridional heat flux); see Chao *et al.* (1996), Smith *et al.* (1998), and Chassignet *et al.* (1998). On the other hand, some features have not yet been seen in OGCM solutions that are present in idealized models at very fine resolution; examples of this are the instability of the western boundary current before separation (Berloff & McWilliams, 1998a), an intrinsic low-frequency variability on decadal periods (Berloff & McWilliams, 1998b), and the emergence of an abundant population of mesoscale and sub-mesoscale coherent vortices (Siegel *et al.*, 1998), no doubt with other phenomena yet to be discovered.

The computational cost scales with horizontal resolution roughly as $1/dx^3$, assuming that the vertical resolution, duration of integration, and

Figure 14. Instantaneous sea surface temperature from a fine-resolution OGCM for the Atlantic basin.

domain size are not varied. This implies roughly a thousand-fold disparity in computation costs for any given problem configuration between OGCMs that resolve the eddies and those that do not. Computer limitations, therefore, cause OGCM usage to be split between these two types of configurations. At present, eddy-resolving models can be used well only for intervals

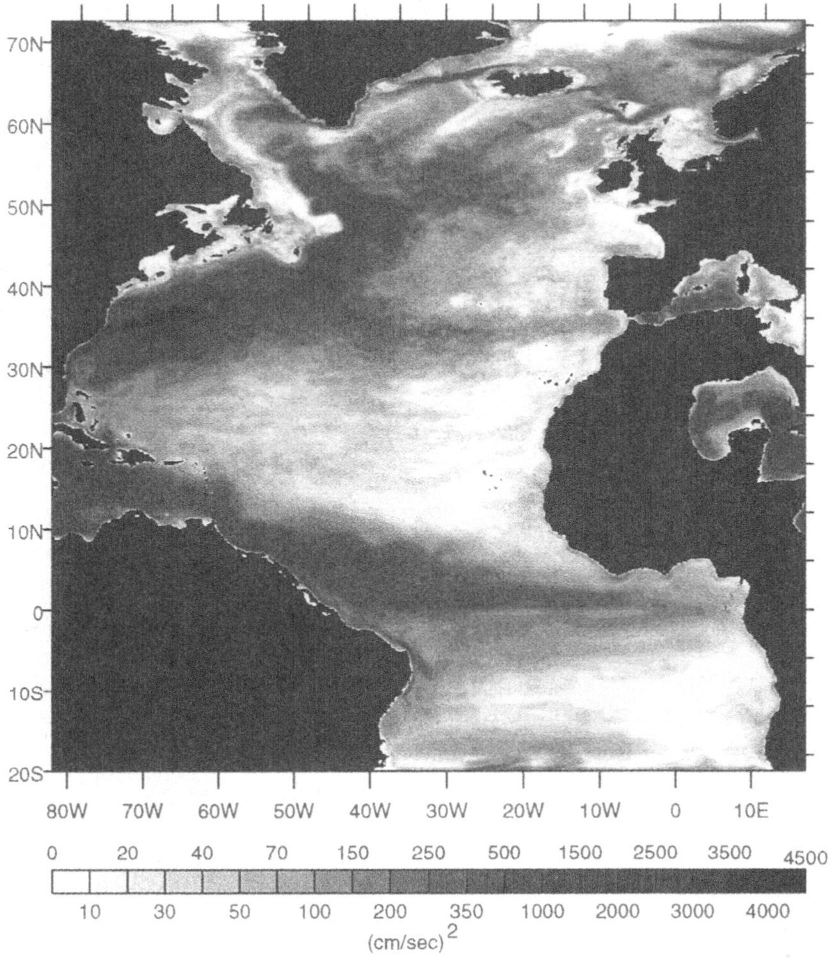

Figure 15. Eddy kinetic energy density at a depth of 46 m from a fine-resolution OGCM for the Atlantic basin.

as long as decades and domains as large as basins, whereas coarse-resolution models are also suitable for centennial and millennial fluctuations and the approach to equilibrium in global domains. Although growing computer power will narrow this division, it will be at least a decade, and perhaps much longer, before it disappears and everyone will prefer an eddy-resolving model. The fact that OGCMs without eddies and with sensible parameterizations can do reasonably well in calculating the large scale thermohaline circulations, heat and water fluxes, and water mass distributions remains somewhat mysterious and thus must be accepted only provisionally; how-

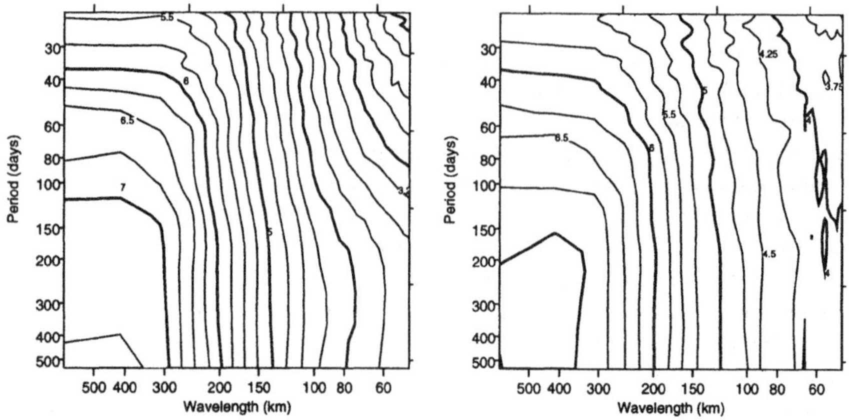

Figure 16. Wavenumber-frequency spectra of sea-surface elevation in the Gulf Stream region, 50-75° W, 32-42° N: (left) from a fine-resolution OGCM for the Atlantic basin; (right) from TOPEX/Poseidon altimeter measurements.

ever, it suggests that there is something of a dynamical decoupling between these phenomena and the mesoscale eddies, strong currents, and other small scale phenomena.

11. The Role of the Ocean in Climate System Models

The scientific problems of ocean circulation and climate have a fundamental mutual dependency. Ocean models have an excessive uncertainty in their surface boundary conditions as specified from observations (Section 3), and the equilibrium climate of the earth strongly depends upon storage and meridional heat transport by the ocean. A more subtle dependency is that fluctuations in either the atmosphere or ocean may elicit a response in the other medium that importantly alters the evolution by feedbacks. ENSO is the best known example of an interannual *coupled mode* of climate variability. However, both the ocean and atmosphere exhibit many other forms of variability than ENSO, and we do not yet know very well how many of these modes are significantly coupled. The paleoclimate record shows many episodes of rapid surface temperature changes of $\mathcal{O}(10)$ K with a probably global extent, most recently during the Alerod–Younger Dryas fluctuations at the end of the last ice age about 12,000 years ago (*e.g.*, Dansgaard *et al.*, 1984). Within the modern era ocean surface temperatures show substantial decadal variability in the regional spatial patterns of atmospheric

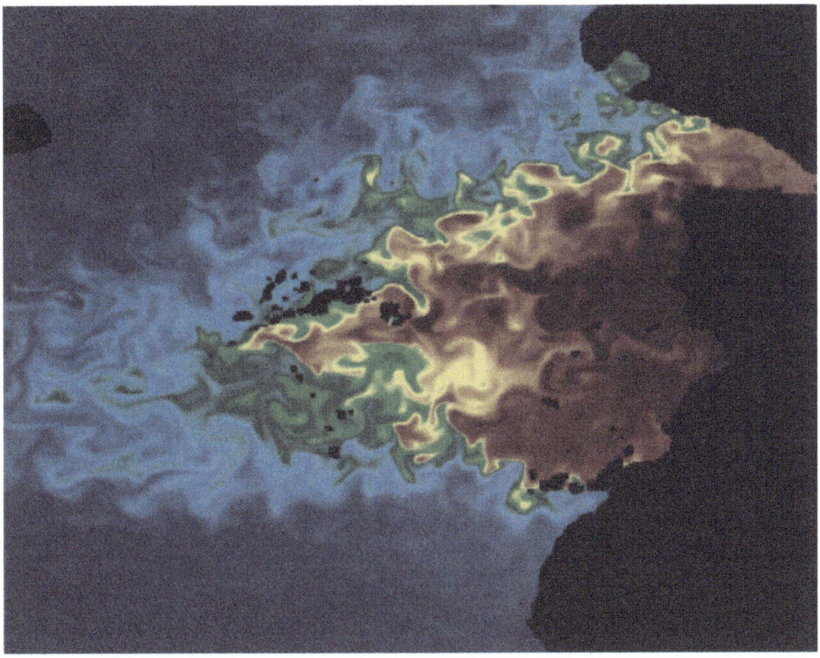

Figure 17. Salinity near the Mediterranean at a depth of 1140 m from a fine-resolution OGCM for the Atlantic basin. Red represents the warm-salty water from the Mediterranean outflow and blue the cold-fresh water of the interior Atlantic Ocean.

"teleconnection" modes such as the North Atlantic Oscillation (*e.g.*, Deser & Blackmon, 1993). My view is that both of these phenomena are related to the delicacy and intrinsic variability of the oceanic thermohaline circulation (Saravanan & McWilliams, 1995, 1997). In addition, a new coupled mode is being excited by the radiative forcing from anthropogenic greenhouse gas emissions, which may further involve the thermohaline circulation because of the associated changes in planetary-scale buoyancy forcing.

A coupled climate model includes a global OGCM, together with models for sea ice, the land surface, and the atmosphere. This type of model has a long history of usage (*e.g.*, Manabe & Bryan, 1969). It has been the principal means for forecasting global warming due to increasing greenhouse gases, and therefore it will undoubtedly become an even more important tool as mankind begins to practice planetary environmental management. One of the least convincing aspects of coupled models has been the prevalent tendency for *climate drift* away from the actual climate. To avoid this the common practice has been to impose surface *flux corrections* or *flux adjustments* as an artificial stabilization technique. This consists of computing equilibrium states for the uncoupled oceanic and atmospheric models (*e.g.*, using surface (S, T) restoring terms and specified sea surface T, respectively), computing the associated heat and water air-sea fluxes for each model, and then imposing the difference between the models' fluxes as spatially and seasonally varying added source terms to one of the models, usually the ocean. This does reduce climate drift in the subsequent coupled solution, and different modeling groups have found differing degrees of drift suppression by this technique.

An obvious goal for coupled modeling, therefore, is to eliminate flux corrections while avoiding excessive climate drift. A notable advance towards this goal is the recent solution for the NSF/NCAR Climate System Model: without flux corrections (other than a global air-sea water balance constraint to compensate for the neglect of river inflow to the ocean), the average air temperature shows no significant drift over an integration of 300 years (Fig. 18; Boville & Gent, 1998). The reasons behind this success appear primarily to be advances in SGS parameterization (clouds and radiation in the atmospheric model, the surface PBL in both models, and isopycnal tracer transport in the ocean model), as well as having an initial condition with an equilibrium, uncoupled ocean solution forced by reasonably accurate surface fluxes. This apparent lack of drift in surface temperature would not continue indefinitely in this model, however, since there is appreciable drift in, for example, the oceanic S field associated with the lack of river inflow and excessive water throughput in the sea-ice model. Furthermore, as always, there are many properties of this coupled solution that are unrealistic and warrant further attention.

Achieving skillful coupled models is an important and difficult scientific challenge. There are so many physical processes that must be included for a physically consistent calculation of climate, and probably there are many failure modes that lead to excessive drift (many of which may not yet have been discovered). Of course, some blemishes in the simulated oceanic behavior will occur because of failures in the other component models in a coupled solution; two examples of known problems of this type are an incorrect location of the sea-ice boundary and excessive surface solar radiation due to underestimation in the atmospheric model of stratus cloud amounts or cloud absorption (*i.e.*, the current controversy over the "anomalous absorption" in nature compared to standard radiative transfer calculations; see Cess *et al.*, 1995).

OGCMs must avoid their own contributions to coupled failure modes. The use of surface restoring terms for T and S in ocean models precludes many of these failure modes. To make progress, then, ocean models must evolve away from using this procedure for the buoyancy forcing. This poses a new requirement that an OGCM yield a correct surface T (necessary for

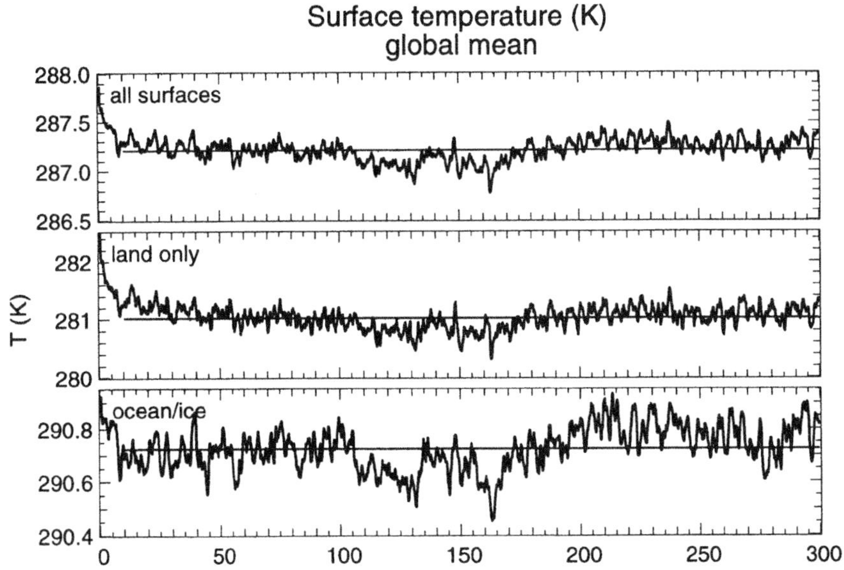

Figure 18. Time series of spatially averaged surface air temperature in a 300-year integration of the Climate System Model (Boville & Gent, 1998).

having correct air-sea fluxes) as a result of the forcing it receives and its own intrinsic variability. In principle, meeting this requirement involves all aspects of the OGCM, but there is probably particular importance to the surface PBL, marginal seas and river inflows, and interactions with sea ice.

For the reasons discussed in Section 10, OGCMs with coarse spatial grids will be most commonly used in coupled calculations, although some exploratory calculations with eddy-resolving models no doubt will be made. However, the prospect of either not resolving eddies or doing so inaccurately is unpalatable. This gives us added incentive to develop more robust methods for accelerating OGCM solutions (Section 4) and more skillful parameterizations for marginally resolved eddying solutions, like those used in Large Eddy Simulations (LES) of small-scale turbulence.

12. Summary and Prospects

Models of the oceanic general circulation are currently in a phase of rapid development and expanding utilization. Within the range of legitimate choices for model formulation and boundary conditions, surveyed above, present model solutions do seem to encompass the major features of the circulation as observed. In the coming decades the primary challenges are to extend and refine the modeling choices and to expand the observational system to reduce the empirical uncertainties for testing the models. The goal, of course, is to end up by achieving a useful degree of consistency between OGCM solution behaviors and nature. The biggest surprises along the way are likely to come in weaning the ocean model from specified, or at least strongly constrained, surface boundary conditions, as it assumes its proper, more fundamental role in modeling the earth's climate system.

Acknowledgements

I thank Gokhan Danabasoglu, Matthew Maltrud, and Richard Smith for their help in preparing many of the figures included here. I thank Eric Chassignet and Jacques Verron for organizing the 1998 Winter School on Ocean Modeling and Parameterization at Les Houches. Support was provided by the National Air and Space Agency (NAG 5-3982) and the National Science Foundation (OCE-9633681).

References

Arakawa, A. (1988) Finite-difference methods in climate modeling. In: *Physically based modeling and simulation of climate and climate change. Part I*, M.E. Schlesinger (ed.), Klewer, 79-168.

Armi, L. (1978) Some evidence for boundary mixing in the deep ocean. *J. Geophys. Res.* **83**, 1971-1979.

Armi, L. and D.M. Farmer (1988) The flow of Mediterranean and Atlantic water through the strait of Gibraltar. *Prog. Oceanogr.* **21**, 1-106.

Barnier, B., L. Siefridt, & P. Marchesiello (1995) Thermal forcing for a global ocean circulation model using a three-year climatology of ECMWF analyses. *J. Mar. Syst.* **6**, 363-380.

Barnier, B., P. Marchesiello, A.P. de Miranda, J.M. Molines, & M. Coulibaly (1998) A sigma-coordinate, Primitive Equation model for studying the circulation in the South Atlantic. Part I: Model configuration with error estimates. *Deep-Sea Res.*, in press.

Beckman, A., & R. Doescher (1997) A method of improved representation of dense water spreading over topography in geopotential-coordinate models. *J. Phys. Ocean.* **27**, 581-591.

Berloff, P., & J.C. McWilliams (1998a) The quasigeostrophic dynamics of western boundary currents. Preprint.

Berloff, P., & J.C. McWilliams (1998b) Low-frequency intrinsic variability in wind-driven gyres. Preprint.

Bleck, R., & E. Chassignet (1994) Simulating the oceanic circulation with isopycnic-coordinate models. In: *The Oceans: Physical-Chemical Dynamics and Human Impact.* S.K. Majumdar, & E.W. Miller, eds., Penn. Acad. Sci., 17-39.

Bleck, R., S. Dean, M. O'Keefe, & A. Sawdey (1995) A comparison of data-parallel and message-passing versions of the Miami Isopycnic coordinate Ocean Model. *Parallel Computing*, submitted.

Boening, C.W. (1989) Influences of a rough bottom topography on flow kinematics in an eddy resolving model. *J. Phys. Ocean.* **19**, 77-97.

Boening, C.W., W.R. Holland, F.O. Bryan, G. Danabasoglu, & J.C. McWilliams (1995) An overlooked problem in model simulations of the thermohaline circulation and heat transport in the Atlantic Ocean. *J. Climate* **8**, 515-523.

Boville, B.A., & P.R. Gent (1998) Century integrations with the NCAR Climate System Model. *J. Climate*, in press.
Bretherton, F. & M. Karweit (1975) Mid-ocean mesoscale modeling. In: *Numerical Models of Ocean Circulation*, R.O. Reid, ed., National Academy Press, Washington, D.C., 237-249.
Bretherton, F. & D.B. Haidvogel (1976) Two-dimensional turbulence above topography. *J. Fluid Mech.* **78**, 129-154.
Bryan, K. (1969) A numerical method for the study of the circulation of the world ocean. *J. Comput. Phys.* **4**, 347-376.
Bryan, K. (1984) Accelerating the convergence to equilibrium of ocean-climate models. *J. Phys. Ocean.* **14**, 666-683.
Bryden, H.L., D.H. Roemmich, & J.A. Church (1991) Ocean heat transport across 24° N in the Pacific. *Deep-Sea Res.* **38**, 297-324.
Chao, Yi, A. Gangopadhyay, F.O. Bryan, & W.R. Holland (1996) Modeling the Gulf Stream system: how far from reality? *Geophys. Res. Lett.* **23**, 3155-3158.
Cess, R.D., & 19 co-authors (1995) Absorption of solar radiation by clouds: observations versus models. *Science* **267**, 496-499.
Chassignet, E.P., Garraffo, Z. and A. Paiva (1998) Fine-mesh (1/12 degree) modeling of the North Atlantic: the spin-up phase. Preprint.
Covey, C. (1995) Global ocean circulation and equator-pole heat transport as a function of ocean GCM resolution. *J. Climate* **11**, 425-437.
Cummins, P.F., G. Holloway, & A.E. Gargett (1990) Sensitivity of the GFDL ocean general circulation model to a parameterization of vertical diffusion. *J. Phys. Ocean.* **20**, 817-830.
Cummins, P.F. (1991) The deep water stratification of ocean general circulation models. *Atmosphere-Ocean* **29**, 563-575.
Danabasoglu, G., J.C. McWilliams, & P.R. Gent (1994) The role of mesoscale tracer transports in the global ocean circulation. *Science* **264**, 1123-1126.
Danabasoglu, G., & J.C. McWilliams (1995) Sensitivity of the global ocean circulation to parameterizations of mesoscale tracer transports. *J. Climate* **8**, 2967-2987.
Danabasoglu, G., J.C. McWilliams, & W.G. Large (1996) Approach to equilibrium in global ocean models. *J. Climate* **9**, 1092-1110.
Dansgaard, W., S.J. Johnson, H.B. Clausen, D. Dahl-Jensen, N. Gundestrup, C.U. Hammer, & H. Oeschger (1984) North Atlantic climatic oscillations revealed by deep Greenland ice cores. In: *Climate Processes and Climate Sensitivity*, J.E. Hansen & T. Takahashi, eds., Geophysical Monographs **29**, AGU Press, 288-298.
Deser, C., & M. Blackmon (1993) Surface climate variations over the North Atlantic Ocean during winter: 1900-1989. *J. Climate* **6**, 1743-1753.
Dukowicz, J.K., R.D. Smith, & R.C. Malone (1993) A reformulation and implementation of the Bryan-Cox-Semtner ocean model on the Connection Machine. *J. Atmos. Ocean. Tech.* **10**, 195-208.
Dukowicz, J.K.,& R.D. Smith (1994) Implicit free-surface method for the Bryan-Cox-Semtner ocean model. *J. Geophys. Res.* **99**, 7991-8014.
DYNAMO (1997) DYnamics of North Atlantic MOdels: Simulation and assimilation with high resolution models, Report No 294, Institut fr Meereskunde, Kiel, Germany.
Ebby, M., & G. Holloway (1994a) Grid transform for incorporating the Arctic in a global ocean model. *Climate Dyn.* **10**, 241-247.
Ebby, M., & G. Holloway (1994b) Sensitivity of a large-scale ocean model to a parameterization of topographic stress. *J. Phys. Ocean.* **24**, 2577-2588.
Ezer, T.,& G.L. Mellor (1994) Diagnostic and prognostic calculations of the North Atlantic circulation and sea level using a sigma coordinate ocean model. *J. Geophys. Res.* **99**, 14159-14171.
Gargett, A.E., & G. Holloway (1992) Sensitivity of the GFDL ocean model to different diffusivities for heat and salt. *J. Phys. Ocean.* **22**, 1158-1177.
Garrett, C. (1991) Marginal mixing theories. *Atmosphere-Ocean* **29**, 313-339.

Gent, P.R., & J.C. McWilliams (1983) Regimes of validity for balanced models. *Dyn. Atmos. Oceans* **7**, 167–183.

Gent, P.R., & J.C. McWilliams (1990) Isopycnal mixing in ocean circulation models. *J. Phys. Ocean.* **20**, 150-155.

Gent, P.R., J. Willebrand, T.J. McDougall, & J.C. McWilliams (1995) Parameterizing eddy-induced tracer transports in ocean circulation models. *J. Phys. Ocean.* **25**, 463-474.

Gent, P.R., F.O. Bryan, G. Danabasoglu, S.C. Doney, W.R. Holland, W.G. Large, & J.C. McWilliams (1998) The NCAR Climate System Model global ocean component. *J. Climate*, in press.

Gnanadesikan, A. (1998) Representing the bottom boundary layer in the GFDL ocean model: Model framework, dynamical impacts, and parameter sensitivity. *J. Phys. Ocean.*, submitted.

Haidvogel, D.B.,& F.O. Bryan (1992) Ocean general circulation modeling. In: *Climate System Modeling*. Trenberth, K. ed., Oxford Press, 371-412.

Haidvogel, D.B., E. Curchitser, M. Iskandarani, R. Hughes, & M. Taylor (1995) Global modeling of the ocean and atmosphere using the spectral element method. *Atmosphere-Ocean*, in press.

Hall, M.M., & H.L. Bryden (1982) Direct estimates and mechanisms of ocean heat transport. *Deep-Sea Res.* **29**, 339-359.

Hirst, A.C., & J.S. Godfrey (1993) The role of Indonesian throughflow in a global ocean GCM. *J. Phys. Ocean.* **23**, 1057-1086.

Holloway, G. (1992) Representing topographic stress for large-scale ocean models. *J. Phys. Ocean.* **22**, 1033-1046.

Hurlburt, H.E., P.J. Hogan, E.J. Metzger, W.J. Schmitz, & A.J. Wallcraft (1995) Dynamics of eddy-resolving models of the Pacific Ocean and the Sea of Japan. In: *Workshop on Numerical Prediction of Oceanographic Conditions*, Japan Meteorological Agency, Tokyo, 51-57.

Ishikawa, I., Y. Yamanaka, & N. Suginohara (1994) Effects of presence of a circumpolar region on buoyancy-driven circulation. *J. Oceanography* **50**, 247-263.

Ivchenko, V.O. & D.P. Stevens (1995) The zonal momentum balance in a realistic eddy resolving general circulation model of the Southern Ocean. *J. Phys. Ocean.*, **26**, 753-774.

Kagimoto, T., & T. Yamagata (1997) Seasonal transport variations of the Kuroshio: An OGCM simulation. *J. Marine Research*, in press.

Kalnay, E., et al. (22 authors) (1996) The NCEP/NCAR 40-year reanalysis project. *BAMS* **77**, 437-471.

Killworth, P.D. (1998) Eddy parameterisation in large scale flow. In *Ocean Modeling and Parameterization*, E.P. Chassignet and J. Verron (Eds.), Kluwer Academic Publishers, 253-269.

Kraus, E.B., & S. Turner (1967) A one-dimensional model of the seasonal thermocline. II: The general theory and its consequences. *Tellus* **19**, 98-106.

Large, W.G., J.C. McWilliams, & S.C. Doney (1994) Oceanic vertical mixing: A review and a model with a non-local K-profile boundary layer parameterization. *Rev. Geophys.* **32**, 363-403.

Large, W.G., G. Danabasoglu, S. Doney, & J.C. McWilliams (1997) Sensitivity to surface forcing and boundary-layer mixing in a global ocean model: annual-mean climatology. *J. Phys. Ocean.*, in press.

Large, W.G., & P.R. Gent (1998) Validation of vertical mixing in an equatorial ocean model using Large-Eddy Simulations and observations. *J. Phys. Ocean.*, submitted.

Leonard, B.P. (1979) A stable and accurate convective modelling procedure based on quadratic upstream interpolation. *Computer Methods in Applied Mechanics and Engineering* **19**, 59-98.

Levitus, S. (1982) Climatological Atlas of the World Ocean. *NOAA Prof. Pap.* **13**, U.S. Gov. Print. Off., 173 p.

Levitus, S., R. Burgett, & T.P. Boyer (1994) World Ocean Atlas 11994: Volume 3, Salinity. *NOAA Atlas NESDES* **3**, U.S. Gov. Print. Off., 99 p.

Li, X., Y. Chao, J.C. McWilliams, and L.-L. Fu (1998) Sensitivity to vertical mixing in a Pacific Ocean General Circulation Model. *Geophys. Res. Lett.*, submitted.

Lunkeit, F., R. Sausen, & J.M. Oberhuber (1996) Climate simulations with a global coupled atmosphere-ocean model ECHAM2/OPYC. Part I: present-day climate and ENSO events. *Climate Dyn.* **12**, 195-212.

MacDonald, A.M., & C. Wunsch (1996) An estimate of global ocean circulation and heat fluxes. *Nature* **382**, 436-439.

Manabe, S., & K. Bryan (1969) Climate calculations with a combined ocean-atmosphere model. *J. Atmos. Sci.* **26**, 786-789.

Marotzke, J. (1997) Boundary mixing and the dynamics of three-dimensional thermohaline circulations. *J. Phys. Ocean.* **27**, 1713-1728.

Marshall, J. (1981) On the parameterization of geostrophic eddies in the ocean. *J. Phys. Ocean.* **11**, 257-271.

Marshall, J., C. Hill, L. Perelman, & A. Adcroft (1997) Hydrostatic, quasi-hydrostatic, and nonhydrostatic ocean modeling. *J. Geophys. Res.* **102**, 5733-5752.

McDougall, T.J. (1987) Neutral surfaces. *J. Phys. Ocean.* **17**, 1950-1964.

McWilliams, J.C., & J.H.S. Chow (1981) Equilibrium geostrophic turbulence: I. A reference solution in a β-plane channel. *J. Phys. Ocean.* **11**, 921-949.

McWilliams, J.C., N.J. Norton, P.R. Gent, & D.B. Haidvogel (1990) A linear balance model of wind-driven, mid-latitude ocean circulation. *J. Phys. Ocean.* **20**, 1349-1378.

McWilliams, J.C. (1996) Modeling the oceanic general circulation. *Annual Rev. of Fluid Mech.* **28**, 1-34.

McWilliams, J.C., G. Danabasoglu, & P.R. Gent (1996) Tracer budgets in the Warm Water Sphere. *Tellus* **48A**, 179-192.

McWilliams, J.C., & J.M. Restrepo (1998) The wave-driven ocean circulation. Preprint.

Mellor, G., & T. Yamada (1982) Development of a turbulence closure model for geophysical fluid problems. *Rev. Geophys. Space Phys.* **20**, 851-875.

Merryfield, W.J., & G. Holloway (1998) Eddy fluxes and topography in stratified quasi-geostrophic models. Preprint.

Munk, W.H., & E. Palmen (1951) Note on the dynamics of the Antarctic Circumpolar Current. *Tellus* **3**, 53-55.

Owens, W.B., & F.P. Bretherton (1978) A numerical study of mid-ocean mesoscale eddies. *Deep-Sea Res.* **25**, 1-14.

Pacanowski, R., & G. Philander (1981) Parameterization of vertical mixing in numerical models of tropical oceans. *J. Geophys. Res.* **11**, 1443-1451.

Pinardi, N., A. Rosati, & R.C. Pacanowski (1995) The sea surface pressure formulation of rigid lid models. Implications for altimetric data assimilation studies. *Jour. Marine Systems* **6**, 109-120.

Polzin, K.L., J.M. Toole, J.R. Ledwell, & R.W. Schmitt (1997) Spatial variability of turbulent mixing in the abyssal ocean. *Science* **276**, 93-96.

Pratt, L.J., ed. (1990) *The Physical Oceanography of Sea Straits*, Dordrecht, Kluwer, 587p.

Price, J.F., & M.O. Beringer (1994) Outflows and deep water production by marginal seas. *Prog. In Ocean.* **33**, 161-200.

Redi, M.H. (1982) Oceanic isopycnal mixing by coordinate rotation. *J. Phys. Ocean* **12**, 1154-1158.

Rhines, P.B. (1977) The dynamics of unsteady currents. In: *The Sea* **VI**, Goldberg, E., ed., New York, Wiley, 189-318.

Rhines, P.B., & W.R. Young (1982) Homogenization of potential vorticity in planetary gyres. *J. Fluid Mech.* **122**, 342-367.

Roberts, M.J., A.L. New, R.A. Wood, & R. Marsh (1995) An intercomparison of a Bryan-Cox type ocean model and an isopycnic ocean model. Part 1: the subpolar gyre and high-latitude processes. *J. Phys. Ocean.*, **26**, 1495-1527.

Roberts, M., & D. Marshall (1997) Do we require adiabatic dissipation schemes in eddy-resolving ocean models? Preprint.

Sakamoto, T., & T. Yamagata (1996) Seasonal transport variations of the wind-driven circulation in a tow-layer planetary-geostrophic model with a continental slope. *J. Mar. Res.* **54**, 261-284.

Salmon, R., G. Holloway, & M.C. Hendershott (1976) The equilibrium statistical mechanics of simple quasi-geostrophic models. *J. Fluid Mech.* **75**, 691-703.

Saravanan, R., & J.C. McWilliams (1995) Multiple equilibria, natural variability, and climate transitions in an idealized ocean-atmosphere model. *J. Climate* **8**, 2296-2323.

Saravanan, R., & J.C. McWilliams (1997) Stochasticity and spatial resonance in interdecadal climate fluctuations. *J. Climate*, in press.

Siegel, A., P. Berloff, J.C. McWilliams, J.B. Weiss, & I. Yavneh (1998) The quasi-geostrophic dynamics of wind-driven ocean gyres at very high resolution. Preprint.

Shchepetkin, A., & J.C. McWilliams (1997) Quasi-monotone advection schemes based on explicit locally adaptive dissipation. *Monthly Weather Review*, in press.

Smith, R.D., S. Kortas, & B. Meltz (1995) Curvilinear coordinates for global ocean models. Los Alamos Technical report, LAUR-95-1146.

Smith, R.D., M.E. Maltrud, M.W. Hecht, & F.O. Bryan (1998) Numerical simulation of the North Atlantic Ocean at $1/10^o$. Preprint.

Sommeria, J. (1998) Statistical mechanics of potential vorticity for parameterizing mesoscale eddies. In *Ocean Modeling and Parameterization*, E.P. Chassignet and J. Verron (Eds.), Kluwer Academic Publishers, 303-326.

Song, Y., & D. Haidvogel (1994) A semi-implicit ocean circulation model using a generalized topography-following coordinate system. *J. Comp. Phys.* **115**, 228-244.

Suginohara, N., & M. Fukasawa (1988) Set-up of deep circulation in multi-level numerical models. *J. Ocean Soc. Japan* **44**, 315-336.

Toole, J. (1998) Turbulent mixing in the ocean: Intensity, causes, and consequences. In *Ocean Modeling and Parameterization*, E.P. Chassignet and J. Verron (Eds.), Kluwer Academic Publishers, 171-191.

Treguier, A.M., & J.C. McWilliams (1990) Topographic influences on wind-driven, stratified flow in a β-plane channel: an idealized model of the Antarctic Circumpolar Current. *J. Phys. Ocean.* **20**, 324-343.

Treguier, A. M., J.K. Dukowicz, & K. Bryan (1996) Properties of nonuniform grids used in ocean general circulation models. *J. Geophys. Res.* **101**, 20877-20881.

Treguier, A.M., I.M. Held, & V.D. Larichev (1997) On the parameterization of quasi-geostrophic eddies in primitive equation ocean models. *J. Phys. Ocean.* **27**, 567-580.

Trenberth, K.E., and A. Solomon (1994) The global heat balance: heat transports in the atmosphere and ocean. *Climate Dyn.* **10**, 107-134.

Visbeck, M., J. Marshall, T. Haine, & M. Spall (1997) Specification of eddy transfer coefficients in coarse-resolution ocean circulation models. *J. Phys. Ocean.* **27**, 381-402.

Wajsowicz, R.C. (1993) A consistent formulation of the anisotropic stress tensor for use in models of the large-scale ocean circulation. *J. Comp. Phys.* **105**, 333-338.

Wijffels, S.E., R.W. Schmitt, H.L. Bryden, & A. Stigebrandt (1992) Transport of freshwater by the oceans. *J. Phys. Ocean.* **22**, 155-162.

Yavneh, I., & J.C. McWilliams (1995) Robust multigrid solution of the shallow-water balance equations. *J. Comp. Physics* **119**, 1-25.

Yin, F.L., & I.Y. Fung (1991) On the net diffusivity in ocean general circulation models. *J. Geophys. Res.* **96**, 10773-10776.

Zhang, J., R.W. Schmitt, & R.-X. Huang (1998) Sensitivity of GFDL Modular Ocean Model to the parameterization of double-diffusive processes. *J. Phys. Ocean.*, in press.

FORCING THE OCEAN

BERNARD BARNIER
Centre National de la Recherche Scientifique,
Laboratoire des Écoulements Géophysiques et Industriels,
Grenoble, France.

1. Introduction

An introduction to an oceanography class could begin with a description of the oceans in these terms: "the ocean is made of saltwater contained in basins ...", and it could continue as: "... and therefore the physical boundaries of the ocean are the bottom of the basins, the coastlines, and the free surface in permanent contact with the atmosphere". Heat entering the ocean through the bottom of the basins is usually considered as negligible. Estimates of local geothermal sources of heat are quite unreliable, since the methods are often based on poorly constrained budgets, but some individual sources have been measured carefully (Converse *et al.*, 1984). These are typically $10^8\,W$, and although this represents one chimney, we can imagine it distributed over the area of the mid-ocean rift valley, to give about $1\,Wm^{-2}$ on average. The flux falls off to values more than three order of magnitude smaller down the flank and onto the abyssal plain. The entire budget for the world ocean is roughly $10^{13}\,W$, which is thought to be carried by the hydrothermal part; the conductive part is a similar order of magnitude. For comparison, the annual mean solar heat flux entering the ocean from the sea surface is above $150\,Wm^{-2}$ on average, corresponding to a net heat input of $50 \times 10^8 W$. In consequence, it is accurate to consider that the main external forcing of the ocean circulation comes from the free surface. Nevertheless, the presence of geothermal heating at the bottom destabilizes the water column, and leads to significant regional flow and mixing (Speer and Helfrich, 1995).

The ocean receives energy through the air-sea interface by exchange of momentum, which represents a source of kinetic energy, and by exchanges of heat and water, which modify the buoyancy into the fluid and so act as a source of potential energy. Forcing an ocean model first requires the knowledge of air-sea exchanges of energy and mass at the ocean surface. One

E. P. Chassignet and J. Verron (eds.), Ocean Modeling and Parameterization, 45–80.

of the first difficulties with which we are confronted is to estimate fluxes through a moving interface, and despite the present technology, measurements at the ocean surface are still a challenge (Figure 1) and air-sea flux estimates are still subject to significant errors and biases. A second major difficulty is that exchanges between the ocean and the atmosphere are by nature interactive. Variations in one fluid will force variations in the other fluid. This raises the question as to whether or not it is conceivable to drive one fluid without considering its coupling with the other. This question can be partially answered by considering the nature of the coupling between the two fluids.

Figure 1. A picture of a NOAA 12-meter diameter platform used to measure atmospheric parameters near the surface of the ocean.

The heating rate of a fluid column (Figure 2) is given by:

$$\frac{\partial T}{\partial t} = \frac{Q}{\rho C_p \Delta z} \tag{1}$$

where T is the fluid temperature, t is time, Q is the incoming heat flux, Δz is the height of the fluid column, ρ is the density, and C_p is the specific heat of the fluid.

Although the specific heats of the ocean and of the atmosphere are not so different (Table 1), the heat capacity per volume (i. e., ρC_p) is much greater for the ocean than for the atmosphere.

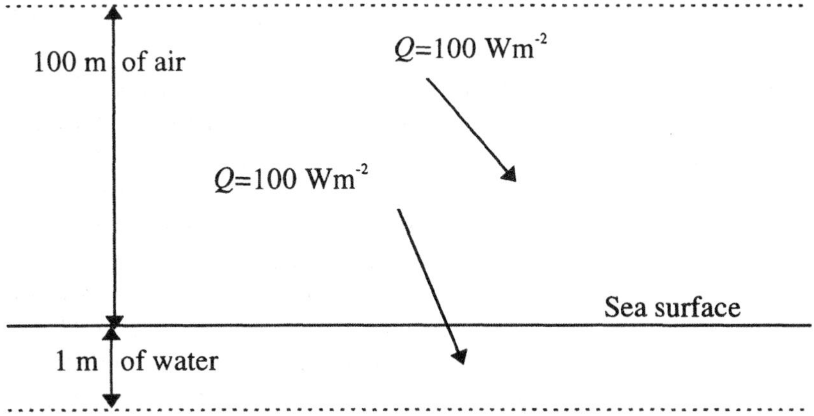

Figure 2. Schematic used to explain the heating rate of the first 100 m of air, and of the first meter of ocean, by an incoming flux of 100 Wm^{-2}.

For a column of air of 100 m heated by a flux of 100 Wm^{-2} the heating rate is

$$\frac{\partial T}{\partial t} = 2.77^o C/h \tag{2}$$

For the first meter of ocean, a heat flux of 100 Wm^{-2} results in a heating rate of

$$\frac{\partial T}{\partial t} = 0.083^o C/h \tag{3}$$

TABLE 1. Parameters entering into the calculation of the heating rate

Atmosphere	Ocean
$C_{p_a} = 1.0048 \times 10^3\ J.kg^{-1}.^oC^{-1}$	$C_{p_w} = 4.18 \times 10^3\ J.kg_{-1}.^oC^{-1}$
$\rho_a = 1.29\ kg.m^{-3}$	$\rho_w = 1.029 \times 10^3\ kg.m^{-3}$

Due to the larger heat capacity (i. e., ρC_p) of sea water, the heating rate is more than 30 time slower for the first meter of ocean than for the first 100 m

of air. Therefore, the ocean is the slow component of the coupled ocean-atmosphere system, and this has several consequences in oceanography and meteorology.

- *Short term weather forecasting (3 to 5 day time scale)*: Atmospheric general circulation models (AGCM) are using a constant sea surface temperature (SST) over the forecast period, and errors induced by this approximation are reasonable due to the much slower evolution of the ocean temperature. Possible exceptions are extreme events such as hurricanes in which perturbations in SST are large over short periods.
- *Seasonal weather forecasting (annual time scale)*: AGCMs require a prognostic SST, which can be provided by a more or less sophisticated ocean model.
- *Long term climate studies*: Fully coupled ocean/atmosphere general circulation models are absolutely necessary.
- *Ocean modeling*: With the objectives of describing the present state of the ocean, and understanding which processes drive the mean circulation and its variability (i. e. the objectives of WOCE and CLIVAR), model simulations of the ocean circulation require long term model integration (over several tens of years). For this purpose, it appears conceivable to drive an ocean general circulation model (OGCM) with a climatology of atmospheric variables obtained over the last 50 years of observations, provided the ocean model does not drift too far from the observed state.

However, there is a paradox in forcing an ocean model with observed atmospheric fluxes, since this attempts to use the fast component of a coupled system to drive the slow component. It is likely that the ocean model will "integrate" the short term variations of the atmospheric forcing, which may appear as a stochastic forcing with regard to the slow variations of the ocean properties. But using fixed observed fluxes (which are independent of the state of the model) usually drives the model toward a state which is unrealistic and incompatible with the forcing function itself, due to inaccuracies in the forcing and the model (Rosatti and Miyakoda, 1988).

For example, overheating of a tropical ocean model could happen due to either too much solar radiation in the forcing data, or too little upwelling in the ocean model because of an inaccurate parameterization of vertical and/or horizontal mixing. Under such conditions, the model SST will increase and may reach values significantly different from the SST used to estimate the air-sea fluxes.

Using observed fluxes only also reduces the degrees of freedom of the ocean model, and may produce an unrealistic model response. The meridional heat transport in a model experiment carried out to a steady state (no heat storage), for example, will be totally controlled by the heat flux imposed at the surface, and will therefore be independent of the model circulation. In this case, allowing a correction which depends on the time evolution of the model SST to be applied to the heat forcing adds one degree of freedom, and allows the model to produce its own estimate for the meridional heat transport (Figure 3).

Two important aspects of the forcing function of an ocean model have been identified in the introduction above: the accuracy of the air-sea flux estimates derived from surface and/or satellite observations, and the definition of a parameterization which accounts for the retroaction of the ocean onto the air-sea exchanges. This paper discusses these aspects. The surface boundary conditions for the primitive equations are described in section 2. The forcing function of an ocean model is defined in section 3, and the accuracy of the various flux fields which are available to force a global model is discussed in section 4.

2. Surface Boundary Conditions for the Primitive Equations

This section provides the formulation of the surface boundary conditions for the primitive equations. Notations are classic and are not described unless we believe it is necessary.

2.1. FLUX BOUNDARY CONDITION

There are several ways to formulate the forcing of an OGCM based on the primitive equations. A formulation which attempts to remain close to the physics of the air-sea interactions is the "flux boundary condition". It is assumed that vertical gradients of the prognostic model variables (horizontal velocity, temperature and salinity) are proportional to the flux across the interface.

2.1.1. *Momentum*

– *Momentum equation*

$$\frac{\partial \vec{u}}{\partial t} + \vec{U}.\vec{\nabla}\vec{u} + f\vec{k} \times \vec{u} = -\frac{1}{\rho_0}\vec{\nabla}P + A_H \nabla^2 \vec{u} + \frac{\partial}{\partial z}(A_V \frac{\partial \vec{u}}{\partial z}) \qquad (4)$$

– *At the surface*

$$(A_V \frac{\partial \vec{u}}{\partial z})_{z=0} = \frac{\vec{\tau}}{\rho_0} \qquad (5)$$

Equation (5) is the flux boundary condition at the ocean surface. It states that the vertical velocity gradient at the surface is a function of the wind

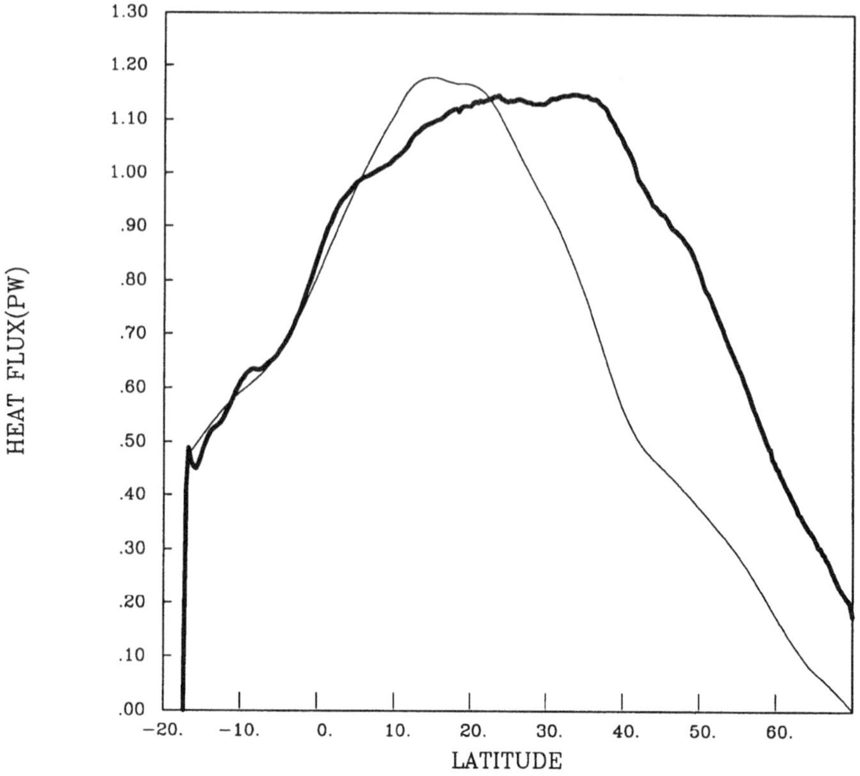

Figure 3. Estimates of the meridional heat transport in the North Atlantic Ocean from a model experiment [here the SPEM model in the Dynamo experiment (Dynamo, 1997)]. The thin curve is the transport estimate obtained from the divergence of the net heat flux forcing the model. It would be the transport given by the model if forced with fixed fluxes. The thick curve is the transport estimate of the dynamical model forced with a retroaction term which depends on the model SST acting on the heat forcing. The difference is what is brought about by the dynamical response of the model.

stress vector and a turbulent vertical viscosity. Note that the knowledge of the vertical viscosity is not necessary to apply this boundary condition in (4).

2.1.2. *Heat*

– *Temperature equation*

$$\frac{\partial T}{\partial t} + \vec{U}.\vec{\nabla} T = K_H \nabla^2 T + \frac{\partial}{\partial z}(K_V \frac{\partial T}{\partial z}) + F_{SOL}(z) \tag{6}$$

– *At the surface*

$$(K_H \frac{\partial T}{\partial z})_{z=0} = \frac{Q_{NSOL}}{\rho_0 C_{p_w}} \tag{7}$$

In this formulation, F_{SOL} is the heating rate (in $^oC/s$) due to the penetration of solar radiation in the upper ocean. It can be calculated from the solar heat flux Q_{SOL} using an absorption law for sea water (Paulson and Simpson, 1977). Equation (7) is the flux boundary condition for the temperature equation. It expresses the vertical temperature gradient at the surface as a function of the non-solar heat flux (Q_{NSOL}, in Wm^{-2}). The solar heating is often not parameterized according to an absorption law in OGCMs, but as a flux boundary condition. In that case, $F_{SOL} = 0$ in (6), and in (7), the non solar heat flux Q_{NSOL} is replaced by the net heat flux Q_{NET}.

2.1.3. *Salt*

The salinity of the ocean, S, is defined as the total amount of dissolved salt (in grams) contained in 1 kg of sea water (Pickard and Emery, 1982). The primitive equations express the conservation of salt in the ocean by an equation of conservation for salinity, and the definition of the surface boundary condition for this equation poses a conceptual problem.

– *Salinity equation*

$$\frac{\partial S}{\partial t} + \vec{U}.\vec{\nabla} S = K_H \nabla^2 S + \frac{\partial}{\partial z}(K_V \frac{\partial S}{\partial z}) + F_{SOL}(z) \qquad (8)$$

– *At the surface*

$$(K_H \frac{\partial S}{\partial z})_{z=0} = S.(E - P - R) \qquad (9)$$

In (9), the surface boundary condition for salinity is formulated in the form of a salinity flux across the air-sea interface, which relates the vertical gradient of salinity at the surface to the surface freshwater flux expressed as the algebraic sum of the evaporation, E, precipitation, P, and river runoff, R, all expressed in m/s. This formulation is the most commonly used in OGCMs. However, it must be kept in mind that it is an approximation which does not conserve salt in the ocean, and is therefore a potential problem in long term ocean and coupled ocean-atmosphere model integrations. This condition is often called the **virtual salinity flux** boundary condition.

2.1.4. *Remark on the virtual salinity flux*

The formulation of the virtual salinity flux can be derived from the local freshwater and salt budgets at the surface of an ocean submitted to a constant evaporation rate E. Considering the control volume just beneath the ocean surface, the evaporation process takes freshwater out across the top surface, and generates a compensating upward flux of salt water across the bottom surface (Figure 4).

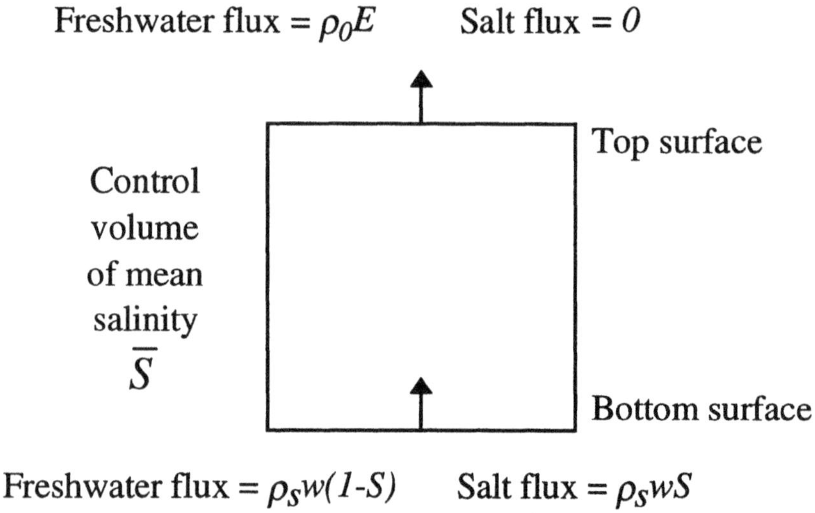

Figure 4. Schematic used for the calculation of the virtual salt flux where E = evaporation rate (m/s) at the top surface of the volume, ρ_0 = density of freshwater, ρ_S = density of saltwater, w = vertical velocity at the bottom surface of the volume and S = salinity of the water at the bottom surface of the volume.

We assume that a steady regime is established. However, vertical fluxes are turbulent by nature, and a budget requires that turbulent transports be considered. Therefore turbulent quantities such as vertical velocity and salinity are split into steady and fluctuating components, according to the usual notations:

$$w = \bar{w} + w', \ \ S = \bar{S} + S' \tag{10}$$

and a salinity flux can be defined as:

$$F_S = \overline{w'S'} \tag{11}$$

Note that F_S is indeed a salinity flux (not a salt flux) and can be used as a surface boundary condition for the salinity equation (8).

Salt budget. There is no flux of salt across the free surface. Therefore, as shown in Figure 4, the salt flux across the top surface of the control volume is identically zero. Considering the conservation of salt in the control volume, the salt budget states that the salt flux across the bottom surface is identically zero:

$$\rho_S \overline{wS} = 0 \tag{12}$$

Using (10) yields

$$\rho_S(\bar{w}\bar{S} + \overline{w'S'}) = 0 \tag{13}$$

and the salinity flux is

$$\overline{w'S'} = -\bar{w}\bar{S} \tag{14}$$

Freshwater budget. The freshwater budget states that the freshwater flux is continuous across the control volume, which yields:

Flux at top surface = flux at bottom surface

$$\rho_0 E = \rho_S \overline{w(1-S)} \tag{15}$$

$$\rho_0 E = \rho_S \overline{(\bar{w} + w')(1 - \bar{S} - S')} \tag{16}$$

and using (14) this yields

$$\rho_0 E = \rho_S \bar{w} \tag{17}$$

An expression for the salinity flux with a combination of (14) and (17) is

$$F_S = \overline{w'S'} = -\frac{\rho_0}{\rho_S} E\bar{S} \tag{18}$$

The flux boundary condition for the salinity equation (8) is then established using the usual approximation of Prandtl which relates the flux and the vertical gradient of the mean quantities:

$$(K_V \frac{\partial \bar{S}}{\partial z})_{z=0} = -\overline{w'S'} \tag{19}$$

Combining (18) and (19), and replacing the evaporation rate E by the full budget including precipitation and river runoff, and dropping the overbar, we obtain a flux boundary condition similar to (9):

$$(K_V \frac{\partial \bar{S}}{\partial z})_{z=0} = \frac{\rho_0}{\rho_S} S.(E - P - R) \tag{20}$$

To obtain the exact expression shown in (9), the ratio of the density of freshwater to the density of salt water is approximated as unity.

We have shown that the virtual salinity flux boundary condition results from several approximations. Furthermore, as we have already mentioned, this formulation is not conservative of the amount of salt in the model. Nevertheless, this formulation is widely used in OGCMs, and is adequate for most model experiments, as long as the integration time (up to several hundred years) remains short compared to the thermohaline equilibrium time scale of the ocean. However, present climate research requires ocean and coupled ocean-atmosphere model integration of several thousand years, attempting to simulate the state of thermohaline equilibrium of the ocean. The formulation of the salinity forcing which we have presented above is not adequate in this case, and other formulations based on the variation of the free surface (thus not adequate for rigid lid models) have been developed, which are conservative of the total amount of salt in the ocean.

2.2. FORCING VARIABLES

We have presented the surface boundary conditions for the primitive equations. These conditions require parameters external to the ocean model: the forcing fields. In the formulation presented above, the necessary forcing fields are:

- the wind stress vector ($\vec{\tau}$ in Nm^{-2}),
- the solar heating (F_{SOL}, in $^oCs^{-1}$) which in fact requires the solar radiation at the surface (Q_{SOL} in Wm^{-2}), the albedo A, and a penetration law for the shortwave radiation,
- the non solar heat flux (Q_{NSOL} in Wm^{-2}), which is the algebraic sum of the net longwave radiation Q_{IR}, latent heat flux Q_{LAT}, and sensible heat flux Q_{SENS},
- the evaporation rate (E in ms^{-1}), the precipitation rate (P in ms^{-1}), and the river runoff (R in ms^{-1}).

The best possible knowledge of these quantities is necessary to define the forcing function of an ocean model. However, as we shall see below, several additional fields are required to fully determine a forcing function which includes a parameterization of the retroaction of the ocean onto air-sea fluxes.

3. Forcing Function of an Ocean Model

This section carefully describes a forcing function for a global ocean model. According to the scientific objectives of the model simulation, different approaches are possible. However, there is an underlying principle that must be consider in all cases: *the ocean is the slowly evolving component of the coupled ocean/atmosphere system, and an Ocean GCM should not be forced without some kind of parameterization to account for the feedback to the atmosphere.*

Our discussion is not intended to be exhaustive and to cover all possible situations. Instead, we consider the case in which we want to drive a primitive equation ocean model with a climatic record of air-sea exchanges (several tens of years). Such a record could be provided, for example, by a monthly mean climatology of surface observations from the Comprehensive Ocean and Atmosphere Data Set (COADS) (Woodruff *et al.*, 1987), or by daily data from the reanalyses carried out at the National Center for Environmental Predictions (NCEP) (Kalnay *et al.*, 1996), or at the European Center for Medium Range Weather Forecast (ECMWF) (Gibson *et al.*, 1997). Three steps are necessary to determine the forcing function. First, we must decide on a formulation for the surface boundary condition. Second, we must define a retroaction term which parameterizes a feedback of the ocean to the atmosphere. And third, we must obtain estimates of the required surface variables (wind stress, heat fluxes, $E-P$, runoff, sea level pressure, ...) with the proper time and space sampling for the application. The first step has been done in section 2 for a flux boundary condition in a rigid lid model. In the following, we discuss a parameterization of the retroaction term and the surface fields, for each component of the forcing (heat, salinity, momentum).

3.1. FORCING FUNCTION FOR THE TEMPERATURE EQUATION

For simplicity in the following, the solar radiation is treated as a surface flux, but the adaptation of what follows to the case of a penetration law is straightforward. At a time t of the model integration, the flux boundary condition for the temperature equation is expressed as:

$$(K_V \frac{\partial T}{\partial z})_{z=0} = \frac{Q_{NET}}{\rho_0 C_{p_w}} \tag{21}$$

where $Q_{NET}(t)$ is the instantaneous forcing function of the ocean, and must include:

- the contribution of the climatic record to the net heat flux at this time, a quantity derived from observations and independent of the model:

$$Q_{NET}^{CLIM}(t) = Q_{SOL}^{CLIM} + Q_{IR}^{CLIM} + Q_{LAT}^{CLIM} + Q_{SENS}^{CLIM} \tag{22}$$

- the contribution of a *model dependent* retroaction term, $Q_{CORR}(t)$, the objectives of which are to account for the feedback of the ocean on the air-sea fluxes, and to compensate for uncertainties in air-sea flux estimates which may have a large impact on model results (since to over-estimate Q_{NET} by only 1 Wm^{-2} would increase the temperature of the upper 100 m of the ocean by 7.5 ^oC after 100 years).

Therefore, the net heat flux applied at the model surface is expressed as follows:

$$Q_{NET}(t) = Q_{NET}^{CLIM}(t) + Q_{CORR}(t) \tag{23}$$

and both components must be fully determined at each model time step. It is shown below how to calculate these two terms.

3.1.1. *Relaxation to a climatic sea surface temperature*
In their early stages, OGCMs have been forced with a simple relaxation of the model surface temperature T_S to a climatic sea surface temperature T_S^{CLIM}, with a time constant, λ, which determines how long an anomaly in surface temperature can persist before it is damped by the heat forcing. In this case, the surface flux is a diagnostic quantity,

$$Q_{NET}(t) = \rho_w C_{p_w} \Delta z \lambda [T_S^{CLIM} - T_S(t)] \tag{24}$$

where Δz is the thickness of the first model level where the relaxation is applied. The forcing function defined in (24) can be written as follows:

$$Q_{NET}^{CLIM}(t) = 0 \text{ and } Q_{CORR}(t) = \rho_w C_{p_w} \Delta z \lambda [T_S^{CLIM} - T_S(t)] \tag{25}$$

With this formulation, the forcing controls the model surface temperature, which tends to remain close to the climatic SST. This formulation has well known shortcomings:

- At equilibrium, a model that simulates an SST equal to the climatic SST (*i.e.*, $T_S = T_S^{CLIM}$) assumes that the net heat flux given by (24) is zero, which is not consistent with what we know of the climate system in which the ocean cannot sustain a state of equilibrium with no exchange with the atmosphere. Note however that in specific regional applications where the net heat flux is expected to be zero, this formulation might be appropriate (Frankignoul *et al.*, 1998).
- A model that diagnoses a realistic heat flux will do so with unrealistic values in sea surface temperature.
- The model SST is controlled by relaxation rather than by ocean dynamics.

3.1.2. *Example of retroaction term*
The underlying idea of the formulation of the retroaction term presented below is that air-sea heat exchanges locally depend upon changes in SST (Haney, 1971). Therefore, the surface heat flux can be expanded in Taylor series about the observed climatic SST used to estimate the heat flux (Barnier *et al.*, 1995):

$$Q_{NET}(T_S) = Q_{NET}(T_S^{CLIM}) + \left(\frac{\partial Q_{NET}}{\partial T}\right)_{T_S^{CLIM}} (T_S^{CLIM} - T_S) \tag{26}$$

where T_S^{CLIM} is the observed climatic surface temperature and T_S is the model surface temperature.

In (26), the term

$$(\frac{\partial Q_{NET}}{\partial T})_{T_S^{CLIM}} \quad (27)$$

is a climatic estimate of the sensitivity of the net heat flux to variations in SST. Compared to the plain relaxation to a climatic SST, this formulation has the advantage of being more consistent with the physics of air-sea interactions. For instance:

- At equilibrium, a model that simulates an SST equal to the climatic SST (*i.e.*, $T_S = T_S^{CLIM}$) assumes that the net heat flux is the observed heat flux climatology.
- As long as the correction term remains small, the model SST is controlled by ocean dynamics rather than by relaxation.

A small value of the sensitivity $\partial Q/\partial T$ does not necessarily mean a weak retroaction because the temperature difference, which is a function of the model errors and an inaccurate estimate of the climatic heat flux, can be large. To estimate the sensitivity, it is necessary to have a formulation of the net heat flux as a function of the sea surface temperature. An interesting aspect of the above formulation, which explicitly separates the climatological forcing flux and the model dependent correction term, is that it allows using independently the best available estimate of the air-sea flux and the best available estimate of the correction term.

3.1.3. *Heat flux bulk formulation*

Direct measurements of the exchange of heat at the air-sea interface are not possible. Fluxes across are the air-sea interface are generally estimated from surface meteorological variables measured at anemometric height (usually 10 m above the sea level) using the standard aerodynamic bulk formulae. We present below a formulation for the various components of the net heat flux. The oceanographic convention is used, so a heat gain for the ocean is positive.

Net solar radiation. The formulation of the net solar radiation presented here can be used for monthly mean estimates over areas at least as large as 100 km^2. Numerical weather prediction centers (NWPCs) are using radiation schemes to work on shorter time scales. The formula for the net solar radiation used by Rosati and Miyakoda (Rosatti and Miyakoda, 1988) to force their numerical world ocean model is:

$$Q_{SOL} = Q_{TOT}(1 - 0.62C + 0.0019\beta)(1 - \alpha) \quad (28)$$

where:

C	is the fractional cloud cover,
β	is the solar noon altitude in degrees,
α	is the albedo of the ocean surface,
Q_{TOT}	is the clear sky radiation at the ocean surface.

Q_{TOT} has two components $\qquad Q_{TOT} = Q_{DIR} + Q_{DIFF}$

t = atmospheric attenuation (t=0.7)	the direct solar radiation at the ocean surface
Z = zenith angle	$Q_{DIR} = Q_{TOA}\tau^{secZ}$
A_a = ozone absorption (A_a = 0.09)	and the isotropic back-scattered radiation
J_0 = solar constant = 135 Wm^{-2}	$Q_{DIFF} = [(1 - A_a)Q_{TOA} - Q_{DIR}]/2$
a = radius of the earth	The radiation at the top of the atmosphere is
D_F = fraction of daylight	$Q_{TOA} = \frac{J_0}{a^2} D_F(\phi, \lambda) cos z$

Net longwave radiation. The formula for the net longwave radiation shown here is the one used by Rosati and Miyakoda (1988) to force their numerical world ocean model. This is very similar to the formulation of Clarke *et al.* (1974) .

$$Q_{IR} = -\epsilon\sigma T_S^4(0.39 - 0.05e_a^{1/2})(1 - BC) - 4\epsilon\sigma T_S^3(T_S - T_a) \qquad (29)$$

where:

e_a	is the atmospheric vapor pressure (mb),
ϵ	is the emissivity of the ocean (0.97),
B	is the cloudiness correction factor (0.5 to 0.8),
C	is the fractional cloud cover used in (28),
T_S	is the ocean surface temperature,
T_a	is the air surface temperature,
σ	is the Stephan-Boltzman constant.

The atmospheric vapor pressure e_a is related to the specific humidity q_S used to estimate the latent heat flux and the saturation pressure vapor. Note that there are different formulations for the cloudiness factor (Large *et al.*, 1977).

Climatological satellite estimates of the net shortwave and longwave radiation obtained from the Surface Radiation Budget (SRB) (Darnell *et al.*,

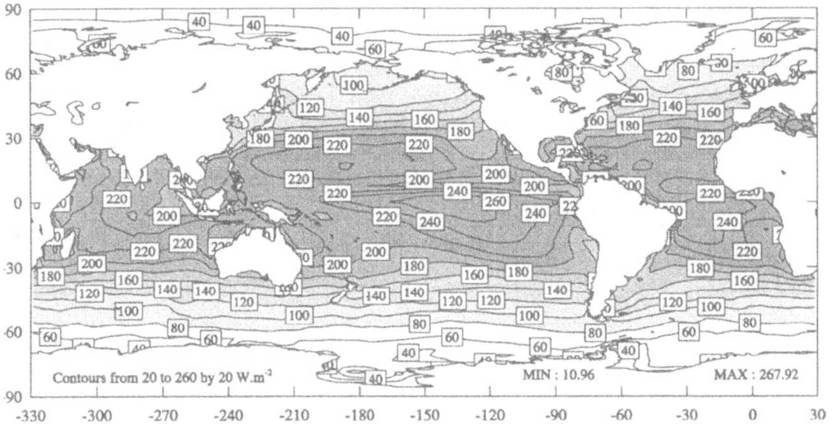

Figure 5. Satellite estimate of the net shortwave radiation (in Wm^{-2}) from the Surface Radiation Budget experiment (Darnell *et al.*, 1996). Contour interval is 20 Wm^{-2}.

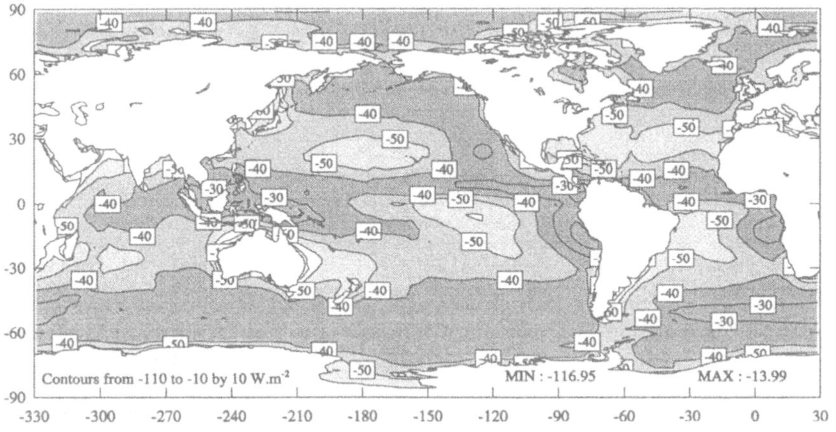

Figure 6. Satellite estimate of the net longwave radiation (in Wm^{-2}) from the Surface Radiation Budget experiment (Darnell *et al.*, 1996). Contour interval is 10 Wm^{-2}.

1996) experiment are shown in Figure 5 and Figure 6. Solar heat flux is zonally distributed outside the tropics, decreasing poleward. It is a maximum at tropical latitudes (it reaches 260 Wm^{-2} in the South Tropical Pacific) and is slightly lower (200 Wm^{-2}) in the intertropical convergence zone. Compared to radiative fluxes from ECMWF (known to be underestimated due to too much absorption from convective clouds), the SRB shortwave is larger in the inter-tropical band (Garnier *et al.*, 1998). In the eastern south tropical Atlantic and Pacific, the shortwave heat flux is lower than in the central tropical oceans (and inversely for the longwave). This feature, related to non-precipitant clouds peculiar to these regions of subsidence, is not found in radiative flux estimates from ECMWF re-analysis, indicating a deficiency in the parameterization of this type of cloud in the atmospheric model.

Sensible heat flux. The classical bulk formula is used in a function of atmospheric variables measured at anemometric height.

$$Q_{SENS} = -\rho_a C_p C_H U_{10}(T_s - T_a) \tag{30}$$

where:

ρ_a	is the air density,
C_p	is the air specific heat,
C_H	is the turbulent exchange coefficient,
U_{10}	is the wind speed at anemometer height,
T_s	is the sea surface temperature,
T_a	is the air temperature at anemometer height.

An estimate derived from COADS in-situ observations (Josey *et al.*, 1998) is shown in Figure 7.

Latent heat flux. The classical expression is used in a function of atmospheric variables measured at anemometric height.

$$Q_{LAT} = -\rho_a C_E L U_{10}(q_s - q_{10}) \tag{31}$$

where:

ρ_a	is the air density,
C_E	is the turbulent exchange coefficient,
L	is the latent heat of vaporization,
U_{10}	is the wind speed at anemometer height,
q_s	is the specific humidity of saturated air,
q_a	is the specific humidity at anemometer height.

An estimate derived from COADS in-situ observations (Josey *et al.*, 1998) is shown in Figure 8.

Net heat flux. The net heat flux is the algebraic sum of the four flux components. A climatological mean estimate of this quantity for the period 1979 to 1993 is presented in Figure 9. It was obtained by averaging 15 years of air sea flux estimates provided by the re-analysis of meteorological data carried out at ECMWF. Major subtropical gyres show considerable heat

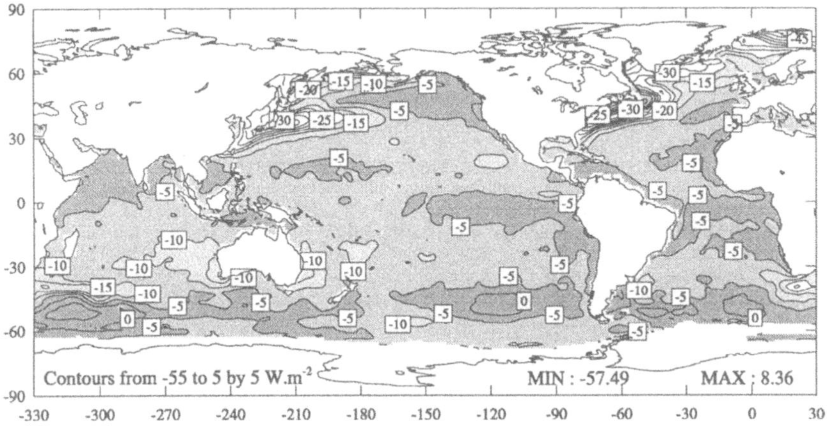

Figure 7. Climatological estimate of the sensible heat flux at the surface of the ocean estimated from the COADS data set (Josey *et al.*, 1998).

loss to the atmosphere (the maximum heat loss is 260 Wm^{-2} in the Gulf Stream), and the great heat gains are found in the equatorial band (150 Wm^{-2} in the eastern Equatorial Pacific) and in areas of strong upwelling.

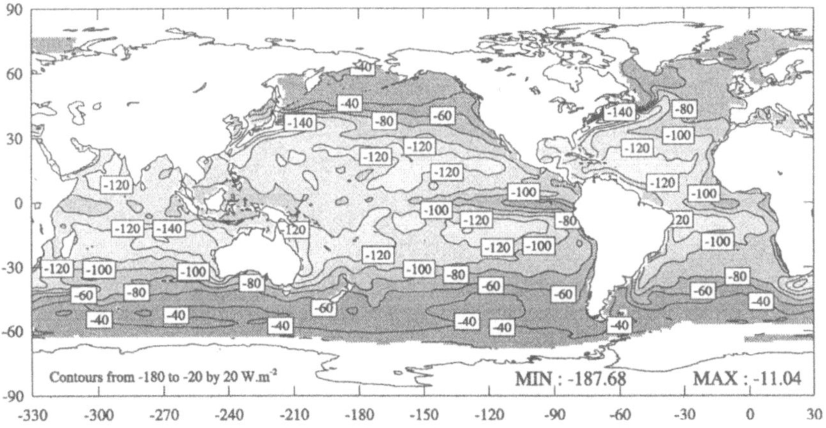

Figure 8. Climatological estimate of the latent heat flux at the surface of the ocean estimated from the COADS data set (Josey *et al.*, 1998).

The bulk formulae presented above intrinsically contain a parameterization of the effect of air-sea feedback on fluxes, due to their dependence on SST, and can be used directly to define the flux boundary condition of an ocean model. In this formulation, which is called the bulk forcing, the sea surface temperature entering the formulae is the surface temperature calculated by the model, and the atmospheric variables entering the formulae are obtained from climatology (Large *et al.*, 1977). This bulk forcing is not discussed in more detail in this paper.

3.1.4. *Formulation of the correction term*

The contribution of each flux component to the correction term $\partial Q/\partial T$ in (27) is obtained by a simple differentiation of the bulk formulae (28), (29), (30), (31), with regard to SST. This calculation has been done by Barnier *et al.* (1995) who proposed the following formulation:

$$\begin{aligned}(\frac{\partial Q_{NET}}{\partial T_S})_{T_S^{CLIM}} &= 4\sigma(T_S^{CLIM})^3 + \rho_a C_p C_H U_{10} \\ &+ \rho_a C_E L U_{10}.2353\ln 10.\frac{q_s(T_S^{CLIM})}{(T_S^{CLIM})^2} \end{aligned} \tag{32}$$

There is no contribution from the solar radiation since it does not directly depend on the ocean surface temperature. The contribution of the long wave radiation, which is only a few $Wm^{-2}K^{-1}$, can be obtained by an exact differentiation of (29), but has been approximated to $4\sigma T^3$.

The calculation of the contribution of the latent heat flux from (31) requires the relationship between the specific humidity of saturated air q_s, and the saturated water vapor e_s,

$$q_s = \frac{0.622}{P_a} e_s(T) \tag{33}$$

The latter quantity is related to the surface temperature by the Clausius-Clapeyron integrated law (P_a is the atmospheric surface pressure).

A 15-year climatological estimate of this correction term, calculated from the ECMWF re-analysis data set, is presented in Figure 10 Values range from 30 $Wm^{-2}K^{-1}$ to 60 $Wm^{-2}K^{-1}$, the largest values being found in the tropics and within the subtropical gyres. This estimate of the retroaction term in very good agreement with the one proposed from three years of ECMWF analyses relative to the period 1986 to 1988 by Barnier *et al.*

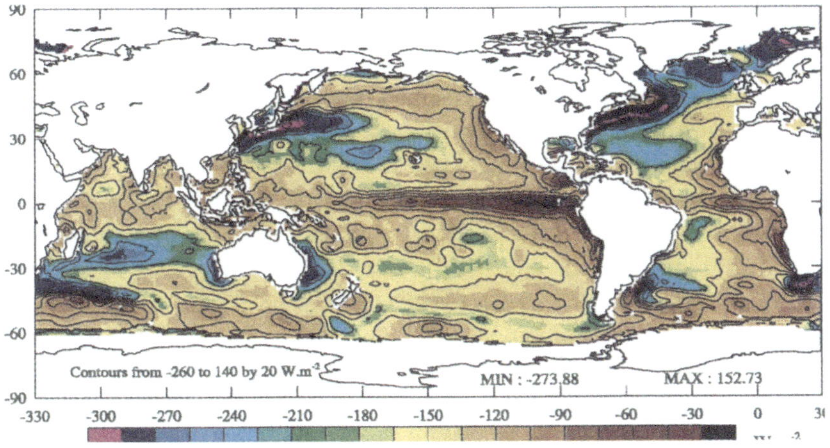

Figure 9. Climatological mean estimate of the neat heat flux at the ocean surface (in Wm^-2) for the period 1979 to 1993 from ECMWF re-analysis. Contour interval is 20 (in Wm^{-2}). Positive values are a gain for the ocean. White areas at high latitudes indicate the extension of sea-ice.

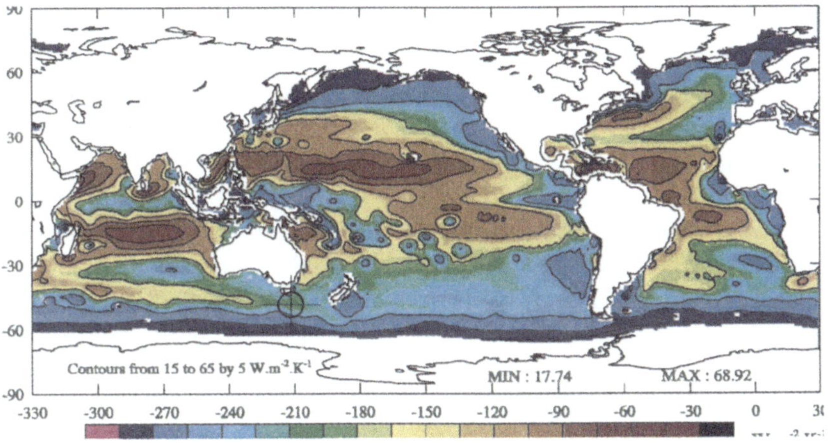

Figure 10. Climatological mean of the net correction term (in $Wm^{-2}K^{-1}$) for the period 1979 to 1993, estimated from ECMWF re-analysis. White areas at high latitudes indicate the extension of sea-ice.

(1995), who also discussed the seasonal cycle of the correction term. The largest variations are found where the monsoon cycle is the dominant signal (with a 30 $Wm^{-2}K^{-1}$ difference between the seasons), and in subtropical gyres where the amplitude of the seasonal cycle can reach 20 $Wm^{-2}K^{-1}$. It is reasonable to consider that the values shown in Figure 10 are maximum estimates of the ocean feedback to the air-sea fluxes. The reason is that, in the formulation of the forcing flux presented in (26) and (32), changes in ocean temperature feed back entirely to the flux, whereas in a coupled system, part of the energy made available by these temperature changes would contribute to change the air temperature of the lower atmosphere, and would feed the dynamics of the atmospheric boundary layer. Therefore, it is likely that the local feedback to the flux would be weaker than what is shown in Figure 10. Regional studies using lag covariance between observed SST and flux anomalies (Frankignoul *et al.*, 1998) suggest that in the North Atlantic subtropical gyre, this estimate of the feedback should be reduced by a factor of 2.

3.2. FORCING THE SALINITY EQUATION

For simplicity in the notation, the salinity forcing is considered as a source term added to the right hand side of the salinity equation:

$$\frac{\partial S}{\partial t} + u\nabla S = K_H \Delta S + \frac{\partial}{\partial z}(K_V \frac{\partial S}{\partial z}) + Source \tag{34}$$

$$Source = \frac{1}{\Delta z} S.(E - P - R) \tag{35}$$

This is a body force formulation which is exactly equivalent to the flux boundary condition formulated in (8) if Δz is the thickness of the first model level.

3.2.1. *Relaxation to a climatic sea surface salinity*

The formulation which has been mostly used in numerical models of the general circulation is a relaxation to a climatic sea surface salinity S^{CLIM}, generally obtained from the atlas of Levitus (1982). The source term is then expressed as:

$$Source = \lambda(S_S^{CLIM} - S_S^{MODEL}) \tag{36}$$

and the fresh water flux is diagnosed as

$$E - P - R = -\lambda \Delta z (1 - \frac{S_S^{CLIM}}{S_S^{MODEL}}) \tag{37}$$

The source term controls the model surface salinity, which tends to remain close to the climatic sea surface salinity (SSS). The shortcomings are similar to those for the temperature, and at equilibrium, a model in which surface salinity is equal to the climatic estimate diagnoses a zero freshwater flux.

3.2.2. *A sample retroaction term*

Following the formulation used for the temperature equation, the source term for the salinity equation can be written as

$$F = \frac{1}{\Delta z} S_S^{MODEL}.(E - P - R) + \lambda(S_S^{CLIM} - S_S^{MODEL}) \tag{38}$$

where:
S_S^{CLIM} is the observed climatic surface salinity,
S_S^{MODEL} is the model surface salinity,
λ is a relaxation time scale (the same as for the SST).

This formulation is more consistent with the physics of the air-sea interactions. For instance;

- at equilibrium, a model that simulates a SSS equal to the climatic SSS (*i.e.*, $S_S^{MODEL} = S_S^{CLIM}$) assumes that the net freshwater flux is the observed freshwater flux climatology.
- as long as the correction term remains small, the model SSS is controlled by ocean dynamics rather than by relaxation.

Heat and salt forcing are coupled through evaporation and latent heat. This is a long term feedback. Considering a given formulation for the heat and salinity forcing, an interesting diagnostic for a simulation of the ocean climate would be to learn if the coupling is correctly parameterized. In any case, a forcing function should attempt to balance with the best possible accuracy the heat and freshwater budgets. An original attempt is described in Large *et al.* (1977) in their formulation of a bulk forcing.

3.2.3. *Formulation of the freshwater flux*

To compute the source term (14), it is necessary to have estimates of the sea surface salinity, the precipitation over the ocean, the river runoff, and the evaporation rate.

Evaporation. In the evaporation process, most of the heat is provided by the ocean, and the evaporation rate can be related to the latent heat flux (31) as follows:

$$E = \frac{Q_{LAT}}{\rho_w L} \tag{39}$$

Uncertainties in the calculation of the latent heat will therefore affect the freshwater flux. A climatological mean estimate of the evaporation field is shown in Figure 11.

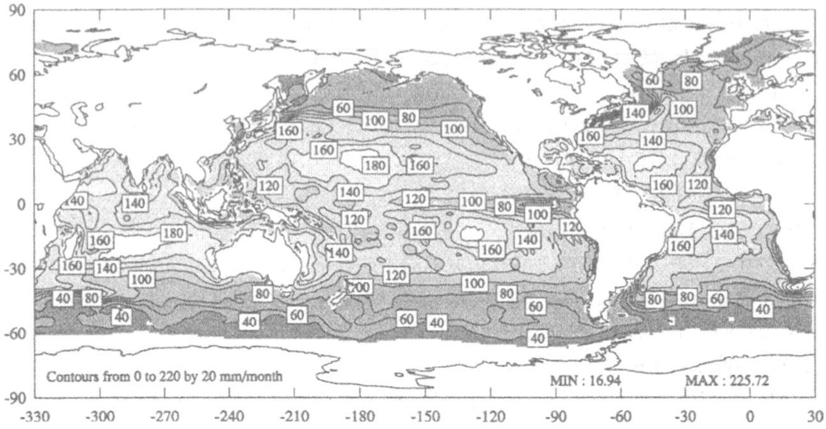

Figure 11. Climatological mean of the evaporation (in $mm/month/m^2$) for the period 1979-1993, estimated from ECMWF re-analyses.

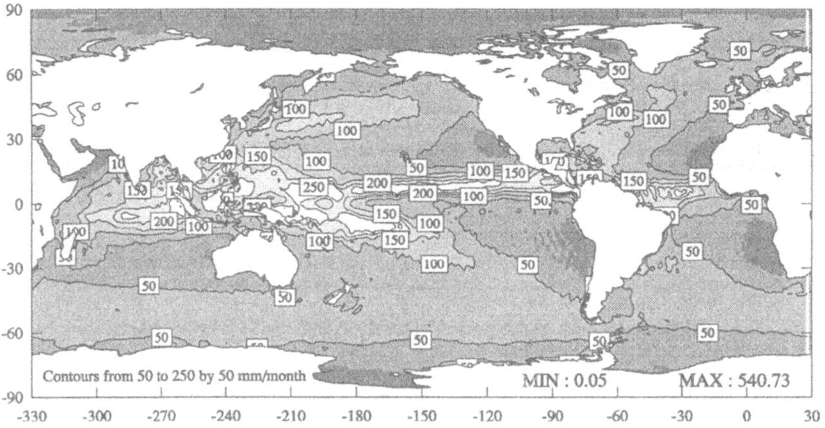

Figure 12. Climatological mean of precipitation (in mm/month/m2) for the period 1979-1993, estimated from ECMWF re-analyses.

Precipitation, river runoff, and sea surface salinity. These quantities must be estimated by direct measurements. Precipitation over the oceans (Figure 12), as well as river runoff, are poorly known, and large discrepancies exist between estimates from different sources, as we show in section 4. Evaporation and precipitation should globally balance to the river runoff.

A mean climatology of the river runoff can be obtained from the annual Baumgartner and Reichel climatology (Baumgartner and Reichel, 1975).

This climatology estimates the total river runoff to $1.26 \times 10^6 m^3 s^{-1}$. For comparison, the $E - P$ balance obtained from the ECMWF re-analysis (Figures 11 and 12) is $2.6 \times 10^6 m^3 s^{-1}$ (Garnier *et al.*, 1998). The UNESCO International Hydrographic Program (IHP) has compiled a data base of the global river discharge for all continents (Vorosmarty *et al.*, 1996). A monthly mean climatology for the major rivers could be derived from this data base.

The sea surface salinity is usually obtained from local or global atlases such as the atlas of Levitus (1982, 1994).

3.3. FORCING THE MOMENTUM EQUATION

The surface boundary condition for the momentum equation (5) requires the knowledge of the wind stress vector $\vec{\tau}$ at the ocean surface. The bulk formulation of the surface stress is:

$$\vec{\tau} = \rho_a C_D |U_{10} - U_{sfc}| \vec{u}_{10} \tag{40}$$

where C_D is a turbulent exchange (drag) coefficient and U_{sfc} is the speed of the ocean surface current. Dynamical variables of the atmosphere and the ocean (*i.e.*, winds and currents) are not as strongly coupled as thermodynamical variables, and no retroaction is considered for the input of momentum.

According to (40), the local retroaction of the ocean on the stress would occur through U_{sfc} and the drag coefficient. The speed of surface currents is on average significantly smaller than the wind speed, and U_{sfc} is generally neglected in (40). This approximation is justified in coarse resolution models which largely underestimate ocean currents, but could be reconsidered in the case of eddy resolving models which exhibit strong western boundary currents and large velocities in the Antarctic Circumpolar Current (ACC). The retroaction of the ocean on the drag coefficient occurs through the surface roughness, which is usually accounted for through the dependence of the coefficient upon wind speed, and through the SST in the case it has an impact on the stability of the air column. It is generally considered that these retroaction effects are negligible compared to the errors made in estimating ocean surface winds.

4. Flux Fields

This section discusses the accuracy of the various flux fields that are available to force an OGCM. Note that in the case of a bulk forcing as in Large

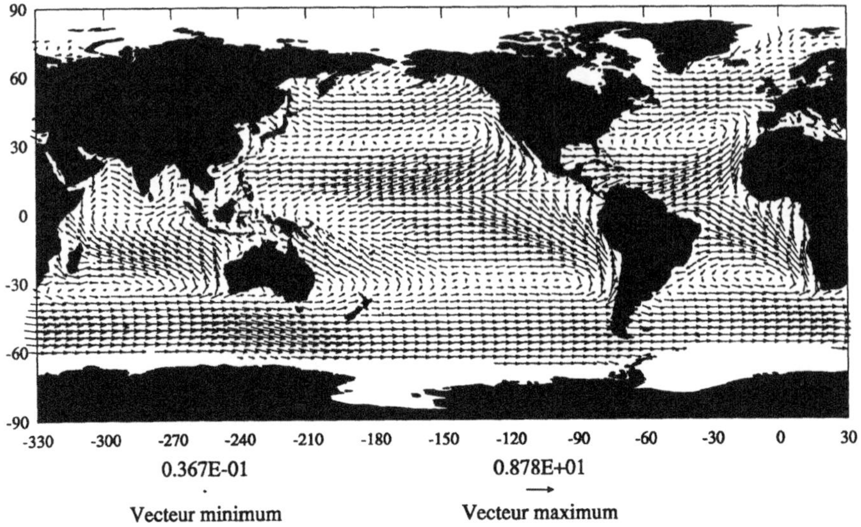

Figure 13. Climatological mean of the wind vector (in m/s) for the period 1979-1993, estimated from ECMWF re-analyses.

et al. (1977), fields representing all variables entering the calculation of the bulk formulae are necessary.

4.1. BASIC FORCING COMPONENTS OF A PROGNOSTIC OGCM

4.1.1. *Fluxes*

The major flux fields required to force an ocean model are listed below. They are:

Q_{sol}	is the net heat input by solar radiation in Wm^{-2},
Q_{nsol}	is the non solar heat flux in Wm^{-2} (sensible, latent and long-wave),
τ_x,τ_y	is the components of the wind stress vector in Nm^{-2},
u_*^3	is the wind stirring in m^3s^{-3} (used in particular mixed layer formulat
E	is the evaporation rate in ms^{-1},
P	is the precipitation rate in ms^{-1},
R	is the river runoff in m^3s^{-1}.

4.1.2. *Other surface variables*

Additional variables are necessary to calculate the retroaction terms. We also mention sea ice, just as a reminder that it is a fundamental component of the ocean forcing at high latitudes. The modeling of sea ice is treated in this volume by Fichefet *et al.* (1998).

SST is the sea surface temperature in ^oC,
U_{10} is the wind speed at anemometer height in ms^{-1},
SSS is the sea surface salinity,
P_a is the mean sea level pressure in Pa,
A is the ocean albedo,
SIC is the sea-ice coverage and growth rate.

4.1.3. *Resolution Issues*
The resolution of the flux fields desired for the various applications of ocean models is presented in Table 2.

TABLE 2. Desired resolution of the flux fields for ocean model application

Model type	Grid resolution	Time resolution	Coverage	Fields
Climate OGCMs	$1^o \times 1^o$	monthly mean	global	all fileds
Eddy resolution OGCMs	better than $1^o \times 1^o$	1 to 3 days	global	all fileds
Diagnostic Inverse Models	$1^o \times 1^o$	monthly to seasonal mean	global	all fields but sea ice growth

4.2. SOURCES OF FLUX ESTIMATES

The three major sources of flux estimates are ship board and buoy (in situ) observations of surface variables, satellite observations, and analyses from numerical weather prediction centers. To compile a catalogue of available surface flux data and flux related data sets, including those becoming available from the various re-analysis projects, and to review the strengths and weaknesses of these data sets, is a task which the present paper does not cover. We give here a somewhat incomplete view of the available air-sea flux climatologies. However, the most important data sets have been cited throughout this paper.

4.2.1. *In situ*
Flux climatologies for all fluxes have been generated historically using ship reports contained in COADS (Woodruff *et al.*, 1987). The most recent global climatologies (monthly mean, 1° x 1° grid), including a correction

for observational biases, are from Da Silva *et al.* (1994), and from Josey *et al.* (1998), who made their data set available at the following sites:

http://www.nodc.noaa.gov

http://www.soc.soton.ac.uk/JRD/MET

In situ flux estimates suffer from a drastic lack of observations in large areas, in particular in the Southern Hemisphere. Their random errors can be quantified, and are evaluated to 15 to 25 Wm^{-2} in the annual mean, according to the field which is considered (Gleckler and Weare, 1997). It is more difficult to evaluate biases. Fluxes are monthly means, averaged over areas of 1° x 1°, although the true resolution is coarser than that, since a significant smoothing has been applied (the turbulent heat fluxes from Josey *et al.* (Josey *et al.*, 1998) shown in Figures 7 and 8 exhibit smooth patterns).

4.2.2. *Satellite*

Fluxes are not directly measured from satellite observations, and sophisticated parameterizations are necessary to derive fluxes from radiative fields effectively observed or from backscattered signals. Another problem with satellite estimates is that records are usually too short to provide a climatology (Schulz *et al.*, 1997).

Shortwave and longwave radiation. It is important to consider both fluxes from the same source so that the cloud error tends to cancel. Satellite estimates have greatly improved in recent years due to the Surface Radiation Budget (SRB) experiment (Darnell *et al.*, 1996). Accuracy is estimated to 10 Wm^{-2} for monthly means over $2.5^o \times 2.5^o$ areas, but comparisons with other sources indicate that they could be overestimated by almost 20 Wm^{-2}. An 8 year monthly mean climatology of shortwave and longwave fluxes is available for the period 1983-1991 from the Langley Research Laboratory (LARC) on a $2.5^o \times 2.5^o$ grid:

http://agni.larc.nasa.gov/srb_homepage.html

Long-term means of both the longwave and shortwave radiation from the LARC (Darnell *et al.*, 1996) have been plotted in Figure 5 and Figure 6.

Turbulent heat fluxes. Satellite estimates of latent and sensible heat fluxes require additional in-situ observations, and are still far from the required accuracy. However, new opportunities are appearing with the possibility of combining information from various sensors. Recent studies, using AVHRR to retrieve the surface temperature, and SSM/I to estimate surface wind speed and surface air humidity, have produced an estimate of the latent heat flux over the world ocean, with a standard error estimated to 30 Wm^{-2} on instantaneous fluxes, decreasing to 15 Wm^{-2} for monthly averages (Schulz *et al.*, 1997).

Precipitation. The most recent climatology of precipitation has been provided for the period 1986 to 1995 by the Global Precipitation Climatology Project (GPCP) (Huffman *et al.*, 1997). This data set is available at:

http://www.dwd.de/research/gpcc

The mean precipitation rate obtained from SSM/I measurements is shown in Figure 14 and should be compared to precipitation from ECMWF (Figure 12).

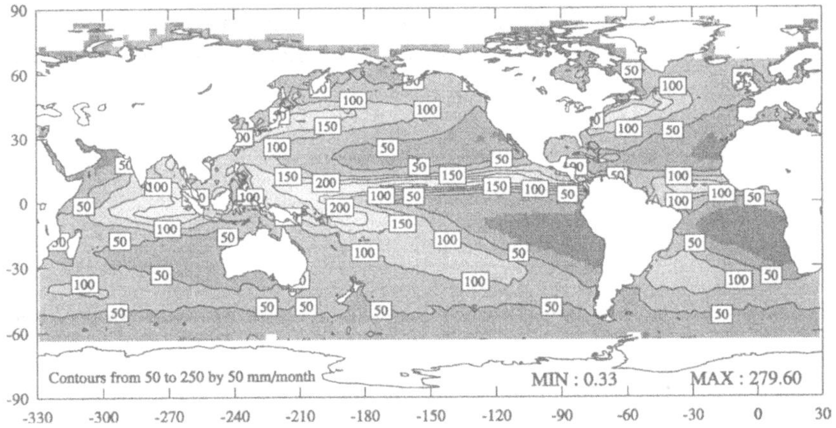

Figure 14. Climatological mean (period 1986 to 1995) of SSM/I precipitation estimates (in mm/month) from Global Precipitation Climatology Project (GPCP) data set (Huffman, 1997).

Wind stress. Satellite estimates of the wind speed and wind vector are recent. Wind speed climatologies from SSM/I and ERS scatterometer are now available. Global fields of the wind vector and wind stress are available on a weekly basis from the European Remote Sensing (ERS) satellite missions since July 1991 (Bentamy *et al.*, 1997) at the following site:

http/www.ifremer.fr/cersat

However, satellite wind products can still be significantly improved. The combination of several satellite products through variational methods is promising (Siefridt, 1998, personal communication, Figure 15), and the NSCAT scatterometer has demonstrated that the global daily wind vector field can be obtained.

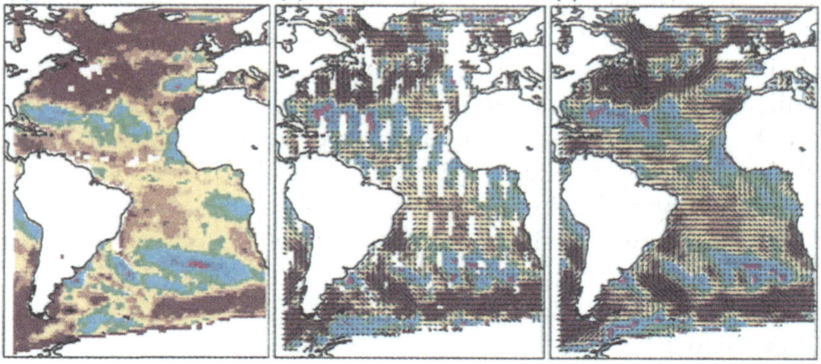

Figure 15. Variational analysis of ERS-1 and SSM/I wind fields for a 3 day period. In this analysis, more weight is put on scatterometer winds (Siefridt, personal communication).

4.2.3. *Numerical weather prediction centers*
Meteorological analyses. For forecast purposes, numerical weather prediction centers such as ECMWF or NCEP perform global analyses of meteorological data every 6 hours. An analysis combines the output of a model forecast and observations (both relative to the same period of 6 hours) in order to produce the state of the model that is the closest to observations in an optimal sense, which depends on the data assimilation technique that is used. Therefore, the analysed fields are the fields described by the prognostic model variables (*i.e.*, surface temperature, air temperature at 2 m, dew point temperature at 2 m, surface pressure, etc.). In the forecast system operating at ECMWF, analysed fields are filtered to produce initialization fields for the next forecast (Figure 16). The first 6 hours of the one day forecast are used to produce the analysis for the next forecast. In consequence, fluxes are not obtained from the analysis, but are accumulated during over the forecast period.

Therefore, the model is not constrained by observations during flux calculations, and fluxes should be considered as forecast quantities, not analysed quantities. A consequence is that spin-up problems in the early phase of the forecast, due to the dynamical adjustment of initial fields, may have an effect on fluxes. At ECMWF, precipitation estimates are known to be very sensitive to the spin-up phase, and large differences are observed in the monthly mean rainfalls estimated from the first 6 hours of the forecast and the monthly mean rainfalls estimated from the last 12 hours of the one day forecast. The impact on heat fluxes is not as important. In ECMWF

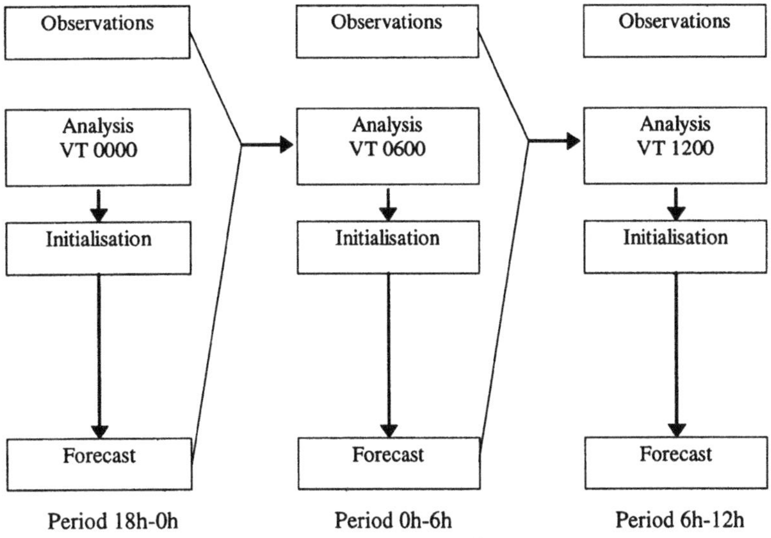

Figure 16. Schematic of the analysis forecast system at ECMWF (Darnell *et al.*, 1996).

re-analyses, this impact has been observed to be 10 Wm^{-2} on the net heat flux (Kallberg, 1997).

Products from NWP centers. Analysed and forecast fields produced by NWP Centers represent a great opportunity to oceanography since they are able to provide fields of surface variables with global, regular, and dense coverage in space and time, which are dynamically consistent through model dynamics and parameterizations. Note that the geophysical consistency of the analysed fields may be deteriorated by the assimilation process which introduces non-physical constraints. At present, only NWP Centers can produce a consistent set of daily fluxes, whereas other sources are limited to monthly means in order to achieve an acceptable accuracy. Like other flux products, NWPC products have biases which are difficult to evaluate. Over long periods, analyses show a variability which is largely induced by changes in the model and the assimilation system (Figure 17) (Siefridt *et al.*, 1998; Béranger *et al.*, 1998). Recent re-analyses performed, over 40 years (1957 to 1996) at NCEP/NCAR (Kalnay *et al.*, 1996), over 15 years (1979 to 1993) at ECMWF (Gibson *et al.*, 1997), and over 15 years (1980 to 1995) for the GEOS-1 re-analysis (Schubert, 1998), have been corrected for this variability, although observational biases remain due to an inhomogeneous data flow over the re-analysed period.

4.3. FLUX REQUIREMENTS

Fluxes used as components of a forcing function of an ocean model should verify several global constraints with some degree of accuracy. Note that, in principle, it is the whole forcing function which should be considered (including the retroaction term), but this is not possible before the model experiment has been terminated. Nevertheless, these constraints are useful diagnostics to perform on the total forcing function after the model integration is finished.

4.3.1. *Global balance*

For oceanographic applications, if we consider a long enough period of time (typically 15 years for ECMWF re-analysis), the heat content of the ocean should be nearly steady. Thermal balance considerations thus require that the long term mean net input of heat be nearly zero for the world ocean. This balance is shown in Figure 17 for ECMWF re-analysis. On average over the 15 years of the re-analysis, the global balance is nearly matched. However, a regional analysis shows that the balance is achieved through compensation of regional biases (Garnier *et al.*, 1998).

4.3.2. *Global transport*

The meridional heat and freshwater transports by the ocean are major components of the climate system. They can be estimated by the divergence of the surface fluxes, and provide a useful diagnostic to assess the accuracy of air-sea exchanges. The meridional heat transport for the world ocean obtained from ECMWF re-analysis is shown in Figure 18. This transport estimate compares well with other sources of data. Divergent heat transport is located at 2.5 S, in agreement with Da Silva (1994). The asymmetry between both hemispheres is also a common feature. However, large differences are found between the heat transport of different sources in individual basins (Garnier *et al.*, 1998), indicating that the agreement for the global ocean results from a compensation of errors.

4.3.3. *Geophysical consistency*

Before running an ocean model, the consistency of the forcing function should be studied, and the geophysical correlation among the various components of forcing should be verified.

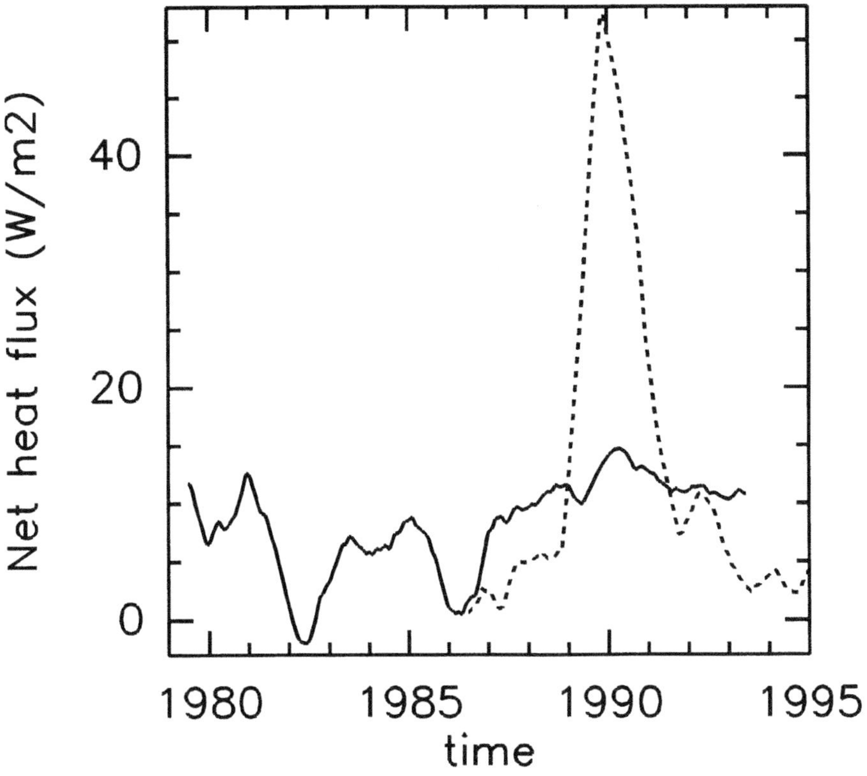

Figure 17. Annual mean net heat flux (in Wm^{-2}) over the global ocean for individual years between 1979 to 1995 from ECMWF re-analysis (full line) and ECMWF analyses (dashed line). (Garnier, 1998, personal communication).

5. Conclusion

This paper has discussed the forcing function of an Ocean General Circulation Model. It was emphasized that the thermohaline forcing requires a model dependent retroaction term to account for the local feedback of the ocean on air-sea fluxes, and to compensate for uncertainties in air-sea flux estimates.

A parameterization of the forcing function has been described (Barnier *et al.*, 1995), which explicitly separates the formulation of the climatic forcing flux and the retroaction term. This makes it easy to use the best available estimates of air-sea fluxes independently of the retroaction term. This formulation has been used (for the temperature equation only) with some success in many WOCE related eddy-permitting model simulations carried out with the objective of simulating the present state of the ocean (DYNAMO, 1997; McLean *et al.*, 1997; Stammer *et al.*, 1996). Note that the above studies used a relaxation condition for salinity.

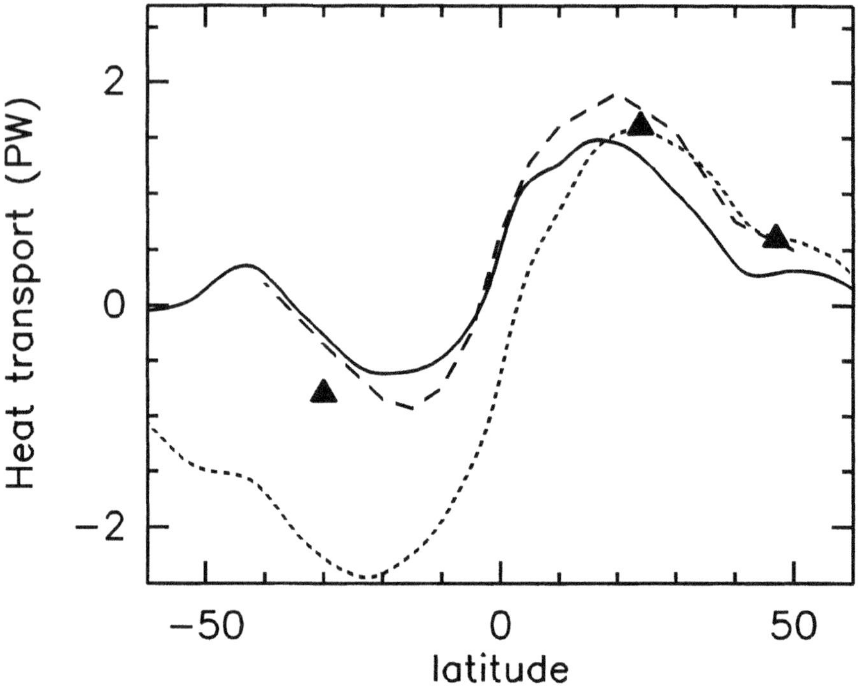

Figure 18. Meridional heat transport (in 1015 W) for the world ocean as a function of latitude, from ECMWF re-analysis (full line), from a combination of turbulent fluxes from ECMWF re-analysis and radiatives fluxes from LARC satellite estimates (Darnell *et al.*, 1996) (dotted line), from Da Silva *et al.* (1994) (dashed line). Hydrographic estimates (McDonald and Wunsch, 1996) of this transport at selected oceanographic sections are indicated by triangles. (Garnier, 1998, personal communication).

Alternative formulations exist which we have not been described in this lecture. As we mentioned, in the bulk forcing (Large *et al.*, 1977), air-sea fluxes are made model dependent by the direct use of the model-computed SST in the bulk formulation. This formulation has been used for long integration of coarse resolution models (Large *et al.*, 1977).

A potential problem with all forcing functions is the geophysical consistency between the thermal and salinity forcing, since heat and freshwater fluxes are linked through evaporation. It is likely that eddy-resolving models are not integrated for long enough periods of time to see their solutions greatly modified. Nevertheless, having balanced heat and freshwater forcing would certainly improve the quantitative aspect of their predictions. This issue is of much greater importance for climate model integration of several hundred to thousands of years. As shown by Large *et al.* (1977), a significant reduction in model drift can be obtained in case of a bulk forcing if the climatology of the atmospheric variables has been adjusted to generate a globally balanced heat and freshwater budget. There is no trivial solution to this problem, and even climatologies of surface fields produced by operational analyses or re-analyses carried out at NWP Centers are still far from matching this balance.

The ideal solution to force an ocean model would be to couple it with a global atmospheric GCM. Coupled models have been developed to carry out climate studies. They include representations of the atmosphere, ocean, cryosphere and land surface. Considerable progress has been made in the last decade in modeling and coupling every component of the system, although the fine dynamical scales are still not considered, due to computational cost and/or inaccurate subgridscale parameterizations. The ability of current coupled models to simulate the observed climate varies with the scales and the variables that are considered (Houghton *et al.*, 1996). For example, models are most accurate at global scale, but skills at regional scales are much lower. In coupled (and atmospheric) models, clouds and radiative fluxes are not well represented, a balanced global freshwater budget is not achieved (in a large part because of an inaccurate hydrological balance over land), and large discrepancies are found in their heat flux estimates over the ocean when compared to other sources of data (Boer, 1996). However, it is likely that the state of the art will rapidly improve, and being able to use, in a coupled mode, a global Atmosphere GCM to force an eddy-resolving Ocean GCM is likely to happen (Barnier *et al.*, 1997).

This lecture also discussed the accuracy of air-sea flux estimates. The main objective of the presentation was to provide information about the data sets which can be used to define a forcing function. Although no critical evaluation of fluxes was made (this can found in the relevant literature), it was shown that large differences are observed between climatic flux estimates from different sources. But the accuracy of fluxes at climatic and seasonal (monthly mean) time scales is not the only issue. Emerging operational oceanography exposes an increasing need of high frequency (daily) products. At present, operational analyses carried-out at NWP Centers provide high frequency fluxes, but their representation of the true variability is not yet accurate enough. For example, the daily winds produced at ECMWF do not represent the observed wind bursts in the western tropical Pacific which are believed to play an important role in the onset of El Niño (Grima *et al.*, 1998). Satellite winds provided by ERS and NSCAT missions seem to represent this variability with sufficient accuracy (Grima *et al.*, 1998), and it is likely that in the near future, satellite flux estimates will be able to compete with meteorological analyses on the sampling issue.

To conclude, it is still necessary to conduct major research efforts to improve the accuracy of air-sea flux estimates from satellite observations and from operational analyses, which are thus far the tools that are the most likely to provide the dense space and time sampling at global scale that ocean models are demanding. Nevertheless, in-situ observations are still of prime importance, at least to calibrate and validate flux calculations from indirect satellite observations or numerical models, and to build independent climatologies that can be used as climate records, since NWPCs analyses and satellite products are subject to large variations related to changes in numerical models and assimilation processes, or to the appearance of new technologies, retrieval algorithms, and chance, which is still a part of the earth observation from space.

Acknowledgments

The author is supported by Center National de la Recherche Scientifique. Je tiens à associer L. Siefridt, K. Béranger et E. Garnier à ce travail, car beaucoup de ce que j'ai appris sur le forçage des modèles d'océan et sur les flux air-mer, je l'ai appris des travaux que nous avons menés ensemble. Je leur exprime ici mes sincères remerciements. During the school, I greatly appreciated discussions with R. Smith on the virtual salinity flux, and with W. Large whose views are so enlightening. Finally, I want to thank J. J. O'Brien, from whom I learned (as a student in his group, too long ago ...), the background of, and got my interest in, the field of air-sea interactions.

References

Barnier, B., Le Provost, C., Madec, G., Tréguier, A.M., Molines, J.M. and Imbard, M. (1997) The CLIPPER Project: High resolution modelling of the ocean circulation in the Atlantic forced by and coupled with the atmosphere, Proceedings of the International Symposium: Monitoring the ocean in the 2000s: an integrated approach, Biarritz, October 13-14, 1997.

Barnier, B., Siefridt, L. and Marchesiello, P. (1995) Thermal forcing for a global ocean circulation model using a three-year climatology of ECMWF analyses, *J. Marine Systems*, **6**, 363-380.

Baumgartner, A. and Reichel E. (1975) The world water balance, Elsevier, New York, 179p.

Bentamy, A., Grima, N., Quilfen, Y., Harscoat, V., Maroni, C. and Pouliquens, S. (1997) An atlas of surface wind from ERS-1 scatterometer measurements 1991-1996, IFREMER, Brest.

Béranger, K., Siefridt, L., Barnier, B., Garnier, E. and Roquet, H. (1998) Evaluation of operational ECMWF surface Freshwater fluxes over oceans during 1991-1997, submitted to *J. Marine Systems.*

Boer, G.J. (1996) Air-sea fluxes from models and their validations, Proceedings of the WCRP workshop on air-sea flux fields for forcing ocean models and validating GCMs, WCRP Report No 762, 157-163.

Clarke, N.E., Eber, L., Laurs, R.M., Renner, J.A. and Saur, J.F.T (1974) Heat exchange between ocean and atmosphere in the eastern North Pacific for 1961–71. NOAA Tech. Rep. NMFS SSRF-682. U.S. Dept. Of Commerce, Washington, DC.

Converse, D.R., Holland H. D., and Edmond J. M. (1984) Flow rates in the axial hot springs of the East Pacific Rise (21° N): implications for the heat budget and the formation of massive sulfide deposits, Earth and Planetary Science Letters 69, 59-175.

Darnell, W.L., Staylor, W.F., Ritchey, N.A., Gupta, S.K. and Wilber, A.C. (1996) Surface radiation budget: A long term global dataset of shortwave and longwave fluxes. EOS Transactions, *Amer. Geophys. Union.*

Da Silva, A., Young, C., and Levitus, S. (1994) Atlas of surface marine data 1994, Volume 4: Anomalies of freshwater fluxes. NOAA atlas NESDIS 9, U.S. Department of Commerce, NOAA, NESDIS.

DYNAMO (1997) Dynamics of North Atlantic models: Simulation and assimilation with high resolution models, Report No 294, Institut fr Meereskunde, Kiel, Germany.

Fichefet, T., Goose, H. and Morales Maqueda, M. (1998) On the large-scale modeling of sea ice and sea ice-ocean interactions. In *Ocean Modeling and Parameterization*, E.P. Chassignet and J. Verron (Eds.), Kluwer Academic Publishers, 399-422.

Frankignoul, C., Czaja, A. and L'Heveder B. (1998) Air-sea feedback in the North Atlantic and surface boundary conditions for ocean models, *J. of Climate*, in press.

Garnier, E., Barnier, B., Siefridt, L. and Béranger, K. (1998) Air-sea fluxes from 15 years of ECMWF reanalysis: A surface boundary condition for OGCMs?, submitted to *J. Climate Dyn.*

Gibson, J.K., Kallberg, P., Uppala, S., Hernandez, A., Nomura, A. and Serrano, E. (1997) ECMWF Re-analysis project, 1. ERA description, Project Report Series, ECMWF report, July 1997.

Gleckler, P.J. and Weare, B.C (1997) Uncertainties in global surface heat flux climatologies derived from ship observations, *J. of Climate*, **10**, 2764-2781.

Grima, N., Bentamy, A., Delecluse, P., Katsaros, K., Levy, C. and Quilfen, Y., (1998) Sensitivity of an oceanic general circulation model forced by satellite wind-stress fields, Submitted to *J. Geophys. Res.*

Haney, R.L. (1971) Surface thermal boundary condition for ocean circulation models, *J. Phys. Oceanogr.*, **1**, 145-167.

Houghton, J.T., Meira Filho L.G., Callander B.A., Harris, N., Kattenberg, A. and Maskell, K. (Eds.) (1996), *Climate Change 1995*, The Science of Climate Change, Contribution of WGI to the second assessment report of the intergovernmental panel on climate change, Cambridge University Press, Cambridge.

Huffman, G.R., Adler, R.F., Arkin, P., Chang, A., Ferraro, R., Gruber, A., Janowiak, J., McNab, A., Rudolf, B. and Schneider, U. (1997) The Global Precipitation Climatology Project (GPCP) combined precipitation data set, *Bull. Amer. Meteor. Soc.*, **78**, 5-20.

Josey, S., Kent, C.E. and Taylor, P. (1998) Ocean-atmosphere heat exchange: a global climatology derived from in-sityu meteorological observations. In preparation for *J. of Climate.*

Kallberg, P. (1997) Fluxes and some other surface characteristics of ECMWF Ranalyses, Proceedings from the Firts International Conference on Reanalyses, SilverSpring, in press.

Kalnay, E., Kanamistu, M., Kistler, R., Collins, W., Deaven, D., Gandin, L., Iredell, M., Saha, S., White, G., Woolen, J., Zhu, Y., Cheliah, M., Ebisuzaki, W., Higgins, W., Janowiak, J., Mo, C.K., Ropelewski, C., Leetma, A., Reynolds, R. and Jenne, R. (1996) The NCEP/NCAR reanalysis project, *Bull. Amer. Meteor. Soc.*, **77**, 437-471.

Large, W.G., Danabasoglu, G., Doney, S. and McWilliams, J.C (1977) Sensitivity to surface forcing and boundary layer mixing in a global ocean model: Annual mean climatology, *J. Phys. Oceanogr.*, **27**, 2418-2447.

Levitus, S. (1982) Climatological atlas of the world ocean, NOAA Prof. Pap. 13, U.S. Department of Commerce, 173p.

Levitus, S., Burgett, R. and Boyer, T.P. (1994) World Ocean Atlas NESDIS. Department of Commerce, NOAA, NESDIS.

McDonald,A. and Wunsch, C. (1996) An estimate of global ocean circulation and heat fluxes, *Letters to Nature*, **382**, 436-439.

McLean, J.L., Semtner, A., and Zlotnicki (1997) Comparisons of mesoscale variability in the Semtner-Chervin 1/4 model, the Los Alamos Parallel Ocean Program 1/6 model, and TOPEX/POSEIDON data, *J. Geophys. Res.*, **102**, 25,203-25,226.

Paulson, C.A and Simpson J.J. (1977) Irradiance measurements in the upper ocean. *J. Phys. Oceanogr.*, **7**, 952-956.

Pickard, G.L.and Emery, W.J. (1982) Descriptive Physical Oceanography, Pergamon press, Oxford.

Rosatti, A. and Miyakoda, K. (1988) A general circulation model for upper ocean simulation, *J. Phys. Oceanogr.*, **18**, 1601-1626.

Schubert, S., The Geos-1 re-analysis: Overview, Proceedings from the First International Conference on Reanalyses, SilverSpring, in press.

Schulz, J., Meywerk, J., Ewald, S. and Schlussel, P. (1997) Evaluation of satellite-derived latent heat fluxes, *J. of Climate*, **10**, 2782-2795.

Siefridt, L., Barnier, B., Béranger, K. and Roquet, H. (1998) Evaluation of ECMWF surface heat fluxes as a thermal ocean forcing in relation to model parameterization changes during the period 1986 to 1995, submitted to *J. Mar. Sys.*

Speer, K.G. and Helfrich, K.R.(1995) Hydrothermal plumes: a review of flow and fluxes. In L.M. Parson, C.L. Walker, and D.R. Dixon, (eds.) Hydrothermal Vents and Processes, Geological Society Special Publication No. 87, pp. 373-385.

Stammer, D.,Tokmakian, R., Semtner, A. and Wunsch, C. (1996) How well does a 1/4 global circulation model simulate large-scale oceanic observations?, *J. Geosphys. Res.*, bf 101, 25,779-25,811.

Vorosmarty C.J., Fekete B., and Tucker B.A., Global river discharge data base, (RivDIS v1.0), IHP-V Theme 1, Unesco 1996.

Woodruff, S.D., Slutz, R.J., Jenne, R.L. and Steurer, P.M. (1987) A comprehensive Ocean Atmosphere data set, *Bull. Amer. Meteor. Soc.*, **68**, 1239-1250.

MODELING AND PARAMETERIZING THE OCEAN PLANETARY BOUNDARY LAYER

WILLIAM G. LARGE
National Center for Atmospheric Research,
Boulder, CO, U.S.A.

1. Introduction

The portion of a geophysical fluid that is directly influenced by the presence of a boundary is referred to as a planetary boundary layer, PBL. The two most notable examples are the atmospheric boundary layer, ABL, where the lower surface is either land or ocean, and the ocean's surface boundary layer, OBL, bounded above by the atmosphere. Although the focus here is the OBL, it should become evident that the observational database is incomplete in some important ways and that the state of modeling is immature relative to that of the atmosphere. Therefore, in the absence of evidence to the contrary, ABL theory, observations and models are presented as being similar to the OBL. However, it is important to identify unique characteristics of the OBL. The ocean's benthic boundary layer and its parameterization are the subjects of Beckmann (1998, this volume). Shallow coastal regions where the OBL and benthic boundary layer interact, and where ocean tidal mixing can be strong are excluded. Also excluded are ocean interior mixing processes, except when they interact with the OBL.

The direct influence of the boundary means that the distance, d, from the boundary is an important length scale of a PBL. Within about a millimeter of the ocean surface, the Reynolds number formed with this length becomes too small ($\simeq$ 1 for a velocity scale of 1 mm/s) to support turbulence and vertical transfer is accomplished through viscous (molecular) processes. This viscous sublayer is not resolved in any models of the OBL and the flux divergences over a millimeter are generally neglibible. Therefore, conceptually, PBL model domains can be regarded as starting at the viscous sublayer. Farther from the boundary the Reynolds number becomes very large and there is fully developed three-dimensional turbulence. The primary function of the PBL, therefore, is the vertical turbulent transport

E. P. Chassignet and J. Verron (eds.), Ocean Modeling and Parameterization, 81–120.

of properties between the more two-dimensional geophysical fluid interior and the viscous transport at the boundary. The challenge for ocean modeling is to faithfully model, or parameterize these turbulent fluxes.

A short bibliography of our subject might begin with Monin and Yaglom (1971), where PBL physics is treated extensively. Some of these topics are updated and explored from a purely atmospheric perspective by articles in Nieustadt and van Dop (1982). A perhaps more accessible treatment of the PBL is Stull's (1988) textbook. For a more oceanographic perspective there are the monograph edited by Kraus (1977) and a recent report by Nurser (1996). The latter provides a comprehensive overview of the current state of the art, and references where deeper examinations can be found.

2. Boundary Layer Physics

What do we know about the turbulent transports that are to be modelled and parameterized? The answer lies in PBL physics, which is distinct from that of geophysical fluid interiors (Wyngaard, 1982). One focus will be on similarity theory, and the other on the turbulent fluxes themselves. Both have important implications for PBL models.

It is convenient to decompose PBL properties into two parts, say X and x. In theory and observations these represent a mean and fluctuation, but in modeling they are the resolved and the unresolved, respectively. Furthermore, by aligning a horizontal coordinate axis in the direction of the mean flow, the mean of the downstream velocity component, U, is the mean speed and the mean cross-stream component is $V = 0$. Other PBL properties that are dynamically important are the potential temperature, Θ, the salinity in the ocean (specific humidity in the atmosphere), S, the pressure, P, and the density, ρ.

However, there are numerous dynamically passive properties that also need to be considered in PBL parameterizations. Critical to biological models are nutrients, species, photosynthetically active radiation and trace metals. Geochemistry involves the whole carbon budget, including carbon dioxide and radioactive carbon, as well as freons and trace gases. These passive scalars tend to behave like salinity, and so are also denoted below with a subscript, 's', as opposed to the subscript 'm' for momentum and 'Θ' for potential temperature when distinguished from other scalars.

A practical equation of state for the ocean is

$$\rho = \rho_r(1 - \alpha(\Theta - \Theta_r) + \beta(S - S_r)), \tag{1}$$

where ρ_r is a reference density at the reference values of Θ_r and S_r. Then the buoyancy per unit volume is defined as $B = g\rho_r^{-1}(\rho_r - \rho)$. In (1) the expansion coefficients for temperature and salt are α and β, respectively.

They are of order $1-2\times10^{-4}K^{-1}$ and $7\times10^{-4}ppt^{-1}$, respectively. Allowing the coefficients to vary with depth is a simple way of incorporating an ocean pressure (depth) effect on density.

Vertical turbulent transports are correlations between property fluctuations and the fluctuating vertical velocity, w, given by

$$\overline{wx} = < uw > = \overline{< wx >}, \tag{2}$$

where the overbar represents a time average at a point and angle brackets a spatial average at a particular time, and equivalency is achieved only for sufficient averaging. The vertical buoyancy flux becomes

$$F = -\overline{wb} = g(-\alpha\overline{w\theta} + \beta\overline{wS}) \tag{3}$$

where $\overline{w\theta}$ is kinematic heat flux and $\overline{wS}$ is the kinematic salt flux in the ocean. Surface values of these kinematic fluxes (denoted as $\overline{wx}_o$) are related the the surface momentum (stress), heat and freshwater fluxes (Barnier, 1998, this volume). Note that these latter fluxes are continuous across the boundary, but the kinematic fluxes, including $\overline{wb_o}$ and F_o are not.

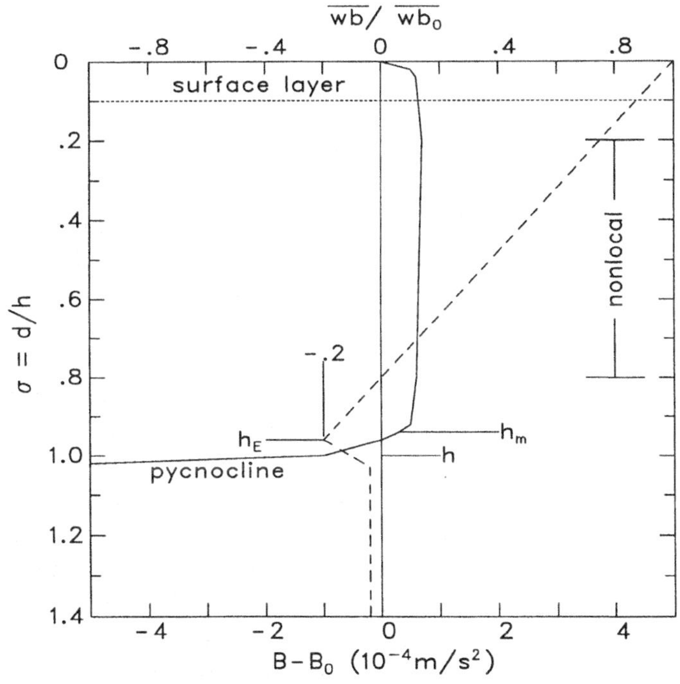

Figure 1. Schematic profiles of buoyancy relative to the surface (solid) and buoyancy flux normalized by the surface flux (dashed) in the upper ocean with purely convective forcing.

Typical property distributions are strikingly similar in the ABL and OBL, supporting the hypothesis that to first order OBL and ABL physics are similar. In pure convection (*e.g.* ocean surface cooling with no wind mixing) experience with the ABL would predict the situation sketched in Fig. 1 for an OBL of depth h. As observed, the buoyancy (temperature in the absence of salt), as well as many other properties are "well mixed" throughout much of the layer down to the mixed layer depth, h_m, which can be very different than the boundary layer depth, h. In the convective ABL there is a near surface layer that is statically unstable. This surface layer extends to about $0.1h$, or to $\sigma = 0.1$ in the nondimensional coordinate $\sigma = d/h$ of Fig. 1. Such an unstable layer has been observed in the ocean too (Anis and Moum, 1992). In Fig. 1 the OBL extends just into a very stable pycnocline (often a thermocline), that is analogous to the an atmospheric inversion aloft.

Comparison of ABL and OBL flux profiles is precluded by the unhappy fact that there are no direct flux measurements in the OBL, except in the special case of under sea-ice (McPhee and Martinson, 1994). The buoyancy flux profile of Fig. 1 is a model result based on ABL experience. It is linear from the surface to the entrainment depth, h_E, where the entrainment flux is prescribed to be a fraction, -0.2, of the surface flux (Ball, 1960). In the region of $0.2 < \sigma < 0.8$, the buoyancy flux is not down-gradient, implying that the local fluxes are driven by nonlocal boundary layer physics. This situation is observed in the ABL, but will it be seen in the OBL?

Now consider a prognostic equation for the U component of momentum in the surface layer of a PBL

$$\begin{aligned} U_t = -UU_x - WU_z &\quad , \quad \text{Advection} \\ +fV &\quad , \quad \text{Coriolis} \\ -\rho P_x &\quad , \quad \text{Pressure Gradient} \\ -(\overline{uw})_z &\quad , \quad \text{Vertical Mixing} \end{aligned} \qquad (4)$$

where f is the Coriolis parameter ($1 \times 10^{-4} s^{-1}$ in mid-latitudes). Near the surface of the ABL, V becomes small and geostrophy is no longer the primary balance. Instead the pressure gradient is balanced by the vertical divergence of the momentum flux, of stress, so that the fluid stress, $\tau(z)$ diminishes nearly linearly with height from its surface value τ_o, so that there is no constant flux layer. Nevertheless, such a term is used by Lumley and Panofsky (1964) for the layer where the loss of flux is less than flux measurement error.

In the OBL the behavior of the near surface stress is not as well known. Unlike the ABL, there are no direct stress measurements in the open ocean to check if the ABL and OBL are similar in this regard. There is no reason to believe that they are because the Coriolis term in (3) vanishes only at the Equator. Elsewhere, there is the potential for a multiterm balance in (4) and a wide range of possible stress profiles. Stress profiles have been inferred from ocean current measurements and from models (*e.g.*, Price, 1998, this volume).

2.1. SIMILARITY THEORY

The semi-empirical Monin-Obukhov similarity theory is the most well tested boundary layer scaling. As such, models should strive to be faithful to its results. The theory argues that in the surface layer near the boundary, but away from the surface roughness elements, the only important turbulence parameters are the distance from the boundary, and the surface kinematic fluxes. The fundamental turbulent parameters that can be formed from these quantities are the friction velocity, u^*, the scales of the turbulent fluctuations of scalars, such as temperature, Θ^*, and ocean salinity (or atmospheric humidity), S^*, and the Monin-Obukhov length scale, L:

$$\begin{aligned} u^{*2} &= \overline{wu}_o = |\vec{\tau}_o|/\rho_r \\ \Theta^* &= -\overline{w\theta}_o/u^* \\ S^* &= -\overline{ws}_o/u^* \\ L &= u^{*3}/(\kappa\, F_o), \end{aligned} \tag{5}$$

where $\kappa = 0.4$ is von Karman's constant. Analysis of the turbulent kinetic energy equation shows that turbulence suppression by stable buoyancy forcing ($F_o > 0$) balances mechanical production at a depth, $d = L$.

Although the surface layer need not be a constant flux layer (Tennekes, 1973), as long as the surface fluxes remain the only important parameters, the following dimensionless profiles are universal functions of the stability parameter $\zeta = d/L$:

$$\begin{aligned} \phi_m(\zeta) &= \kappa\, d\, u^{*-1}\ \partial_z U \\ \phi_s(\zeta) &= \kappa\, d\, \Theta^{*-1}\ \partial_z \Theta = \kappa\, d\, S^{*-1}\ \partial_z S. \end{aligned} \tag{6}$$

These functions have been empirically determined from measurements in the ABL, but there is no consensus about their precise functional form (Högström, 1988). However, theoretical considerations dictate that the turbulent velocity scales, $w_x = \kappa u_*/\phi_x$ (w_m for momentum and w_s for scalars) should become dependent on the convective velocity scale, $w^* = (-F_o h)^{\frac{1}{3}}$, in the convective limit of ζ approaching minus infinity ($u^* = 0$ with unstable forcing $F_o < 0$). This is satisfied if in this limit

$$\begin{aligned} \phi_m &= (a_m - c_m\zeta)^{-1/3} \\ \phi_s &= (a_s - c_s\zeta)^{-1/3}, \end{aligned} \tag{7}$$

where a_x and c_x are empirical constants (*e.g.*, Large *et al.*, 1994).

Integration of (6) gives the familiar logarithmic mean property profiles,

$$\begin{aligned} U(d) &= U_o + \frac{u^*}{\kappa} \left[ln\left(\frac{d}{z_m}\right) - \psi_m(\zeta) \right] \\ \Theta(d) &= \Theta_o + \frac{\Theta^*}{\kappa} \left[ln\left(\frac{d}{z_\theta}\right) - \psi_s(\zeta) \right] \\ S(d) &= S_o + \frac{S^*}{\kappa} \left[ln\left(\frac{d}{z_s}\right) - \psi_s(\zeta) \right] \text{, with} \\ \psi_x(\zeta) &= \int_{Z_x/L}^{\zeta} \frac{[1 - \phi_x(\zeta')]}{\zeta'} \, d\,\zeta', \end{aligned} \tag{8}$$

where the constants of integration, z_m, z_θ and z_s, are the roughness lengths. The ψ_x functions give the stability corrrection. They have been integrated for some common analytic forms of the ϕ_x functions by Paulson (1970). The surface layer does not extend all the way to the surface (Monin and Yaglom, 1971; Lui *et al.*, 1979), because at a finite distance surface roughness elements and molecular processes begin to directly influence the turbulence. Therefore, neither (6), nor (8) can be expected to hold as d approaches z_m.

2.2. TURBULENT FLUXES

The vertical transport of properties through a PBL is accomplished by three dimensional turbulent eddies. At a fixed point their frequency is related to the downstream radian wavenumber, k, by Taylor's frozen turbulence hypothesis, $k = 2\pi f/U$. The time average of a vertical flux of property X at a point is the integral over frequency, f, of the cospectrum, $\Phi_{wx}(f)$, of the property fluctuations and the vertical velocity. Similarly, the spatial average at a point in time is the integral of the cospectrum over k. In practice the eddies occupy a finite range of wavenumber space, say k_1 to k_2, and hence frequency space, say f_1 to f_2, so the fluxes become

$$\begin{aligned} \overline{wx} &= \int_{f_1}^{f_2} \Phi_{wx}(f)\, df \\ < uw > &= \int_{k_1}^{k_2} \Phi_{wx}(k)\, dk \end{aligned}$$

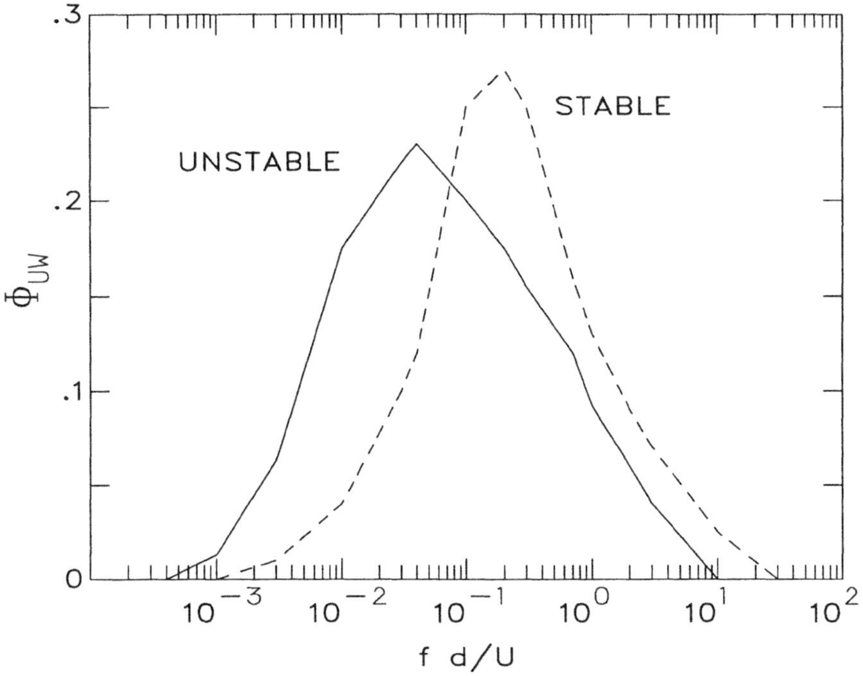

Figure 2. Observed momentum flux cospectra at 13 m height in the ABL for both stable and unstable conditions. The ordinate is normalized and is variance preserved by muliplying by f/u^{*2}. The nondimensional abscissa is natural frequency, $fd/U = d/\lambda = kd/(2\pi)$.

$$< \overline{uw} > \quad = \quad \int_{f_1}^{f_2} \int_{k_1}^{k_2} \Phi_{wx}(f,k)\, dk\, df. \tag{9}$$

Momentum flux cospectra, Φ_{uw}, observed at about 13m height in the marine ABL are shown in Fig. 2. for both stable and unstable forcing conditions. As the boundary is approached the flux transporting eddies become confined to "fit" within the height, d, and so become smaller in wavelength, λ (larger, k), with higher frequency temporal fluctuations. This effect is removed by plotting the cospectra as functions of $fd/U = kd/(2\pi) = d/\lambda$. The lower limits of integration for Φ_{uw} are $f_1 = 4 \times 10^{-4}U/d$ and $k_1 = 8\pi \times 10^{-4}/d$. The upper limits are $f_2 = 30U/d$ and $k_2 = 60\pi/d$. The heat flux cospectrum, $\Phi_{w\theta}$, is similar, with the same stability dependence (*e.g.*, Large and Pond, 1982). Thus at a given height the same five decades of eddy sizes transport both the heat and the momentum vertically.

3. Resolved Turbulence Modeling

Computationally, it is not possible for a model to resolve all scales from the global ($10^8 m$) to the smallest three-dimensional turbulent (order 1

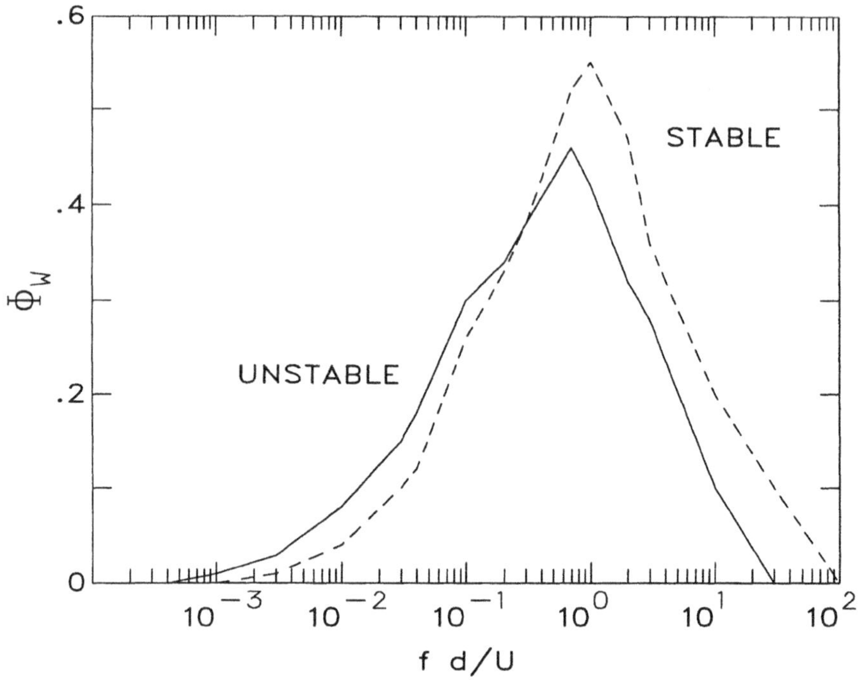

Figure 3. As for Fig. 2, but for observed vertical velocity spectra.

mm). General circulation models (GCMs) resolve only the larger scales of this range. They may resolve up to 3 decades of motions with horizontal grid spacing of order $10^5 m$, and relatively few vertical grid point. At the other end of the spectrum are models that resolve the high wavenumbers of fully turbulent motions. From Fig. 2, it is apparent that there are three-dimensional turbulent eddies ranging over 4 decades in size that are important to the momentum flux. To completely model such a range requires in excess of 20000 grid points in each of the three coordinates, or a total of nearly 10^{13} grid points, which is not computationally feasible. There are codes which contain roughly 512 points in each direction and this allows only about 2.5 decades to spanned.

Direct numerical simulation (DNS) attempts to resolve the smaller scales (higher wavenumbers) of turbulence. In particular, it tries to include the energy containing eddies. The vertical velocity spectra of Fig. 3 indicate that spanning from 0.1 to 30 in λ/d space may be adequate. Unfortunately, such a range excludes most of the flux transporting eddies (Fig. 2), especially in unstable conditions. Therefore, DNS is not applicable to the problem of turbulent transport through a PBL. In contrast Large Eddy Simulation (LES) centers the available 2.5 decades near the cospectral peaks, so that the largest eddies responsible for the bulk of the turbulent transport are resolved. As shown by Fig. 3, the higher wavenumbers can be made to include the peak of the kinetic energy spectrum, so that the unresolved small scale motions are all in a relatively well understood, theoretically, portion of spectral space. In this case, relatively well based sub-grid-scale (SGS) parameterizations are available, and since most of the energy containing and flux transporting eddies are resolved the solutions are relatively insensitive to the parameterization.

The Large Eddy Simulation technique was pioneered by Deardorff (1970; 1972) for numerical investigations of laboratory and atmospheric convection. It has been successfully applied to many atmospheric studies (*e.g.*, Wyngaard and Brost, 1984; Moeng and Wyngaard, 1984), including evaluations of boundary layer turbulence closures (Moeng and Wyngaard, 1989; Ayotte *et al.*, 1996). Recently, LES models have been applied to ocean problems including Langmuir circulations (Skyllingstad and Denbo, 1995; McWilliams and Sullivan, 1997), and turbulent mixing in the upper equatorial ocean (Wang *et al.*, 1996; Wang *et al.*, 1998; Skyllingstad *et al.*, 1997). These equatorial studies addressed the diurnal cycle of solar heating, the horizontal component of the earth's rotation, eddy viscosity, the diurnal cycle of deep turbulence, and turbulent dissipation during a westerly wind burst.

Accepting that LES greatly reduces the SGS parameterization problem, there are still several issues that need to be addressed. First, with only 2.5 decades of model resolution, the domain must be kept artifically small, so that larger scale motions, including some turbulent transport eddies (Fig. 2) are not contained. Second, the scales that need to be resolved change with height. A fixed grid appropriate for the middle of a PBL will not resolve the energy containing and flux transporting eddies close to the surface, where the solution becomes more dependent on the SGS parameterization. A similar situation arises near the atmospheric inversion and the ocean mixed layer base where local stratification tends to reduce the scales of important motions. One approach to ameliorating this situation is to modify the SGS parameterizations in these regions, but this is not straightforward because the theoretical base is weaker. Another is to refine the grid locally (Sullivan *et al.*, 1996). Figure 2 also serves to illustrate the difficulties with simulating a diurnal cycle in either the ABL or PBL. The peak of the cospectra tend to be at larger scales, by as much as a decade, in unstable conditions (daytime in the ABL and nighttime in the OBL), than in stable conditions. Thus, the model range may become inadequate when trying to model both conditions, as required for a diurnal cycle.

3.1. AN LES MODEL

A basic LES model has been developed for the unstable ABL by Moeng (1984). The surface boundary conditions are flux type for momentum and temperature. The boundary conditions with the interior are radiation type to allow propagating internal waves to leave the system (Klemp and Durran 1983). Wang *et al.* (1998), hereafter WML, modify this model for application to the equatorial ocean boundary layer, where their neglect of salinity is not catastrophic. Their most important development is the inclusion of the effects of large scale flow field, which they require in order to capture the essential features of equatorial upper ocean dynamics. The resulting LES model equations for the prognostic variables $\mathbf{u} = (u = u_1, v = u_2, w = u_3)$, the resolved velocity, Θ and e, the subgrid turbulent kinetic energy, are :

$$\begin{aligned}
\mathbf{u}_t + \mathbf{u}\cdot\nabla\mathbf{u} + 2\boldsymbol{\Omega}\times\mathbf{u} &= -\nabla p - \alpha\mathbf{g}\Theta + \nabla\cdot\vec{\tau} + \mathbf{G}, \\
\nabla\cdot\mathbf{u} &= 0, \\
\Theta_t + \mathbf{u}\cdot\nabla\Theta &= \nabla\cdot\mathbf{q} + \frac{1}{\rho_o c_p}\partial_z I(t,z) + H_\Theta \\
e_t + \mathbf{u}\cdot\nabla e + W e_z &= \vec{\tau} : \nabla\mathbf{u} - \alpha g q_3 - \epsilon + \nabla\cdot 2\mathbf{K}\cdot\nabla e,
\end{aligned} \tag{10}$$

where $\boldsymbol{\Omega}$ is the Earth's rotation vector and ϵ the turbulent dissipation. The absorption of solar radiation with depth in the ocean gives the radiation profile, $I(t,z)$. The colon in (10) represents the inner product operator between two tensors. The parameterization of the unresolved subgrid Reynolds stress tensor, $\vec{\tau}$, of the eddy viscosity tensor, $\mathbf{K}$, and of the subgrid heat flux vector, $\mathbf{q}$, are the classic problems of LES modeling. These issues are detailed in WML and will not be discussed further here.

The novel aspect of (10) is the specification of the large scale interaction terms ($\mathbf{G}$, H_Θ and W). The non-zero terms of these functions are :

$$\begin{aligned}
G_U &= -u\partial_x U - W\partial_z u - \partial_x P + F_U \\
G_V &= -v\partial_y V - W\,\partial_z v \\
G_w &= -W\partial_z w \\
H_\Theta &= -u\partial_x\Theta - W\partial_z\Theta + F_\Theta \quad .
\end{aligned} \tag{11}$$

They are determined by prescribing profiles of the large scale zonal pressure gradient $\partial_x P$, the zonal gradient of zonal velocity $\partial_x U$, the meridional gradient of meridional velocity $\partial_y V$, and the meso-scale eddy contributions to the zonal momentum and heat equations, F_U and F_Θ, respectively. These profiles are estimated by WML from ocean observations (Bryden and Brady, 1989) and eddy resolving tropical basin models (*e.g.*, Gent and Cane, 1989). They show profiles of each, as well as of the following derivatives. First, continuity is used to compute the large scale vertical gradient of vertical velocity $\partial_z W$, which is then integrated to give the $W(z)$ of (10) and (11). Finally, the large scale horizontal temperature gradients are determined by the hydrostatic relation

$$\partial_x \Theta = \frac{1}{\alpha g} \partial_{xz} P \ . \tag{12}$$

What can be learned from such an LES model? As examples, consider what the results of WML reveal about the vertical profiles of fluxes, the turbulent kinetic energy budget and the relationship between mixing coefficients (viscosity and diffusivity) and local gradient Richardson number,

$$Ri_g = \frac{\partial_z B}{(\partial_z U)^2 + (\partial_z V)^2} \ . \tag{13}$$

This LES examined the equatorial ocean response to a diurnal cycle of solar heating, with a steady surface cooling of 200 W/m^2 and a noon insolation of 776 W/m^2, such that the daily average heating was 47 W/m^2. The wind wind stress was constant and zonal at $-0.042\, N/m^2$, such that the daily turbulent scales were $u^* = 0.0064\, m/s$ and $\theta^* = -0.0018\, K$.

Figure 4 shows the daily average flux profiles. There is some curvature of the stress profile, but a linear decrease of the momentum flux with d, is not a poor approximation, even though the boundary layer cycles between about $5\,m$ and $40\,m$ depth over the diurnal cycle. However, a linear buoyancy (temperature) flux profile is not evident at any depth. The wind forcing and daytime heating are evidently sufficient to dominate over the linear flux profile expected during windless, nighttime convection (Fig. 1). Until ocean flux profiles are measured directly, the profiles of Fig. 4 provide one of our better pictures of the variation of flux in the OBL. Pictures applicable to other ocean situations could be provided by appropriate LES modeling.

The turbulent kinetic energy (TKE) equation relates the time tendency of $e = q^2/2$, where $q^2 = u^2 + v^2 + w^2$, to the production of e by shear, $P_S = u^{*2} \partial_z U$, to its suppression by buoyancy, $F = -\overline{wb}$, to the divergence of the turbulent transport of e, $D_T = \partial_z \overline{we}$, plus the work done by the fluctuating pressure, $D_P = \rho_r^{-1} \partial_z \overline{wp}$, and to turbulent dissipation, ϵ:

$$\partial_t \, e = P_S - F - D_T - D_P - \epsilon. \tag{14}$$

WML find that the dominant balance is between P_S, F and ϵ, as is often assumed, but find that the balance of these terms has a distinct diurnal cycle. Furthermore, contrary to some arguements (*e.g.*, Osborn, 1980; Peters *et al.* 1988), the flux Richardson number, $Ri_f = -F/P_S$, is neither small, nor constant. Between 30 m and 80 m depth it varies between 0.2 and 0.5. Since buoyancy flux suppresses turbulence, $F > 0$, in the equatorial upper

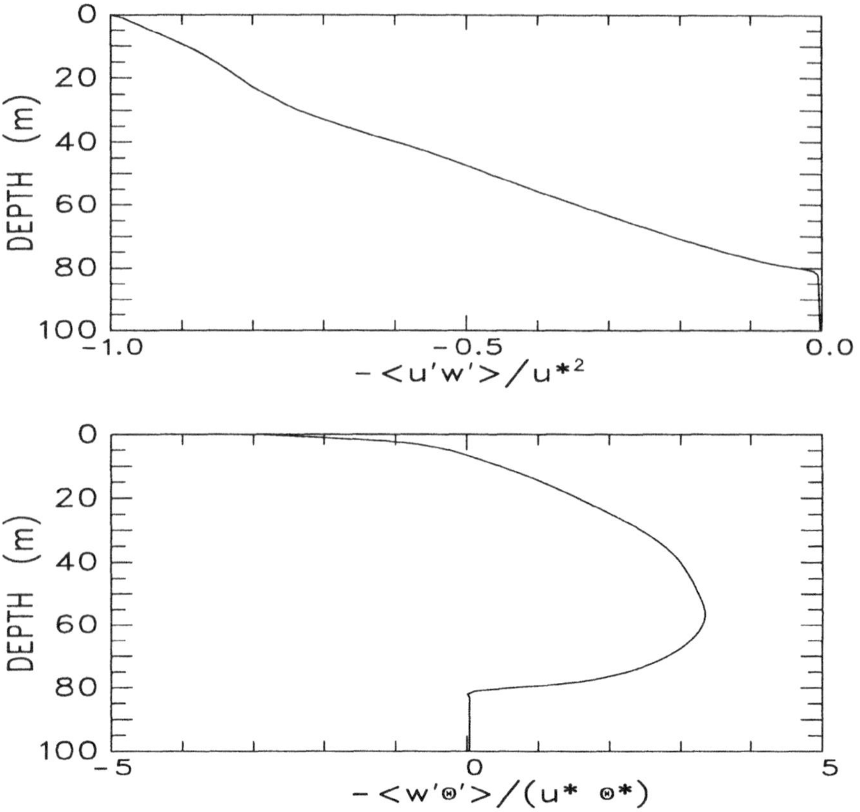

Figure 4. Daily averaged profiles of kinematic momentum flux and heat flux from the LES solution of Wang et al. (1998). The normalization factors are $u^{*2} = 4.1 \times 10^{-5} m^2 s^{-2}$ and of $u^*\theta^* = 1.15 \times 10^{-5} ms^{-1}\, K$, respectively.

ocean, its neglect by Peters *et al.* 1988 would lead to underestimates of P_S, viscosity and diffusivity that are based on ϵ measurements.

Figure 5 (dashed lines) shows the Peters *et al.* (1988) fit of their estimated mixing coefficients as functions of Ri_g (13). The LES results (see Fig. 16 of WML) do suggest such an underestimate. Furthermore, they indicate similar values for viscosity and diffusivity for $Ri_g < 0.6$ and that at a given value of Ri_g these coefficients can range over 3 orders of magnitudes. Thus, a universal parameterization of mixing as a function of Ri_g may not be possible. However, if the LES results are separated into boundary layer and interior then the latter do show a functional relationship. In particular, coefficients between $0.001\, m^2/s$ and $0.006\, m^2/s$ are found for $Ri_g < 0.1$, while viscosity and diffusivity are of order $0.0001\, m^2/s$ and $0.00001\, m^2/s$, respectively for $Ri_g > 1$. Although, these features are not reflected in the Peters *et al.* (1988) fits, they are present in the two parameterizations also

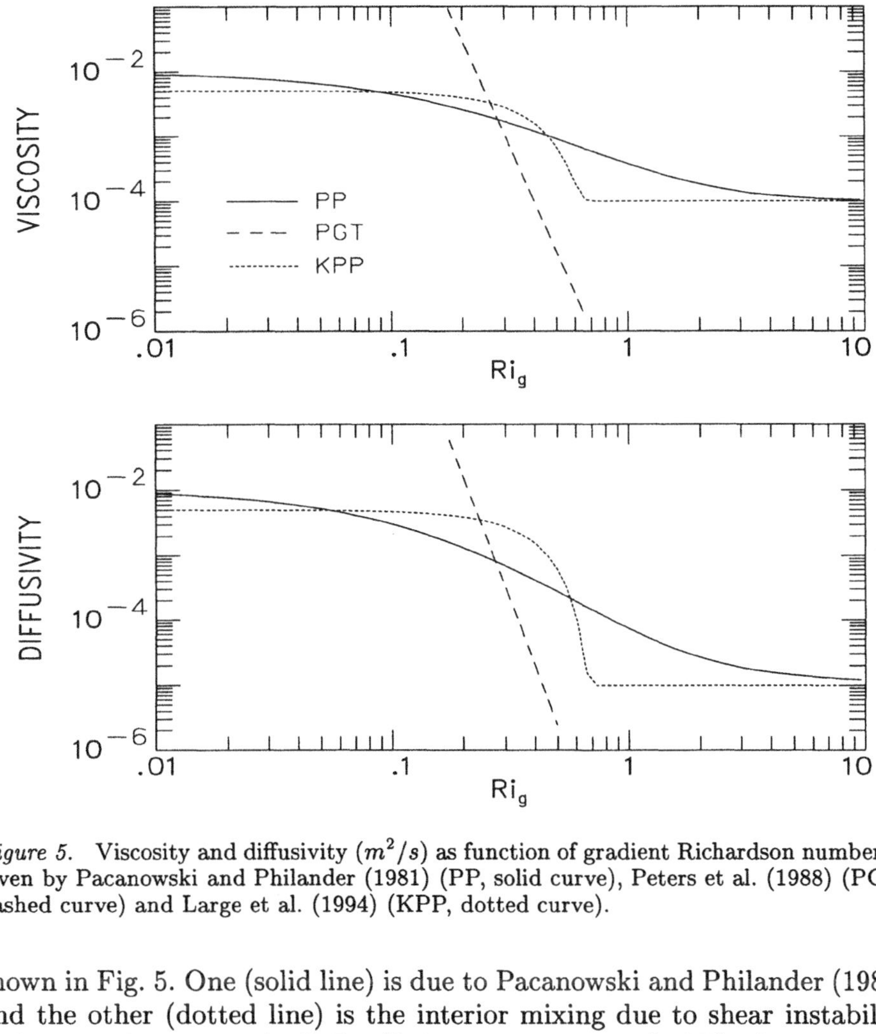

Figure 5. Viscosity and diffusivity (m^2/s) as function of gradient Richardson number as given by Pacanowski and Philander (1981) (PP, solid curve), Peters et al. (1988) (PGT, dashed curve) and Large et al. (1994) (KPP, dotted curve).

shown in Fig. 5. One (solid line) is due to Pacanowski and Philander (1981) and the other (dotted line) is the interior mixing due to shear instability and internal waves in the K-Profile Parameterization (KPP) of Large *et al.* (1994). However, in the former viscosity is significantly larger than diffusivity in the range $0.1 < Ri_g < 0.5$.

4. Parameterizations (1-D Models)

A one-dimensional numerical model of the ocean solves prognostic equations for the vertical profiles of potential temperature, Θ, salinity, S, and the components (U, V) of horizontal velocity, $\vec{U}$. These equations are, respectively,

$$\begin{aligned}\partial_t U &= \partial_z[-\overline{wu}] + f\, V + G_U \\ \partial_t V &= \partial_z[-\overline{wv}] - f\, U + G_V\end{aligned}$$

$$\partial_t \Theta \;=\; \partial_z[-\overline{w\theta}] \;+\; (\rho_o C_p)^{-1}\, \partial_z I(z,t) + H_\Theta. \tag{15}$$

where the surface values of the kinematic fluxes need to be specified for closure. In (15) the terms G_U, G_V, and H_Θ can be specified to represent any large scale flow not explicity modeled, and can be made consistent with the large scale flow of an LES model (11).

Models can be classified according to how they deal with the kinematic flux terms. If we assume that the fluxes are the result of a number of processes and that the turbulent motions associated with each process, p, are separated by spectral gaps in both frequency and wavenumbers from other processes, then from (9)

$$< \overline{wx} >= \sum_p \int_{k_{p1}}^{k_{p2}} \int_{f_{p1}}^{f_{p2}} \phi_{wx}(f,k) df\, dk. \tag{16}$$

A complete parameterization would then consist of the summation of the parameterizations for each process, p. It is clear that a parameterization should depend on what portion of frequency (f_{p1} to f_{p2}) and wavenumber (k_{p1} to k_{p2}) space is being accounted for. This becomes fundamental when the lower limits of integration are set by grid resolution, Δ, and model timestep, Δt:

$$k_{p1} = (2\Delta)^{-1} \qquad\qquad f_{p1} = (2\Delta t)^{-1}. \tag{17}$$

A simple summation of parameterizations is complicated by possible interaction between processes and by processes that overlap in spectral space. However, usual practice in modeling three-dimensional PBL turbulence is to use a single parameterization encompassing all the vertical transport scales.

The two general categories of such parameterization are Mixed-Layer (Bulk) Models and Turbulence Closure Models (Fig. 6). Mixed-layer models have been the more common and have had some successes. They all assume, *a priori*, the existence of a perfectly well-mixed layer of depth, h_m. Observations of deep isothermal layers have clearly provided the foundation for this approach, but it is not a universally correct assumption. These models can further be classified according to whether h_m evolves prognostically or diagnostically. An example of the former is presented in Section 4.2, while the classic example of the latter (Price *et al.*, 1982) is referred to by Price (1998, this volume).

Turbulence closure models in the ocean are very much like those in the atmosphere. They are all attempts at overcoming the fundamental turbulence "closure" problem. The problem is that the number of equations needed to describe turbulent flow is always less than the number of unknowns. A three-equation set like (15), has 6 unknowns, the 3 means plus

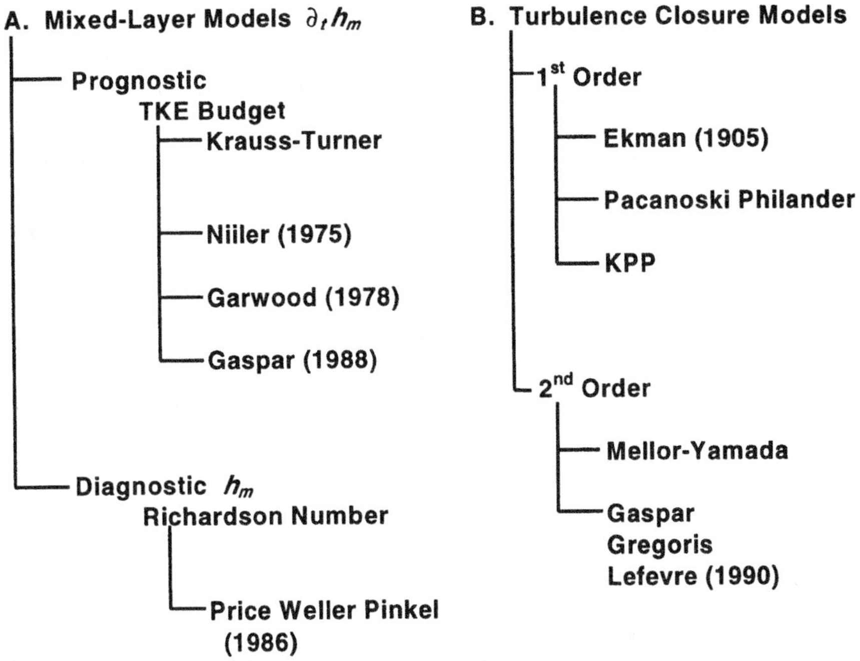

Figure 6. Classification of OBL mixing schemes into Mixed-Layer and Turbulence Closure models.

the three double correlations, or second moments ($\overline{wx}$). Closure can be obtained by expressing these unknowns in terms of the means. The order of a turbulence closure is defined to be equal to the highest prognostic moment, so this would be a first order closure. Alternatively, prognostic equations for the three second order moments can be derived to give a system of six equations, but then the number of unknowns increases to 10, because of the triple correlations in the second moment equations. Writing these third moments in terms of first and/or second moments results in a second order closure. In a third order closure there are 15 unknowns and the fourth order moments are generally written in terms of the second order moments. Convergence to the "correct" solution as the order of a closure is increased should not necessarily be expected, since the number of unknowns increases faster than the number of equations.

4.1. PACANOWSKI-PHILANDER

The simplest first order closure of (15) is to assume down-gradient, Fickian diffusion :

$$\overline{wx} = - K_x \, \partial_z X \quad , \tag{18}$$

with prescribed diffusivity, K_s, and viscosity, K_m. Price (1998, this volume) discusses the Ekman solution for x = horizontal velocity. A more sophisticated approach that has been applied at all ocean depths is to make K_s and K_m functions of Ri_g (13). Although originally proposed by Pacanowski and Philander (1981) for equatorial oceans, it has been used more widely. Their general forms are

$$\begin{aligned} K_m &= \frac{\nu_0}{(1 + a\ R_i)^n} + \nu_m \\ K_s &= \frac{K_m}{(1 + a\ R_i)} + \nu_s\ , \end{aligned} \tag{19}$$

where a fit of equatorial model results to observations was used to determine the constants as $a = 5$, $n = 2$, $\nu_m = 1.0\ cm^2/s$, $\nu_s = 0.1\ cm^2/s$, $\nu_0 = 100\ cm^2/s$. These functions are plotted as solid curves in Fig. 5. As noted in Section 3.1, (19) is consistent with LES results from the highly sheared ocean interior between the Equatorial Undercurrent and the bottom of the boundary layer. It would be expected to be successful only in ocean regions where this interior mixing dominates OBL mixing.

The parameterization (19) basically says that in such a region of large shear, there is a tendency for instabilty as indicated by a small Ri_g, and hence large K_s and K_m values and large momentum and heat fluxes. This picture is consistent with observations of the equatorial ocean. However, as the ocean becomes unstratified, $\partial_z B = 0$, as near the surface in the OBL, Ri_g goes to zero and the mixing coefficients become constant with depth at ν_0, like the simplest Ekman theory.

Now consider non-linear mixing length theory (L-Theory) of turbulence. Prandtl (1925) proposed that the fundamental mixing length, L_m, should be proportional to the standard deviation of vertical particle displacements, from which Heisenburg (1948) arrived at

$$K_x = L_m^{4/3}\ \epsilon^{1/3}. \tag{20}$$

Following Blackadar (1962), dissipation is equated to production of TKE:

$$\epsilon = P_S = u^{*2}\ \partial_z U = L_m^2 (\partial_z U)^3 \tag{21}$$

from which the viscosity becomes

$$K_m = L_m^2\ \partial_z U \tag{22}$$

$$L_m = \kappa d(1 + (\kappa d)/L_o)^{-1}, \tag{23}$$

where L_o is empirically determined from measurements in the free atmosphere, as was ν_m in (19).

The expression (23) is formulated to agree also with similarity theory in neutral conditions near the boundary where $(\kappa d)/L_o$ is much less than 1. Then (6) when squared with $\phi_m(\zeta = 0) = 1$, and (22) both give a viscosity

$$K_m = (\kappa d)^2 \partial_z U = \kappa \; d \; u^*. \tag{24}$$

Therefore, in this situation Packanowski and Philander (1981) mixing (19) and its constant diffusivity, is inconsistent with similarity theory and the L-Theory prediction of the dependence of viscosity on d and u^*.

4.2. KRAUS-TURNER MIXED LAYER

Mixed-layer models where h_m evolves prognostically generally follow Kraus and Turner (1967), and utilize the TKE equation (14), with zero time tendency and divergent transport, $D_T = 0$. Vertically integrating from $z = -h_m$ to the surface yields

$$h_m \overline{P_S} - h_m \overline{F} - \rho_r^{-1} \overline{wp}_o + \rho_r^{-1} \overline{wp}(-h_m) = h_m \overline{\epsilon}, \tag{25}$$

where the overbar represents the vertical average. Unique to the OBL is the $-\rho_r^{-1}\overline{wp}_o$ term, which represents the net surface flux of turbulent kinetic energy due gravity wave and atmospheric pressure correlations. It is commonly parameterized as $m_D \; {u^*}^3$. The shear production term is also parameterized as being proportional to the cube of the friction velocity: $h_m \overline{P_S} = m_p \; {u^*}^3$.

Assuming that the buoyancy flux is linear between its surface value, F_o, and its value at $z = -h_m$, F_{-h_m}, gives

$$\overline{F} = \frac{1}{2}(F_{-h_m} + F_o). \tag{26}$$

Corrections to (25) for solar absorption are possible. Implicit in Kraus-Turner mixed layer models is another assumption about property profiles, namely; there are differences in mixed layer values, X_m, and values at $z = -h_m$, such that property jumps across the mixed layer base $\Delta X = X_m - X(-h)$ are significantly non zero. For downward fluxes positive, the fluxes across $z = -h_m$ are then equated to $w_e \Delta X$, where the entrainment velocity is defined as

$$\begin{aligned} w_e &= \partial_t \; h_m \qquad \partial_t h_m > 0 \\ &= 0 \qquad\quad\; \partial_t h_m < 0, \end{aligned} \tag{27}$$

and $F_{-h_m} = w_e \Delta B$. The flux of energy at the mixed-layer base, $\rho_r^{-1}\overline{wp}(-h_m)$, is generally negligible so that (25) becomes

$$m_p \; {u^*}^3 + m_D \; {u^*}^3 - \frac{h_m}{2}(F_o + w_e \; \Delta B) = h_m \overline{\epsilon}. \tag{28}$$

Mixed layer models differ primarily in their parameterization of dissipation. A simple way (Niiler, 1975) assumes that a constant fraction of each production term is dissipated. The two u^* terms can then be combined, the surface buoyancy flux is a production term only if it is negative, and the entrainment term is never a production term. Therefore, (28) is closed with

$$\begin{aligned} h_m \, \overline{\epsilon} &= m_1 \, u^{*^3} + \frac{m_2 h_m (|F_o| - F_o)}{4} : \\ h_m \, w_e \, \Delta B &= 2 m_0 \, u^{*^3} - \frac{h_m}{2} \Big[m_2 |F_o| + (2 - m_2) F_o \Big], \end{aligned} \tag{29}$$

where $m_0 = m_p + m_D - m_1$. In deepening entrainment situations (29) is the desired prognostic equation for $w_e = \partial_t h_m$. Shallowing situations, $w_e = 0$, result in a diagnostic equation, which for $F_o > 0$ becomes

$$h_m = 2 m_0 L \; . \tag{30}$$

Generally accepted values are $m_2 = 0.8$ and $0.4 < m_0 < 0.5$.

Which observations do not support this bulk mixed layer model? First, the OBL generally is not uniformly mixed in momentum, temperature, and salinity (Lukas and Lindstrom, 1991). Second, discontinuities are not observed at the base of the mixed-layer (Kantha and Clayson, 1994). Third, buoyancy flux profiles may not always be linear (Fig. 4). Fourth, accounting for $-\rho_r^{-1} \overline{wp}(-h_m)$ tends to improve model behavior. Finally, the assumed constants may not be universal, and may have neglected functional dependencies. For example, Crawford and Large (1996) find that the ratio m_p is highly variable, depending on the time history of wind direction and on the mixed layer depth. This issue is clouded by combining the individual coefficients into $m_0 = m_p + m_D - m_1$, which may explain the adjustments that have been made to m_0 and m_2. For example, Garwood (1977) suggests $m_o = 7.6$, but Martin (1985) and Gaspar (1988) tune the same model and find $m_o = 4.5$ and 3.5, respectively.

4.3. SECOND ORDER CLOSURE

A second-order closure model directly computes the time evolution of at least one second-moment quantity. Full closure of all the second-moment equations carries the substantial computing burden of integrating many prognostic equations. For example, a system with three velocity components, potential temperature, and one conservative scalar requires 15 prognostic second moments (Donaldson, 1973). Mellor and Yamada (1974; 1982) develop a hierarchy of second-moment closures for a u, v, w, and θ system. At each degree of approximation the "level" of the hierarchy decreases (from 4 to 1). There are 10 prognostic equations for the second moments in their full, level 4 closure. To date it has not been demonstrated that the results from such models justify the additional complexity and computations, even if simplified by coordinate choice to have $\overline{uv} = \overline{vw} = \overline{v\theta} = 0$.

At level 3, Mellor and Yamada solve prognostic equations for only two second moments; temperature variance, $\overline{\theta^2}$, and TKE. This simplification arises from expressing the other second-order moments as diagnostic equations depending only on themselves; mean quantities, including TKE; and empirical length scales. The vertical fluxes, for example, are complex expressions:

$$\overline{uw} = \frac{-3A_1\ell}{q}\left[(\overline{w^2} - C_1 q^2)\partial_z U - \alpha g \overline{u\Theta}\right]$$
$$\overline{w\Theta} = \frac{-3A_2\ell}{q}\left[\overline{w^2}\partial_z \Theta - (1 - C_3)\alpha g \overline{\Theta^2}\right], \qquad (31)$$

involving second moments, "universal" constants, and length scales, ℓ.

Similarly, expressing $\overline{\theta^2}$ as a diagnostic equation results in the intermediate level 2.5 degree of approximation, with q^2 the only prognostic second moment.

$$\frac{\partial q^2}{\partial t} = \frac{\partial}{\partial z}\left[\ell\ q\ S_q\ \frac{\partial q^2}{\partial z}\right] + 2(P_S - F - \epsilon). \qquad (32)$$

Since all the length scales are related to the scale of energy containing turbulence, they are usually made proportional to a master length scale, such as L_m. Then L-Theory (20) and the parameterization $\epsilon = q^3/(B_1\ell)$ with $\ell \propto L_m$ give $K_x = S_x\ q\ \ell$, with S_x an empirical constant. Now (32) can be recognized as a form of the TKE equation (14), with the pressure work effectively ignored by combining it with the turbulent transport in the square brackets, then expressing this term as $\overline{wq^2} = K_{q^2}\ \partial_z q^2 = S_{q^2}\ q\ \ell\ \partial_z q^2$. This balance is different than assumed by Kraus-Turner models.

The level 2 approximation assumes that shear plus buoyant production of TKE is balanced locally by turbulent dissipation to give a diagnostic equation for q^2 as well. At this point in the hierarchy there are no second moment prognostic equations, and the closure is first order and akin to L-theory.

The most popular member of the Mellor–Yamada hierarchy is the level 2.5. However, these schemes are known to have the general trait of too little entrainment during convection and the related problem of too little mixing across stabilizing density gradients (Kantha and Clayson, 1994). There are several possible contributors. Realistic entrainment requires buoyancy effects to be accounted for in the closure of the turbulent transport of third-order fluxes. Also, the use of a master length scale, which sets all processes proportional to a single scale, is recognized as a major weakness of the model (Mellor and Yamada, 1982). An evaluation of second-order closures (Moeng and Wyngaard, 1989) reveals two more objections to usual practice; namely, the down-gradient transport assumption is inadequate in the convective ABL because it neglects nonlocal effects, and neglect of buoyancy effects on the pressure-scalar covariance closure. This study also finds that the Mellor–Yamada parameterization of the eddy diffusivity for TKE is more than an order of magnitude less than that computed from LES data. A remark of Nurser (1996) is that the constants are determined from neutral laboratory flows, and do not seem to apply to convective boundary layers.

There have been several responses to these criticisms. A prognostic equation for the product of q^2 times the master length scale has been added (Mellor, 1989). Gaspar *et al.* (1990) specify two independent diagnostic length scales. Also, an enhanced mixing in the shear region below the ocean mixed-layer has been added (Kantha and Clayson, 1994).

4.4. NONLOCAL K-PROFILE PARAMETERIZATION

The K-Profile Parameterization (KPP) is similar to existing nonlocal models of the atmospheric boundary layer (Troen and Mahrt, 1986; Holtslag *et al.*, 1990). Particular attention is given to the distinct physics of the boundary layer, which are characterized by Fickian diffusion (18) plus nonlocal transport. The crucial parameter is the boundary layer depth, h, which is diagnosed from the surface fluxes and the velocity and buoyancy (temperature and salinity) profiles. There is no *a priori* assumption of the existence of a well mixed layer. However, as with observations, a near isothermal layer usually develops and its depth, h_m, depends on definition (Lukas and Lindstrom, 1991). As observed (*e.g.*, Davis *et al.*, 1981), significant velocity shear and stratification can exist within this mixed layer. In the interior, the eddy mixing coefficients are used to parameterize the combined effects of double diffusion, unresolved internal waves and resolved scale shear (Richardson number) instability. These parameterizations are detailed by Large *et al.* (1994), and the sum of the latter two effects are shown in Fig. 5.

In the boundary layer the kinematic fluxes are given by

$$\overline{wx}(z) = -K_x\ (\partial_z X - \gamma_x) \qquad 0 < \sigma \leq 1. \tag{33}$$

The K_x values are nonlocal, depending on h, $\sigma = d/h$ and the surface forcing, but act on the local property gradients to produce down-gradient transports. Only in the case of heat and salt with convective (unstable) forcing do the diffusivities also multiply the nonlocal term, γ_x, which can produce counter-gradient fluxes (*e.g.*, Fig. 1).

It is important to appreciate that the h is not directly related to h_m. Essentially, it is the depth to which turbulent boundary layer eddies can penetrate vertically before being stopped by the stratification. When such a structure has a lot of kinetic energy due to shear between where it is formed (usually the surface) and the thermocline, the penetration can be substantial. At such times h is very much greater than the mixed layer depth, and there is vigorous property exchange between the boundary layer and interior. The boundary layer depth depends on the surface forcing, on the oceanic velocity profile $\vec{U}(z)$, on the buoyancy profile and on a parameterized turbulent velocity shear. It is the shallowest depth where a bulk Richardson number equals a critical value:

$$Ri_c = \frac{(B_r - B(h))h}{|\vec{U}_r - \vec{U}(h)|^2 + V_t^2} \quad , \tag{34}$$

where B_r and $\vec{U}_r$ are averages over the surface layer and empirically $Ri_c = 0.3$. The unresolved shear leads to the V_t^2 term and is most important in situations of little or no wind to produce mean shear. It is formulated (Large et al, 1994) to make the model entrainment agree with the theoretical arguments (Ball, 1960) and the observational data (Deardorff *et al.*, 1969; Tennekes, 1973; Moeng and Wyngaard, 1989). These suggest that with only convective surface forcing, the ratio of the entrainment to the surface buoyancy flux equals a constant, -0.2 (Fig. 1), independent of the stratification in the pycnocline.

The profile of boundary layer diffusivity is expressed as the product of the depth dependent turbulent velocity scale, w_x, and a non-dimensional vertical shape function $G(\sigma)$:

$$K_x(\sigma) = \; h \; w_x(\sigma) \; G(\sigma), \tag{35}$$

The dependence on h reflects the ability of deeper boundary layers to contain larger, more efficient turbulent eddies. Particular examples of $G(\sigma)$ and $w_x(\sigma)$ profiles are shown in Fig. 7. The shape function is assumed to be a cubic polynomial (O'Brien, 1970),

$$G(\sigma) = a_0 + a_1\sigma + a_2\sigma^2 + a_3\sigma^3. \tag{36}$$

Turbulent eddies do not cross the surface, so the necessary condition, $K_x(0) = 0$, is imposed by setting $a_0 = 0$. In the surface layer, $\sigma < 0.1$, where Monin Obukhov similarity theory applies, $G(\sigma) \simeq \sigma a_1$ and from (6)

$$\phi_x = \frac{\kappa \; d \; u^*}{K_x} = \frac{\kappa \; u^*}{w_x \; a_1}, \tag{37}$$

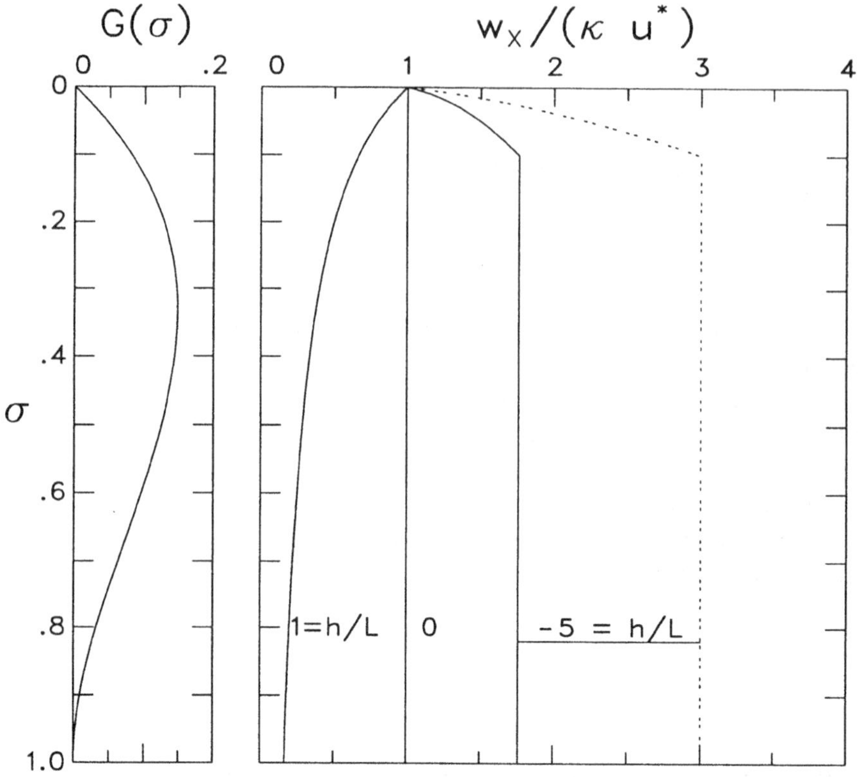

Figure 7. Left panel: Vertical profile of the shape function $G(\sigma)$, where $\sigma = d/h$, in the special case of $G(1) = \partial_\sigma G(1) = 0$. Right panel: Vertical profiles of the normalized turbulent velocity scale, $w_x(\sigma)/(\kappa u^*)$, for the cases of $h/L = 1$, 0 and –5. In unstable conditions $w_s(\sigma)$ (dashed traces) is greater than $w_m(\sigma)$ (solid traces) at all depths, but for stable forcing, $h/L \geq 0$, the two velocity scales are equal at all depths.

which is satisfied by $w_x = \kappa\, u^*\, \phi_x^{-1}$ and $a_1 = 1$. Also, as discussed earlier, the velocity scales become proportional to the convective velocity scale in pure convection. As argued in Troen and Mahrt (1986), this formulation is assumed to be valid everywhere in the stably forced boundary layer. In unstable conditions, the turbulent velocity scales beyond the surface layer are assumed to remain constant at their $\sigma = 0.1$ values (Fig. 7).

In the KPP scheme the ocean interior is allowed to force the OBL. Specifically, a_2 and a_3 are chosen so that interior and boundary layer mixing coefficients and their vertical derivatives match at $d = h$. In this way flux divergences are continuous at $d = h$, even though the mixing rules change.

The nonlocal transport term in (33) is non-zero only for scalars in unstable (convective) forcing conditions. Following Deardorff (1972b), and

Mailhôt and Benoit (1982) it is parameterized by

$$\gamma_s = C_x \frac{\overline{ws_o}}{w_s(\sigma)\ h} \ , \quad \gamma_\theta = C_s \frac{(\overline{w\theta_o} + \overline{w\theta_R})}{w_s(\sigma)\ h} \ ; \ \zeta < 0 \tag{38}$$

with $C_s = 10\ \kappa(c_s \kappa \epsilon)^{1/3}$ in order to be consistent in the convective limit outside the surface layer. The $\overline{w\theta_R}$ term represents the amount of radiative heat absorbed in the boundary layer that effectively contributes to the nonlocal transport of heat.

There is considerable evidence from observations (Kaimal *et al.*, 1976; Lenschow *et al.*, 1980) and LES (Deardorff, 1972a) of very small velocity gradients in the central portion of the convective ABL. This implies that momentum fluxes in this region are also maintained by nonlocal transport. However, such a term cannot be derived as Deardorff (1972b) did for the heat flux, so it is unlikely to have the form of (38). Until a suitable form has been determined nonlocal momentum flux is not parameterized ($\gamma_m = 0$).

5. Evaluation Case Studies

A thorough evaluation of OBL models is important both to developers seeking ways to improve the parameterizations and to users whose research depends on the model results. Ideally, the basis of evaluation would be a full suite of observations obtained over a full range of parameter space. Unfortunately, the inherent difficulty of measuring turbulent flows has resulted in an inadequate database for the purpose. As models become more complex and are applied to a wider range of research questions, this problem worsens. It is particularly acute in the ocean where surface-wave contamination has prevented any direct flux measurements. There are an appreciable number of turbulent dissipation estimates, but even these are indirect. They are inferred from measurements of the vertical gradients of horizontal velocity, and temperature (Toole, 1998, this volume).

5.1. AN EVALUATION OF ABL PARAMETERIZATIONS

An alternative, more widely used for the ABL than OBL, is to augment the database with turbulence simulations, such as LES. For example, Ayotte *et al.* 1996 use LES (Moeng, 1994) to evaluate the behavior of turbulent closure schemes over a range of ABL conditions. In this code the flow variables of interest are u, v, w, θ and two passive, conservative tracers referred to as B and C scalars. The B scalar mimics humidity in the atmosphere and salinity in the ocean, if effects on density are ignored, in that there are nonzero fluxes both at the surface and the inversion. In the atmosphere the C scalar is "top-down", with zero surface flux, while in the ocean it would be "to-surface" and mimic nutrients being entrained into the mixed layer from below.

Once LES has been chosen as the standard of comparison, the next issue is which schemes to evaluate. Ayotte *et al.* (1996), hereafter A96, choose seven :

1. $K(Ri)$: a first order closure (Louis, 1981) where, like Pacanowski and Philander (1981), the diffusivity and viscosity are functions of Richardson number.
2. MY2.0 the level 2 second order closure of Mellor and Yamada (1982).
3. MY2.5 the level 2.5 second order closure of Mellor and Yamada (1982).
4. KPROF: the nonlocal K-profile model of Troen and Mahrt (1986), which was the basis for KPP.
5. MIX: a mixed-layer model based on the (Driedonks, 1982) TKE equation at the base of the inversion, rather than integrated over the layer as in Kraus and Turner.
6. STAB: the stability bounds scheme originally developed for the ocean by Price *et al.* (1986).
7. TRAN: the transilient model of Stull (1984), where a set of rules are applied that allow a given level to mix with every other level.

After the schemes are selected and numerically implemented in a common experimental framework, no tuning of parameters is allowed. An evaluation is not an optimization exercise or sensitivity study.

The next issue is to select a number of cases that span the parameter range of interest. Since LES are only now beginning to simulate the stable boundary layer, A96 is restricted to the neutral and unstable boundary layers. They use 10 cases for simulation by LES and the 7 schemes;

1. EKM: an idealized case of Ekman flow where there is no surface heating ($L = \infty$), no stratification, and forced with a uniform geostrophic wind of $10\,m/s$ ($u^* = .39\,m/s$).

2.,3. OOWC and O5WC: two cases with a weak capping inversion at about 1075 m and forced with a uniform geostrophic wind of 15 m/s ($u^* \simeq .67\,m/s$). The first case has no buoyancy forcing and the second has moderate surface heating given by $u^*\theta^* = 0.05\,K\,ms^{-1}$, ($L = -481\,m$).

4.,5.,6.,7. 00SC, 03SC, 05SC, 24SC: four cases with a strong capping inversion at 450 m, 490 m, 480 m, and 1025 m, respectively. In all cases the geostrophic wind is uniform at 15 m/s ($u^* = .50$ to $.62\,m/s$). The surface heating is ($u^*\theta^*$) is systematically increased over the 4 cases (0 to .03 to .05 to .24 $K\,ms^{-1}$), with corresponding changes in L(∞ to -380 to -370 to -56 m).

8. 24F: a free convection ABL, where there is no mean wind and a strong capping inversion at 1000 m to limit the growth of the ABL.

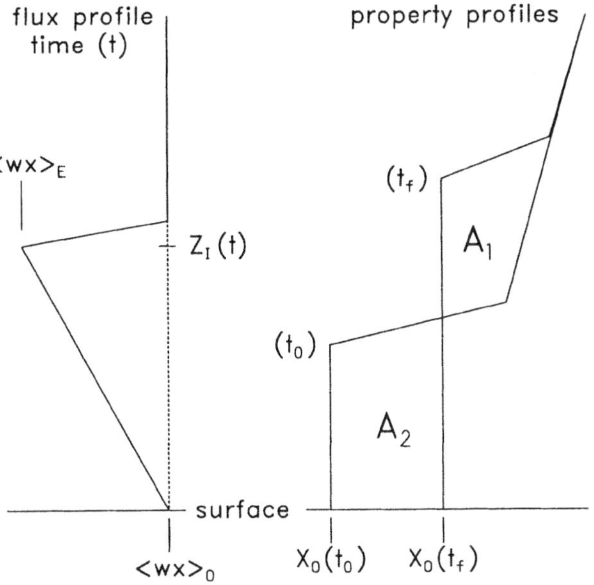

Figure 8. Idealized sketch of entrainment during a one-dimensional model integration from time t_o to t_f, showing the evolution of a property profile. The flux profile at an intermediate time defines the inversion height, $Z_I(t)$. The evaluation measures of Ayotte et al. (1996) are the entrainment, A_1, the sum of entrainment and surface flux, A_2, the final mixed-layer property values, $X(t_f)$, the surface flux, $< wx >_o$, and the entrainment flux $< wx >_E$.

9.,10. 15B and 24B: two baroclinic cases produced by varying the speed and direction of the geostrophic wind with height ($u^* \sim .63m/s$), the surface heat flux is $u^*\theta^* = .15\,K\,ms^{-1}(L = -125\,m)$ in the first case, and $.24\ km\,s^{-1}(L = -83\,m)$ in the second. Both cases have a strong capping inversion at about 1000 m.

In each case the LES is run until quasi-steady turbulence is developed. At this time, t_o, the scalars are introduced, the LES continued and the PBL model integration started, using LES initial conditions. All solutions are then integrated for about 10 eddy turnover times until time, t_f.

Evaluation requires quantitative measures of a parameterization's performance relative to LES. This issue is not straightforward and somewhat subjective as different schemes will perform the "best" under different circumstances depending on the measure. The important thing is to discover systematic biases and evaluate overall performance. Figure 8 is a schematic of the six measures employed by A96. They are tabulated in Fig. 9, along with the properties they are applied to. They are used for two purposes. First, they are evaluated for each property, for each of the 10 cases and used for case by case evaluation. Second, normalized measures are summed

Error Measure	Property, X			
	(u, v)	θ	B	C
A_I Entrainment		C		C
$A_2 =$ $A_I +$ Surface Flux			C	
$\frac{A_1}{A_2 - A_1}$		C		
Mean Over Mixed Layer	R_I	R_I	R_I	C R_I
Surface Flux $\langle wx \rangle_o$	C R_2	Given	R_2	0
Entrainment Flux $\langle wx \rangle_E$		R_2		R_2

Figure 9. Table showing which error measures are applied to which properties in both the case by case evaluations, C, and the resolution sensitivity studies of Ayotte et al. (1996). For the latter, R_1 denotes final mixed-layer property values, $X(t_f)$, and R_2 denotes surface and entrainment fluxes.

over the 10 cases, yielding a single number for each of a range of resolutions. Figure 9 denotes the measures and properties used for case by case evaluations by a "C", and those used for resolution sensitivity by R_1 for propery averages over the boundary layer and by R_2 for fluxes.

Two of the these six measures are designed to assess overall ability to reproduce the evolution of mixed-layer means by producing the correct fluxes. The first, A_1, is an overall measure of entrainment over the integration time and the second, A_2, is a similar measure of entrainment plus surface flux. As shown by Fig. 8, they are defined as

$$A_1 = -\int_{Z_I(t_o)}^{H} \frac{(X(z,t_f) - X(z,t_o))}{(t_f - t_o)} dz$$
$$A_2 = \int_{0}^{Z_I(t_o)} \frac{(X(z,t_f) - X(z,t_o))}{(t_f - t_o)} dz, \quad (39)$$

where H is the model top. The entrainment as measured by A_1 is just the time integral of the vertical property flux at $z = Z_I(t_o)$, and is used for θ and C-scalar. The difference between this time integral and the integrated surface flux is A_2, which is used for B-scalar only. A third measure of the relative strengths of entrainment and surface flux, $A_1/(A_2 - A_1)$, is used only for θ (Fig. 9). The other three measures (Fig. 8) are evaluated at the final time, t_f. They are the mean mixed-layer values (wind speed, θ, B-scalar, and C-scalar), the surface fluxes (momentum, and B-scalar), and finally the entrainment fluxes (θ and C-scalar).

The case by case evaluations can be found in A96, so only the highlights will be described here. The two Mellor-Yamada results are very similar, indicating that the level 2.5 is not an improvement over 2.0. They both systematically entrain too little. There is a universal lack of C-scalar entrainment in the baroclinic cases. The mixed layer scheme tends to over-entrain C-scalar and θ in the barotropic cases. Some models perform very differently with a weak capped inversion than a strong capped (*e.g.*, $K(Ri)$, KPROF).

As an overall assessment of a model's ability to reproduce the LES mixed-layer mean properties and fluxes is provided by the resolution sensitivity plots in A96. The resolution is defined as the number of grid points in the boundary-layer and varies between 3 and 15. An ideal parameterization would give the least accumulated error for both the means and the fluxes and have little resolution dependency. Then the best model performance possible would be achieved regardless of how the boundary layer depth varied in time and space. There is no such model. Only the mixed layer model error is independent of resolution, but the aggregate flux error is larger than any of the other schemes, even at coarse resolution. The major contributor is the entrainment heat flux $< w\theta >_E$. Overall KPROF shows the least aggregate error in both mixed layer means and in fluxes, at all resolutions. However, in entrainment heat flux at low resolution it does not do as well as $K(R_i)$, for example. The TRAN, STAB, MY20 and MY25 show more resolution depedency than the other models.

Alternatively, one can pick a "typical" resolution and decide which errors are most serious for a particular research project and select the "best" model for the problem. On the other hand these A96 results show model developers how their model compares with others and in what respects it does poorly or well. Poor performance should motivate efforts at improvement. For example, the parameterized turbulent shear in KPP (34) that produces stability independent entrainment was a direct response to the poorer performance of KPROF with a weaked capped inversion (00WC and 05WC) compared to strong capping cases (00SC, 03SC, 05SC and 24SC).

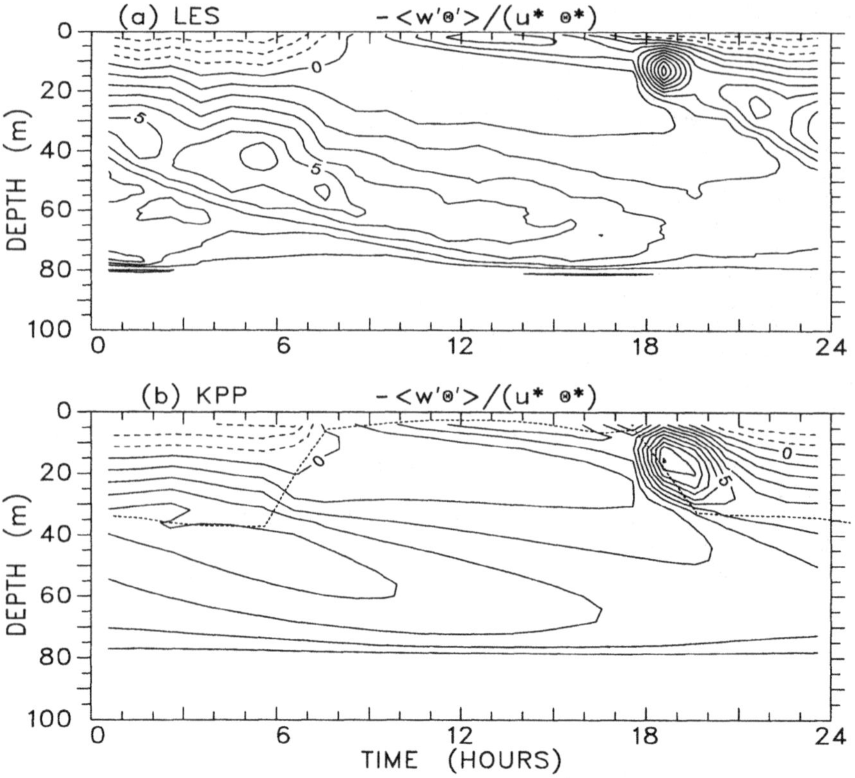

Figure 10. Diurnal cycle of heat flux normalized by $u^*\theta^* = 1.15 \times 10^{-5} ms^{-1}K$, with a contour interval of 1. a) LES solution and b) KPP solution, with boundary layer depth shown as the dotted trace.

5.2. EQUATORIAL OBL

Large and Gent (1998) compare the K-profile parameterization (KPP) to LES solutions (WML) of the diurnal cycle of the OBL at the equator. The LES model is given by (10) and the KPP model by (15). Both allow for specified large scale flow (11) and (12). The comparison is clean because both the surface forcing and the large scale flow are identical in the two models. The comparison is direct because the parameterized turbulent flux profiles are explicitly computed in LES. They demonstrate the fidelity of KPP vertical mixing in the equatorial regime, where observational limitations had precluded such evaluations. The comparison of the diurnal cycle of heat flux in the upper 100 m is shown in Fig. 10. All the features seen in the LES, both in and below the boundary layer, are reproduced by 1-dimensional KPP mixing, despite its coarser resolution (4m versus 1m). Important examples in the boundary layer are the nighttime region of $- < w\theta >< 0$ (dashed contours), including the implicit countergradient fluxes, and the rapid deepening with large fluxes at the onset of convection following sunset (18:30 h). Below the boundary layer KPP mixing is governed the coefficents in Fig. 4 and the heat flux is similar to LES and consistent with the observed deep diurnal cycle of turbulence (*e.g.*, Lien *et al.*, 1995).

In contrast, Large and Gent (1998) find that a control mixing scheme that is like Peters *et al.* (1986) for $Ri_g < 0.4$ (Fig. 5) and Pacanowski and Philander (1981) elsewhere fails to reproduce the LES results in some important ways. Most notably, there is no evidence of intense convection at sunset. Nighttime cooling does appear as a core of high momentum and heat flux, but these tend to be shallower than the LES feature throughout the diurnal cycle. The heat flux at depth is also much weaker. Because of the delayed onset of increased heat fluxes until about 21:00h, the high daily SSTs persist too long. The region of negative heat fluxes is less extensive than in both LES and KPP. Such fluxes can only occur where there is a temperature inversion, because the control only has down-gradient heat flux. In general, the diurnal cycle of momentum flux in the control is much more similar to LES than is the heat flux.

From this comparison it can be concluded that KPP represents a more complete suite of physics than the control. This view is not inconsistent with Pacanowski and Philander (1981) being a good description of observations in the shear region below the equatorial boundary layer, but not applicable to the boundary layer itself (see Chen *et al.*, 1994). However, when the two schemes are implemented in a region equatorial Pacific model (Gent and Cane, 1989) and the results compared to observations, there is no clear "better" performer. Furthermore, systematic biases relative to the observations are evident with both mixing schemes.

These results illustrate the danger of using observations to evaluate a piece of model physics. The observations are a response to true forcing and ocean flow. Model forcing can be very inaccurate and the model ocean flow is probably always defficient. Depending on how these problems interact, it is very possible for improved physics to degrade a model solution relative to observations and visa versa. These problems are not nearly so severe when an LES solution is used as the standard of comparison.

LES solutions are also very useful for determining parameter values and model sensitivity to these values. For example, a parameter, ν_0, gives the viscosity and diffusivity for $Ri_g \leq 0$ for the interior mixing of KPP. The Fig. 5 curves are for $\nu_0 = 50 \times 10^{-4} m^2/s$, which was the value that produced the favorable comparison of Fig. 10. Figure 11 shows the same diurnal cycle of heat flux, but for $\nu_0 = 40, 60,$ and $70 \times 10^{-4} m^2/s$. The sensitivity is surprising and dramatic. As ν_0 is increased, the onset of the vigorous nighttime convection and rapid deepening of the boundary layer

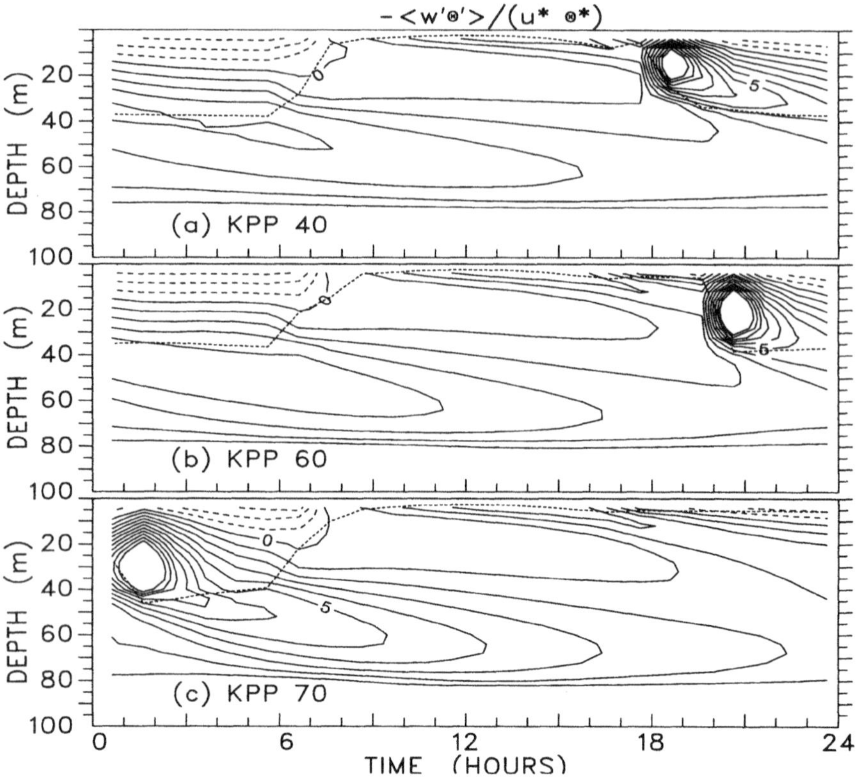

Figure 11. As for Fig. 10 except only for KPP solutions with ν_0 equal (a) 40, (b) 60 and (c) $70{\times}10^{-4}m^2/s$.

depth is delayed later and later. For $\nu_0 = 40 \times 10^{-4}m^2/s$ the onset is around 18:00h, as for $\nu_0 = 50 \times 10^{-4}m^2/s$ and in LES (Fig. 10). However, at $\nu^0 = 60 \times 10^{-4}m^2/s$ the rapid deepening doesn't begin until after 19:30 (Fig. 11). A further increase to $\nu^0 = 70 \times 10^{-4}m^2/s$ delays the onset and deepening to just after midnight. Thus, as ν_0 is increased the larger interior mixing appears to be capable of balancing the convective nighttime surface cooling longer, before convective deepening is required. However, when it does occur the magnitude of the entrainment flux remains about the same.

6. The OBL in Ocean GCMs

The hypothesis guiding the global ocean model integrations of Large *et al.* (1997) is that more realistic and less constraining surface boundary conditions, and more physical SGS vertical mixing, including boundary layer processes, will result in improved equilibrium OGCM solutions. Their study first examines the global, volume-average temperatures and salinities for

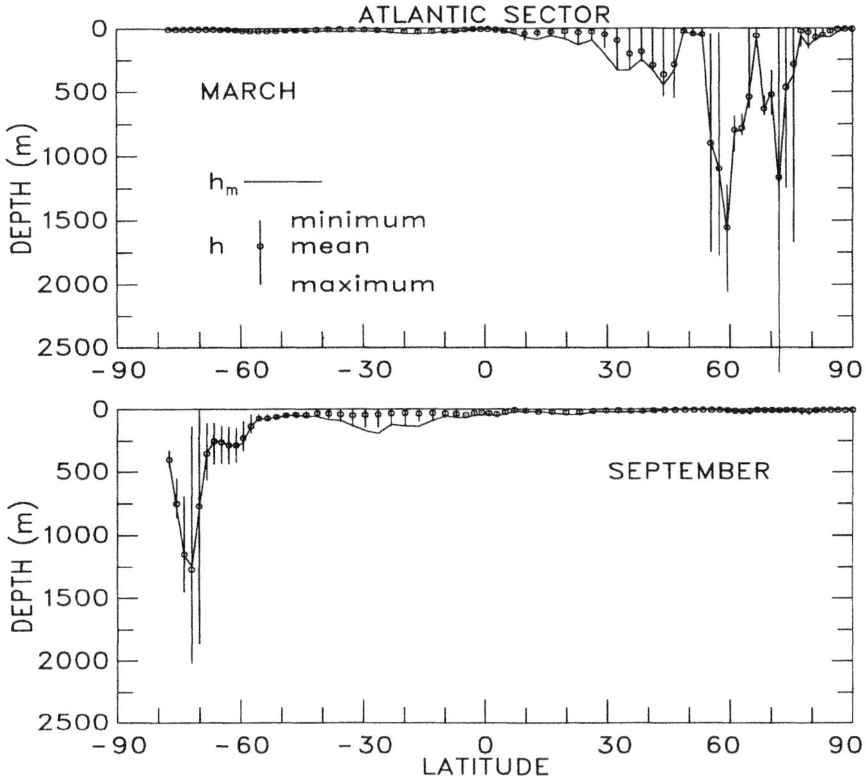

Figure 12. March and September monthly mean mixed layer depth (solid line), and boundary-layer depths (circles) along a 1200 km wide (3 grid points) meridional section of a model Atlantic Ocean. Vertical lines range from the minimum to the maximum of boundary layer depth during the month.

four model equilibrium solutions and the Levitus (1982) climatology. The displacements from climatology reflect the integrated net surface heat and freshwater. As in Danabasoglu and McWilliams (1995), the combination of surface restoring and the Gent and McWilliams (1990) mesoscale eddy transport parameterization leads to a global mean state that is only about 0.2°C too cold, but nearly 0.15 ppt too fresh. Without the eddy transport of tracers, the mean state is much too warm and still too fresh. In contrast, the bulk flux forcing cases of Large *et al.* (1997) fall much closer to Levitus. There is somewhat less net drift when KPP is used instead of enhanced mixing to 50 m depth, but clearly the major impact on the improved global, volume-average temperature and salinity is the change in surface forcing.

In contrast to a uniform 50 m mixed-layer depth, KPP mixing develops boundary layers that are highly variable both spatially and in time. This variability is chiefly a response to the surface forcing and is illustrated by the equilibrium results of Fig. 12. Monthly averages over 10.8 degrees of longitude (3 mid-basin grid cells) of both h (circles) and h_m (solid line) are much shallower in the summer hemisphere than the winter. In the winter hemisphere the maximum boundary layer depth can reach the model's bottom, and it is often much deeper than the average of h. In some places like the subtropical gyres this characteristic contributes to average mixed-layers that are deeper than average boundary layers. In regions of known deep convection, the Weddell Sea, the sub-polar gyre of the North Atlantic and the Nordic Seas, the boundary-layer with its large mixing coefficients reaches well below the mixed-layer to beyond 2000 m depth in winter. This penetrative mixing is KPP's parameterization of the entrainment caused by convective elements that extend throughout the boundary-layer. Is it sufficient, or is another parameterization of deep convection needed? Note that the boundary layer is not near its maximum depth long enough for the mixed layer to even approach these depths. Consistent with observations, the winter-time mixed-layer depths in the subtropical gyres are about 400 m in the North Atlantic and about 200 m deep in the South Atlantic.

The KPP boundary layer scheme has several features not commonly found in OGCM parameterizations. Three examples are the contribution of unresolved shear to the diagnosis of h, the forcing by the interior, and the nonlocal term in the heat and salt fluxes. Are these important, or unnecessary complications? To examine this question, we analyse the Large *et al.* (1997) equilibrium solutions. Given the dearth of observations, especially in the Southern ocean, such analyses represent our only global picture of how the OBL functions throughout the annual cycle. However, without more widespread and direct (ie. flux measurements) observations to constrain the modeling results, aspects of this picture could be either misleading, or at worst, completely incorrect.

The diagnostic equation for h (34) can be written as

$$h = Ri_c \left(\frac{\mid \vec{U}_r - \vec{U}(h) \mid^2}{(B_r - B(h))} + \frac{V_t^2(h)}{(B_r - B(h))} \right) . \tag{40}$$

The first term gives the contribution of the resolved shear and the second the parameterized unresolved shear. The ratio, R_h, of the former to h is shown in Fig. 13. In order to give a flavour of the annual cycle, results from the extreme months (December and six months later in June) are shown. Regions of $R_h > 0.8$ are where the resolved shear is dominant. In December these areas cover the whole of the ACC, and most of the Tropical ocean south of the equator, but little of the winter hemisphere. Again in June little of the winter hemisphere is covered, but resolved shear appears dominant in much of the OBL between the equator and about 30°N.

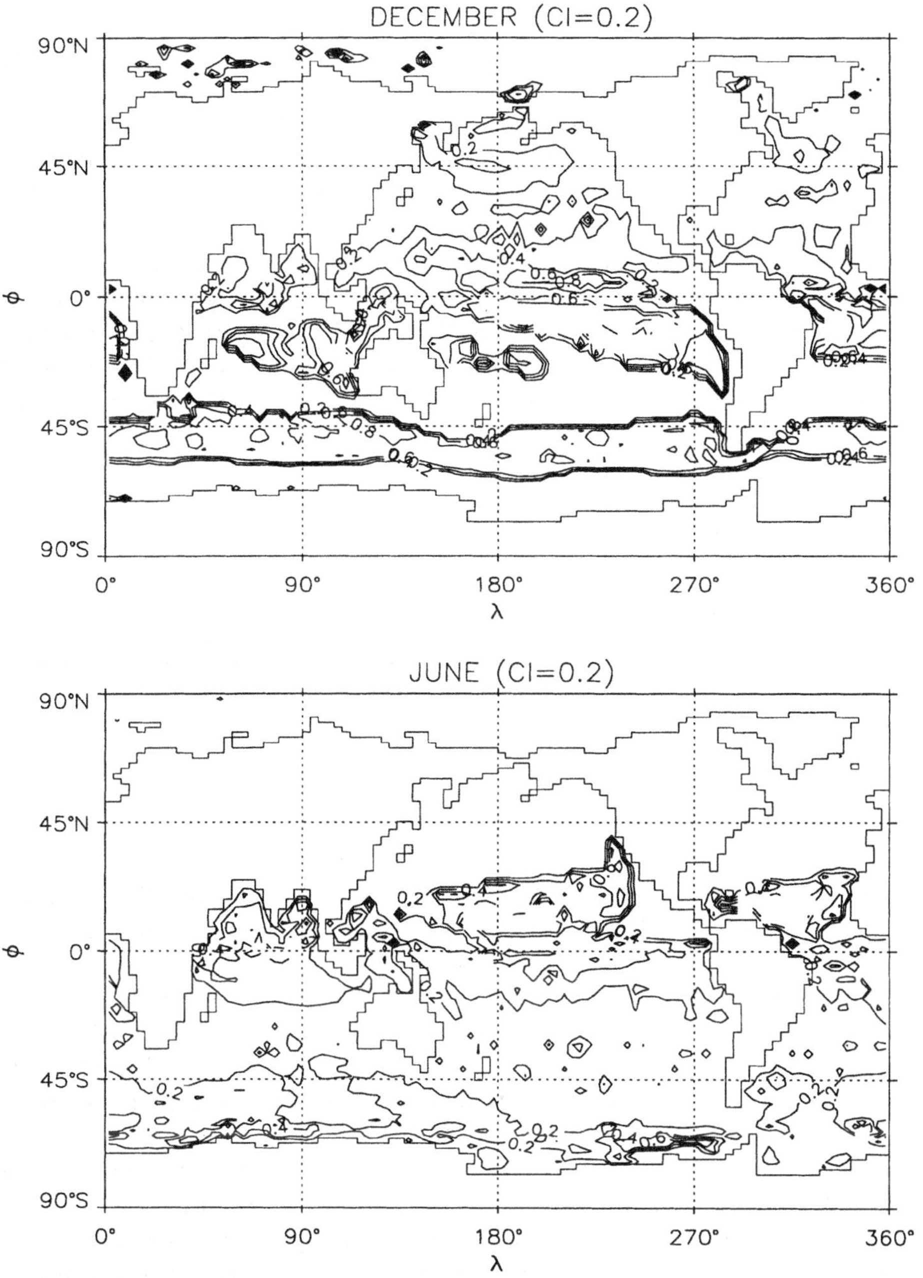

Figure 13. December and June global distributions of the ratio R_h; the fractional contribution of the resolved shear to the boundary layer depth. The remaining fraction is due to the parameterized, unresolved shear.

The parameterizaton of unresolved shear was motivated by the desire to improve the performance of KPROF models in weak capped inversions without degrading the strong capped inversion results of Ayotte *et al.* (1996). It was based on making the ratio of entrainment to surface buoyancy flux, in pure convection, equal to -0.2 (Fig. 1), independent of stratification. Figure 13 shows that in the general conditions of both wind and buoyancy forcing this parameterization plays an important role in determing the boundary layer depth. In particular, there are large regions of $R_h < 0.2$, where it dominates the resolved shear contribution in (40). Such an area is found everywhere north of about $30° N$ in June, and most of this area in other months too. Another is between about $45° S$ and $15° S$ in June, and most of a narrower band ($40° S$ to $20° S$) in December.

Figure 7 shows $G(\sigma)$ for the special case of no interior forcing. The function maximum is $(4/27)$ at $(\sigma = \frac{1}{3})$. Interior forcing increases this function at all depths and the ratio, $R_G = G(\sigma = 1/3)/(4/27)$ is shown in Fig. 14. Again two extreme months (January and July) are shown. At most times and places the ratio is less than 1.5 and the interior forcing does not overly influence mixing within the OBL. However, there are coherent regions where the mixing coefficients are more than doubled ($R_G > 2.0$; thick contour), which would significantly reduce the diffusive time scale and result in a more well mixed OBL. There are a number of small areas of $R_G > 10$ (third contour).

Interior forcing appears to be important where the mixed layer is shallow. This characterization includes the following general regions; poleward of the ACC in January, in the poleward branches of the subtropical gyres of the summer hemisphere, in northern subpolar gyres in July and in the Arctic throughout the year. It is also important throughout the year in the equatorial oceans, although there is a bias north of the equator in July. A contributor to this feature along the equator is likely the shear between the Equatorial Undercurrent and the surface. As shown in Fig. 10 there are large fluxes below the boundary layer, because the shear leads to small Richardson numbers and hence to large mixing coefficients (Fig. 5).

In the KPP scheme (33) there are two scalar flux components: Fickian diffusion and nonlocal transport. The former is always down-gradient, but the latter can sometimes be counter-gradient (Fig. 1). In order to assess nonlocal behavior globally, we construct a function, Λ, that expresses the relative contribution of the nonlocal flux to the total flux as well as its relation to the local gradient. If the flux components are plotted with Fickian the ordinate and nonlocal the abscissa, Λ is a function of the angle, ψ :

$$\begin{aligned} \Lambda &= 1 - \psi/90 \qquad 0 \leq \psi < 180 \\ &= 3 - \psi/90 \qquad 180 \leq \psi < 360. \end{aligned} \tag{41}$$

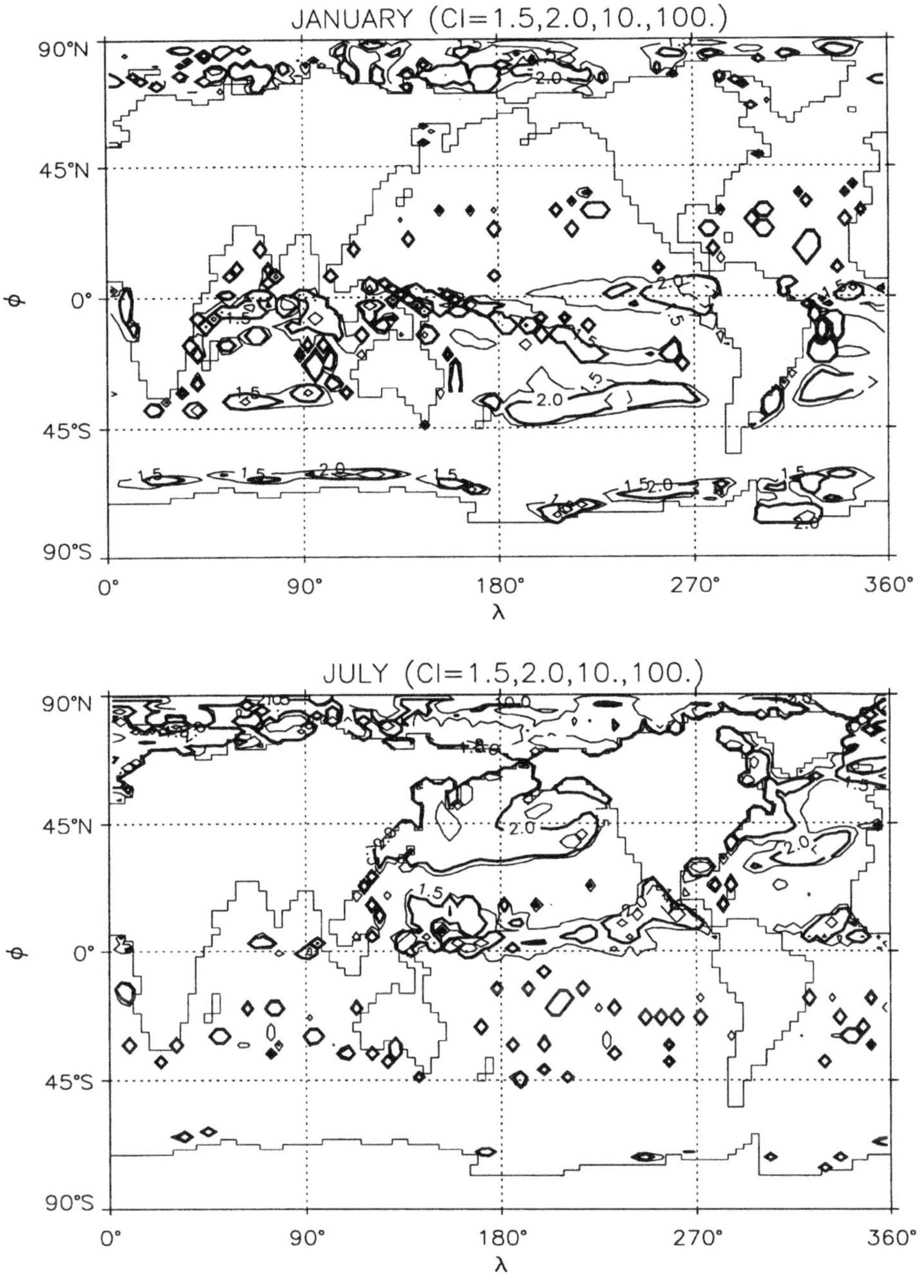

Figure 14. January and July global distributions of the ratio $R_G = G(\sigma = 1/3)/(4/27)$; the enhancement of viscosity and diffusivity due to interior forcing of the ocean boundary layer.

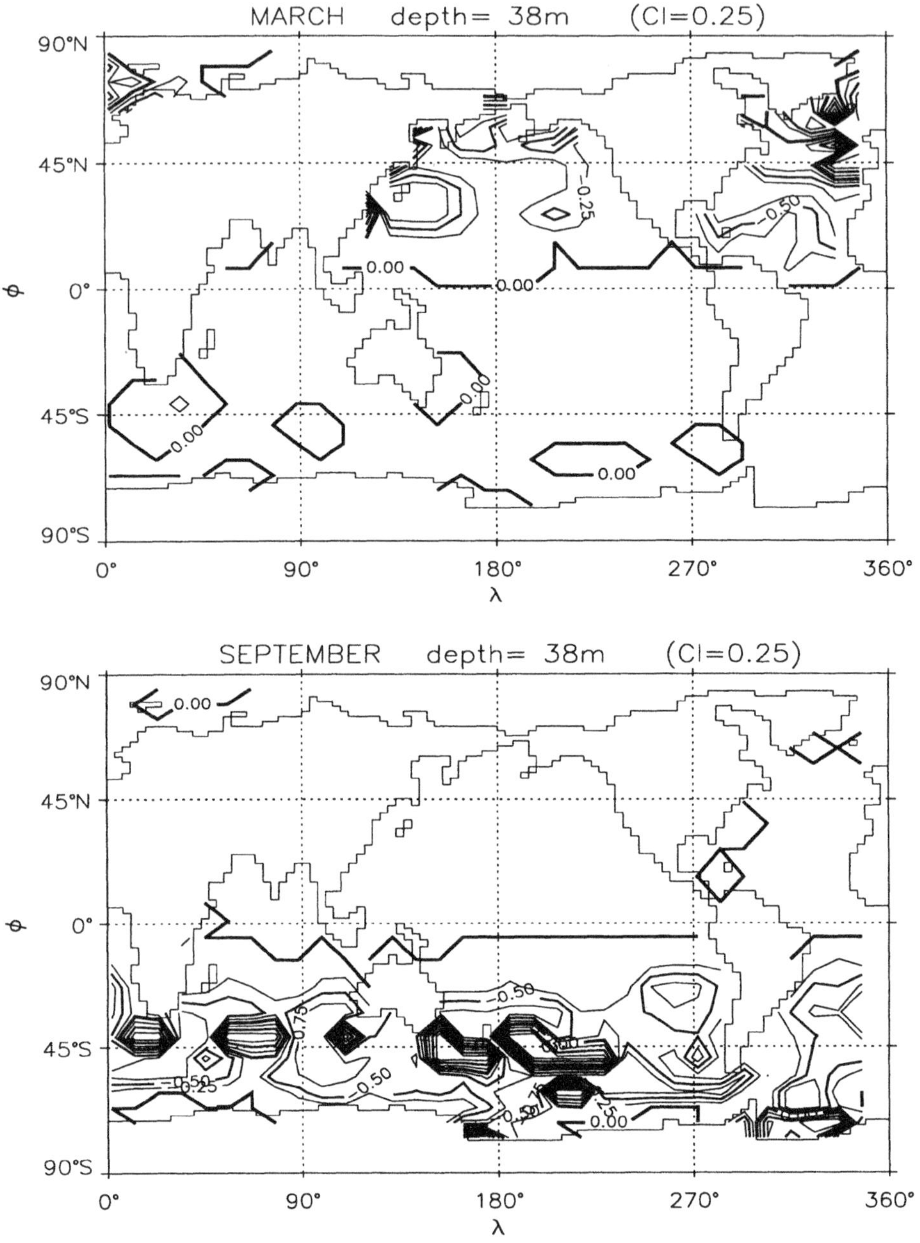

Figure 15. March and September global distributions of the function $\Lambda(\psi)$ at the $38\,m$ model interface. The relative contribution of nonlocal flux is indicated by the absolute value, while negative (positive) values indicate that this flux is counter-gradient (down-gradient).

When a point lies in the first or third quadrants the nonlocal flux is down-gradient as represented by $\Lambda > 0$. A negative value of Λ is used to indicate a counter-gradient nonlocal flux (second and fourth quadrants). Note that in the case of buoyancy (Fig. 1), the nonlocal flux is nonzero only for $\overline{wb_o} < 0$, so the nonlocal flux must be negative too and points can only lie in the second and third quadrants. This characteristic does not hold for passive traces, such as CO_2.

At an angle, ψ, of 45 or 225° there is an equal contribution and $\Lambda = 1/2$. At ψ equal 135 or 315° the conter-gradient nonlocal term is equal, but opposite to the Fickian, so $\Lambda = -1/2$ and the total flux is zero. Values of Λ near zero indicate that Fickian diffusion is dominant. Dominant nonlocal flux that is down-gradient, or counter-gradient is shown by Λ approaching 1, or -1, respectively. In such a situation the gradient is small and can easily change signs, so rapid transition between $\Lambda = 1$ and $\Lambda = -1$ are possible in both time and space.

To illustrate what the function $\Lambda(\psi)$ looks like spatially and temporally, Fig. 15 shows it at the model interface at $38m$ depth, globally, in the extreme months of March and September. In the summer hemisphere this depth is usually below the boundary layer, so there is zero nonlocal flux and $\Lambda = 0$ is most places. However, the nonlocal is important at mid to high latitudes in the winter hemisphere. In March there is a thicker contour at $\Lambda = -1/2$ along which it is balancing Fickian diffusion and there is no net flux at this level. There are also large areas of more negative values where the nonlocal flux is strong enough to produce a net counter-gradient flux. In the deep water mass formation regions of the North Atlantic there are areas where the vertical gradients and hence Fickian fluxes are small, compared to the nonlocal fluxes, but these can be either down-gradient or counter-gradient, with rapid transition between the two. In either case the net buoyancy flux would be $\overline{wb}(35m) < 0$.

These transitions are even more evident in the Southern hemisphere in September. Clearly the nonlocal fluxes are very active everywhere poleward of about 20°S. In particular, in the Weddell Sea they are dominant and down-gradient, so the net flux is much enhanced over the Fickian.

References

Anis, A., and J.N. Moum (1992) The superadiabatic surface layer of the ocean during convection. *J. Phys. Oceanogr.*, **22**, 1221–1227.

Ayotte, K.W., P.P. Sullivan, A. Andren, S.C. Doney, A.A.M. Holtslag, W.G. Large, J.C. McWilliams, C.-H. Moeng, M.J. Otte, J.J. Tribbia, and J.C. Wyngaard, 1996: An evaluation of neutral and convective planetary boundary-layer parameterizations relative to large eddy simulations. *Bound. Layer Meteorol.*, **79**, 131–175.

Ball, F.K., 1960: Control of inversion height by surface heating. *Quart. J. Roy. Meteor. Soc.*, **86**, 483–494.

Barnier, B., 1998: Forcing the oceans. In *Ocean Modeling and Parameterization*, E.P. Chassignet and J. Verron (Eds.), Kluwer Academic Publishers, 45-80.
Beckmann, A., 1998: Representation of bottom boundary layer processes in numerical ocean circulation models. In *Ocean Modeling and Parameterization*, E.P. Chassignet and J. Verron (Eds.), Kluwer Academic Publishers, 135-154.
Blackadar, A.K., 1962: The vertical distribution of wind and turbulent exchange in a neutral atmosphere. *J. Geophy. Res.*, **67**, 3095–3102.
Bryden, H.I., and E.C. Brady, 1989: Eddy momentum and heat fluxes and their effects on the circulation of the equatorial Pacific ocean. *J. Mar. Res.*, **47**, 55–79.
Chen, D., L.M. Rothstein, and A.J. Busalacchi, 1994: A hybrid vertical mixing scheme and its application to tropical ocean models. *J. Phys. Oceanogr.*, **24**, 2156–2179.
Crawford, G.C., and W.G. Large, 1996: A numerical investigation of ocean inertial resonant response to wind events. *J. Phys. Oceanogr.*, **26** , 873–891.
Danabasoglu, G., and J. C. McWilliams, 1995: Sensitivity of the global ocean circulation to parameterizations of mesoscale tracer transports. *J. Climate*, **8**, 2967–2987.
Davis, R.E., R. deSzoeke, D. Halpern, and P. Niiler, 1981: Variability in the upper ocean during MILE. Part I: The heat and momentum balances. *Deep-Sea Res.*, **28A**, 1427–1451.
Deardorff, J.W., 1970: A numerical study of three-dimensional channel flow at large Reynolds numbers. *J. Fluid Mech.*, **41**, 453–480.
Deardorff, J.W., 1972a: Numerical investigation of neutral and unstable planetary boundary layers. *J. Atmos. Sci.*, **29**, 91–115.
Deardorff, J.W., 1972b: Theoretical expression for the counter-gradient vertical heat flux. *J. Geophys. Res.*, **77**, 5900–5904.
Deardorff, J.W., G.E. Willis, and D.K. Lilly, 1969: Laboratory investigation of non-steady penetrative convection. *J. Fluid Mech.*, **35**, 7–31.
Driedonks, A.G.M., 1982: Models and observations of the growth of the atmospheric boundary layer. *Bound. Layer Meteorol.*, **23**, 283–386.
Donaldson, C., 1973: Construction of a dynamic model of the production of atmospheric turbulence and the dispersal of atmospheric pollutants. In *Workshop on Micrometeorology*, D.A. Haugen, ed., AMS, Boston, 313–392.
Ekman, V.W., 1905: On the influence of the earth's rotation on ocean currents. *Arkiv. Math. Astron. O. Fysik*, **2**, 11.
Gaspar, P., 1988: Modeling the seasonal cycle of the upper ocean. *J. Phys. Oceanogr.*, **18**, 161–180.
Gaspar, P., Y. Gregoris, and J.-M. Lefevre, 1990: A simple eddy kinetic energy model for simulations of the oceanic vertical mixing: Tests at station Papa and Long-Term Upper Ocean Study Site. *J. Geophys. Res.*, **95**, 16,179–16,193.
Garwood, R.W., 1977: An oceanic mixed layer model capable of simulating cyclic states. *J. Phys. Oceanogr.*, **7**, 455–471.
Gent, P.R. and M.A. Cane, 1989: A reduced gravity, primitive equation model of the upper equatorial ocean. *J. Comput. Phys.*, **81**, 444–480.
Gent, P. R., and J. C. McWilliams, 1990: Isopycnal mixing in ocean circulation models. *J. Phys. Oceanogr.*, **20**, 150–155.
Heisenburg, W., 1948: On the theory of statistical and isotropic turbulence. *Proc. Roy. Soc. A*, **195**, 402.
Högström, U., 1988: Non-dimensional wind and temperature profiles in the atmospheric surface layer: A re-evaluation. *Boundary-Layer Meteorology*, **42**, 55–78.
Holtslag, A.A.M., E.I.F. de Bruijn, and H.-L. Pan, 1990: A high resolution air mass transformation model for short-range weather forecasting. *Mon. Wea. Rev.*, **118**, 1561–1565.
Kaimal, J.C., J.C. Wyngaard, D.A. Haugen, O.R. Cote, Y. Izumi, S.J. Caughey, and C.J. Readings, 1976: Turbulence structure in the convective boundary layer. *J. Atmos. Sci.*, **33**, 2152–2169.
Kantha, L.H., and C.A. Clayson, 1994: An improved mixed layer model for geophysical

applications. *J. Geophys. Res.*, **99**, 25,235–25,266.

Klemp, J.B, and D.R. Durran, 1983: An upper boundary condition permitting internal gravity wave radiation in numerical mesoscale models. *J. Atmos. Sci.*, **111**, 430–444.

Kraus, E.B., and J.S. Turner, 1967: A one-dimensional model of the seasonal thermocline; II. The general theory and its consequences. *Tellus*, **19**, 98–105.

Large, W.G., and S. Pond, 1982: Sensible and latent heat flux measurements over the ocean. *J. Phys. Oceanogr.*, **12**, 464–482.

Large, W.G., J.C. McWilliams, and S.C. Doney, 1994: Oceanic vertical mixing : A review and a model with a nonlocal boundary layer parameterization. *Rev. Geophys.*, **32**, 363–403.

Large, W.G., G. Danabasoglu, S.C. Doney, and J.C. McWilliams, 1997: Sensitivity to surface forcing and boundary layer mixing in a global ocean model: Annual mean climatology. *J. Phys. Oceanogr.*, **27**, 2418–2447.

Lenschow, D.H., J.C. Wyngaard, and W.T. Pennell, 1980: Mean-field and second-moment budgets in a baroclinic convective boundary layer. *J. Atmos. Sci.*, **37**, 1313–1326.

Lien, R.-C., D.R. Caldwell, M.C. Gregg, and J.N. Moum, 1995: Turbulence variability at the equator in the central Pacific at the beginning of the 1991–1993 El Nino. *J. Geophys. Res.*, **100**, 6881–6898.

Liu, W.T., K.B. Katsaros, and J.A. Businger, 1979: Bulk parameterization of air-sea exchanges of heat and water vapor including the molecular constraints at the interface. *J. Atmos. Sci.*, **36**, 1722–1735.

Louis, J.F., 1979: A parametric model of vertical eddy fluxes in the atmosphere. *Boundary-Layer Meteorol.*, **17**, 187–202.

Lukas, R., and E. Lindstrom, 1991: The mixed layer of the western equatorial Pacific ocean. *J. Geophys. Res.*, **96**, 3343–3357.

Lumley, J.A., and H.A. Panofsky, 1964: *The Structure of Atmospheric Turbulence.* Wiley and Sons, 239 pp.

Mailhôt, J., and R. Benoit, 1982: A finite-element model of the atmospheric boundary layer suitable for use with numerical weather prediction models. *J. Atmos. Sci.*, **39**, 2249–2266.

Martin, P.J., 1985: Simulation of the ocean mixed layer at OWS November and Papa with several models. *J. Geophys. Res.*, **90**, 903–916.

McPhee, M.G., and D.G. Martinson, 1994 : Turbulent mixing under drifting pack ice in the Weddell Sea. *Science*, **263**, 218–221.

McWilliams, J.C., P.P. Sullivan, and C.-H. Moeng, 1997: Langmuir turbulence in the ocean. *J. Fluid Mech.*, **334**, 1–30.

Mellor, G.L. (1989): Retrospect on oceanic boundary layer modeling and second moment closure. In *Parameterization of Small-Scale Processes, Proceedings of the 'Aha Huliko'a Hawaiian Winter Workshop, January 1989*, University of Hawaii at Manoa, P. Muller and D. Henderson (Eds.), Honolulu, Hawaii.

Mellor, G.L., and T. Yamada, 1974: A hierarchy of turbulent closure models for planetary boundary layers. *J. Atmos. Sci.*, **31**, 1791–1806.

Mellor, G.L., and T. Yamada, 1982: Development of a turbulence closure model for geophysical fluid problems. *Rev. Geophys. Space Phys.*, **20**, 851–875.

Moeng, C.-H., 1984: A large eddy simulation model for the study of planetary boundary-layer turbulence. *J. Atmos. Sci.*, **41**, 2052–2062.

Moeng, C.-H., and J.C. Wyngaard, 1984: Statistics of conservative scalars in the convective boundary layer. *J. Atmos. Sci.*, **41**, 3161–3169.

Moeng, C.-H., and J.C. Wyngaard, 1989: Evaluation of turbulent transport and dissipation closures in second-order modeling. *J. Atmos. Sci.*, **46**, 2311–2330.

Monin, A.S., and A.M. Yaglom, 1971: *Statistical Fluid Mechanics, vol. 1*, The MIT Press, 769 pp.

Nieuwstadt, F.T.M., and H. van Dop, (Eds.), 1982: *Atmospheric Turbulence and Air Pollution Modeling*, Reidel, Dordrecht, Holland, 358 pp.

Niiler, P.P., 1975: Deepening of the wind mixed layer. *J. Mar. Res.*, **33**, 405–422.

Niiler, P.P., and E.B. Kraus, 1977: One-dimensional models of the upper ocean. In *Modeling and Prediction of the Upper Layers of the Ocean*, E.B. Kraus, Ed., Pergamon.
O'Brien, J.J., 1970: A note on the vertical structure of the eddy exchange coefficient in the planetary boundary layer. *J. Atmos. Sci.*, **27**, 1213–1215.
Osborn, T.R., 1980: Estimates of the local rate of vertical diffusion from dissipation measurements. *J. Phys. Oceanogr.*, **10**, 83–89.
Pacanowski, R.C., and S.G.H. Philander, 1981: Parameterization of vertical mixing in numerical models of the tropical oceans. *J. Phys. Oceanogr.*, **11**, 1443–1451.
Paulson, C.A., 1970: Representation of wind speed and temperature profiles in the unstable atmospheric surface layer. *J. Appl. Meteorol.*, **9**, 857–861.
Peters, H., M.C. Gregg, and J.M. Toole, 1988: On the parameterization of equatorial turbulence. *J. Geophys. Res.*, **93**, 1199–1218.
Prandtl, L., 1925: Bericht uber Untersuchungen sur augogebildeten Turbulenz, *Z. Angev. Math. Mech.*, **5**, 136–139.
Price, J.F., 1998: Parameterization of the fair weather Ekman layer. In *Ocean Modeling and Parameterization*, E.P. Chassignet and J. Verron (Eds.), Kluwer Academic Publishers, 121-134.
Price, J.F., R.A. Weller, and R. Pinkel, 1986: Diurnal cycling: Observations and models of the upper ocean response to diurnal heating, cooling and wind mixing. *J. Geophys. Res.*, **91**, 8411–8427.
Skyllinstad, E.D., and D.W. Denbo, 1995: An ocean large-eddy simulation of Langmuir circulations and convection in the surface mixed layer. *J. Geophys. Res.*, **100**, 8501–8522.
Skyllinstad, E.D., W.D. Smyth, J.N. Moum, and H. Wijesekera, 1997: Turbulent dissipation during a westerly wind burst: A comparison of large-eddy simulation results and microstructure measurements. *J. Phys. Oceanogr.*, **27**,
Stull, R.B., 1984: Transilient turbulence theory. Part I: The concept of eddy-mixing across finite differences. *J. Atmos. Sci.*, **41**, 3351–3367.
Stull, R.B., 1988: *An Introduction To Boundary Layer Meteorology*, Kluwer, Boston, 666 pp.
Sullivan, P.P, J.C. McWilliams, C.-H. Moeng, 1996: A grid nesting method for large-eddy simulation of the planetary boundary layer flows. *Boundary-Layer Meteor.*, **80**, 167–202.
Tennekes, H., 1973 : The logarithmic wind profile. *J. Atmos. Sci.*, **30**, 234–238.
Toole, J.M., 1998: Turbulent mixing in the ocean: Intensity, causes and consequences. In *Ocean Modeling and Parameterization*, E.P. Chassignet and J. Verron (Eds.), Kluwer Academic Publishers, 171-190.
Troen, I.B., and L. Mahrt, 1986: A simple model of the atmospheric boundary layer; sensitivity to surface evaporation. *Boundary-Layer Meteor.*, **37**, 129–148.
Wang, D., W.G. Large, and J.C. McWilliams, 1996: Diurnal cycling, eddy viscosity and horizontal rotation effects in equatorial ocean boundary layers. *J. Geophys. Res.*, **101**, 3649–3662.
Wang, D., J.C. McWilliams, and W.G. Large, 1998: Large eddy simulation of the diurnal cycle of deep equatorial turbulence. *J. Phys. Oceanogr.*, **28**, 129–148.
Wyngaard, J.C., 1982: Lectures on the planetary boundary layer. In *Mesoscale Meteorology — Theories, Observations and Models*, D.K. Lilly and T. Gal-Chen (Eds.), NATO ASI, Reidel, 781 pp.
Wyngaard, J.C., and R.A. Brost, 1984: Top-down and bottom-up diffusion in the convective boundary layer. *J. Atmos. Sci.*, **41**, 102–112.

PARAMETERIZATION OF THE FAIR WEATHER EKMAN LAYER

JAMES F. PRICE
Woods Hole Oceanographic Institution,
Woods Hole, MA, U.S.A.

Abstract

Most models of the upper ocean Ekman layer require a parameterization of turbulent mixing. Two very simple forms of parameterization are considered here and compared with fair weather data sets. A steady classical diffusion model is found to give reasonably good simulations with familiar values of diffusivity. However, the solutions show a systematic error in current direction that arises, apparently, from a non-parallel relation between shear and stress in the observed profiles.

Inspection of these and other data suggests that the transient, surface-trapped stratification associated with the diurnal cycle is likely to be important in fair weather conditions. The process of diurnal cycling can be treated in a layered model to yield a closed solution for Ekman layer currents. This solution has some of the characteristics of a useful parameterization for use in an OGCM, though it covers only a part of the relevant parameter range.

1. The Upper Ocean Ekman Layer

The upper ocean Ekman layer presents two significant problems – to observe and describe the currents due to an imposed wind stress, and to develop accurate and physically consistent models of wind-driven currents that can be applied over a wide range of conditions. In this note we examine some recent upper ocean observations, and consider some issues of Ekman layer parameterization. Readers looking for a broader perspective on the Ekman layer should see Kraus and Businger (1994), and for a review of upper ocean models, Nurser (1996).

The essential dynamics of wind-driven currents were identified almost 100 years ago in Walfrid Ekman's (1905) landmark analysis. From some

E. P. Chassignet and J. Verron (eds.), Ocean Modeling and Parameterization, 121–134.

very limited observations of wind-driven ice motion, Ekman realized that (1) the Coriolis acceleration due to earth's rotation was crucially important for wind-driven currents, as was already understood well for geostrophic currents, and (2) wind stress must be absorbed over a surface boundary layer having a thickness of order several tens of meters. The latter implies that wind stress must be transmitted downward from the surface as a turbulent momentum flux, $\boldsymbol{\tau}$, that cannot be observed directly in most open ocean conditions (see McPhee and Martinson, 1994, for an example of turbulent stress measured under ice). Away from boundaries and other strong ocean currents, the momentum balance of a steady wind-driven current is expected to be largely between the Coriolis and wind stress accelerations,

$$f\boldsymbol{V}\exp(i\frac{\pi}{2}) = \frac{1}{\rho}\frac{\partial \boldsymbol{\tau}}{\partial z}, \tag{1}$$

where f is the Coriolis parameter, ρ is the nominal density of sea water, $\boldsymbol{V}$ is the steady or time-averaged current (with stress averaged the same way and bold symbols denoting complex variables). Boundary conditions are presumed known on $\boldsymbol{\tau}$: that it equals the given wind stress at the surface, and that it vanishes along with the current at the depth of semi-permanent stratification, $z = z_r$, also presumed to be given from observations. The crux of the theoretical half of the Ekman layer problem is to calculate the stress within the water column above z_r so that the current profile $\boldsymbol{V}(z)$ can also be calculated. This is a classical problem in turbulent parameterization that has been studied intensively since Ekman (1905) and yet is still unsolved in important ways today.

The observational half of the Ekman layer problem requires accurate current measurements near the sea surface, which has been possible only since about 1980 (Weller and Davis, 1980; Weller, 1981). Two recent data sets (Price *et al.*, 1987, and Chereskin, 1995; Figure 1) provide what appears to be a reliable view of the upper ocean Ekman layer in fair weather, subtropical conditions. The current profiles have a spiral shape in which the current speed decays with depth as the current vector rotates to the right — this is very similar to a classical Ekman spiral.

This note addresses two specific questions: **What processes set the thickness and the shape of these current profiles?** and, **How can the relevant processes be parameterized within simple models?** There is now a full spectrum of upper ocean models that are relevant to the Ekman layer problem, and it may be helpful to point out where the present "simple" models will fall, and also to clarify a second way in which the term 'parameterization' will be used. At the greatest resolution are numerical turbulence models that compute explicitly nearly all of the stress-carrying part of the turbulence and require parameterization of only

TABLE 1. Some relevant external variables and depth scales estimated from the current profiles. Q is the average daily maximum heat flux. Current e-folding is the e-folding scale of the current speed estimated by fitting an exponential to the speed profile. Current turning is the depth over which the current turns through one radian, estimated by fitting a straight line to the direction profile.

Data set	f ($10^{-5}s^{-1}$)	τ (Pa)	Q (W m^{-2})	Current e-folding (m)	Current turning (m)
LOTUS3	8.36	0.07	630	10	18
EBC	8.77	0.09	570	16	66

the smallest, dissipative scales (McWilliams *et al.*, 1997). These models are enormously powerful research tools, but they are also very demanding computationally and for that reason they generally can not be used to represent, or parameterize, upper ocean processes within OGCMs. At a step down in resolution, there is a wide range of intermediate numerical models that have sufficient temporal and spatial resolution to be run with synoptic surface fluxes and that can simulate the evolution of the seasonal and diurnal thermocline (Mellor and Durbin, 1975; Price *et al.*, 1986; Large *et al.*, 1994; Large, 1998, discusses some of these models). These intermediate models parameterize nearly all aspects of turbulence, and they are generally affordable for use as the upper ocean component of three-dimensional circulation models. At still another major step down in resolution are the kind of truly simple models that are emphasized here, a familiar example being Ekman's (1905) classical diffusion model. These models too must parameterize all aspects of turbulence, and in addition they can accept only limited surface flux information, e.g., time-averaged wind stress in place of the time series of wind stress that was possible at the intermediate level. Hence their turbulence parameterizations may be very different from that used at the previous model level. These models make almost no computational demand, and in some limits they can be solved explicitly. These most-reduced models and their solutions may be suitable as upper ocean parameterizations in some circumstances (e.g., when only climatological surface flux data is available) but their intrinsic value is that they may help us reach a deeper understanding of Ekman layer phenomenon.

The two questions noted above could be taken in either order, but given the obvious similarity of the observed spirals to classical Ekman spirals we begin with the turbulent parameterization question and the classical diffusion theory.

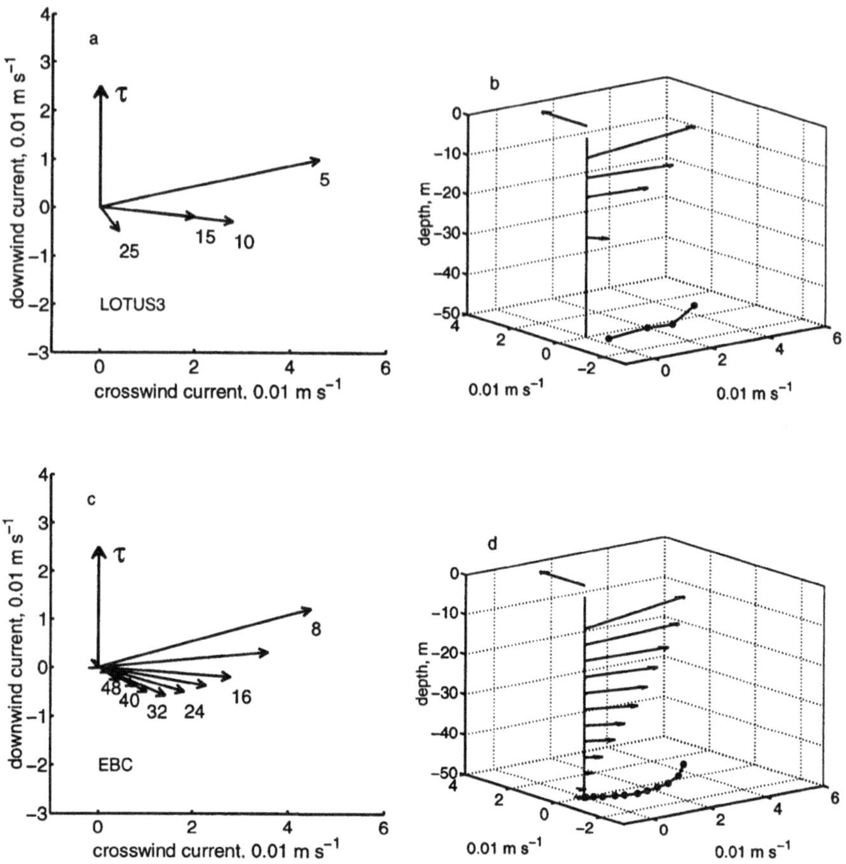

Figure 1. Hodographs and three-dimensional profiles of time-averaged upper ocean currents and wind stress. (a, b) Observations from LOTUS3 in the western Sagrasso Sea (after Price *et al.*, 1987). (c, d) Observations from the Eastern Boundary Current Experiment (after Chereskin, 1995, hereafter, EBC). The LOTUS3 data were collected over the summer half of the year at 35N in the western Sargasso Sea (Table 1). The EBC mooring was at 37N in the eastern North Pacific also during the summer. The reference frames have been rotated so that wind stress points northward in hodographs (the τ vector indicates direction only) and the depth in meters is written just below the tip of the current vectors. Note that these two current profiles are very similar, which is reassuring since the external conditions were also very similar (Table 1).

2. Diffusion Theory

As a first model of the turbulent momentum flux, Ekman adopted Boussinesq's hypothesis of ca. 1870 that a turbulent stress might be treated by analogy with a laminar (molecular) stress,

$$\tau/\rho = K\frac{\partial \boldsymbol{V}}{\partial z}, \tag{2}$$

where K, the eddy diffusivity, represents the stirring effects of turbulence acting upon the vertical shear of the time-mean (or steady) wind-driven current, $\boldsymbol{V}$ (a concise review of the eddy diffusivity concept is in Frisch, 1995).

This parameterization places nearly the entire burden of the problem upon the specification of K, and indeed it is not too strong to say that it turns the Ekman layer problem into the task of finding a suitable K. This would be an inspired step if it turned out that K was in any sense simpler than the current profile itself, i.e., if K had simpler depth-dependence or reduced parameter dependence. Experimental and theoretical efforts to determine K in this and other contexts have formed a major research program on geophysical turbulence, though it is not evident that there has been a convergence on any particular form (Pollard, 1975; Huang, 1979). This vagueness serves to inoculate the classical diffusion theory against critical test, since any particular version of K will appear to be no more than a strawman.

2.1. THE LAMINAR DIFFUSION MODEL

The classical "laminar" diffusion model, CLDM, presumes that K is constant with depth and yields the Ekman spiral, which we have already noted is visually similar to the observed spirals. The CLDM can be solved given a steady wind stress, taken "north" (or imaginary), and boundary conditions as noted above (with $z_r = -H$). Denoting the diffusive length scale within the upper ocean by

$$D_K = \sqrt{2K/f}, \tag{3}$$

and normalizing depth by H,

$$z' = z/H,$$

the solution is (Gonella, 1971),

$$\boldsymbol{V}(z) = U_n\frac{\alpha}{\sqrt{2}}[a\exp(-(1+i)z'\alpha) + b\exp((1+i)z'\alpha)], \tag{4}$$

where

$$\begin{aligned} a &= \exp(i\pi/4)(\exp(2\alpha) - \exp(i2\alpha))/c, \\ b &= \exp(i\pi/4)(\exp(-i2\alpha) - \exp(-2\alpha))/c, \\ c &= \cosh(2\alpha) - \cos(2\alpha). \end{aligned}$$

TABLE 2. Statistical measures of the CLDM and SEL4 solutions (the SEL4 is developed in Section 3). The best fit diffusivities for the former (K_b) and the rms vector error (Misfit) and the percent variance accounted for (PV) are shown for each model.

Data set	rms V	CLDM solution K_b (10^{-4} m^2 s^{-1})	Misfit (m s^{-1})	PV	SEL4 solution Misfit (m s^{-1})	PV
LOTUS3	0.026	100 ± 20	0.011	83	0.008	90
EBC	0.021	175 ± 25	0.007	89	0.005	94

The scale

$$U_n = \frac{U_*^2}{Hf}$$

is called the "neutral" velocity scale, and the only non-dimensional parameter is

$$\alpha = H/D_K.$$

This solution simplifies considerably in two important limits. In the limit $\alpha \ll 1$, say because K is very large, then the solution is just

$$V(z) = U_n \tag{5}$$

in the depth range $-H < z < 0$, and vanishing below. The Ekman layer current is then depth-independent (i.e., slab-like) and flows at right angles to the wind stress. In the other limit, when $\alpha \gg 1$, say because K is small, then this solution reduces to the classical Ekman spiral for infinite water depth,

$$V(z) = U_n \alpha \exp(z'\alpha) \exp(iz'\alpha)(1+i).$$

The surface current is a well-known result

$$V(0) = \frac{U_*^2}{\sqrt{fK}} \exp(i\frac{\pi}{4}). \tag{6}$$

This CLDM solution is parametric in K, which must be specified for any practical use. In the absence of an accepted form, the best fit K, K_b, was found by minimizing the root mean square vector misfit between the CLDM solution (4) and the observed currents (Table 2). In both cases there was a distinct minimum of the misfit, and thus a well defined K_b and a corresponding solution that looks fairly good (Figure 2a,b). These estimated K_b are well within the (large) range inferred from other analyses

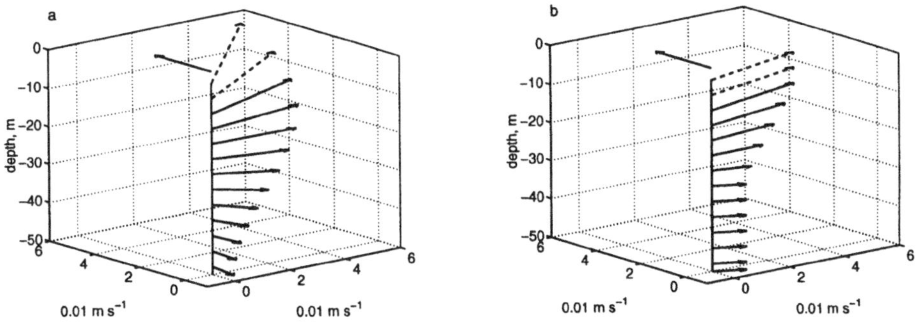

Figure 2. Three-dimensional profiles of the horizontal current computed by the best fit CLDM solution (a), and by the SEL4 solution (b). External parameters were those of the EBC data set (Table 1) and the solution is plotted at the nominal 4 m depth interval of the EBC observations; dashed current vectors at $z = 0$ and -4 m are shallower than the available observations (cf. Figure 1d).

(Pollard, 1975). In the EBC case the CLDM solution had an rms misfit of only about 0.007 m s^{-1}, and accounted for 89% of the variance of the observed currents. A solution this accurate would suffice for many practical purposes, assuming that one could predict K. These two cases are nearly identical and give little clue to K dependence, but they are not far from the neutral, turbulent Ekman layer scale, $h_E = c_1 U_*/f$, where h_E is the e-folding depth scale (Table 1) from which a diffusivity can be inferred. In these cases $c_1 = 0.1$, while the nominal value is $c_1 = 0.25$ (Coleman *et al.*, 1990; McPhee and Martinson, 1994).

On closer inspection, there is seen to be an error in the current direction and in the shape of the spiral. The error is small near the surface and increases with depth to roughly 45 degrees. For many purposes this directional error might not be significant since it occurs mainly at depth. Such a small error would probably not be considered a reliable indicator if it occurred in only one example. However, a very similar directional error is evident in the LOTUS3 case as well, and a closely related phenomenon is the reported mismatch of diffusivity estimated from current direction and speed (the former being larger, Weller, 1981; Chereskin, 1995).

2.2. A DIAGNOSED DIFFUSIVITY

An unconstrained diffusivity can be diagnosed from the observed current profile. The EBC data is best suited for this purpose since it has very good vertical resolution. The stress profile was estimated by integrating the steady momentum balance (1) upward from the reference depth $z_r = -H$, where the wind-driven current and stress were presumed to vanish (Figure

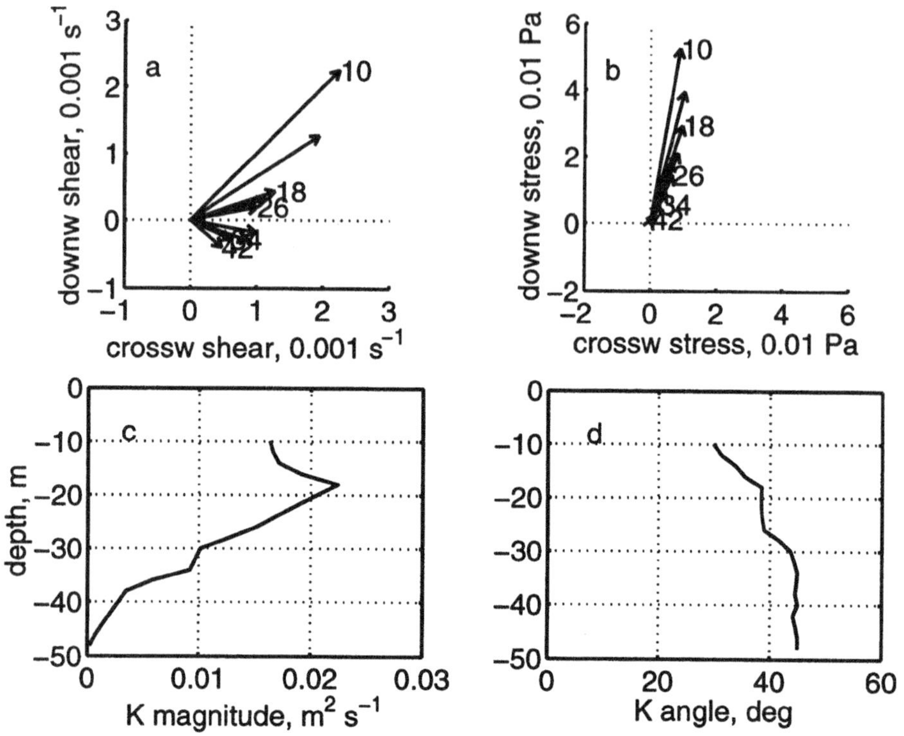

Figure 3. (a) Shear profile estimated from the EBC current profile. (b) Stress profile inferred from the EBC current profile and assuming a steady momentum balance throughout the water column. (c) The amplitude of the complex diffusivity estimated from (a) and (b). (d) The angle of the complex-valued diffusivity. Angles greater than zero indicate that the stress vector at a given depth is to the left of the corresponding shear vector, as can be seen comparing (a) and (b).

3, and see also Chereskin, 1995). The resulting stress profile extrapolates to approximately the wind-derived estimate of stress at the sea surface, as expected from Chereskin's (1995) analysis of the transport. The vertical shear is readily computed, and the stress and shear may be compared in hodographs (Figure 3a and b). The shear and the stress are evidently not parallel; the stress vector points more nearly downwind than does the shear vector by roughly 40 degrees, but depending upon depth. This non-parallel relation between stress and shear is the root of the direction error made by the CLDM. A non-parallel shear/stress relation has been found also in numerical calculations of turbulent Ekman layers (Coleman *et al.*, 1990) where it evidently arises from the large eddy structure of Ekman layer turbulence, and in the analysis of surface drifter data (Krauss, 1993).

The classical diffusion model could accommodate this non-parallel relation between stress and shear by admitting a complex-valued $K(z)$, (Figure 3c and d). But this complex K is no simpler than the current profile itself, and the prospects of finding or developing a universal form of K are not brightened (and indeed Rossby similarity theory, e.g., Kraus and Businger, 1994, seems just as apt). While diffusion theory offers ample room for development, it may be useful to consider what might have been overlooked at the starting point, Eq (2).

3. A Model of the Stratified Ekman Layer

Under the fair weather conditions that held during both observation periods, the upper ocean density profile is often stratified and remixed on a diurnal cycle. This is accompanied by corresponding large variations in the thickness of the turbulent surface layer and in the current profile (Brainerd and Gregg, 1993; Price *et al.*, 1986). The effects of this diurnal variability can be assessed within a layered model that leads to a closed solution for the time-averaged, wind-driven current.

3.1. A LAYERED MODEL

The mixed layer depth is presumed to vary with a top-hat time-dependence over the course of a day, with the shallow, daytime mixed layer depth being the trapping depth defined by Price *et al.* (1986; hereafter, PWP),

$$D_Q = \frac{U_*^2 P_T}{\sqrt{Q_* P_Q / 2.}}, \tag{7}$$

where $P_T = (1/f)\sqrt{2 - 2\cos(fP_Q/2)}$, P_Q is the interval during which the surface heat flux is warming, and $Q_* = g\gamma Q/\rho C_p$, with Q the amplitude of the daily maximum heat flux, γ the thermal expansion coefficient and C_p the heat capacity of sea water. The idea behind D_Q is that the thickness of the diurnal warmed layer is set by a Richardson number condition based upon the temperature difference (a proxy for the density difference) and the vertical shear of the wind-driven current (the diurnal jet). The time-averaged two layer wind-driven current can then be computed from the time-dependent momentum balances (details in Price and Sundermeyer, 1998). Within the upper layer 1, $z \geq -D_Q$,

$$\mathbf{V_1} = U_n \left[1 + (\alpha - 1)\sqrt{\Psi_r^2 + \Psi_i^2}\exp(i\theta)\right], \tag{8}$$

while in the lower layer 2, $-H < z < -D_Q$,

$$\mathbf{V_2} = U_n \left[1 - \sqrt{\Psi_r^2 + \Psi_i^2}\exp(i\theta)\right]. \tag{9}$$

The scale $U_n = \frac{U_*^2}{fH}$, exactly as before for the CLDM solution, and in this context,

$$\alpha = H/D_Q.$$

This is analogous to the ratio H/D_K of the CLDM, but note that this α depends only upon external variables. The angle

$$\theta = \text{atan}(\frac{\Psi_i}{\Psi_r}),$$

where the $\Psi_{r,i}$ are functions of the phases, fP_Q and fP_{24}, which do not have a counterpart in the CLDM,

$$\Psi_r = \frac{fP_Q - \sin(fP_Q)}{fP_{24}}, \quad \text{and} \quad \Psi_i = \frac{1 - \cos(fP_Q)}{fP_{24}},$$

where P_{24} is the length of a day.

3.2. EMBELLISHMENTS AND A CONSISTENCY CHECK

These are the essential results for the layered model, but if the full profile of the Ekman layer current is required then the profile structure must be considered further. The two layer model taken literally indicates a velocity jump between the layers. There is, of course, no such velocity jump found in the ocean, nor in any highly resolved upper ocean model (Figure 4). To simulate a continuous profile a third, intermediate layer of thickness D_Q is therefore inserted into the solution on what is admittedly an *ad hoc* basis. Velocity in this intermediate layer is computed by a linear interpolation in the depth range $-1.5D_Q < z < -0.5D_Q$. Another velocity jump occurs at the bottom of layer 2, at $z = -H$. At subtropical or higher latitudes this velocity jump will usually be quite small. However, at lower latitudes the deep Ekman layer currents can be fairly large, O(0.1 m s^{-1}), and cause significant mixing. To simulate this feature a fourth (and final) layer has been added, whose thickness is set by a critical gradient Richardson number condition. Thus the velocity profile may show up to four distinct layers, two vertically uniform and two vertically sheared, and the solution is termed SEL4 (four layer stratified Ekman layer).

The SEL4 solution is free from adjustable parameters, and can be readily evaluated given the surface fluxes and H as before, plus the daily maximum surface heating and the duration of heating. The SEL4 profiles look reasonably good (compare Figure 2b with 1d) and the percent variance accounted for is $\geq 90\%$, or slightly better than the CLDM solution (Table 2).

The most obvious error is at depth (Figure 2), where both the CLDM and SEL4 solutions overestimate the speed of the observed current. This can

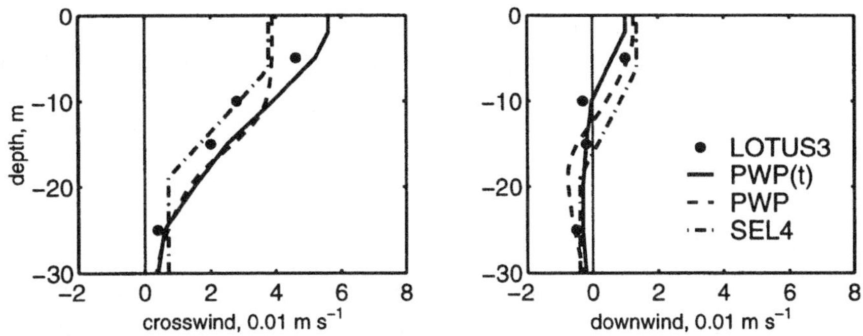

Figure 4. Crosswind and downwind Ekman layer currents from the LOTUS3 data set (discrete points) and as computed by the PWP numerical model run with the hourly surface fluxes (solid line, PWP(t), after Price *et al.*, 1987), by the PWP numerical model run with time-averaged fluxes (dash-dot line, PWP) and by the SEL4 solution (dashed line, SEL4). This comparison serves to check several approximations made during the development of the SEL4 solution and in particular the greatly simplified time-dependence of mixed layer depth (SEL4 compared with PWP). It also shows that the use of time-averaged fluxes in place of hourly fluxes does not introduce large errors (PWP (and SEL4) compared with PWP(t)).

be attributed at least partly to the assumption that the depth of semipermanent stratification, H, was constant. Within the LOTUS3 data set, the advection by internal wave motions alone was about 15 m peak-peak at 50 m depth. This would have a significant smearing effect on the mean current (Davis *et al.*, 1981) that is not represented in the models, and that greatly exceeds the layer four thickness of the SEL4 model.

3.3. PARAMETER DEPENDENCE OF THE SURFACE CURRENT

The parameter dependence of the surface current reveals some of the character of the SEL4 solution and can be compared directly with that of the CLDM solution. First, if D_Q approaches H, say because heating is negligible or wind stress is strong, then diurnal cycling is of no consequence. This arises when

$$\frac{Q_* P_Q H^2}{U_*^4 P_\tau^2} \leq 1, \tag{10}$$

which is the condition that defines weak heating. In that case the surface current goes as the neutral limit,

$$V_1 = U_n, \tag{11}$$

exactly as does the CLDM in the limit of large K, Eq (5). This is not an interesting result for this model, and neither is it likely to be a realistic solution for a neutral Ekman layer. At the other extreme when $D_Q \ll H$, analogous to the CLDM limit of very small K, Eq (6), the surface current

$$\mathbf{V_1} = U_h \exp(i\theta), \tag{12}$$

where

$$U_h = \frac{\sqrt{Q_* P_Q/2}}{f P_\tau} \sqrt{\Psi_r^2 + \Psi_i^2}$$

is termed the "heating" scale. This is an interesting result in as much as U_h is independent of the stress. The EBC and LOTUS3 cases approach this limit and $U_h \approx 0.04$ m s^{-1}, which is consistent with the shallowest measured speeds (Figure 1). Most conditions, i.e., most days and most places, will fall between these two extremes and the surface Ekman current computed by SEL4 will depend upon both H and D_Q. The stress-dependence of the surface current will then be intermediate as well.

The direction of the surface current varies with latitude and heating, but typically the surface current is well to the right of the wind ($f \geq 0$) and the current profiles are "flat" compared to a classical Ekman spiral. In the large α limit, the surface current direction (for EBC or LOTUS3 latitude and P_Q =13 hrs) is $\theta = \text{atan}(\Psi_i/\Psi_r) = 15$ degrees, or 75 degrees to the right of the wind stress (Figure 1). In this regard the SEL4 solutions tend to be a little more realistic than are the CLDM solutions.

4. Remarks

A checklist of desirable qualities of a theory, model, or parameterization might include accuracy, consistency, simplicity, scope, and fruitfulness (Kuhn, 1977), to which we add falsifiability. Simplicity is clearly a virtue of the models considered here, and fruitfulness is an issue for the future.

Accuracy compared to some absolute standard, here taken to be the LOTUS3 and EBC observations. Both the SEL4 and the CLDM solutions exhibit reasonably good accuracy (measured by rms misfit and compared to uncertainties in the data themselves), though CLDM had the advantage of an adjustable parameter.

Consistency with an accepted or more fundamental theory, here measured only for SEL4 and only by comparison with the PWP numerical model results, which would not count as an accepted standard. This comparison did serve to test the conjecture that surface fluxes could be represented by their average values, and found to be valid at least for the LOTUS3 case where wind stress was especially variable. Comparisons with much more sophisticated turbulence-resolving models (as in McWilliams *et al.*, 1997) could show whether the observed flat spiral shape could arise even in the absence of diurnal cycling.

Scope sufficient for the required applications and ideally beyond the domain of construction. If the task were to estimate Ekman layer currents at an arbitrary location and season, then the SEL4 solution would not be a good candidate, since it is applicable only to fair weather conditions. The range of application is at least known, being defined by Eq (10). It is clear that the phenomena included in SEL4 are only a subset of possible Ekman layer dynamics and that more comprehensive models, not so constrained in their formulation, are needed for practical applications.

Falsifiability in the face of conflicting, relevant observations. The greatest virtue of SEL4 when compared with classical diffusion models is that it is closed, and without adjustable parameters. The SEL4 solution can be evaluated unambiguously, and if it truly fails, then it can be declared truly dead in a way that the classical diffusion models probably never can be.

5. Access and Acknowledgements

The models described here are available from J. Price (jprice@whoi.edu). At the time of this writing, they can also be retrieved from the anonymous ftp site 128.128.29.54, cd pub/ekman. Most of this note is taken from Price and Sundermeyer (1998), which is also accessible from the same directory.

JFP's research on upper ocean dynamics has been supported by the US Office of Naval Research under contract N00014-95-1-0105.

References

Brainerd, K. E. and M. C. Gregg (1993) Diurnal restratification and turbulence in the oceanic surface mixed layer 1. Observations. *Journal of Geophysical Research*, **98**, 22645–22656.

Chereskin, T. K. (1995) Direct evidence for an Ekman balance in the California Current. *Journal of Geophysical Research*, **100**, 18261–18269.

Coleman, G. N., J. H. Ferziger and P. R. Spalart (1990) A numerical study of the turbulent Ekman layer. *Journal of Fluid Mechanics*, **213**, 313–348.

Davis, R. E., R DeSzoeke, D. Halpern and P. Niiler (1981) Variability in the upper ocean during MILE, I: The heat and momentum balances. *Deep-Sea Research*, **28**, 1427–1451.

Ekman, V. W. (1905) On the influence of the earth's rotation on ocean currents. *Arch. Math. Astron. Phys.*, **2**, 1–52.

Frisch, U. (1995) *Turbulence*. Cambridge University Press, Cambridge, England.

Gonella, J. (1971) A local study of inertial oscillations in the upper layers of the ocean. *Deep-Sea Research*, **18**, 775–788.

Huang, N. E. (1979) On surface drift currents in the ocean. *Journal of Fluid Mechanics*, **91**, 191–208.

Kraus, E. B., and J. A. Businger (1994) *Atmosphere-Ocean Interaction*. 2nd Ed. Oxford University Press, New York. 362 pp.

Krauss, W. (1993) Ekman drift in homogeneous waters. *Journal of Geophysical Research*, **98**(C11), 20187–20209.

Kuhn, T. S. (1977) Objectivity, value judgment and theory choice. In *The Essential Tension*. The University of Chicago Press. Chicago.

Large, W. G., J. C. McWilliams and S. C. Doney (1994) Oceanic vertical mixing: A review and a model with nonlocal boundary layer parameterization. *Review of Geophysics*, **32**, 363–403.

Large, W. G. (1998) Modeling and parameterizing ocean planetary boundary layers. In *Ocean Modeling and Parameterization*, E.P. Chassignet and J. Verron (Eds.), Kluwer Academic Publishers, 81–121.

McPhee, M. G. and D. G. Martinson (1994) Turbulent mixing under drifting pack ice in the Weddell Sea. *Science*, **263**, 218–221.

McWilliams, J. C., P. S. Sullivan and C.-H. Moeng (1997) Langmuir turbulence in the ocean. *Journal of Fluid Mechanics*, **334**, 1–30.

Mellor, G. L. and P. A. Durbin (1975) The structure and dynamics of the ocean surface mixed layer. *Journal of Physical Oceanography*, **5**, 718–728.

Nurser, A. J. G. (1996) A review of models and observations of the oceanic mixed layer. Internal Document 14, Southampton Oceanographic Center. Southampton, England.

Pollard, R. T. (1975) Observations and models of the structure of the upper ocean. In: *Modelling and Prediction of the Upper Layers of the Ocean*, E. B. Kraus, Editor, Pergamon Press, Oxford, England.

Price, J. F., R. A. Weller and R. Pinkel (1986) Diurnal cycling: Observations and models of the upper ocean response to heating, cooling and wind mixing. *Journal of Geophysical Research*, **91**, 8411–8427.

Price, J. F., R. A. Weller and R. R. Schudlich (1987) Wind-driven ocean currents and Ekman transport. *Science*, **238**, 1534–1538.

Price, J. F. and M. A. Sundermeyer (1998) Stratified Ekman layers. *Journal of Geophysical Research*, submitted.

Weller, R. A. (1981) Observations of the velocity response to wind forcing in the upper ocean. *Journal of Geophysical Research*, **86**, 1969–1977.

Weller, R. A., and R. E. Davis (1980) A vector measuring current meter. *Deep-Sea Res.*, **27**, 565–582.

THE REPRESENTATION OF BOTTOM BOUNDARY LAYER PROCESSES IN NUMERICAL OCEAN CIRCULATION MODELS

AIKE BECKMANN
Alfred-Wegener-Institute for Polar and Marine Research, Bremerhaven, Germany

1. Introduction

Processes in the oceanic bottom boundary layer (BBL) have received much less attention in ocean modeling than surface mixed layer processes. Since the ocean is mainly driven from the surface, high vertical resolution is traditionally placed in the upper ocean. Consequently, the only effect of the lower boundary was considered to be its effect as a sink for momentum and energy. But BBLs are also the place for enhanced diapycnal mixing (especially over steep and rough topography), wind-driven cross-slope transports of tracers and particles (including sediment), and last but not least, gravity-driven dense water spreading. Thus, they play an important role for both local and large-scale ocean dynamics.

The representation of these processes in state-of-the-art numerical ocean general circulation models (OGCMs) has not been investigated systematically. But results from the recent DYNAMO project (DYNAMO group, 1997) can be used to get an idea of how different the resulting flow fields and tracer distributions can be, and to estimate their relevance for the wind-driven and thermohaline circulations. As an illustration, the time-mean near-bottom velocity fields in the subpolar North Atlantic as produced by the three DYNAMO models are shown in Figures 1a-1c. These flow fields are the combined result of steady wind-forced motion (up- and downwelling at the upper continental margin), the time-average of fluctuating currents, and gravity-driven cross-slope flows. We expect strong down-slope outflows south of the Greenland-Iceland-Scotland Ridge, an equatorward deep western boundary current (DWBC) along the continental margin, and a general cyclonic flow around deep basins (in the direction of propagating topographic waves). Considerable disagreement is found in the patterns,

E. P. Chassignet and J. Verron (eds.), Ocean Modeling and Parameterization, 135–154.

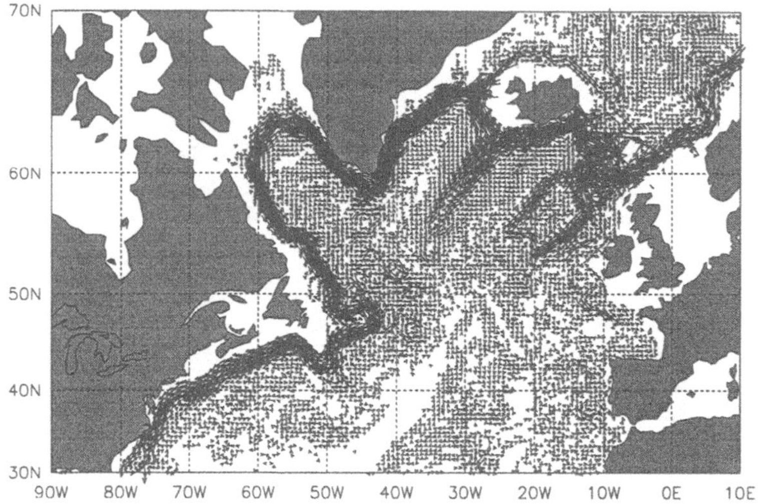

Figure 1a. Time-mean bottom layer circulation in the subpolar North Atlantic from the 1/3 degree DYNAMO-SPEM. Vectors are only plotted for depths greater than 500 m and velocities exceeding 0.5 $\mathrm{cm\,s^{-1}}$ (after DYNAMO group, 1997).

strength, and direction of the deep flow: the SPEM results (Figure 1a) are dominated by a strong, coherent along-slope flow along the upper continental slope (1000-2000m); MICOM, on the other hand, has the strongest cross-slope component of deep flow, notably in the Irminger and Iceland Basins (Figure 1b); finally, MOM has generally weak flows and no continuous DWBC in the bottom layer (Figure 1c). Not surprisingly, these differences have direct consequences for the bottom layer density distribution, the deep water mass characteristics, and the large-scale meridional overturning.

Obviously, these differences are the result of differing (coarse) vertical resolution near the lower boundary, inadequate sub-grid scale parameterizations, and the peculiarities of the vertical coordinate and the treatment of topography. To improve our understanding of BBL processes and their representation in numerical models, a combination of theoretical investigations, high resolution modeling, systematic model testing, and validation with observations is necessary.

This paper will give an overview of BBL processes and parameterizations over sloping topography. It is organized as follows: Section 2 describes the standard treatment of the lower boundary in ocean models. The properties and effects of externally forced boundary layers are considered in Section 3. This includes both steady and periodically forced regimes. Section 4 addresses the modeling of internally driven BBL flows (gravity plumes).

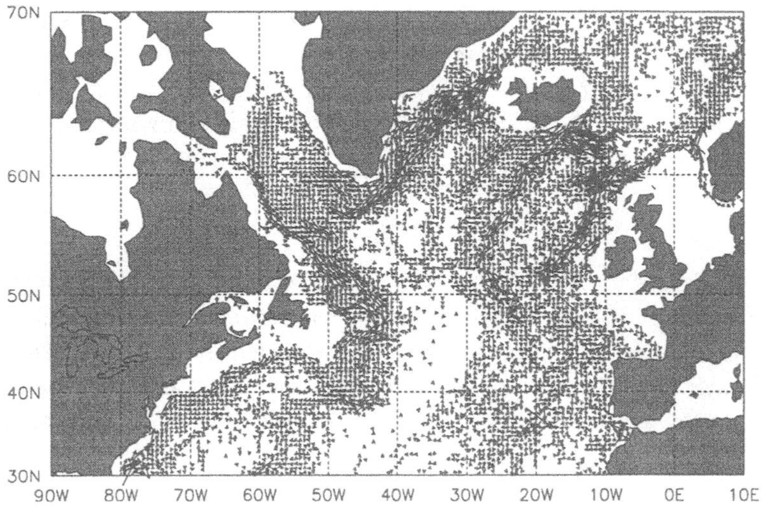

Figure 1b. As Figure 1a, but for DYNAMO-MICOM (after DYNAMO group, 1997).

Section 5 briefly considers sediment transport modeling from an oceanographic perspective. A summary is presented in Section 6.

2. Standard Bottom Boundary Layer Treatment

Today's large-scale ocean circulation models do not resolve the BBL adequately (with the exception of sigma-coordinate models in shallow areas). And even the currently used parameterizations give a very crude representation of the various bottom boundary sublayers (viscous layer, constant stress layer, logarithmic layer, Ekman layer; see, e.g., Wimbush and Munk, 1970). Vertical turbulent mixing for momentum and tracers is approximated by Laplacian terms

$$\frac{\partial \vec{u}}{\partial t} = \frac{\partial}{\partial z}\left(A_{vM}\frac{\partial \vec{u}}{\partial z}\right), \quad \frac{\partial T}{\partial t} = \frac{\partial}{\partial z}\left(A_{vT}\frac{\partial T}{\partial z}\right), \tag{1}$$

often simply with constant vertical eddy coefficients A_v [typically $\mathcal{O}(10^{-3})$ m^2s^{-1} for momentum and $\mathcal{O}(10^{-5})$ m^2s^{-1} for tracers]. The boundary layer scale is then

$$\delta = \sqrt{\frac{2A_v}{|f|}}, \tag{2}$$

which is $\mathcal{O}(10)$ m, and at least one grid point per δ is necessary to capture the main effects of this layer. To account for increased mixing near

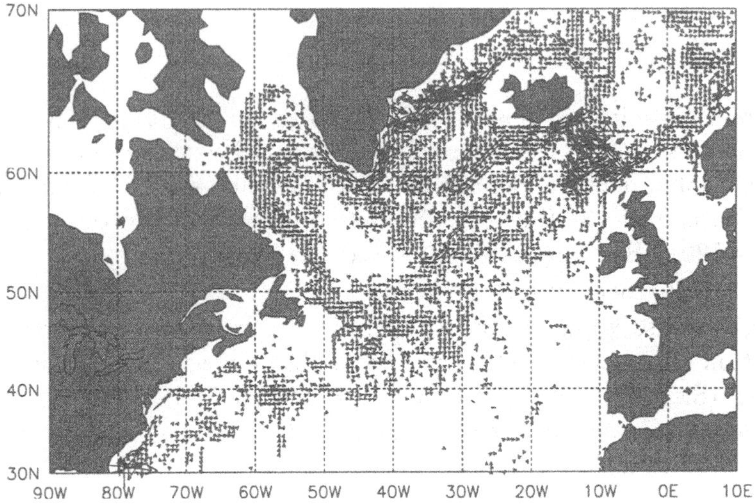

Figure 1c. **As Figure 1a, but for DYNAMO-MOM (after DYNAMO group, 1997).**

the bottom due to bottom roughness, vertically varying (bottom intensified) coefficients have been used occasionally. But the BBL thickness also depends on the stratification and current shear (see, e.g., Weatherly and Martin, 1978; Middleton and Ramsden, 1996); therefore time-dependent (stability or Richardson number based) coefficients or higher order turbulent closure schemes (Mellor and Yamada, 1982; Large *et al.*, 1994; see also Large, 1998, this volume) are more appropriate.

In addition, bottom boundary conditions are needed. Usually, a bottom stress term is used, which in its most general form includes a linear and a quadratic term, as well as a term associated with the effects of tidal motion and designed to increase the drag for weak bottom flows. The stress at the sea bed is then written as

$$\vec{\tau}_b = A_v \frac{\partial \vec{u}_b}{\partial z} = r_b \vec{u}_b + c_D \vec{u}_b \sqrt{U_T^2 + u_b^2 + v_b^2} \tag{3}$$

where $\vec{u}_b$ is the prognostic near-bottom velocity (usually taken at the lowest grid point), $r_b[\mathrm{m\,s^{-1}}]$ is the linear and c_D the nonlinear bottom drag coefficient, and U_T is a prescribed tidal residual velocity. Typical values are $r_b = \mathcal{O}(10^{-4})\ \mathrm{m\,s^{-1}}$, $c_D = \mathcal{O}(10^{-3})$, and $U_T = \mathcal{O}(0.01)\ \mathrm{m\,s^{-1}}$. For tracers, the no-flux condition

$$\frac{\partial T}{\partial z} = 0 \tag{4}$$

is assumed at the bottom. (Geothermal heat fluxes at the bottom are small and limited to mid-ocean ridges and other geologically active zones.)

As is common in the treatment of the surface mixed layer, parameterizations of BBL dynamics are one-dimensional (i.e., vertically local); lateral effects (e.g., due to a sloping bottom) are ignored. This assumption may work quite well for high resolution (especially in terrain-following) models, but they fail to reproduce the BBL dynamics in coarse resolution applications.

3. Externally Driven Boundary Layer Flow

The response of a BBL to a prescribed external flow has been the focus of several idealized studies, both for steady and time-dependent forcing.

3.1. STEADY WIND-DRIVEN UP- AND DOWNWELLING

Studies on the effects of a sloping bottom under steady or quasi-steady forcing have identified the following processes:

- the asymmetry of up- and downwelling circulations (Trowbridge and Lentz, 1991; Ramsden, 1995a; Ramsden, 1995b; Middleton and Ramsden, 1996; Klinck, 1996); with thin (thick) boundary layers for up-slope (down-slope) flow;
- the process of *boundary layer shutdown* for stratified flows (MacCready and Rhines, 1993), in which the near-bottom flow is largely reduced due to an adverse density gradient; and
- the evolution of *shelf break fronts* (Gawarkiewicz and Chapman, 1992; Chapman and Gawarkiewicz, 1995) due to a convergence of shelf and deep ocean flow.

To illustrate some of these effects, a fully prognostic, high resolution simulation of BBL dynamics in a case of steady up- and downwelling is presented here. The experiment investigates wind-driven flow over sloping topography in a two-dimensional (y,z) configuration on an f-plane. A "double-shelf" topography is used to include both up- and downwelling processes simultaneously in one single model run. The numerical sigma-coordinate model (SPEM; Haidvogel *et al.*, 1991) is used with 1 km horizontal resolution and 50 levels in 20 to 200 m deep water. To further increase the vertical resolution in the surface and bottom boundary layers, a nonlinear stretching is applied (Song and Haidvogel, 1994). The topographic slopes are moderate ($\nabla h_{max} = 0.012$) and the Coriolis parameter is constant ($f_o = 10^{-4}$ s^{-1}). Note that the sigma-coordinate pressure gradient error is negligible for this very high resolution application. The model is initialized with an exponential stratification $\rho(z) = 28 - 2 \cdot e^{z/50\text{m}}$ $\sigma-$units, and forced with an along-channel wind stress τ^x (10^{-4} Pa kg $^{-1}$ m^3) that increases smoothly in time with a one day time scale. Vertical mixing is based on a Richardson

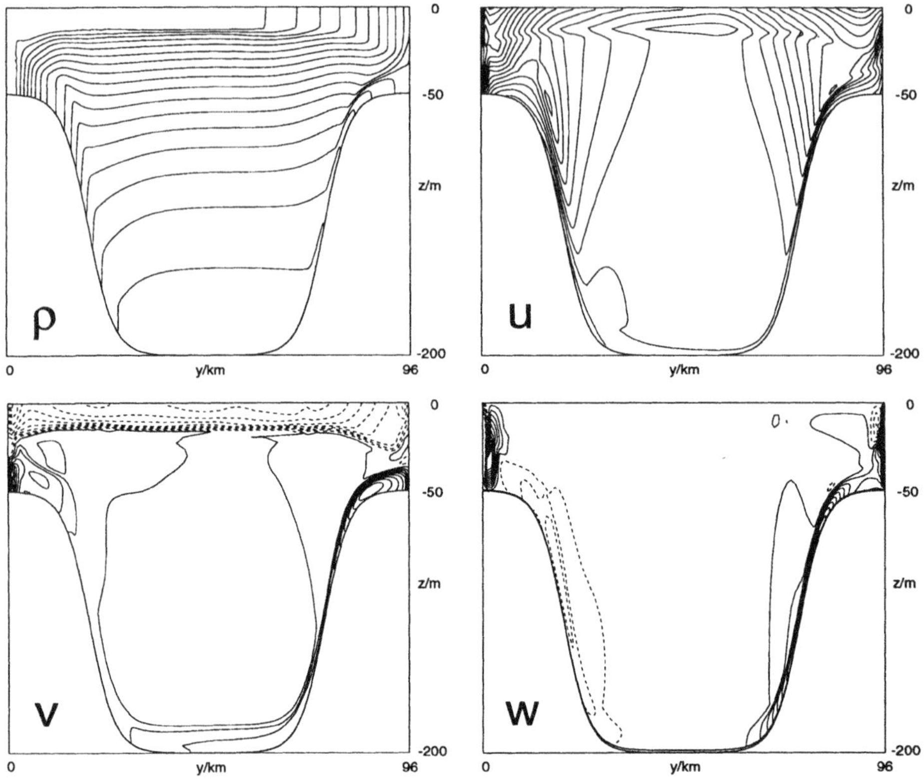

Figure 2. Model results at day 5 for steady wind forcing: density (ρ), contour interval: 0.1 σ-units; along-shelf velocity (u), contour interval: 0.03 $\mathrm{m\,s^{-1}}$; cross-shelf velocity (v), contour interval: 0.01 $\mathrm{m\,s^{-1}}$; vertical velocity (w), contour interval: $2{\cdot}10^{-4}$ $\mathrm{m\,s^{-1}}$

number dependent vertical viscosity/diffusivity; lateral mixing is harmonic with A_h = 20 $\mathrm{m^2s^{-1}}$. Finally, the bottom drag term is quadratic, with c_D = $2{\cdot}10^{-3}$ $\mathrm{m\,s^{-1}}$.

After 5 days, a typical duration of an upwelling event, the simulation shows a significant asymmetry between up- and downwelling shelves (Figure 2). The surface Ekman layer flow (directed to the left) has induced compensating currents below, which have caused characteristic changes in the density field within the BBL. At the bottom, vertically homogenized layers have developed, relatively thick ($\sim$ 30 m) in downwelling, much thinner ($\sim$ 10 m) in upwelling situations. The differences can be explained by the cross-isobath Ekman transport of buoyancy. Upslope density transport increases the density difference between the BBL and the interior, thus suppressing turbulent entrainment at the upper BBL interface, which leads to a thinner boundary layer with higher velocities. Downslope buoyancy transports, on the other hand, cause intense convective adjustment, because lighter water

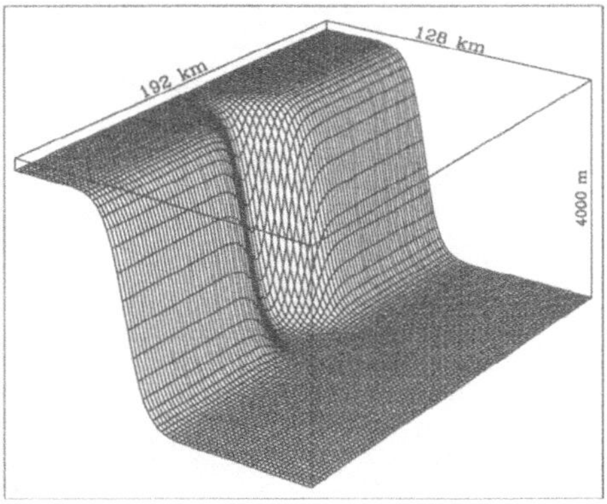

Figure 3. Idealized canyon model topography.

is advected under denser water. The along-shelf flow exhibits the signature of *boundary layer shut-down*, with largely reduced velocities in and maxima just above the BBL. It is important to note that the use of a dynamically adjustable vertical mixing appears to be important, even at this very high resolution. A justification of the chosen Richardson number dependence, however, must come from a comparison with observations.

3.2. PERIODICALLY FORCED BOUNDARY LAYER FLOW

The asymmetry of upwelling and downwelling circulations as described in the previous subsection may lead to a substantial net circulation when periodic forcing is applied. This residual flow is well known to exist in areas with large tidal amplitudes, but any periodic or stochastic forcing will generate such a rectified flow, especially for steep and irregular topography: a rectified boundary layer evolves. A number of idealized laboratory and numerical experiments on this issue have been conducted over the last few years, featuring

- a shelf with undulations (Haidvogel and Brink, 1986);
- a shelf with canyon (Haidvogel and Beckmann, 1998);
- a coastal bank (Verron *et al.*, 1995); and
- an isolated seamount (Haidvogel *et al.*, 1993).

Here, we will look specifically at the role of the BBL in the canyon and the seamount cases.

3.2.1. *Shelf with canyon*

The basic model set-up for the canyon experiments features a periodic channel of 192 km length in the along-shelf direction, bounded by solid (inshore/offshore) walls 128 km apart. The horizontal resolution was chosen

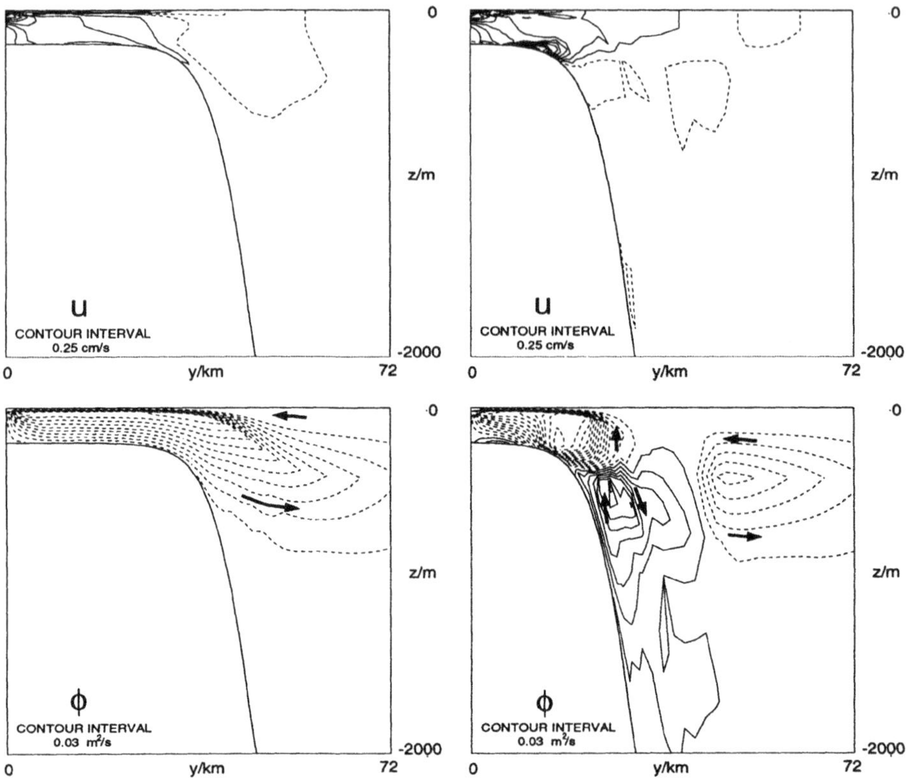

Figure 4. Time-mean along-channel averages of the along-channel velocity (upper panels) and secondary (overturning) circulation in the cross-channel/vertical plane (lower panels) for a straight continental shelf (left), and the canyon topography (right). Solid lines for u denote flow in the direction of coastally trapped waves. Dashed contours for ϕ indicate clockwise, i.e. downwelling, motion. Only the upper 2000 m and coastal 72 km of the model domain are shown.

to be 2 km uniformly across the whole model domain; the constant Coriolis parameter of $f_o = 10^{-4}s^{-1}$ corresponds to about 45°N. Again, we use the terrain-following coordinate model SPEM.

The idealized shape of the topography was chosen to accommodate several features of realistic shelf/slope bathymetries: the continental slope is a steep transition from a 200 m deep shelf to a 4000 m deep abyssal ocean. The canyon intersects the otherwise straight continental margin, roughly 30 km wide at the mouth, the same order of magnitude as that of a typical mid-latitude Rossby radius of deformation. Its functional form is similar to that of Haidvogel and Beckmann (1998). The vertical coordinate is stretched to yield higher resolution near both the surface and the bottom. The topography (see Figure 3) varies between 200 and 4000 m, and the minimum grid spacing for 20 sigma-levels is 10 m on the shelf and 40 m in the deep ocean.

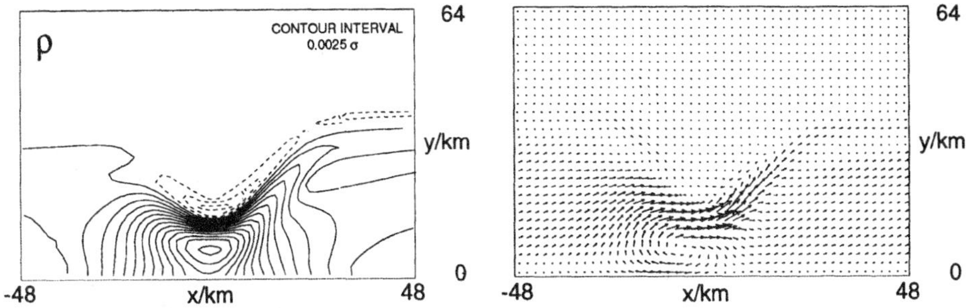

Figure 5. Time-mean density anomaly and horizontal flow in the bottom-most layer for the canyon case. The maximum vector is 3.2 $\mathrm{cm\,s^{-1}}$. Only the central 96×64 km of the model domain are shown.

An exponential initial stratification typical of midlatitude oceans was prescribed as $\rho(z) = 28 - 2 \cdot e^{z/1000\mathrm{m}}$ σ−units. The canyon flow is driven by an along-channel wind stress τ^x (10^{-4} Pa kg^{-1} m^3) that varies sinusoidally with a ten day period. The viscous and diffusive processes in the model are parameterized by a quadratic bottom friction of $c_D = 2 \cdot 10^{-3}$ and a harmonic lateral viscosity and diffusivity along geopotential surfaces ($\nu_{uv\rho} = 20$ m^2 s^{-1}). The prescribed vertical viscosity profile was assumed to decrease exponentially from the $5{\times}10^{-3}$ m^2 s^{-1} at the surface and the bottom into the fluid interior on a 50 m exponential scale. To obtain the time-mean fields, the quasi-stationary model fields from day 90 to day 100 were averaged. Longer test runs have shown that the transients from the model spin-up phase are weak by that time.

As a reference, an experiment without the canyon was performed. The main diagnostic quantities are the along-channel averages of along-channel velocity and overturning motion. The time-mean along-channel velocity in this experiment (Figure 4, left panels) is prograde (i.e., in the direction of topographic waves, with the shallow water to their right in the northern hemisphere). The maximum occurs in the lateral boundary layer at the coast, but also in the bottom layer extending offshore to the shelf break. There is a weak retrograde flow over the slope, possible due to remaining transients from the initialization process. The corresponding secondary circulation in the cross-shelf/vertical plane, presented as an overturning streamfunction ϕ, shows a large-scale downwelling pattern, with on-shore flow in a thin surface layer and broad off-shore return flow. This overturning cell extends down to about 600-800 m depth.

With the canyon present (Figure 4, right panels), the along-channel flow shows a substantial bottom-intensified prograde flow above the shelf break. The secondary circulation is now *two-celled*: the presence of the canyon leads to an increase in downwelling strength on the shelf and adds

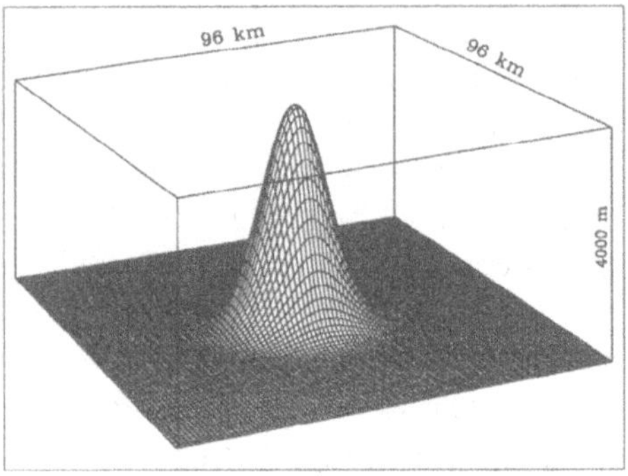

Figure 6. Idealized seamount model topography.

a counterrotating cell in the canyon, forcing the downward flow to leave the bottom (a *detached boundary layer*), thus forming a shelf break front. The horizontal density and flow structure within the BBL (Figure 5) show a positive anomaly on the shelf, the result of the alternating up- and downwelling circulations. A strong current flowing on the downstream flank of the canyon to greater depth dominates the boundary layer flow. There is a convergence at the upstream rim of the canyon. Since all continental shelf-slope regions have topographic irregularities, it can be expected that this is a typical circulation pattern.

3.2.2. *An isolated seamount*

A rectification process observed in a different physical context is tidally forced flow at isolated seamounts. Here, the resonant excitation of seamount trapped waves (Brink, 1989) will lead to the generation of a time-mean anticyclonic flow about the seamount. The propagation of bottom intensified topographic waves depends to a large degree on BBL dynamics.

In an idealized experiment, a Gaussian seamount (Figure 6) is placed in the center of a periodic f-plane channel. The seamount topography was chosen to be tall and steep (3500 m in 4000 m water depth, with 10% maximum slope), typical of many real seamounts. An along-channel barotropic diurnal tidal forcing of 1 cm s^{-1} is applied. All other model parameters are identical to those in the previous subsection.

After 20 days of integration, the rectified time-mean flow in the BBL has reached a steady state with about 10 cm s^{-1} anticyclonic flow above the seamount summit (Figure 7). This vortex is associated with a cap of dense water on top of the seamount. The overturning circulation shows downwelling over the seamount summit, and a second counterrotating cell at the upper flanks.

The circulation patterns presented above seem to be typical for real

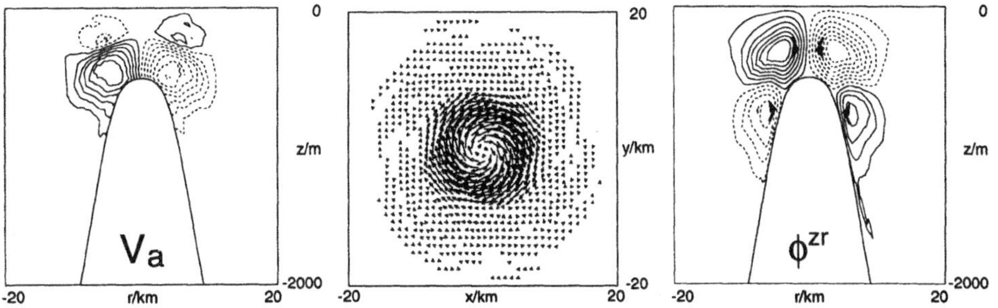

Figure 7. Results from the isolated seamount experiment: (left) time-mean azimuthal velocity along a radial section through the center of the seamount (negative flow is directed into the paper); (center) time-mean horizontal flow in the BBL (maximum vector length 10 $cm\,s^{-1}$; (right) time-mean secondary circulation in the radial/vertical plane (positive contours denote clockwise motion). The transects show only the upper 1000 m and central 40 km of the model domain.

seamounts under tidal forcing and have been observed, e.g., at Fieberling Guyot in the Northeast Pacific (see Brink, 1995; Kunze and Toole, 1997). In a very high resolution sigma-coordinate model study, Beckmann and Haidvogel (1997) have simulated the flow at Fieberling Guyot and validated their results with the large amount of available observational data. The model was shown to reproduce both spatial structure and amplitude of the time mean state quite reliably.

3.2.3. *Parameterization of rectified mean flows*

Both examples of small-scale flow–topography interaction, each with a different forcing mechanism, indicate that three-dimensional time-mean flow fields generated by topographic stress are generated under a wide variety of circumstances. Their effect on large-scale flows and tracer distributions may be substantial, and they need to be included in simulations of large-scale ocean circulation.

A parameterization for the barotropic component of topographically induced mean flows (the *Neptune effect*) was proposed by Holloway (1992), based on principles of mechanical statistics. According to this theory, the rectified barotropic flow is proportional to the gradients of the topography. A more detailed description is given in this volume by Alvarez and Tintoré (1998). This parameterization has been applied in coarse resolution ocean models and has been proven to increase the realism of the large-scale circulation patterns (Eby and Holloway, 1994; Alvarez *et al.*, 1994). However, it neglects the vertical structure of the flow and the corresponding density anomalies (as seen in the experiments above). Consequently, improvements are desirable, especially for coastal ocean tracer transport modeling studies.

4. Internally Driven Boundary Layer Flow

In contrast to the situation assumed in the previous section, the BBL may not only respond to the exterior flow, i.e., the flow in the interior of the ocean, but may also develop internal dynamics. In case of lateral density contrasts within the BBL, gravitational adjustment will occur; over variable topography *gravity plumes* will descend downslope.

Prominent examples of these processes are the outflows from marginal seas (Denmark Strait, Straits of Gibraltar; Price and O'Neil Baringer, 1994; see also Price and Yang (1998), this volume), and also the deep water formation areas with brine-enriched shelf water masses on the Arctic and Antarctic shelves.

4.1. BOTTOM BOUNDARY LAYER MODELS

Numerical modeling of these processes requires more sophisticated methods than the relatively simple one-dimensional parameterizations presented in section 2. Aside from resolution issues, the representation of topography plays a major role here. For example, downslope flow in z-coordinate models is represented as an alternating sequence of lateral advection and vertical convection events. This procedure includes a large amount of (artificial) mixing, such that dense bottom water masses lose their density contrast quite rapidly. Although known for quite some time, this issue has been addressed only recently by Beckmann and Döscher (1997) and Winton and Hallberg (1998), who have compared z- vs. sigma-, and z- vs. isopycnic coordinates, respectively. Obviously, geopotential coordinate models may need improvements most urgently, but coarse vertical resolution terrain-following and isopycnic models will also benefit from a BBL submodel. Several different strategies are currently being applied, and are briefly described below.

4.1.1. *High resolution sigma-coordinate models*

Sigma coordinates seem well suited for bottom boundary layer studies. Consequently, a number of investigations of coastal dynamics have used high resolution primitive equation models with terrain-following coordinates. Process studies on the formation and maintenance of shelf break fronts (Gawarkiewicz and Chapman, 1992), and on the offshore transport of dense shelf water masses over sloping topography (Gawarkiewicz and Chapman, 1995; Chapman and Gawarkiewicz, 1995; Jiang and Garwood, 1995; Jiang and Garwood, 1996) have been carried out. Applications also include a regional study on the Mediterranean outflow (Jungclaus and Mellor, 1998). However, very high resolution near the bottom is necessary to suppress the pressure gradient error and high horizontal resolution is required to avoid hydrostatic inconsistency.

4.1.2. *Reduced gravity models*

Jungclaus and Backhaus (1994) have developed a (stand-alone) reduced gravity model, which predicts the evolution of two-dimensional velocity and tracers in a vertically homogeneous BBL beneath a resting ambient fluid. Intrinsic disadvantages are the neglect of a barotropic mode, of any upper layer flow, and of the process of detrainment. Adjustable parameters for such a model are the ambient density, the bottom drag coefficient, the turbulent exchange coefficient at the upper boundary of the plume, and the turbulent Schmidt number. The model has been successfully applied in stand-alone mode to the Denmark Strait overflow (Jungclaus and Backhaus, 1994) and the Storfjord (Jungclaus *et al.*, 1995). In principle, it can be coupled to a fully three-dimensional primitive equation model. Attempts are currently underway (Beckers and Backhaus, pers. comm.).

4.1.3. *Thin shell models*

Dietrich *et al.*'s (1987) thin shell model is designed as a sub-model for a general circulation model but can also be used in stand-alone mode. It contains several vertical levels and predicts three-dimensional velocity and tracer fields, based on a turbulent kinetic energy (TKE) closure. However, bottom slopes are not explicitly included. Additional adjustable parameters are the vertical resolution of the sub-model, lateral viscosity and diffusivity within the BBL, and the empirical coefficients for the TKE equations.

4.1.4. *Slab layer models*

Also designed as a sub-model, the Killworth and Edwards (1998) approach assumes a vertically homogeneous layer *within* the lowest grid box of a discrete numerical model. Prognostic quantities are horizontal velocities, tracers, and the layer thickness. Entrainment and detrainment are treated in a separate step, using an empirical formula based on results of large eddy simulation (LES) experiments (Zilitinkevich and Mironov, 1996). Adjustable parameters are a bottom drag coefficient, lateral viscosity and diffusivity within the BBL, as well as two coefficients for the determination of the layer thickness. Limitations are the presence of a pressure gradient error in the BBL and that the maximum BBL thickness cannot exceed half the size of the lowermost model grid box.

A similar method by Gnanadesikan (1998) assumes an additional layer of prescribed thickness *below* the standard model grid. The large number of adjustable parameters includes Rayleigh friction and bottom drag in the BBL, viscosity at the upper boundary of the BBL, and the (temporally invariant) BBL thickness. This method includes the evaluation of the lateral pressure gradients as a sum of an along-slope and a hydrostatic correction term (as in sigma-coordinate models).

4.1.5. *A hybrid approach for z-coordinate models*

Another, simpler and somewhat more technical approach was proposed by Beckmann and Döscher (1997). Their method predicts the tracer evolution at the lowermost model points due to lateral (topography-following) diffusive and advective processes. No momentum equation for the BBL is solved, which eliminates the problem of pressure gradient errors. While lateral diffusion connects the bottom tracer points in a straightforward way, topography-following advection requires the redirection of horizontal velocities into neighboring bottom boxes, and modifications of the vertical velocity field to ensure mass conservation. The lateral diffusivity A_l^σ in the BBL can be prescribed with a spatial dependence, e.g., in the conditional form

$$A_l^\sigma(x,y,t) = \begin{cases} \text{large} & \text{if } \nabla\rho\nabla h < 0 \\ 0 & \text{otherwise} \end{cases} \tag{5}$$

which may serve as a parameterization of slope convection. The only other additional adjustable parameter is a weighting factor to specify the degree of boundary parallel advection.

4.2. MODEL TESTING AND VALIDATION

It is currently not clear which of the methods works best under the widest variety of (physical and numerical) circumstances. A large number of model tests, model-model comparisons, and validation studies will be necessary to identify the optimal BBL model.

One such test problem features a self-advected dense water plume on a linearly sloping shelf. This basic configuration has been adopted repeatedly, e.g., by Jungclaus and Backhaus (1994), Jiang and Garwood (1996) and Beckmann and Döscher (1997), and is used here to illustrate the effect of BBL models. In a basin of 320 $\times$ 320 km size, the water depth decreases northward with a 1% linear slope. The resolution is uniformly 10 km in the horizontal and 100 m in the vertical. A density anomaly of 0.5 σ-units is prescribed (and held fixed) in a 30 km wide and 100 m thick opening at the shallower boundary. The standard z-coordinate result after 50 days (Figure 8a) shows a strong dilution of the density signal, a short downslope penetration, and generally very weak flows. In contrast, both BBL models predict a stronger and more penetrative plume (Figure 8b,c). And while both seem to be an improvement over the standard result, it is not clear which of the three solutions is the most realistic. Theoretically, the localized density anomaly at the upper boundary should begin to slide

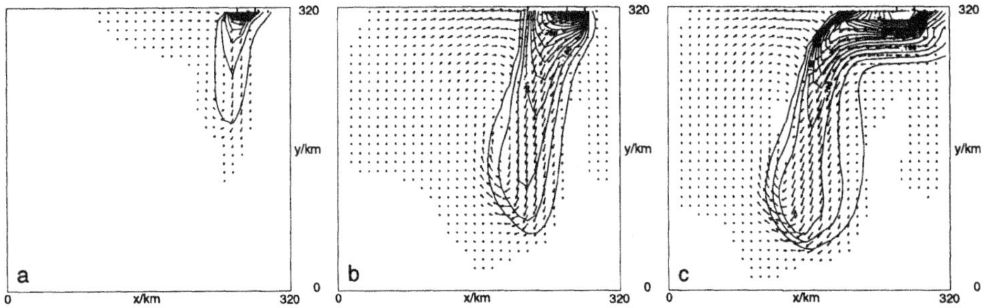

Figure 8. Simulated gravity plume; bottom layer density anomaly and flow field after 50 days: (a) standard z-coordinate model, (b) Beckmann and Döscher (1997); (c) Killworth and Edwards (1998).

down the slope and turn into an along-slope flow under the influence of the Coriolis force. However, without detailed knowledge of entrainment rates at the head of such a plume, the exact solution remains unknown. High resolution isopycnic two-layer models may be used to obtain the minimum entrainment solution for this problem, but comparison to observations is necessary to validate the model results.

It should also be noted that the above gravity plume configuration may be characteristic for episodic shelf water drainage; however, it seems less applicable to the quasi-steady outflows from marginal seas. The main reason is that a plume descending into a resting fluid will also have to spin up the overlying fluid. The relevant dynamics are presumably quite different from those of persistent down-slope flows.

4.3. PARAMETERIZATION OF SLOPE CONVECTION

Although first proposed in the context of Beckmann and Döscher's (1997) BBL model for z-coordinates, the slope convection parameterization (Equation 5) can be used independently. The basic idea is to apply some form of convective adjustment between neighboring bottom tracer cells also, if the deeper one is less dense.

The impact of this method is shown using the results of a relatively simple, non-eddy resolving ($2^o \times 2^o$) simulation of the North Atlantic circulation. The model domain covers the entire basin from the equator to 80°N and is discretized in 28 vertical levels of increasing thickness (50 - 300 m). Versions with and without the BBL model were integrated for ten years each, long enough to reach a quasi-steady state. An artificial passive tracer was released north of Denmark Strait to indicate the southward spreading of water masses of Arctic origin.

Figure 9 shows the distribution of this passive tracer on a meridional section along 31°W. The influence of water of Arctic origin is clearly limited to the upper 2500 m, despite the fact that the density of the overflowing water mass corresponds to the 3000 m level. Instead, the tracer spreads into the ocean's interior with maximum concentrations between 1000 m and 2000 m depth (Figure 9a). The application of the slope convection parameterization causes changes throughout the water column: while concentrations at mid-depth are reduced significantly, a wedge-like intrusion of higher concentration is found at the bottom (Figure 9b), in better agreement with theoretical considerations and observations. There is a similar effect on the near-bottom temperature and salinity fields (not shown). The resulting density changes lead to an increase in horizontal mass transport in the subpolar gyre; at the same time, the NADW cell of the meridional overturning (Figure 10) is stronger and extends deeper, both characteristics being improvements in the simulation. These preliminary results indicate that the thermohaline component of the circulation in coarse resolution climate ocean models can be improved effectively and efficiently with this simple parameterization. Further tests in coarse resolution applications are presently underway (Döscher and Beckmann, pers. comm.).

5. Sediment Transport Effects

The erosion, transport, and deposition of sediment is one of the main interests of marine geology. Consequently, coupled physical-geological studies have been driven by geological questions. There is, however, an important process relevant for ocean dynamics: sediment plumes (turbidity currents) may contribute significantly to the deep water formation at continental margins.

Down-slope flows, as described in the previous section, can reach substantial velocities at the bottom, depending on the density contrast relative to the ambient water masses, the steepness of the topographic slope, and the roughness of the bottom. When the bottom velocity exceeds a critical value, erosion of sediment occurs. This critical velocity depends on the grain size of the sediment, but is typically between 10 and 20 $\mathrm{cm\,s^{-1}}$ for grain sizes between 10 and 100 μm. The additional suspended material increases the density of the fluid; Fohrmann *et al.* (1998) cite values of more than 1 density unit for medium density suspension flows. Converted to salinity equivalents, the sediment load might add up to 1 psu to the density of the moving, turbulent BBL. Thus it seems plausible that these plumes may gain additional speed from an even larger density contrast to the ambient field, leading to deeper penetration into the abyss. Sediment-laden bottom plumes are thus potentially important for setting the deep water characteristics; a decelerating plume will lose its sediment load and *upward*

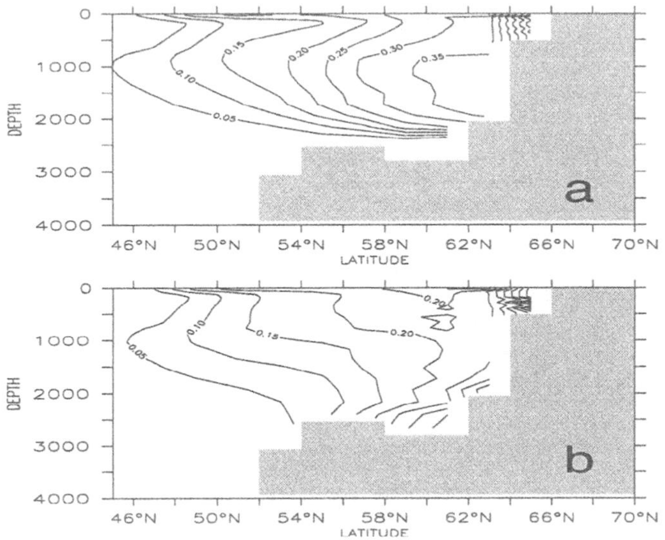

Figure 9. Distribution of an artificial passive tracer on a meridional section at 31°W in ten-year coarse resolution z-coordinate model simulations of the North Atlantic circulation. (a) the standard model; (b) an experiment with the parameterization of slope convection as proposed by Beckmann and Döscher (1997).

convection may occur.

Fohrmann *et al.* (1998) have shown that these effects can be modeled by an advection-diffusion equation for sediment C in the bottom *nepheloid layer* with special source and sink terms:

$$\frac{\partial}{\partial t}C + \frac{\partial}{\partial x}(u_b C) + \frac{\partial}{\partial y}(v_b C) = A\nabla^2 (hC) + \text{sources} - \text{sinks}, \qquad (6)$$

where u_b and v_b are divergent BBL velocities. The source terms are based on critical velocities for erosion (specified individually for each sediment class); the sink terms are modeled similarly by prescribing settling velocities and deposition probabilities. Fohrmann *et al.* (1998) have added such a sediment transport model to the reduced gravity model of Jungclaus and Backhaus (1994), and find substantial effects on the down-slope penetration of individual plumes. It has been speculated that sediment plumes are important for determining the properties of deep water masses, and for the supply of the deep water with oxygen and other nutrients.

6. Summary and Conclusions

Detailed understanding and adequate representation of BBL processes in numerical ocean circulation models is necessary for

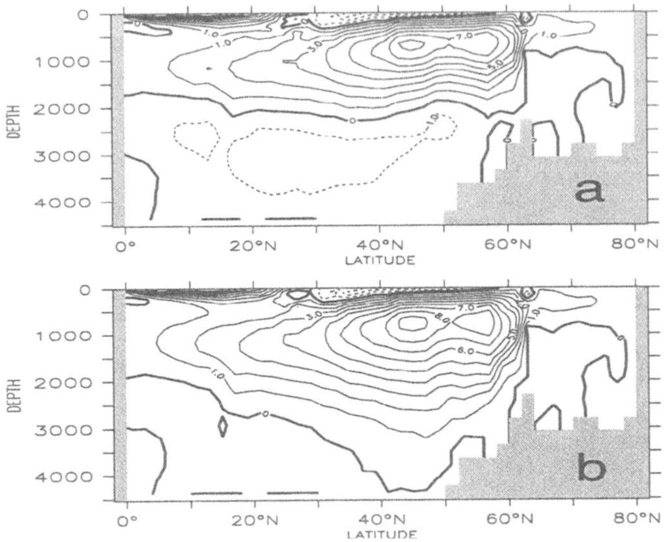

Figure 10. Meridional overturning streamfunction in the coarse resolution simulation of the North Atlantic (a) without, (b) with slope convection parameterization.

- regional coastal studies (BBL dynamics determine how the coastal ocean redistributes physically, biologically, and chemically important tracers and material);
- global (thermohaline) circulation and climate variability studies (across-slope transport and entrainment in the BBL control the rate of deep water formation; rectified along-slope transports are relevant for the spreading of tracers and signal transmission along the continental margins); and
- sediment transport and paleoceanographic studies (near-bottom flows interact with the uppermost layers of the sediment).

The development of BBL models for OGCMs is still in its infancy. More systematic investigations of BBL processes and their representation in ocean models, including quantitative validation against observations, are needed. The bottom boundary layer in numerical ocean circulation models deserves as much attention as the surface mixed layer.

Acknowledgments

Helpful comments by Jim Price, Bernard Barnier, and the students of the 1998 Les Houches Winter School on an earlier version of the manuscript were appreciated. The manuscript also benefited from discussions with René Redler. Ralf Döscher kindly provided the coarse resolution model figures.

References

Alvarez, A. and J. Tintoré, 1998: Topographic stress: Importance and parameterization. In *Ocean Modeling and Parameterization*, E.P. Chassignet and J. Verron (Eds.), Kluwer Academic Publishers, 327-350.

Alvarez, A., J. Tintoré, G. Holloway, M. Eby and J.-M. Beckers, 1994: Effect of topographic stress on circulation in the western Mediterranean. *J. Geophys. Res.*, **99**, 16053-16064.

Beckmann, A. and R. Döscher, 1997: A method for improved representation of dense water spreading over topography in geopotential-coordinate models. *J. Phys. Oceanogr.*, **27**, 581-591.

Beckmann, A. and D.B. Haidvogel, 1997: A numerical simulation of flow at Fieberling Guyot. *J. Geophys. Res.*, **102**, 5595-5613.

Brink, K.H., 1989: The effect of stratification on seamount-trapped waves, *Deep-Sea Res.*, **36**, 825-844.

Brink, K.H., 1995: Tidal and lower frequency currents above Fieberling Guyot, *J. Geophys. Res.*, **100**, 10817-10832.

Chapman, D.C. and G. Gawarkiewicz, 1995: Offshore transport of dense water in the presence of a submarine canyon. *J. Geophys. Res.*, **100**, 13373-13387.

Dietrich, D.E., M.G. Marietta and P.J. Roache, 1987: An ocean modelling system with turbulent boundary layers and topography). a model description. *Int. J. Num. Meth. Fluids*, **7**,, 833-855.

DYNAMO group (Barnard, S., B. Barnier, A. Beckmann, C.W. Böning, M. Coulibaly, D'A. DeCuevas, J. Dengg, Ch. Dieterich, U. Ernst, P. Herrmann, Y. Jia, P.D. Killworth, J. Kröger, M.-M. Lee, Ch. LeProvost, J.-M. Molines, A.L. New, A. Oschlies, T. Reynaud, L.J. West, J. Willebrand), 1997: DYNAMO - Dynamics of North Atlantic Models: Simulation and assimilation with high resolution models. *Ber. Inst. f. Meereskunde Kiel*, **294**, 333 pp.

Eby, M. and G. Holloway, 1994: Sensitivity of a large-scale ocean model to a parameterization of topographic stress. *J. Phys. Oceanogr.*, **24**, 2577-2588.

Fohrmann, H., J.O. Backhaus, F. Blaume and J Rumohr, 1998: Sediments in bottom arrested gravity plumes - numerical case studies. Submitted to *J. Phys. Oceanogr.*

Gawarkiewicz, G. and D.C. Chapman, 1992: The role of stratification in the formation and maintenance of shelf-break fronts. *J. Phys. Oceanogr.*, **22**, 753-772.

Gawarkiewicz, G. and D.C. Chapman, 1995: A numerical study of dense water formation and transport on a shallow, sloping continental shelf. *J. Geophys. Res.*, **100**, 4489-4507.

Gnanadesikan, A., 1998: Representing the bottom boundary layer in the GFDL ocean model: Model framework, dynamical impacts, and parameter sensitivity. Submitted to *J. Phys. Oceanogr.*

Haidvogel, D.B. and K.H. Brink, 1986: Mean currents driven by topographic drag over the continental shelf and slope. *J. Phys. Oceanogr.*, **16**, 2159-2171.

Haidvogel, D.B., J.L. Wilkin and R.E. Young, 1991: A semi-spectral primitive equation ocean circulation model using vertical sigma and orthogonal curvilinear horizontal coordinates. *J. Comp. Phys.*, **94**, 151-185.

Haidvogel, D.B., A. Beckmann, D.C. Chapman and R.-Q. Lin, 1993: Numerical simulation of flow around a tall isolated seamount: Part II: Resonant generation of trapped waves. *J. Phys. Oceanogr.*, **23**, 2373-2391.

Haidvogel, D.B. and A. Beckmann, 1998: Numerical modeling of the coastal ocean. In: Brink, K.H. and A.R. Robinson (Eds.): *The Sea*, **Vol. 10**, 457-482.

Holloway, G., 1992: Representing topographic stress for large-scale ocean models. *J. Phys. Oceanogr.*, **22**, 1033-1046.

Jiang, L. and R.W. Garwood, 1995: A numerical study of three-dimensional dense water bottom plumes on a Southern Ocean continental slope. *J. Geophys. Res.*, **100**, 18471-18488.

Jiang, L. and R.W. Garwood, 1996: Three-dimensional simulations of overflows on continental slopes. *J. Phys. Oceanogr.*, **26**, 1214-1233.
Jungclaus, J.H. and J.O. Backhaus, 1994: Application of a transient reduced gravity plume model to the Denmark Strait Overflow. *J. Geophys. Res.*, **99**, 12375-12396.
Jungclaus, J.H., J.O. Backhaus and H. Fohrmann, 1995: Outflow of dense water from the Storfjord in Svalbard: A numerical model study. *J. Geophys. Res.*, **100**, 24719-24728.
Jungclaus, J.H. and G.L. Mellor, 1998: A three-dimensional model study of the Mediterranean outflow. Submitted to *Journal of Marine Systems.*
Killworth, P.D. and N.R. Edwards, 1998: A turbulent bottom boundary layer code for use in numerical ocean models. Submitted to *J. Phys. Oceanogr.*.
Klinck, J.M., 1996: Circulation near submarine canyons: a modeling study. *J. Geophys. Res.*, **101**, 1211-1223.
Kunze, E. and J.M. Toole, 1997: Tidally driven vorticity, diurnal shear, and turbulence atop Fieberling seamount. *J. Phys. Oceanogr.*, **27**, 2663-2693.
Large, W., 1998: Modeling and parameterizing ocean planetary boundary layers. In *Ocean Modeling and Parameterization*, E.P. Chassignet and J. Verron (Eds.), Kluwer Academic Publishers, 81-120.
Large, W.G., McWilliams, J.C. and S.C. Doney, 1994: Oceanic vertical mixing: a review and a model with a nonlocal boundary layer parameterization. *Rev. Geophys.*, **32**, 363-403.
MacCready, P. and P.B. Rhines, 1993: Slippery bottom boundary layers on a slope. *J. Phys. Oceanogr.*, **23**, 5-22.
Mellor, G. L. and T. Yamada, T., 1982: Development of a turbulent closure model for geophysical fluid problems, *Rev. Geophys.*, **20**, 851-875.
Middleton, J.F. and D. Ramsden, 1996: The evolution of the bottom boundary layer on the sloping continental shelf: a numerical study. *J. Geophys. Res.*, **101**, 18061-18077..
Price, J.F. and M. O'Neil Baringer, 1994: Outflows and deep water productions by marginal seas. *Prog. Oceanogr.*, **33**, 161-200.
Price, J.F., and J. Yang, 1998: Marginal sea overflows for climate simulations. In *Ocean Modeling and Parameterization*, E.P. Chassignet and J. Verron (Eds.), Kluwer Academic Publishers, 155-170.
Ramsden, D., 1995a: Response of an oceanic bottom boundary layer on a slope to interior flow. Part I: Time-independent interior flow. *J. Phys. Oceanogr.*, **25**, 1672-1687.
Ramsden, D., 1995b: Response of an oceanic bottom boundary layer on a slope to interior flow. Part II: Time-dependent interior flow. *J. Phys. Oceanogr.*, **25**, 1688-1695.
Song, Y. and D.B. Haidvogel, 1994: A semi-implicit ocean circulation model using a generalized topography-following coordinate. *J. Comp. Phys.*, **115**, 228-244..
Trowbridge, J.H., and S.J. Lentz, 1991: Asymmetric behavior of an oceanic boundary layer above sloping bottom. *J. Phys. Oceanogr.*, **21**, 1171-1185.
Verron, J., D. Renouard, D. L. Boyer, G. Chabert d'Hières, T. Nguyen and H. Didelle, 1995b: Rectified flow over an elongated topographic feature along a vertical wall. *J. Phys. Oceanogr.*, **25**, 2185–2203.
Weatherly, G.L., and P.J. Martin, 1978: On the structure and dynamics of the oceanic bottom boundary layer. *J. Phys. Oceanogr.*, **8**, 557-570.
White, M., 1994: Tidal and subtidal variability in the sloping benthic boundary layer. *J. Geophys. Res.*, **99**, 7851.-7864
Wimbush, M. and W. Munk, 1970: The benthic boundary layer. In: A.E. Maxwell (Ed.), *The Sea: Ideas and Observations on Progress in the Study of the Seas*, **4**, 731-758.
Winton, M. and R. Hallberg, 1998: Simulation of density-driven frictional downslope flow on z-coordinate ocean models. Submitted to *J. Phys. Oceanogr.*
Zilitinkevich, S. and D.V. Mironov, 1996: A multi-limit formulation for the equilibrium depth of a stably stratified boundary layer. *Bound.-Layer Meteor.*, **81**, 325-351.

MARGINAL SEA OVERFLOWS FOR CLIMATE SIMULATIONS

JAMES F. PRICE AND JIAYAN YANG
Woods Hole Oceanographic Institution,
Woods Hole, MA, U.S.A.

Abstract

This note describes a very simple parameterization of water mass transformation by marginal seas. This parameterization attempts to collapse marginal sea processes into a sidewall boundary condition suitable for an OGCM. Exchange dynamics are treated by hydraulic control models, and descent and mixing are treated by a model of a rotating, entraining density current. This parameterization has been tested by comparison to some well-measured overflows, and it has been implemented in a z-level OGCM to see what effects marginal sea processes have on the deep circulation of an Atlantic-sized basin. Compared to an OGCM solution without a marginal sea, but one which still has a vigorous thermohaline circulation, the combined model solution has much better deep water properties and a rather different circulation in the northern basin.

1. Water Mass Transformation by Marginal Seas

The source waters of the deep circulation can be traced to a few marginal seas that produce a distinctive, dense water type by virtue of their restricted exchange with the open ocean (Whitehead *et al.*, 1974; Warren, 1981). In the present climate, the most important sources of deep water are two overflows from the Nordic Seas — one through Denmark Strait and another through Faroe Bank Channel (Dickson *et al.*, 1990; Borenäs and Lundberg, 1988) — and from the marginal seas and shelf regions surrounding the Antarctic continent (principally the southern Weddell Sea and the Ross Sea; Foldvik *et al.*, 1985). These marginal seas serve as concentration basins (Stommel and Bryden, 1984) that take in oceanic surface water and convert it to a much denser source water that is returned as an overflow. Several important intermediate water types are produced in a similar way. The

E. P. Chassignet and J. Verron (eds.), Ocean Modeling and Parameterization, 155–170.

effects of marginal sea overflows are thus imprinted directly onto the water mass properties and the circulation of the deep ocean (Reid and Lynn, 1971).

For some purposes, e.g., decadal simulation of the present North Atlantic circulation (DYNAMO, 1997), it may suffice to specify the important marginal sea overflows at their present values and then leave them fixed. For climate simulations, and especially those involving a time-changing deep ocean (e.g., Duplessy *et al.*, 1988), it seems essential to represent water mass transformation by marginal seas in a physically consistent and plausible way. This presents a considerable challenge.

Overflows are bottom-trapped density currents, and are thus strongly affected by bottom topography. They have small vertical scales, being typically about 100 - 200 m thick (Figure 1), and they have horizontal scales (width) of from 10 to several hundred kilometers. Their explicit representation would thus require much greater vertical and horizontal resolution than is otherwise desirable in climate models. As they descend the continental slope, overflows may mix intensely with oceanic water, typically doubling their initial volume transport (Smith, 1975; Price and Baringer, 1994). As a consequence of these small spatial scales and energetic mixing dynamics, realistic marginal sea overflows do not arise spontaneously in climate-scale OGCMs.

Here we report on an ongoing effort to make a very simple parameterization of the water mass transformation process by marginal seas. From an oceanic perspective, this process occurs in a small region near the connecting strait, roughly within one grid cell of a large-scale OGCM. Thus it seems appropriate to collapse the process into a sidewall boundary condition on an OGCM, and the present version is called the Marginal Sea Boundary Condition, or MSBC. The MSBC is built from coupled models of marginal sea/ocean exchange, and descent and entrainment of a density current (Section 2) (see Speer and Tziperman, 1990, for a similar approach). Although the MSBC is not completely satisfactory in its present form (some tests are in Section 3), we have nevertheless begun to examine what consequences an explicit representation of marginal sea processes might have on the deep ocean climate and circulation of an idealized Atlantic-size basin (Section 4). In closing (Section 5), we indicate a few of the many areas where this parameterization could be improved, and we also consider other possibilities for the marginal sea problem.

2. A Marginal Sea Boundary Condition

The process of water mass transformation by a marginal sea is imagined to occur in four steps (Figure 2):

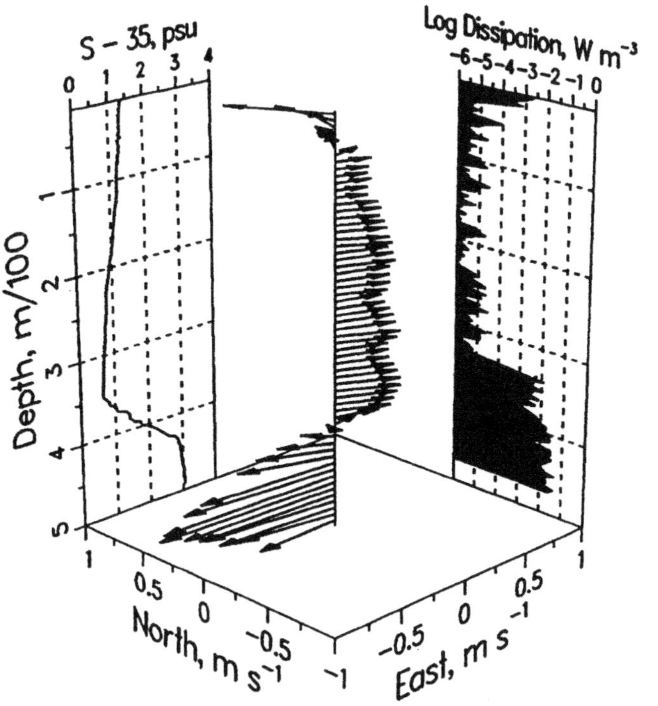

Figure 1. Profiles of salinity, horizontal current, and turbulent dissipation measured near the center of the Mediterranean overflow and just seaward of the shelf-slope break in the Eastern Gulf of Cádiz (from Price *et al.* 1993). The view is toward the northeast. Saline Mediterranean overflow water was flowing westward into the Gulf of Cádiz while the overlying North Atlantic Central water was flowing eastward into the Mediterranean Sea. The dissipation measured within the overflow was about four orders of magnitude larger than in the Atlantic water (which has dissipation values typical of the mid-ocean thermocline). Inferred bottom stress was about 3 Pa, and there was intense vertical mixing between the Mediterranean overflow water and Atlantic water.

First, there must be a buoyancy loss to the atmosphere over the marginal sea. The heat and fresh water flux over the marginal sea are presumed given either from a climatology, as in the test cases described below, or in some future version by an atmospheric GCM. The marginal sea is modeled as a single box having uniform and steady properties.

Second, the dense 'source' water fills the marginal sea up to the depth of the sill within the strait connecting the marginal sea to the open ocean, and overflows steadily into the open ocean. This sets up a two-way exchange flow in which the volume transport of overflowing source water, M_s, is compensated by a nearly equal 'inflow' of oceanic water, M_i, where M is a volume transport. The properties of the inflowing oceanic water are given from climatology in the test cases below, or from the OGCM. The volume,

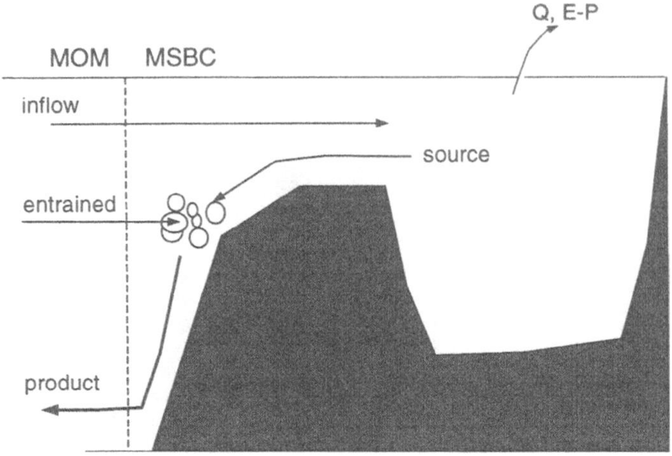

Figure 2. A schematic of the water mass transformation process envisioned to occur in marginal seas that produce deep or intermediate waters. The dashed vertical line at left is the sidewall boundary of an OGCM (the Modular Ocean Model, MOM, in this study) through which the MSBC withdraws, modifies, and reinjects water. The OGCM is unaware of the source water *per se*.

heat, and salt balances for the marginal sea are then

$$M_s + M_i = (E - P)A, \tag{1}$$

$$T_s M_s + T_i M_i = QA/(\rho C_p), \tag{2}$$

$$S_s M_s + S_i M_i = S_s(E - P)A, \tag{3}$$

where A is the surface area of the marginal sea subject to the heat and fresh water fluxes Q and $E - P$. In practice the net freshwater flux through the sea surface is much less than the exchange terms, and to avoid having to treat a barotropic flow through the boundary, the approximation $M_s = -M_i$ is made in the volume budget implemented in the numerical model (Section 4).

To calculate the magnitude of the exchange, M_s, we have used hydraulic models of density-driven exchange. If the strait is narrow compared to the radius of deformation and if the surface inflow has speeds comparable to that of the overflow (as in the Mediterranean overflow through the Strait of Gibraltar), then we use the Bryden and Stommel (1984) model updated by Bryden and Kinder (1991),

$$M_s = 0.07 g'^{1/2} d^{3/2} W, \tag{4}$$

where $g' = g(\rho_s - \rho_i)$ is the buoyancy anomaly of the source water with respect to the adjacent oceanic water, d is the sill depth, and W is the

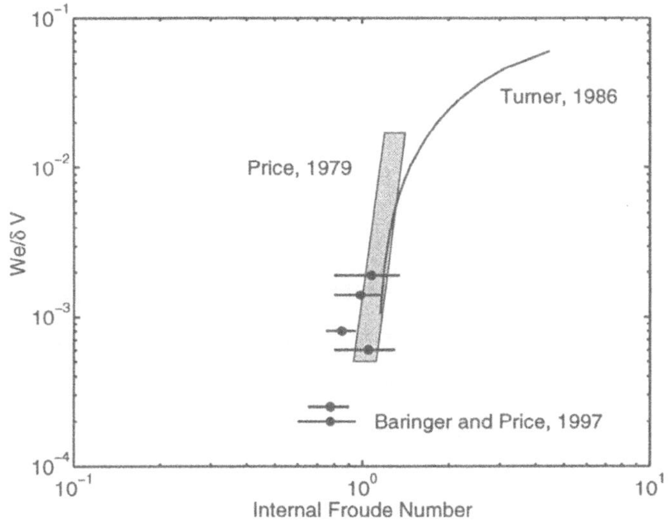

Figure 3. Observed entrainment rate, W_e, normalized by the velocity difference δV, and plotted against the internal Froude number. Data are from laboratory experiments analyzed by Price (1979) (the shaded rectangle), from laboratory experiments analyzed by Turner (1986) (the thin solid line), and from estimates of the Mediterranean overflow in the eastern Gulf of Cádiz by Baringer and Price (1997a,b) (the six discrete points that represent averages across the overflow). The last-named data are evidently the only such oceanic estimates, and it is encouraging (for laboratory studies) that they appear to be consistent with the laboratory-derived estimates.

width of the strait. If the strait is wider than the radius of deformation and if the inflow is not geometrically constrained (as in the Nordic Sea exchange with the North Atlantic), then we use the Whitehead *et al.* (1974) model for maximal geostrophic flow through a strait,

$$M_s = \frac{g' h_u^2}{2f}, \tag{5}$$

where f is the usual Coriolis parameter, and h_u is the thickness of the source water layer above sill depth (more on this below). The oceanic inflow is not explicit in this one-layer exchange theory, but we have assumed that the volume budget (1) still holds.

Third, the overflow forms a density-driven bottom current which descends the continental shelf and slope as it flows into the open ocean. Along the way, most marginal sea overflows double or triple their volume transport by entraining oceanic water which will generally have a considerably different temperature and salinity. Entrainment thus causes a substantial change of the overflow properties. To model the entrainment process, we use the end point model of an entraining, rotating density current developed by

case	latitude	A, 10^6 km^2	Q, W m^{-2}	E-P, m yr^{-1}
Mediterranean	36	2.5	0	0.7
Denmark Strait	62	3.0	-30	-0.08
Nordic Sea	62	3.0	-60	-0.15

TABLE 1. Some relevant external variables for simulation of the Mediterranean Sea, Denmark Strait, and combined Nordic Sea overflows. Area is ice-free area, Q is the annual average heat flux (negative indicates heat loss from the marginal sea) (from Bunker, 1980), and E-P is the evaporation minus precipitation (positive indicates excess evaporation) (after Bryden and Kinder, 1991, and Peixoto and Oort, 1992). For the Denmark Strait case the Q and E-P are reduced by half from their full values in order to give reasonable heat and salt flux through Denmark Strait alone.

Price and Baringer (1994). This model assumes that entrainment occurs in a localized region just over the shelf-slope break and that it depends upon an internal Froude number of the overflow (Figure 3). The entrained oceanic water is taken to be the oceanic water at the depth of the shelf-slope break ('entrained' in Figure 2). The product water properties are then

$$M_p = M_s \frac{1}{1 - \Phi}, \tag{6}$$

$$T_p = T_s - (T_s - T_e)\Phi, \tag{7}$$

$$S_p = S_s - (S_s - S_e)\Phi, \tag{8}$$

where the entrainment parameter

$$\Phi = 1 - F_{geo}^{-2/3} \tag{9}$$

depends upon the geostrophic Froude number, $F_{geo} = U_{geo}/\sqrt{g'h_{geo}}$, evaluated at the shelf-slope break. The geostrophic speed, $U_{geo} = g'\alpha/f$, where α is the slope of the continental slope, and the thickness, h_{geo}, depends upon the spreading of the overflow from the strait to the shelf-slope break (further details are in Price and Baringer, 1994).

Fourth, the mixed overflow water, called 'product' water, is presumed to descend the continental slope without further mixing. This product water is the net result of the water mass transformation process and is a named water type, e.g, Mediterranean Water, and Eastern and Western North Atlantic Deep Water from the Nordic Sea overflows through Faroe Bank Channel and the Denmark Strait. The product water is passed over to the OGCM at the depth where it is either equilibrated in the mid water

case	d, m	h_u, m	W, km	slope	depth ssb, m
Mediterranean	300	-	20	0.012	400
Denmark Strait	600	400	40	0.028	1000
Nordic Sea	700	500	50	0.028	1000

TABLE 2. External variables continued. d is sill depth, W is strait width (for the Mediterranean case, at the surface), h_u is the thickness of the source water within the marginal sea (not required for the Mediterranean), width is strait width, slope is the bottom slope seaward of the shelf-slope break, and depth ssb is the depth of the shelf-slope break.

column or settled onto the bottom as a density current. This last step can be problematic; in some cases where the product settles at mid-depth (e.g., the Mediterranean), we have found that the OGCM may be unable to carry the product water away from the boundary at a sufficient rate, with the result being a literal (and then a figurative) blowup within the OGCM thermocline. In the cases shown in Section 4, the product water is dense enough to reach the sea floor and can readily advect away as a density current.

3. Some Tests and Experiments

3.1. THE MEDITERRANEAN OVERFLOW

This scheme has been tested by comparison to observations of the Mediterranean overflow through the Strait of Gibraltar, which is the best-observed and most intensively analyzed overflow (Ambar and Howe, 1979; Armi and Farmer, 1988; Bryden and Kinder, 1991; Ochoa and Bray, 1991; Baringer and Price, 1997a,b). External parameters needed to run the MSBC in a stand-alone mode are in Tables 1 and 2; the inflow and entrained water properties were taken from observations in the western Gulf of Cádiz.

The Mediterranean case goes through readily (Table 3), and the MSBC gives transports and T/S predictions of the source and product waters that are reasonable compared to observed values. In particular, the product water is predicted to have a salinity of about 36.5 and a transport of a little more than 3 Sv, both of which compare well to the observed Mediterranean overflow in the western Gulf of Cádiz (Ochoa and Bray, 1991).

variable	mid-depth, m	T, C	S	density, sigma	M, Sv
inflow	70	14.58	36.25	27.03	-.9
source	190	14.58	38.62	28.87	.9
entrained	400	12.10	35.70	27.11	-2.6
product	915	12.74	36.46	27.58	3.5

TABLE 3. A simulation of the Mediterranean overflow by MSBC. Inflow is surface oceanic water that flows into the marginal sea (properties taken from an observed profile), source is the source water that overflows from the marginal sea, entrained is the oceanic water entrained into the overflow (properties from an observed profile), and product is the equilibrated product water that is passed to the OGCM. mid-depth is the mid-depth of the layer, density is potential density, and positive transports are directed into the OGCM.

3.2. THE NORDIC SEA OVERFLOWS

The MSBC has also been tested against observations of the Nordic Sea overflows, which are more interesting (problematic) on several counts. The transport and water mass properties of these overflows are well known (Warren, 1981; Swift, 1984; Dickson *et al.*, 1990; Saunders, 1990), though the exchange and mixing dynamics have not been studied intensively and appear to be quite complex. There are two major overflows from the Nordic Seas, one through Denmark Strait and one through Faroe Bank Channel, and there is a significant exchange with the Arctic Ocean through Fram Strait. Thus the heat and salt budgets for the Nordic Sea are not closed using the overflows only. We will nevertheless attempt both a combined Nordic Sea overflow simulation, as we will describe in a moment, and a Denmark Strait case separately since the oceanic conditions are somewhat different for the two overflows. To run the Denmark Strait case, the air/sea fluxes were reduced by half from their nominal values on the assumption that about half the total Nordic Sea overflow comes through Denmark Strait (Table 1). This gives a reasonable density contrast between the source and oceanic waters, about 0.5 kg m^{-3}, and thus makes a useful test of the exchange and entrainment pieces of the MSBC.

The estimated exchange through Denmark Strait is reasonable, $\approx$2.9 Sv, as are the source and product water properties. However, to obtain this result we had to insert an additional key piece of information, namely that the upstream thickness of the source water layer was less than the sill depth by 200 m (the h_u of Table 2; Whitehead, 1995). This is a fair approximation of the actual stratification in the Nordic Sea, and had we not done this, i.e., had we applied the box model approximation consistently,

variable	mid-depth, m	T, C	S	density, sigma	M, Sv
inflow	150	6.4	34.98	27.48	-2.9
source	550	-0.3	34.89	27.93	2.9
entrained	1000	4.1	34.91	27.71	-2.1
product	4000	1.5	34.90	27.93	5.0

TABLE 4. A (partial) simulation of the Denmark Strait overflow. The product water was predicted to reach the sea floor at 4000 m depth.

then the exchange theory would have given too much transport by about a factor of two (also see Saunders, 1990; Killworth, 1994; Whitehead, 1995). This important effect of the marginal sea stratification could perhaps be computed in a future version of this model that includes a resolved marginal sea. For now it must be taken from observations, which leaves this result less than fully predictive.

3.3. A COMBINED, COUPLED NORDIC OVERLOW

In the so-called Nordic Sea overflow case, the MSBC was configured to represent the combined Denmark Strait and Faroe Bank Channel overflows (Tables 1 and 2), and the air/sea fluxes were set to their full values. The MSBC was fully coupled to the northern end of a square, Atlantic-sized basin represented by the Modular Ocean Model, MOM. Inflow and entrained water properties were taken from the northern end of the ocean basin (and the h_u from above was retained). The result was a considerably stronger overflow, having a product water transport of ~11.4 Sv (Table 5), of which about half came from entrainment. This product water transport is about 10% greater than the southward flow of North Atlantic Deep Water (NADW) observed along the eastern continental slope of Greenland (Dickson *et al.*, 1990). The temperature, salinity, and density of the product water (2.2^{o}C, 34.93, and 27.90 kg m^{-3}) are reasonable as an average of newly formed NADW, but note that the present MSBC is unable to give an estimate of the considerable range of temperature and salinity within the real NADW.

3.4. SENSITIVITY TO SURFACE FLUXES

The MSBC development would be otiose if it reproduced only known cases. Its value is that it can make definite (and provocative, we hope) predictions of the change in marginal sea overflows that might occur as part of a climate change scenario. A straightforward example to consider is the effect of an

variable	mid-depth, m	T, C	S	density, sigma	M, Sv
inflow	150	6.3	34.96	27.48	-5.9
source	650	-1.7	34.86	28.06	5.9
entrained	1000	6.3	34.96	27.48	-5.5
product	4000	2.2	34.93	27.90	11.4

TABLE 5. Output from a simulation of the combined Nordic Sea overflow implemented in MOM. In this case the inflow and entrained water properties came from the northern boundary of MOM.

increased (but steady) E-P over the Mediterranean basin (and holding all other external variables fixed at their nominal values). The salinity and the transport of the source water both increase with an increased E-P, and in a comparable way (Figure 4). This change in source water properties is directly attributable to the Stommel and Bryden (1984) exchange dynamics. The source water then must go through the descent and entrainment step before it creates the mixed product water that reaches the open ocean. The MSBC predicts that the transport of product water will be increased substantially while the salinity will remain almost unchanged. This is a direct consequence of the entrainment formulation of Price and Baringer (1994). The upshot is that for all but a very large decrease of E-P from present values, the Mediterranean overflow is predicted to respond to changed E-P with a changed transport of product water having a nearly constant salinity. A rather similar result comes from changing the heat flux over the Nordic Sea (Figure 5).

Are these results realistic? We are not aware of an empirical basis upon which they can be judged (a point we will return to at the end).

4. Does It Matter How or Whether We Treat Marginal Seas?

To see whether an explicit treatment of marginal seas makes a difference in climate-scale simulations, we have run some twin experiments in which the ocean is a square Atlantic-size basin represented by MOM. The surface boundary conditions were restored to averaged T/S, and there was an applied wind stress. Over a small strip on the northern edge of the basin, the boundary conditions were an imposed heat flux and fresh water flux; these were either spread over an open ocean domain, in which case we will call it the ocean-only model, or absorbed by the MSBC configured to represent the combined Nordic Sea overflow. (For a different take on this question, see Gerdes and Koberle, 1995, and also Roberts and Wood, 1997)

The ocean-only model makes a vigorous thermohaline circulation (THC)

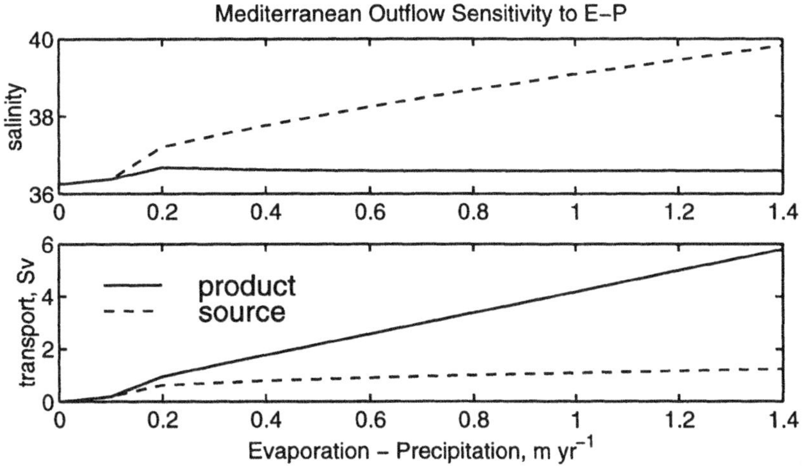

Figure 4. The MSBC simulation of the Mediterranean overflow as a function of E-P over the Mediterranean basin. The upper curves are the salinities of the source water (dashed line) and the product water (solid line). The lower set of curves are the transports. Note that for small changes of E-P about the present value ($\approx$ 0.7 m yr^{-1}), the result is a nearly constant product water salinity, though with a changed transport.

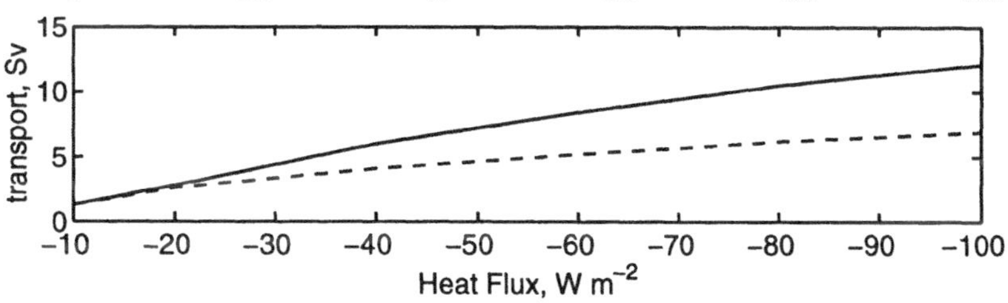

Figure 5. The MSBC simulation of the combined Nordic overflow as a function of heat flux over the basin. The upper curves are the densities of the source water (dashed line) and product water (solid line). The lower set of curves are the corresponding transports. As before, the largest changes appear to be in the transport rather than in the density of the product water (though even rather small changes of density could be important when compared to the density of the southern source water).

driven by deep, open ocean convection. The amplitude of the THC is at least as large as we expect for the true North Atlantic (Figure 6a; the absolute maximum is roughly 25 Sv). With the MSBC present (Figure 6b), the amplitude of the THC in the far field (at the equator, say) is about the same as before, but the absolute maximum is reduced to about 17 Sv. Some significant features of the THC, especially the strength and the direction of the western boundary current, also change with the presence or absence of the MSBC. These changes are due in part to different sources of vorticity, as described further by Yang and Price (1998).

The stratification along the northern boundary is affected markedly by the deep water source (Figure 7). In the ocean-only model, the deep water comes directly from the sea surface by deep convection. The northern end of the ocean is thus very weakly stratified, and the abyssal ocean is much too warm. With the MSBC present, the deep water in the northern basin comes from the marginal sea as a deep overflow, and the concentration basin effect noted at the beginning of this note is evident as a much colder and much more realistic abyssal ocean.

The southern ocean does not have a marginal sea process in these simulations. A shift between mostly northern or mostly southern deep water can thus be fairly pronounced when the MSBC is applied or removed from the northern boundary.

5. Closing Remarks

Two of the obvious shortcomings of the MSBC — the inability to treat time-changing conditions and the need to insert additional information about stratification within the Nordic Sea — stem from the box model approximation of the marginal sea. The box model was adopted purely for convenience, and could be replaced by a resolved marginal sea while still retaining the exchange and entrainment dynamics (which would then be applied on both sides of the ocean/marginal sea boundary).

We are aware of another shortcoming that might not have been apparent since we emphasized bulk properties, e.g., the average temperature and the net transport of the overflows. These bulk properties are, of course, passed consistently from one model stage to the next (i.e., they are conserved). However, real overflows have significant across-stream variability of temperature, salinity and currents (Baringer and Price, 1997a) that should also be matched consistently through the entire model (see Pratt and Smith 1997). This may be of some importance to the ocean climatology and to the dynamics of the deep circulation because overflows are sources not only of heat and volume, as accounted for here, but also of thermal and haline variability and of vorticity.

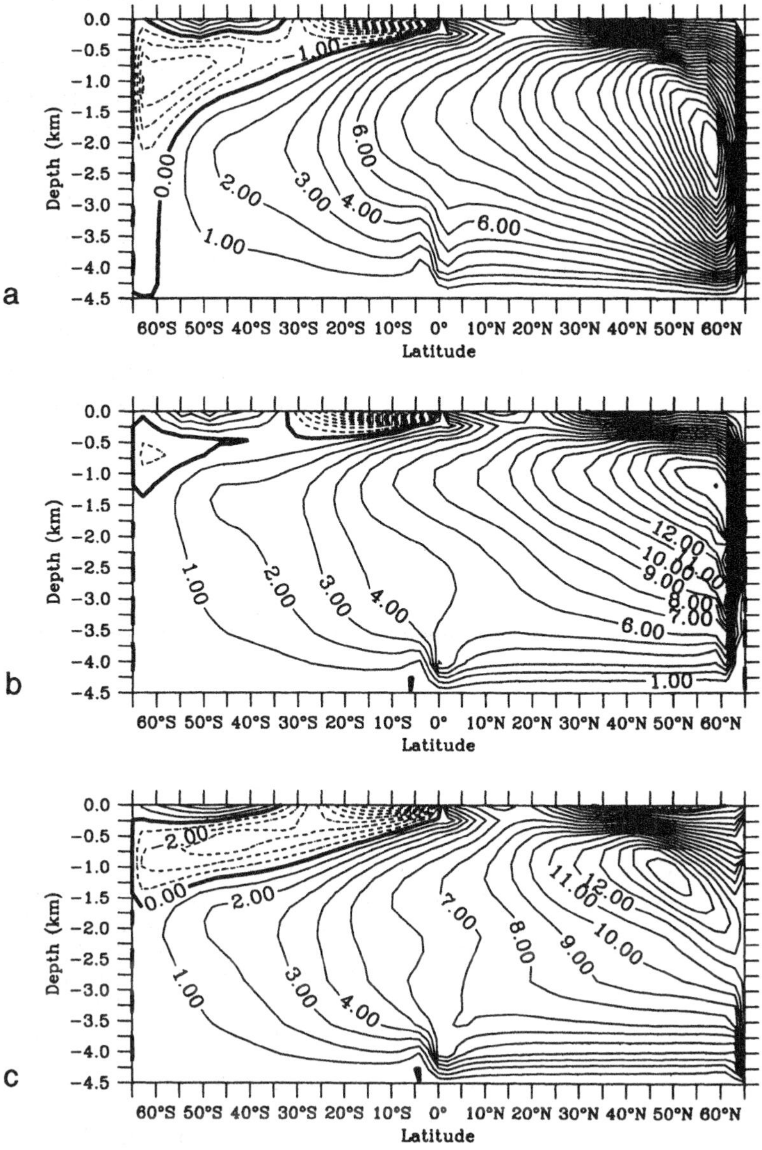

Figure 6. The meridional overturning streamfunction from MOM simulations of an Atlantic size basin. (a) MOM only. (b) MOM with a northern sponge layer that restores the northern boundary T/S toward observed values. (c) MOM with a Nordic Sea-like MSBC. Note that in the latter solution many, but not all, streamlines originate on the northern boundary where there is flow out of the ocean and into the marginal sea at depths above about 1 km, and flow into the ocean at depths greater than about 4 km. The amplitude of the far field THC is little affected by the MSBC, but the distribution of northern/southern source deep water is somewhat altered, as is the deep stratification.

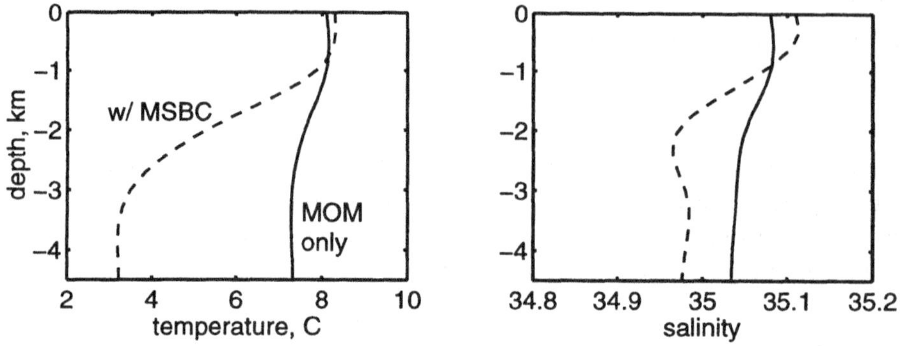

Figure 7. T/S profiles averaged over the northern end of the ocean basin. The solid lines are from MOM only, while dashed lines are from MOM with the MSBC. In the former, deep water comes directly from the sea surface by convection, and hence the northern basin is nearly homogeneous vertically.

An alternative approach to representing overflows in OGCMs is to work toward a model of bottom boundary layer processes that will suffice to treat all bottom currents, including marginal sea overflows (see Beckmann, 1998). This unified approach is sufficiently different from the present patchwork approach that it is difficult to make comparisons, but we will hazard a couple of comments. First of all, a bottom boundary layer model that successfully subsumes overflow dynamics should, of course, be the long range goal. The reason has less to do with marginal sea overflows *per se*, than with related phenomenon — control and mixing at deep passages (Whitehead, 1995) and seasonal deep water production on continental shelves — which such a model could also represent. Exchange dynamics will probably remain as a distinct problem even after bottom boundary layer processes are a settled issue, if only because of the small spatial scales of the connecting straits. Second, it is interesting to consider whether the kind or quality of understanding needed to build and test a unified model is different from that required to build and test a patchwork model. The construction of a patchwork model requires a greater upfront understanding, or at any rate, more explicit information. However, if the goal is to make climate simulations in a parameter regime outside of modern values, and to do so with some confidence, then the level of understanding needed to test a model must be very high regardless of the way the model was built. For example, the MSBC can readily generate predictions of the Mediterranean overflow under conditions of changed E-P over the Mediterranean basin. As is true with many climate predictions, there are no observations able to refute the result directly (Figure 4). In that case the most sensitive contributor to the prediction is the entrainment formulation, Eqs. (6)–(9). We can have confidence in the result only to the extent that the physical processes that cause entrainment in the Mediterranean overflow are understood from observations, and are represented faithfully within the model. A slightly different example arises from the Nordic Sea exchange problem of Section 3.2 where the issue was the geostrophic exchange model, Eq. (5). Progress toward useful models of the thermohaline circulation is likely to be paced as much by our ability to observe and describe the dynamics of the major overflows, especially those in the Antarctic and Nordic Seas, as by our ingenuity at model building.

6. Access and Acknowledgements

At the time of this writing, the marginal sea parameterization is available from the anonymous ftp site: 128.128.29.54, directory pub/oflow. It is also available from J. Price (jprice@whoi.edu).

JFP's research on the dynamics of marginal sea overflows has been supported by the National Oceanic and Atmospheric Administration contract no. NA47GP0188 and by the National Science Foundation under award OCE94-01300. JY's research has been supported by the National Science Foundation under award OCE96-16951.

References

Ambar, I. and M. R. Howe, 1979. Observations of the Mediterranean outflow: I. Mixing in the Mediterranean outflow. *Deep-Sea Research,* **26A**, 535–554.

Armi, L. and D. M. Farmer, 1988. The flow of Mediterranean water through the Strait of Gibraltar. *Progress in Oceanography,*, **21**, 1–105.

Baringer, M. O., and J. F. Price, 1997a. Mixing and spreading of the Mediterranean outflow. *Journal of Physical Oceanography,* **27**, 1654–1677.

Baringer, M. O., and J. F. Price, 1997b. Momentum and energy balance of the Mediterranean outflow. *Journal of Physical Oceanography,* **27**, 1678–1692.

Beckman, A., 1998. The representation of bottom boundary layer processes in numerical ocean circulation models. In *Ocean Modeling and Parameterization,* E.P. Chassignet and J. Verron (Eds.), Kluwer Academic Publishers, 135–154.

Borenäs, K. M. and P. A. Lundberg, 1988. On the deep-water flow through the Faroe Bank Channel. *Journal of Geophysical Research,* **93**, 1281–1292.

Bryden, H. L., and H. M. Stommel, 1984. Limiting processes that determine basic features of the circulation in the Mediterranean Sea. *Oceanologica Acta,* **7**, 289–296.

Bryden, H. L. and T. H. Kinder, 1991. Steady two-layer exchange through the Strait of Gibraltar. *Deep-Sea Research,* **38**, Supplement 1A, S445–S464.

Bunker, A. F., 1980. Trends of variables and energy fluxes over the Atlantic Ocean from 1948 to 1972. *Monthly Weather Review,* **108**, 720–732.

Dickson, R. R., E. M. Gmitrowicz, and A. J. Watson, 1990. Deep water renewal in the northern North Atlantic. *Nature,* **344**, 848–850.

Duplessy, J. C., N. J. Shackleton, R. G. Fairbanks, L. Labeyrie, D. Oppo and N. Kallel, 1988. Deepwater source variations during the last climatic cycle and their impact on the global deepwater circulation. *Paleoceanography,* **3**, 343–360.

DYNAMO, 1997. DYnamics of North Atlantic MOdels: Simulation and assimilation with high resolution models. Report No 294, Institut fr Meereskunde, Kiel, Germany.

Foldvik, A., T. Kvinge and T. Torresen, 1985. Bottom currents near the continental shelf break in the Weddell Sea. *Oceanology of the Antarctic Continental Shelf,* Antarctica Res. Ser., **43**, 21–34.

Gerdes, R. and C. Koberle, 1995. On the influence of DSOW in a numerical model of the North Atlantic general circulation. *Journal of Physical Oceanography*, **25**, 2624–2642.

Killworth, P. D., 1994. On reduced gravity flow through sills. *Geophys. Astrophys. Fluid Dynamics*, **75**, 91–106.

Ochoa, J. and N. A. Bray, 1991. Water mass exchange in the Gulf of Cádiz. *Deep-Sea Research*, **39**, 1553–1572.

Peixoto, J. P. and A. H. Oort, 1992. *Physics of Climate*. American Institute of Physics, New York, NY.

Pratt, L. J. and S. G. Smith, 1997. Hydraulically drained flows in rotating basins. Part I; Method. *Journal of Physical Oceanography*, **27**, 2509–2521.

Price, J. F., 1979. On the scaling of stress-driven entrainment experiments. *Journal of Fluid Mechanics*, **90**, 509–529.

Price, J. F., M. O'Neil Baringer, R. G. Lueck, G. C. Johnson, I. Ambar, G. Parrilla, A. Cantos, M. A. Kennelly, and T. B. Sanford, 1993. Mediterranean outflow mixing and dynamics. *Science*, **259**, 1277–1282.

Price, J. F., and M. O'Neil Baringer, 1994. Outflows and deep water production by marginal seas. *Progress in Oceanography*, **33**, 161–200.

Reid, J. L., and R. J. Lynn, 1971. On the influence of the Norwegian-Greenland and Weddell seas upon the bottom waters of the Indian North Pacific oceans. *Deep-Sea Research*, **18**, 1063–1088.

Roberts, M. J. and R. A. Wood, 1997. Topographic sensitivity studies with a Bryan-Cox-type model. *Journal of Physical Oceanography*, **27**, 823–836.

Saunders, P. M., 1990. Cold outflow from the Faroe Bank Channel. *Journal of Physical Oceanography*, **20**, 29–43.

Smith, P. C., 1975. A streamtube model for bottom boundary currents in the ocean. *Deep-Sea Research*, **22**, 853–873.

Speer, K. and E. Tziperman, 1990. Convection from a source in an ocean basin. *Deep-Sea Research*, **37**, 431–446.

Swift, J. H., 1984. The circulation of the Denmark Strait and Iceland-Scotland overflow waters in the North Atlantic. *Deep-Sea Research*, **31**, 1339–1355.

Turner, J. S., 1986. Turbulent entrainment: the development of the entrainment assumption and its application to geophysical flows. *Journal of Fluid Mechanics*, **173**, 431–471.

Warren, B. A., 1981. Deep circulation of the world ocean. In: *Evolution of Physical Oceanography, Scientific Surveys in Honor of Henry Stommel,* B. A. Warren and C. Wunsch, editors; The MIT Press, Cambridge, MA.

Whitehead, J. A., A. Leetmaa, and R. A. Knox, 1974. Rotating hydraulics of strait and sill flows. *Geophysical Fluid Dynamics*, **6**, 101–125.

Whitehead, J. A., 1995. Critical control by topography - Deep passages, straits and shelf fronts. In *Topographic Effects in the Ocean*, Proceedings of the Hawaii Winter Workshop. University of Hawaii, Manoa, Hawaii. 141–156.

Yang, J. and J. F. Price, 1998. Water mass formation and vorticity balance in an abyssal ocean circulation. *Journal of Physical Oceanography*, submitted.

TURBULENT MIXING IN THE OCEAN

Intensity, causes, and consequences

JOHN M. TOOLE
Woods Hole Oceanographic Institution,
Woods Hole, MA, U.S.A.

1. Introduction

The distributions of flows and water properties in the world's oceans exhibit structure on a vast range of space and time scales. The largest scales (of order 10^7m) are limited chiefly by the size of ocean basins and ultimately of the earth. On the other end of the spectrum, gradients in velocity and water properties exist to sub-meter scales (see below). The total range of spatial scales is thus some nine or more orders of magnitude. The temperature and velocity structures on scales of a few centimeters evolve on a time scale of seconds to minutes (Dillon, 1984); scientists conducting climate studies seek to understand ocean variability on periods of decades to centuries or longer. This gives us in time, a span of at least nine orders of magnitude too. Between these extremes is of course a continuous wavenumber–frequency spectrum of ocean kinetic and potential energy encompassing a host of physical processes (Figure 1). Given these space and time-scale ranges, it is obviously impossible to develop global-scale numerical models that simultaneously resolve all possible oceanographic motions over long times.

The commonly used technique for proceeding beyond this resolution roadblock involves simulations that resolve only part of the frequency-wavenumber space of ocean flows, coupled to schemes that deduce the effects of motions occurring at the excluded scales. Driven by interest in the general circulation, ocean models are commonly set up to resolve regional-, basin- or global-scale flows by truncating horizontal resolution at several kilometers and vertical resolution at tens-to-hundreds of meters. But it is at the excluded small scales that turbulent dissipation and irreversible mixing occur, processes that are basic to establishing the intensity of the ocean's circulation and distribution of water properties.

E. P. Chassignet and J. Verron (eds.), Ocean Modeling and Parameterization, 171–190.

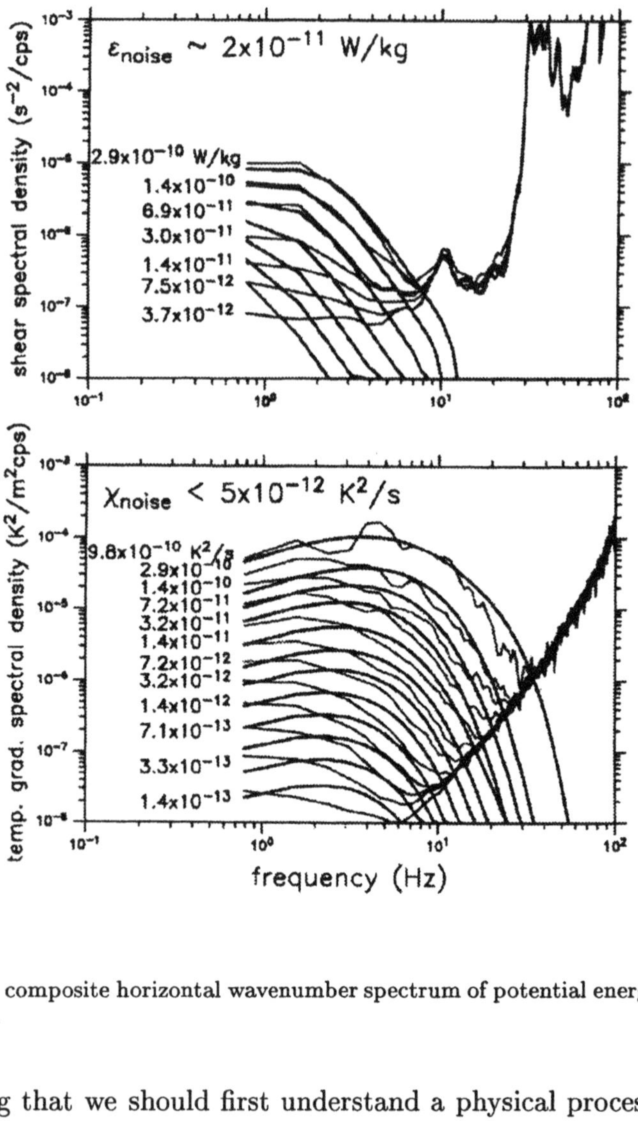

Figure 1. A composite horizontal wavenumber spectrum of potential energy from Dugan *et al.* (1986).

Arguing that we should first understand a physical process before developing, applying, or accepting a parameterization for it, I will in the following review what we now understand about mixing processes in the ocean interior, and some simple thoughts on how these processes might be parameterized. As the processes occurring within the planetary boundary layers at the ocean's top and bottom are discussed in subsequent chapters, the present section concentrates on mixing mechanisms in the remaining ~95% of the water column. And, as they are arguably the best understood, I will concentrate on the diapycnal mixing of scalars (temperature, salinity, density, tracers...), touching only briefly on the questions of momentum fluxes and stress, and epipycnal fluxes.

Basic to my presentation will be the differentiation between mixing and stirring. Formally, mixing is the irreversible action of molecular viscosity, heat conduction, and solvent diffusivity responsible for the destruction of velocity and water property gradients. Owing to seawater's relatively small coefficients of molecular viscosity (ν), and of heat (D_T) and salt (D_S) diffusivity [$\nu \sim O(10^{-6}$ m^2 s^{-1}), $D_T \sim O(10^{-7}$ m^2 s^{-1}), $D_S \sim O(10^{-9}$ m^2 s^{-1})], mixing and the frictional dissipation of kinetic energy is only effective at centimeter-to-millimeter scales in the ocean (see below). Interactions between larger scales of motion act to flux energy to the these small scales where viscous dissipation is effective. Similarly, straining of scalar property distributions by larger scale flows (stirring) enhances gradients on small scales where diffusion acts. Thus, eddies in the ocean stir, and by themselves do not flux water across density or water property surfaces. The key to the parameterization question is whether the fluxes of energy and scalar variance to smaller scales where mixing occurs can be determined in terms of the larger-scale fields resolved by a numerical simulation. As I will discuss, such a link appears roughly in hand between the internal wave field and the turbulent dissipation. We are presently investigating how the internal wave field varies in space and time, and with respect to the mesoscale flow field.

The smallest velocity scales seen in a turbulent fluid are about the Kolmogoroff scale (L_k) given by dimensional argument as $2\pi(\nu^3/\epsilon)^{\frac{1}{4}}$ where ν is the molecular viscosity of sea water and ϵ the rate of turbulent kinetic energy dissipation. For typical ocean thermocline values of ν ($\sim 10^{-6}$ m^2/s) and ϵ ($\sim 10^{-9}$ W/kg $\simeq$ the energy dissipated by one hairdryer per cubic kilometer of ocean) L_k is about 4 cm. We do not as yet have dynamical theory for the shape of the turbulent wavenumber shear spectrum, but Nasmyth (1970) and Oakey (1982) have given us the next best thing: an empirical spectral form which scales with ϵ (Figure 2). The molecular diffusivity of heat and dissolved scalars in sea water (*i.e.*, salinity) are smaller than the viscosity. This means that temperature and salinity gradients exist on smaller spatial scales than do velocity gradients. Building on this fact, Batchelor (1949) developed a theoretical description of the small-scale wavenumber spectrum of a passive tracer. An assumption here is that at these scales, buoyancy effects are not important. The relevant spatial scale from this analysis is $L_b \sim 2\pi(D^2\nu/\epsilon)^{\frac{1}{4}}$. For temperature, with a molecular diffusivity of order 10^{-7} m^2/s, L_b is about 1 cm; for salinity with a diffusivity 100 times smaller, the scale is order 1 mm (again for $\epsilon = 10^{-9}$ W/kg).

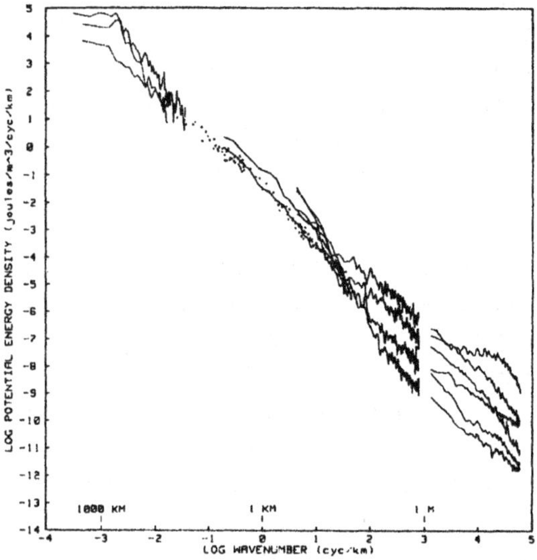

Figure 2. Vertical wavenumber spectra of velocity shear (top) and temperature gradient (bottom) binned by dissipation rate. For reference are shown the Nasmyth (1970) empirical shear spectrum and the Batchelor temperature gradient spectrum. Frequency may be converted to vertical wavenumber through the nominal instrument fall-rate of 0.5 m/s. (Thus, 10 Hz corresponds to 20 cpm or a 5-cm vertical wavelength.) Departure of the data from the theoretical curves at high frequency is caused by mechanical and electronic noise.

1.1. IMPLICATIONS OF OCEAN MIXING

But why do we even care that there is turbulent mixing and centimeter-scale structures in the oceans? Fundamentally, it has been argued that the existence of a vigorous mean overturning circulation is a direct consequence of ocean mixing. To overcome friction in a steady ocean circulation, there must be an input of mechanical energy over each closed streamline. This concept led Sandstrom (1916) to postulate that a closed steady circulation can be maintained only if the heating source is situated at a lower level than the cooling source (Huang, 1998). Sandstrom carried out simple laboratory experiments that appeared to confirm this idea. But this presents a problem for the real ocean circulation, as sea level in the tropics where the oceans are heated is some 1 m higher (relative to a geopotential surface) than it is in the polar oceans where cooling occurs. Why then does a mean thermohaline circulation exist in the ocean? Jeffreys (1925) suggested that mixing is capable of redistributing density, leading to a pressure field of the sense to sustain an overturning circulation in the face of frictional damping. Huang

(1998) has developed a simple closed-tube model for the thermohaline circulation that nicely demonstrates how the flow is controlled principally by diffusion of density when heating occurs at a higher potential than cooling. His model suggests that both Sandstrom and Jeffreys were right: when the heating source lies above the cooling source, an overturning is still possible because of mixing; but when that mixing is solely the result of molecular diffusion (as would virtually be the case in a careful laboratory experiment), the flow is very slow (and so difficult to observe in a laboratory setting.)

Simple scaling arguments for the overturning circulation also imply a dependence of the overturning circulation on buoyancy diffusion. Following Bryan (1987) and others, the continuity and thermal wind equations, combined with the vertical advective-diffusive balance for the density yield the scaling $M = VHL \sim K^{\frac{2}{3}}$ where M is the meridional overturning volume transport (given by the product of the meridional velocity scale V, the basin width L, and the scale height of the pycnocline H) and K is the (assumed uniform) diapycnal diffusivity. Numerical box-model experiments using the GFDL MOM2 code (Bryan, 1969; Pacanowski, 1995) with the Gent and McWilliams (1990) eddy transport parameterization carried out by Zhang *et al.* (1998) exhibited this scaling with little dependence on the strength of the wind driving. Interestingly, Bryan's (1987) earlier study using a model developed with horizontal and vertical diffusion (as opposed to epi- and diapycnal) showed a weaker dependence on K, possibly due to cross-isopycnal mixing by horizontal diffusion in regions of sloping isopycnals (the so-called Veronis effect), and/or because the model was not run sufficiently long to achieve steady state.

Ocean observations also provide evidence, albeit indirect, that there is turbulent mixing taking place in the ocean at rates that have significance for the ocean water properties and circulation. Water mass property-property curves tend to be tightly defined despite the fact that the source water formation mechanisms are temporally variable. J. Reid, for example, is able without much difficulty, to draw smoothly varying property fields on isopycnals on basin to global scales (*e.g.*, Reid, 1998). Presumably if mechanisms for straining and ultimately dissipating property anomalies were very weak, the fields of water properties would be "lumpy" or even uncontourable.

Ocean circulation schemes, particular for the Indian and Pacific Oceans (*e.g.*, Wunsch *et al.*, 1983; Wijffels *et al.*, 1996; Roemmich and McCalister, 1989; Robbins and Toole, 1997) have implicit need for diapycnal mixing in the ocean interior. The meridional overturning circulation deduced for these oceans includes net northward flow of Circumpolar Deep Water at the bottom and a mid-depth southward return of Indian/Pacific Deep Water. As the isopycnals associated with these water masses do not outcrop anywhere in these oceans (north of the Southern Ocean), these circulations imply interior diapycnal mixing and flow. Dense overflows from marginal seas also demonstrate existence of mixing. For example, the overflow through the Strait of Gibraltar is sufficiently dense that were it allowed to sink adiabatically down the European continental slope, it would easily make it to the bottom, filling the eastern North Atlantic with a very warm, saline bottom water. In fact, the overflow does not sink this deep. Intense entrainment mixes ambient fluid with the Mediterranean overflow as it descends to produce a water that is neutrally buoyant at around 1000-m depth (Price and O'Neil Barringer, 1994).

1.2. INDIRECT ESTIMATES OF MIXING RATES

So exactly how intense is mixing in the ocean anyway? The seminal paper that infers the ocean mixing rate is Munk's (1966) "Abyssal Recipe" work. From a simple vertical advective-diffusive balance for the main thermocline: $w\theta_z = K\theta_{zz}$, combined with observations of the typical e-folding scale for $\theta(z)$ and a guess at the average vertical upwelling rate (10^{-7} m s^{-1}: equivalent to a global bottom water formation rate of some 25 Sv uniformly upwelling into the thermocline) this model gives $K \sim 10^{-4}$ m^2/s. Admittedly, this is a crude model; the ocean has a fully three-dimensional circulation. (Munk himself suggested that boundary mixing followed by lateral exchange with the interior might be the actual mechanism sustaining the deep thermocline.) However, diathermal heat fluxes in the abyssal ocean estimated from heat budgets for semi-enclosed deep basins, when expressed in terms of an implied diffusivity also return values of this magnitude (Whitehead and Worthington, 1982; Hogg *et al.*, 1982; Saunders, 1987; Roemmich *et al.*, 1996; Morris *et al.*, 1997). Interestingly, the comparable type of analyses for upper-ocean layers bound by isotherms (pycnals) and the free surface find similar need for significant diathermal fluxes (*e.g.*, Speer, 1998; Zhang and Talley, 1998). These latter estimates are far more complicated than the abyssal ones because the upper surface is not insulating.

The budget calculations, while valuable for setting bounds, do not of course shed light on the possible mechanisms supporting the mixing or where within each control volume it occurs. Indeed, the ocean mixing community has been engaged in a 30-year effort to find mechanisms capable of supporting diapycnal fluxes consistent with $O(10^{-4}$ m^2/s) diffusivities.

2. Analysis of Ocean Microstructure

Unlike in the atmosphere, the seemingly straightforward direct observation of turbulent ocean heat and momentum fluxes (*e.g.*, $< w'\theta' >$, $< u'w' >$) has been attempted only rarely in oceanography. This is in part a consequence of our sensors for measuring dissipation-scale velocities. The airfoil "shear" probe (Osborn and Crawford, 1980), deployed on an instrument platform moving through the ocean, is sensitive to turbulent flows at right-angles to its direction of travel. As the majority of instrument systems profile vertically, these sensors cannot be used to observe vertical turbulent flow. Moum (1990) has had some success measuring turbulent-scale vertical velocities with a Pitot tube in energetic flows, but accumulating a large sample at any one depth interval (needed to develop statistically significant flux estimates) with a vertical profiling instrument is difficult, and the technique is not sufficiently sensitive to measure the weak fluxes in the thermocline. Horizontal sampling with shear probes mounted on towed platforms (Osborn and Lueck, 1985a) and submarines (Gargett *et al.*, 1984; Osborn and Lueck, 1985b) has yielded some interesting results (where extreme care was taken to address vibrational noise contamination), but the techniques have not seen widespread application.

Rather, the common approach has been to adopt statistical models for the turbulent temperature variance and kinetic energy, and from these deduce turbulent fluxes and "eddy diffusivities." The basic assumption is of a statistical steady state balance between the production of variance by eddies acting on a mean, or background gradient and the dissipation of variance by molecular viscosity/diffusion. Osborn and Cox (1972) derived an expression relating the production of temperature variance by turbulent eddies working on the mean (vertical) temperature gradient to its dissipation by molecular diffusion: $< w'\theta' > \overline{\theta}_z = -\chi$. Here $-\chi$ is the dissipation rate of temperature variance, given for isotropic turbulence as $\chi = 6D_T < \theta_z'^2 >$. The averaging operator $<>$ is frequently thought of as over time, though Winters and D'Asaro (1996) suggest that the above holds instantaneously if one first sorts the temperature field to a monotonic profile. Given the shape of the temperature gradient spectrum (Figure 1), it is apparent that estimation of χ involves resolving temperature gradients down to L_b.

The Osborn–Cox "eddy diffusivity" parameterizing of the turbulent heat flux consists of expressing the flux as an eddy diffusivity, K_T, times the mean temperature gradient $< w'\theta' > \; = \; K_T\overline{\theta}_z$. This gives a relation for K_T as $K_T = \chi/\overline{\theta}_z^2 = 3D_T < \theta_z'^2 > /\overline{\theta}_z^2$ (the latter again assuming isotropic turbulence). The governing temperature equation for the background temperature field becomes: $\overline{\theta}_t + u \cdot \nabla\overline{\theta} = \partial_z(K_T\overline{\theta}_z)$.

In similar fashion, Osborn (1980) details the development of a dominant 3-term balance in the turbulent kinetic energy equation:

$$< u'w' > \overline{U}_z = -\epsilon - g < w'\rho' > /\rho_0 \quad .$$

The dissipation rate of turbulent kinetic energy, ϵ, for isotropic turbulence, is given by $\epsilon = 7.5\nu < u_z'^2 >$ and, analogous to χ, involves integrating shear wavenumber spectra to beyond L_k. In words, the above equation says that the production of turbulent kinetic energy by Reynolds stresses acting on the mean shear is balanced in the mean by the viscous dissipation of energy and/or an increase in the mean potential energy by a turbulent vertical buoyancy flux. Visual simplicity is gained through definition of the flux Richardson number:

$$R_f = -(g/\rho_0) < w'\rho' > [< u'w' > \overline{U}_z]^{-1}$$

yielding:

$$(g/\rho_0) < w'\rho' >= [R_f/(1-R_f)]\epsilon \ .$$

Though itself a statistical quantity, R_f is thought to have values of about 0.2 in shear-generated turbulence within a stratified water column (Oakey, 1982), and is frequently taken as constant.

Again adopting a parameterization of the form: $< w'\rho' >= K_\rho \overline{\rho}_z$ gives us an expression for an eddy diffusivity for buoyancy:

$$K_\rho = R_f/(1-R_f)\epsilon/N^2 = \Gamma\epsilon/N^2 \ .$$

As with the Osborn–Cox model, K_ρ then appears in the scalar conservation equations in place of molecular diffusivity. Note that by equating K_T and K_ρ, one is able to investigate the size of the mixing efficiency Γ (Oakey, 1982). This analysis is particularly instructive where double diffusion is possible (see Schmitt, 1998; St. Laurent and Schmitt, 1998).

Though less widely employed, a turbulent viscosity parameterization has also been developed from the above turbulent energy equation. Based on the assumption that the background shear U_z is supplying the energy for the turbulence, one writes: $< u'w' >= A\overline{U}_z$. Again adopting the R_f "simplification", we arrive at a formula for the eddy viscosity A (Gregg *et al.*, 1985):

$$A = (1-\Gamma)\epsilon/\overline{U}_z^2 \ .$$

3. Observations of Ocean Finestructure and Microstructure

Dissipation and mixing in the ocean interior is thought to be principally the result of shear instability. Linear theory gives us the necessary condition for instability that the gradient Richardson number $Ri = N^2/\overline{U}_z^2 < 1/4$. Experimentally, we find this to be a sufficient condition as well in most cases. An example of shear instability in the ocean interior comes from atop Fieberling Guyot, a seamount in the eastern North Pacific (Figure 3) (Kunze and Toole, 1997). As demonstrated by this example, Kelvin–Helmholtz-type instability waves that lead to turbulence and mixing are sensitive to ocean shears on vertical scales of around ten meters. Finescale

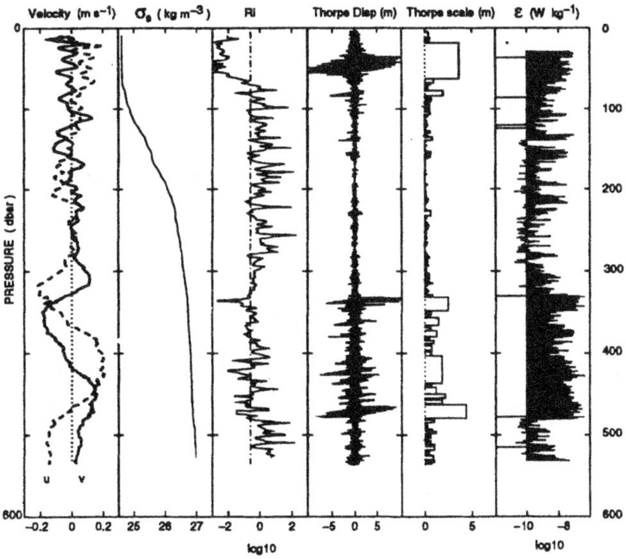

Figure 3. Vertical profile of horizontal velocity, potential density, Richardson number, vertical displacements of overturns, the Thorpe scale, and turbulent dissipation rate from atop Fieberling Guyot (Kunze and Toole, 1997). Note the association of large vertical overturns and large dissipation rate with low Richardson number.

shears on these vertical scales are chiefly the product of the ocean's internal wave field (Gargett *et al.*, 1981). Indeed, the gradient Ri of major ocean currents like the Gulf Stream are invariably much greater than 1. The chief exceptions to this are the equatorial undercurrents, discussed below. Understanding mixing in the ocean interior is thus intimately linked to understanding the finescale shear field associated with the internal waves.

In the 1970's, a basic spectral description of the background field of internal waves was developed by Garrett and Munk (1972, commonly referred to as the GM description). With updates (Garrett and Munk, 1975; Cairns and Williams, 1976), this model remains valid today. Mixing associated with the wave field is believed to be the consequence of random superposition of waves leading to intervals of supercritical Ri, shear instability, turbulence and mixing. Several wave-wave interaction models have been developed to investigate the flux of energy through wavenumber space towards small vertical scale where shear instability takes place and ultimately, energy is dissipated. Using the mixing efficiency arguments discussed above, these models may be used to develop parameterizations for the turbulent diapycnal buoyancy flux ($\Gamma\epsilon$), or if you prefer, the diapycnal eddy diffusivity K_ρ. Polzin *et al.* (1995) discuss several such models in their examination of ocean microstructure and finestructure observations and find greatest support for the model by Henyey *et al.* (1986), as was previously observed by Gregg (1989). This model's prediction for the turbulent dissipation associated with the energy flux through the GM internal wave field goes like: $\epsilon \sim N^2$. That means that the predicted diapycnal diffusivity, K_ρ, is independent of stratification (again for the background internal wave field.) Moreover, the prediction for the diffusivity associated with the background wave field is small: $K_{\rho\ GM} = 7 \times 10^{-6}$ m^2 s^{-1}.

Ocean microstructure data taken chiefly in the main thermoclines of the North Atlantic and Pacific Oceans imply that diapycnal diffusivity is indeed at least a factor of 10 less than what indirect analyses had inferred (Gregg, 1987). This seeming discrepancy caused some to question the traditional interpretation of ocean microstructure observations (Gibson, 1982; Baker and Gibson, 1987). But weak diapycnal mixing of order 10^{-5} m^2 s^{-1} was confirmed in the thermocline by Ledwell *et al.* (1993; 1998) with a tracer release experiment.

How then can we rationalize the large diapycnal fluxes inferred from thermocline-layer heat budget analyses of Speer (1998) and colleagues? Nurser *et al.* (1998) and Marshall *et al.* (1998) suggest that much of this inferred mixing occurs about the base of the time-varying surface mixed layer. For example, the former authors obtained apparent layer-averaged diapycnal diffusivities of 10^{-4} m^2/s out of an ideal thermocline model with mixing limited to mixed layer processes.

More intense mixing may also be associated with special ocean regions characterized by an energized internal wave field (more energetic than the GM background). The Henyey *et al.* model suggests that the energy flux through the wave field (and thus the dissipation rate) increases with the finescale shear level of the internal waves. (An adjustment is additionally suggested for the average wave frequency: Henyey, 1991; Polzin *et al.*, 1995). Comparison of this model to microstructure data find agreement to about a factor of two (Figure 4), a quite reasonable fit given the statistical uncertainty in the turbulent dissipation estimates. Thus, to identify areas of enhanced mixing in the ocean interior one might equivalently search for regions of energetic internal waves.

Ocean areas characterized by negative relative vorticity are often sites of enhanced internal wave energy density. As discussed by Kunze (1985), low-frequency internal waves sense an effective inertial period that includes the relative vorticity of the large-scale flow. Flows with negative relative vorticity are capable of forming wave trapping zones in the horizontal and critical layers in the vertical. At the latter, wave energy accumulates and eventually dissipates. Kunze *et al.* (1995) and Kunze and Toole (1997) offer examples of this process within a warm core ring of the Gulf Stream and

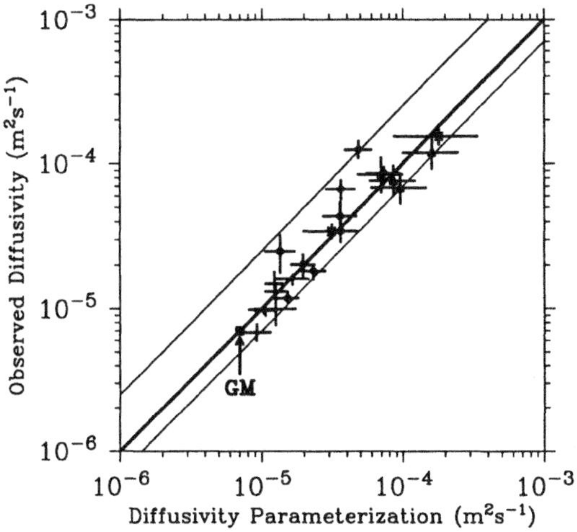

Figure 4. Comparison of turbulent dissipation rate predicted from internal wave characteristics with observed dissipation (Polzin *et al.*, 1995). The thick line denotes perfect agreement; the thin lines bracket the outliers. The data displayed come from a variety of geographic sites characterized by differing internal wave climatologies and background stratification.

within a rectified anticyclonic flow atop a mid-latitude seamount, respectively. In the latter case, the diapycnal buoyancy flux above the seamount was some 200 times that of the background interior (on a per unit area basis). Lueck and Mudge (1997) also found intense mixing about Cobb Seamount.

Internal wave reflection from sloping bathymetry has also been suggested as causing enhanced wave energy density and mixing (*e.g.*, Eriksen, 1985). Linear reflection theory breaks down about a critical bottom slope where infinite reflected wave energy is predicted. For a given incident internal wave, this slope parallels the group velocity of the reflected wave (which is thus unable to carry energy away from the bottom). Enhanced energy on reflection is predicted for a range of slopes about the critical value. Equivalently, for a given slope, there will be a range of incident waves that are predicted to exhibit enhanced energy density on reflection. Evidence of enhanced kinetic energy about estimated critical frequencies has been seen in some moored current meter records obtained above sloping bathymetry (Eriksen, 1985; 1998). With enhanced wave energy density, the wave dissipation parameterization implies enhanced mixing, and indeed, the observa-

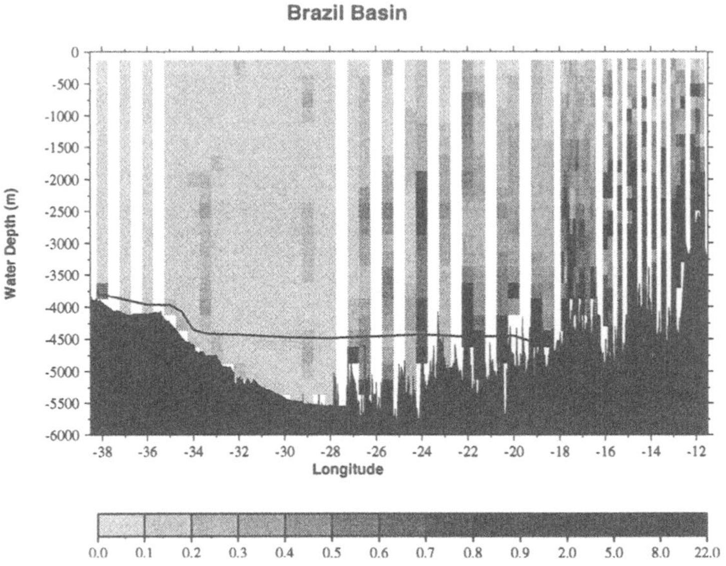

Figure 5. Spatial distribution of diapycnal diffusivity in the Brazil Basin deduced from velocity microstructure (Polzin *et al.*, 1997). This composite zonal section runs from the continental slope off Brazil to the crest of the mid-Atlantic Ridge. High-resolution underway bathymetry from one transect is also displayed.

tions of Toole *et al.* (1997) reveal diffusivities of $\sim 5 \times 10^{-4}$ m^2 s^{-1} within about 500 m vertically of the flanks of Fieberling Guyot. One key point is that due to the ability of waves to propagate vertically, the turbulent dissipation can occur in the stratified water column well above any homogenized bottom layer and thus support vertical mixing of water properties.

Motivated by the large diapycnal diffusivities inferred in the deep, semi-enclosed basins, K. Polzin, R. Schmitt, J. Ledwell and I initiated in 1996 a joint finestructure/microstructure and tracer release study in the deep Brazil Basin. Full water column profiles of temperature, salinity, and horizontal velocity along with turbulent dissipation rate information have now been obtained on two cruises spanning the zonal interval between the 3800-m isobath on the continental slope east of Brazil and the mid-Atlantic Ridge crest (Polzin *et al.*, 1997). Those stations occupied in the western part of the Basin were characterized by background-intensity internal waves and weak turbulent mixing, including those taken over the smoothly sloping South American continental rise (Figure 5).

In contrast, above the rough flanks of the mid-Atlantic Ridge, internal wave energy was enhanced and accompanied by energetic turbulence supporting diffusivity estimates of order 10^{-3} m^2/s. A site over the Ridge flank was selected for the tracer release.

A rough estimate based on these turbulence data of the heat flux across the 0.8°C potential temperature surface, one of the control surfaces used by Hogg *et al.* (1982), comes close to that derived from their Vema Channel transport observations (and thus comes close to closing the abyssal heat budget). The tracer program is still underway, but results of the initial survey carried out in 1997 suggests that the average diapycnal diffusivity at the tracer depth is somewhat larger than was estimated from the turbulence data (J. Ledwell, personal communication, 1997). This slight discrepancy between the microstructure and tracer results is likely a result of aliasing and undersampling the microstructure about the tracer patch.

Polzin *et al.* (1997) postulated that the source of the enhanced internal wave energy, and thus the strong turbulence, is the generation of waves by flow over the rough bathymetry of the mid-ocean ridge. In particular, barotropic tidal currents were implicated, mean and mesoscale flows being thought too weak in the eastern Brazil Basin. Interestingly, Munk and Wunsch (1998) are independently exploring this idea from an astronomical viewpoint. From the change in the moon's orbit, a global tidal dissipation estimate may be derived. While a large fraction of the tidal energy dissipation may happen in shallow seas within bottom boundary layers, perhaps as much as 10^{12} W of energy is supplied to open-ocean tides that, in turn, dissipates and supports diapycnal mixing in the ocean interior.

The observed profiles of K_ρ over the rough topography have an interesting consequence for the diapycnal velocity. Neglecting (demonstratively small) consequences of the nonlinear equation of state, the density equation in a coordinate system aligned with the isopycnals is $w^* \rho_z = \partial_z(K_\rho \rho_z) \sim \partial_z(\Gamma \epsilon)$. Above the spurs of the mid-Atlantic Ridge, ϵ is observed to increase with depth below ~2500 m. This implies negative w^*. But mass conservation for the bottom waters, in light of the flow entering the Brazil Basin through the Vema Channel, requires net upwelling across isopycnals (positive average w^*). We suspect that mass balance is achieved by strong "upward" directed flow across isopycnals within the many fracture zones of the ridge. How one accounts for this in a model that is unable to resolve the narrow fracture zones is a challenging issue.

In the Southern Ocean south of Australia, Polzin and Firing (1997) thought the enhanced internal wave shears observed there might be the product of eddy currents, not tides. Based on the internal-wave-energy parameterization, they inferred $K_\rho \sim 5 \times 10^{-4}$ m^2 s^{-1} below 1000 m within the Antarctic Circumpolar Current. In addition to supporting enhanced mixing through much of the water column, the surmised bottom generated internal waves may also carry momentum upwards and thus distribute the bottom stress up into the water column. If so, an Ekman-type balance might support net meridional flow above the height of the bathymetric sills of the zonal flow in the Southern Ocean.

Before departing the deep ocean, it is important to call attention to another potentially significant mixing process associated with abyssal flow through bathymetric constrictions. Under some conditions, waters may be accelerated through such passages as occurs in hydraulics problems. Those gaps that include a vertical sill can also result in downstream flow acceleration by gravity. Such flows may internally produce low *Ri* and develop shear instability. Internal lee waves and hydraulic jumps might also play a mixing role within such currents. One example of this is the flow through the Romance Fracture Zone in the equatorial Atlantic (Polzin *et al.*, 1996). Mixing downstream of the main sill warms the coldest waters passing over the sill from about 0.9°C to nearly 1.5°C. Turbulence on centimeter scales appears as important as topographic blocking in setting the bottom water properties of basins in the Eastern Atlantic! These processes are discussed further in this volume by Price and Yang (1998).

As noted earlier, another ocean current system that internally develops supercritical *Ri* is found about the equator. Strong shear is found between the westward-directed South Equatorial Current (SEC) and the eastward Equatorial Undercurrent (EUC). Frictional damping and diapycnal heat fluxes are believed to be of first order importance in the momentum and heat balances. A series of cruises conducted in the 1980's examined the turbulence in this environment (*e.g.*, Moum *et al.*, 1989; Gregg *et al.*, 1985; Lien *et al.*, 1995). Intense turbulent dissipation was observed above the undercurrent core. The microstructure was strongly modulated diurnally with the cycle of air–sea buoyancy exchange, with the strength of the wind stress, and the ambient shear between the SEC and EUC. Large (1998, this volume) discusses these observations in the context of a large eddy simulation of the equatorial surface waters. Averaging over these fluctuations, mean diapycnal diffusivities in excess of 10^{-3} m^2/s^{-1} were obtained. Mean eddy viscosities were also estimated in this flow with comparably large values. At greater depths, a remarkable array of zonal flows on or adjacent to the equator has been observed (Firing, 1987). These flows support greater shear than does the GM internal wave field, and are characterized by enhanced mixing (Gregg *et al.*, 1996). As of yet we do not have a general relationship between these near-equatorial sheared flows and the turbulent dissipation.

4. Parameterizing Diapycnal Mixing in the Ocean Interior

The Garrett–Munk description of the internal wave field is a widely accepted baseline for waves in the ocean interior (away from boundaries and regions of forcing and localized dissipation). Interactions of waves in this environment lead to an up-vertical-wavenumber flux of energy and eventually to wave breaking and dissipation. A model for wave-induced dissipation at Garrett–Munk wave energy levels, verified by observations spanning more than two orders of magnitude in N^2, predicts that $\epsilon \sim N^2$ (or equivalently K_ρ is independent of N^2). Moreover, the diapycnal diffusivity predicted for the background wave field is small, ($< 10^{-5}$ m^2 s^{-1}). This suggests a baseline diffusivity for model calculations of $K_\rho \sim 10^{-5}$ m^2 s^{-1}.

Low-frequency internal waves may be trapped to regions of negative relative vorticity associated with the mesoscale flow field. If baroclinic, these flows may present critical layers to the waves (at the level where the wave frequency equals the effective Coriolis frequency). As wave energy accumulates on approach to such critical layers, breaking and dissipation result. This suggests a K_ρ dependence on the relative vorticity of the background flow field, though no formal effort has yet been made to parameterize this effect.

Observations clearly document enhanced internal wave energy density and turbulent dissipation adjacent to rough, sloping bathymetry. Eriksen (1985) and Garrett and Gilbert (1988) have thought about what turbulent dissipation and mixing may result from near-critical wave reflection from sloping bathymetry. At Fieberling Guyot, K_ρ values of about 5×10^{-4} m^2 s^{-1} were observed within ~500 m vertically of these steep slopes. Wave reflection may be in part responsible for this, but bottom generated waves have also been implicated.

Polzin (1998) is working presently to relate flow over bathymetry to wave generation, and in turn, the vertical propagation of these waves and eventual dissipation to the vertical profile of turbulent dissipation and diapycnal diffusivity. The work is beginning with the early models of Bell (1975a,b) of internal tidal and internal lee wave generation, and the wave–wave interaction work demonstrated above to have skill predicting dissipation. The early results show promise that, given finescale bathymetric information and an estimate of the bottom tidal and/or mesoscale currents, a prediction for $K(z)$ may eventually be had.

Accelerated flows in deep channel flows and gravity currents are known sites of intense dissipation. Here some *Ri*-based parameterization may be fruitful, perhaps with additional consideration of hydraulic jumps. J. Price will discuss these in detail. The equatorial current system is another characterized by large vertical shear. Peters *et al.* (1988) fit hourly-averaged eddy diffusivity and eddy viscosity estimates to power-laws in Ri. Great scatter was found in the observations, and the exponential dependence on Ri for $Ri < 0.5$ was large: both limiting confidence in the result. Indeed, Moum *et al.* (1989) chose not to attempt such a fit to their observations taken nearby. It may prove more profitable for modelers to employ schemes that instantaneously mix to neutral Ri as discussed by Schudlich and Price (1992). In any case, the scale dependence of Ri should be considered before blindly applying any scheme based on Ri. Large (1998, this volume) discusses the equatorial upper ocean in greater detail in his presentation.

Lastly we must not forget the boundary layers at the top and bottom of the ocean. These constitute the dominant sites of turbulent dissipation in the sea. Relevant presentations in this volume include those of Large (1998), Price (1998), Send and Käse (1998) and Beckmann (1998).

5. A Few Thoughts on Epipycnal Diffusion

I would be remiss in not at least commenting on mechanisms of epipycnal diffusion in the ocean. Fundamentally, mixing is mixing and the actual modification of a water parcel's properties laterally is no different than vertically: occurring on the Batchelor and Kolmogoroff scales. Lateral stirring and straining, being unconstrained by buoyancy forces are much more effective than in the vertical though, generally giving rise to larger coefficients of epipycnal diffusivity.

Depending on the horizontal resolution of a model, the adopted epipycnal diffusivity must account for different physical processes. Generally speaking, the greater the horizontal scale employed, the greater the diffusivity required. The North Atlantic Tracer Release Experiment (Ledwell *et al.*, 1993, 1998) very nicely demonstrates this. With time, the injected tracer spread epipycnally as well as diapycnally. From the time rate of change of the tracer patch's area, one obtains an estimate of the lateral diffusivity. Ledwell *et al.* (1998) found that the effective diffusivity associated with the lateral spread of the tracer patch increased with time quadratically.

Initially, when the patch(es) had horizontal scales of 100–1000 m, the tracer dispersed with an estimated lateral diffusivity of $\sim$0.07 m^2/s. This rate is consistent with shear dispersion due to near-inertial internal waves (Young *et al.*, 1982). After about six months into the experiment, the tracer had been drawn into long streaks with horizontal widths of 1–30 km. The width-to-length ratio of the streaks suggested rather stronger lateral diffusion on these scales, $\sim$ 2 m^2/s. Polzin *et al.* (1998) propose that vortical mode motions (finescale structures with non-zero Ertel potential vorticity) might be responsible for the mixing on these scales. When it achieved a horizontal scale of 30–300 km, the tracer patch had comparable size to individual mesoscale eddies. Ledwell *et al.* felt an eddy parameterization on this scale was inappropriate; patch growth was chiefly dictated by the eddy strain. Eventually though, as the tracer patch broadened further to scales of 300–1000 km, an eddy diffusivity was again possible to estimate. Ledwell *et al.* report a zonal diffusivity of 2300 (1000–4000) m^2/s and a meridional diffusivity of 650 (200–1200) m^2/s. This North–South/East–West difference may reflect strain by the mean anticyclonic gyre flow.

Finally, I should note that a special form of lateral diffusion is possible at water mass fronts: thermohaline intrusions. Joyce (1977) discussed how interleaving of water masses at fronts can greatly increase the surface area between waters and thus enhance the mixing. Additionally, double-diffusion may play a role in this mixing. This aspect is disscussed in this volume by Schmitt (1998) and Richards (1998).

Acknowledgments

Support for my research was provided by the U.S. National Science Foundation and the Office of Naval Research, for which I am most grateful. I wish to particularly thank Kurt Polzin and Raymond Schmitt for their major contributions to the WHOI-based research that I cited, and our support team of Ellyn Montgomery and David Wellwood.

References

Baker, M.A. and C.H. Gibson (1987) Sampling turbulence in the stratified ocean: Statistical consequences of strong intermittency. *J. Physical Oceanography* **17**, 1817–1836.

Batchelor, G.K. (1949) Small-scale variation of convected quantities like temperature in a turbulent fluid, Pt 1, *J. Fluid Mechanics* **5**, 113–133.

Bell, T.H. (1975a) Topographically generated internal waves in the open ocean, *J. Geophysical Research* **80**, 320–327, 1975a.

Bell, T.H. (1975b) Lee waves in stratified flows with simple harmonic time dependence, *J. Fluid Mechanics* **67**, 705–722.

Beckmann, A., 1998: Representation of bottom boundary layer processes in numerical ocean circulation models. In *Ocean Modeling and Parameterization*, E.P. Chassignet and J. Verron (Eds.), Kluwer Academic Publishers, 135-154.

Bryan, K. (1969) A numerical method for the study of the circulation of the world ocean, *J. Computational Physics* **4**, 347–376.

Bryan, F. (1987) Parameter sensitivity of primitive equation ocean general circulation models, *J. Physical Oceanography* **17**, 970–985.

Cairns, J.L. and G.O. Williams (1976) Internal wave observations from a midwater Float, 2, *J. Geophysical Research* **81**, 1943–1950.

Dillon, T.M. (1984) The energetics of overturning structures: Implications for the theory of fossil turbulence, *J. Physical Oceanography* **14**, 541–549.

Dugan, J.P., W. D. Morris, and B.S. Okawa (1986) Horizontal wavenumber distribution of potential energy in the ocean, *J. Geophysical Research* **91**, 12,993–13,000.

Eriksen, C.C. (1985) Implications of ocean bottom reflection for internal wave Spectra and mixing, *J. Physical Oceanography* **15**, 1145–1156.

Eriksen, C.C. (1998) Internal wave reflection and mixing at Fieberling Guyot, *J. Geo-*

physical Research, in press.

Firing, E. (1987) Deep zonal currents in the central equatorial Pacific, *J. Marine Research* **45**, 791–812.

Gargett, A.E., P.J. Hendricks, T.B. Sanford, T.R. Osborn, and A.J. Williams III (1981) A composite spectrum of vertical shear in the ocean, *J. Physical Oceanography* **11**, 1258–1271.

Gargett, A.E., T.R. Osborn, and P.W. Nasmyth (1984) Local isotropy and the decay of turbulence in a stratified fluid, *J. Fluid Mechanics* **144**, 231–280.

Garrett, C. and D. Gilbert (1988) Estimates of vertical mixing by internal waves reflected off a sloping bottom, in: J.C.J. Nihoul and B.M. Jamart, (eds.), *Small-Scale Turbulence and Mixing in the Ocean*, Proceedings of the 19th International Liege Colloquium on Ocean Hydrodynamics, Elsevier, New York, pp. 405–424.

Garrett, C. and W. H. Munk (1972) Space time scales of internal waves, *Geophysical Fluid Dynamics* **2**, 225–264.

Garrett, C. and W. H. Munk (1975) Space time scales of internal waves: A progress report, *J. Geophysical Research* **80**, 291–297.

Gent, P.R. and J.C. McWilliams (1990) Isopycnal mixing in ocean circulation models, *J. Physical Oceanography* **20**, 9686–9698.

Gibson, C.H. (1982) Alternative interpretations for microstructure patches in the thermocline, *J. Physical Oceanography* **12**, 374–383.

Gregg, M.C., H. Peters, J.C. Wesson, N.S. Oakey, and T.J. Shay (1985) Intensive measurements of turbulence and shear in the equatorial undercurrent, *Nature* **318**, 140–144.

Gregg, M.C. (1987) Diapycnal mixing in the thermocline: A review, *J. Geophysical Research* **92**, 5249–5286.

Gregg, M.C. (1989) Scaling turbulent dissipation in the thermocline, *J. Geophysical Research* **94**, 9686–9698.

Gregg, M.C., D.P. Winkle, T.B. Sanford, and H. Peters (1996) Turbulence produced by internal waves in the oceanic thermocline at mid- and low latitudes, *Dynamics of the Atmosphere and Oceans* **24**, 1–14.

Henyey, F.S., J. Wright, and S.M. Flatte (1986) Energy and action flow through the Internal wave field: An Eikonal approach, *J. Geophysical Research* **91**, 8487–8495.

Henyey, F.S. (1991) Scaling of internal wave model predictions for the dynamics of oceanic internal waves, in: P. Muller and D. Henderson (eds.) *Proceedings of the 'Aha Juliko'a Hawaiian Winter Workshop*, University of Hawaii at Manoa, Honolulu, pp. 233–236.

Hogg, N.G., P. Biscaye, W. Gardner, and W.J. Schmitz (1982) On the transport and modification of antarctic bottom water in the Vema Channel, *J. Marine Research* **40**, Suppl., 231–363.

Huang, R.X., Mixing and energetics of the oceanic thermohaline circulation, *J. Physical Oceanography*, submitted, 1998.

Jeffreys, H. (1925) On Fluid motions produced by differences of temperature and humidity, *Quarterly J. Royal Meteorological Society* **51**, No. 216, 347–356.

Joyce, T.M. (1977) A note on the lateral mixing of water masses, *J. Physical Oceanography* **7**, 626–629.

Kunze, E. (1985) Near-inertial wave propagation in geostrophic shear, *J. Physical Oceanography* **15**, 544–565.

Kunze, E., R.W. Schmitt, and J.M. Toole (1995) The energy balance in a warm-core ring's near-inertial critical layer, *J. Physical Oceanography* **25**, 942–957.

Kunze, E. and J.M. Toole (1997) Tidally-driven vorticity, diurnal shear, and turbulence atop fieberling seamount, *J. Physical Oceanography* **27**, 2663–2693.

Large, W., 1998: Modeling and parameterizing ocean planetary boundary layers. In *Ocean Modeling and Parameterization*, E.P. Chassignet and J. Verron (Eds.), Kluwer Academic Publishers, 81-120.

Ledwell, J.R, A.J. Watson, and C.B. Law (1993) Evidence for slow mixing across the pycnocline from an open-ocean tracer-release experiment, *Nature* **364**, 701–703.

Ledwell, J.R, A.J. Watson, and C.B. Law, Mixing of a tracer in the pycnocline, *J. Geophysical Research*, submitted, 1998.

Lien, R.-C, D.A. Caldwell, M.C. Gregg, and J.N. Moum (1995) Turbulence variability at the equator in the central Pacific at the beginning of the 1991–93 El Niño, *J. Geophysical Research* **100**, 6881–6898.

Lueck, R. and R. Reid (1984) On the production and dissipation of mechanical energy in the ocean, *J. Geophysical Research* **89**, 3439–3445.

Lueck, R. and T. Mudge (1997) Topographically-induced mixing around a shallow seamount, *Science* **276**, 1831–1833.

Marshall, J., D. Jamous, and J. Nilsson, Reconciling 'thermodynamic' and 'dynamic' methods of computation of water-mass transformation rates, Manuscript in preparation, 1998.

Morris, M., N. Hogg, and W.B. Owens (1997) Diapycnal mixing estimated from advective budgets in the deep Brazil Basin, *International WOCE Newsletter* **28**, 23–25.

Moum, J.N., D.A. Caldwell, and C.A. Paulson (1989) Mixing in the equatorial surface layer and thermocline, *J. Geophysical Research* **94**, 2005–2021.

Moum, J.N. (1990) Profiler measurements of vertical velocity microstructure in the ocean, *J. Atmospheric and Oceanic Technology* **7**, 323–333.

Munk, W.H. (1966) Abyssal recipes, *Deep-Sea Research* **13**, 707–730.

Munk, W.H. and C. Wunsch, The moon and mixing: Abyssal recipes II, Manuscript in preparation, 1998.

Nasmyth, P. (1970) Oceanic turbulence. Ph.D. thesis, Institute of Oceanography, University of British Columbia, 69 pp.

Nurser, A.J.G. R. Marsh, and R.G. Williams, Diagnosing water mass formation from air–sea fluxes and surface mixing, Manuscript in preparation, 1998.

Oakey, N.S. (1982) Determination of the rate of dissipation of turbulent energy from simultaneous temperature and velocity shear microstructure measurements, *J. Physical Oceanography* **12**, 256–271.

Osborn, T.R. and C.B. Cox (1972) Oceanic finestructure, *Geophysical Fluid Dynamics* **3**, 321–345.

Osborn, T.R. (1980) Estimates of the local rate of diffusion from dissipation measurements, *J. Physical Oceanography* **10**, 83–89.

Osborn, T.R. and W.R. Crawford (1980) Turbulent velocity measurements with an airfoil probe, in L. Hasse, F. Dobson and R. Davis (eds.), *Instruments and Methods of Air–Sea Interaction*, Plenum Press.

Osborn, T.R. and R.G. Lueck (1985a) Turbulence measurements with a towed body, *J. Atmospheric and Oceanic Technology* **2**, 517–527.

Osborn, T.R. and R.G. Lueck (1985b) Turbulence measurements with a submarine, *J. Physical Oceanography* **15**, 1502–1520.

Pacanowski, R.C. (1995) MOM2 documentation, user's guide and reference manual, *GFDL Ocean Group Technical Report* **3**, Princeton, NJ.

Peters, H., M.C. Gregg, and J.M. Toole (1988) On the parameterization of equatorial turbulence, *J. Geophysical Research* **93**, 1199–1218.

Polzin, K.L., J.M. Toole, and R.W. Schmitt (1995) Finescale parameterizations of turbulent dissipation, *J. Physical Oceanography* **25**, 306–328.

Polzin, K.L. and E. Firing (1997) Estimates of diapycnal mixing from I8s, *International WOCE Newsletter* **29**, 39–41.

Polzin, K.L., Abyssal Mixing: Inertial sub-range solution for the energy balance of the finescale internal wave field, Manuscript in preparation, 1998.

Polzin, K.L., E. Kunze, J.M. Toole, and R.W. Schmitt, The partition of small-scale energy into internal waves and geostrophic motions, Manuscript in preparation, 1998.

Polzin, K.L., K.G. Speer, J.M. Toole, and R.W. Schmitt (1996) Intense mixing of Antarctic Bottom Water in the equatorial Atlantic Ocean, *Nature* **380**, 54–57.

Polzin, K.L., J.M. Toole, G.R Ledwell, and R.W. Schmitt (1997) Spatial variability of turbulent mixing in the abyssal ocean, *Science* **276**, 93–96, 1997.

Price, J.F., 1998: Parameterization of the fair weather Ekman layer. In *Ocean Modeling and Parameterization*, E.P. Chassignet and J. Verron (Eds.), Kluwer Academic Publishers, 121-134.

Price, J.F., and J. Yang, 1998: Marginal sea overflows for climate simulations. In *Ocean Modeling and Parameterization*, E.P. Chassignet and J. Verron (Eds.), Kluwer Academic Publishers, 155-170.

Price, J.F., and M. O'Neil Baringer (1994) Outflows and deep water production by marginal seas, *Progress in Oceanography* **33**, 161–200.

Reid, J.L., On the total geostrophic circulation of the Pacific Ocean: Flow patterns, tracers and transports, Manuscript in preparation, 1998.

Richards, K., 1998: Interleaving at the Equator. In *Ocean Modeling and Parameterization*, E.P. Chassignet and J. Verron (Eds.), Kluwer Academic Publishers, 235-252.

Robbins, P.E., and J.M. Toole (1997) The dissolved silica budget as a constraint on the meridional overturning circulation of the Indian Ocean, *Deep-Sea Research I* **41**, 879–906.

Roemmich, D., and T. McCalister (1989) Large-scale circulation of the North Pacific Ocean. *Progress. in Oceanography* **22**, 171–204, 1989.

Roemmich, D., S. Hautala, and D. Rudnick (1996) Northward abyssal transport through the Samoan passage and adjacent regions, *J. Geophysical Research* **101**, 14,039–14,056.

Sandstrom, J.W. (1916) Meteorologische studien im Schwedischen hochgebirge, *Goteborgs K. Ventenskaps-och Vitterhetssamhalles Handl., Ser. 4* **22**, No. 2, 48 pp.

Saunders, P.M. (1987) Flow through Discovery Gap, *J. Physical Oceanography* **17**, 631–643.

Schmitt, R., 1998: Double-diffusive convection: Its role in ocean mixing and parameterization schemes for large scale modeling. In *Ocean Modeling and Parameterization*, E.P. Chassignet and J. Verron (Eds.), Kluwer Academic Publishers, 215-234.

Schudlich, R.R. and J.F. Price (1992) Diurnal cycles of turbulent dissipation, current and temperature in a model of the equatorial upper ocean, *J. Physical Oceanography* **97**, 5409–5422.

Send, U., and R. Käse, 1998: Processes and effects of deep convection. In *Ocean Modeling and Parameterization*, E.P. Chassignet and J. Verron (Eds.), Kluwer Academic Publishers, 191-214.

Speer, K.G. (1998) A Note on average cross-isopycnal mixing in the North Atlantic Ocean, *Deep-Sea Research I* **44**, in press.

St. Laurent, L. and R.W. Schmitt, The contribution of salt fingers to vertical mixing in the North Atlantic Tracer Release Experiment, Manuscript in preparation, 1998.

Toole, J.M., R.W. Schmitt, K.L. Polzin, and E. Kunze (1997) Near-boundary mixing above the flanks of a mid-latitude seamount, *J. Geophysical Research* **102**, 947–959.

Whitehead, J.A. and L.V. Worthington (1982) The flux and mixing rates of Antarctic Bottom Water within the North Atlantic, *J. Geophysical Research* **87**, 7903–7924.

Wijffels, S.E., J.M. Toole, H.L. Bryden, R.A. Fine, W.J. Jenkins, and J.L. Bullister (1996) The Water Masses and Circulation at 10°N, *Deep-Sea Research I* **43**, 501–544.

Winters, K.B. and E.A. D'Asaro (1996) Diascalar flux and the rate of fluid mixing, *J. Fluid Dynamics* **317**, 179–193.

Wunsch, C., D.-X. Hu, and B. Grant (1983) Mass, heat, salt, and nutrient fluxes in the South Pacific Ocean, *J. Physical Oceanography* **13**, 725–753.

Young, W.R., P.B. Rhines, and C.J.R. Garrett (1982) Shear flow dispersion, internal waves and horizontal mixing in the ocean, *J. Physical Oceanography* **12**, 515–517.

Zhang, H.-M., and L.D. Talley (1998) Heat and buoyancy budgets and mixing rates in the upper thermocline. *J. Physical Oceanography*, accepted.

Zhang, Z., R.W. Schmitt, and R.X. Huang, The relative influence of diapycnal mixing and hydrologic forcing on the stability of the thermohaline circulation, *J. Physical Oceanography*, submitted, 1998.

PARAMETERIZATION OF PROCESSES IN DEEP CONVECTION REGIMES

UWE SEND AND ROLF H. KÄSE
Institut für Meereskunde,
Kiel, Germany

1. Introduction

Convection to large depth is the generation mechanism for the deep water masses circulating in the world ocean, and the few known sites of deep convection can thus be regarded as the sources for the lower branch of the thermohaline circulation. Convection activity therefore has an essential influence on the water mass properties and their volumes, and possibly also on the forcing of the circulation. For this reason we seek to understand the processes and consequences of deep convection and to correctly represent them in numerical circulation models.

There is a wealth of literature on many aspects of deep convection, including two review papers (Killworth, 1983; Marshall and Schott, 1998), the latter of which presents the state of the art in observations, theory, and modelling. In view of this, the short overview given here will emphasize a personal perspective of what might be relevant for progress in representing and parameterizing deep convection in models.

The three known and well-studied sites of open-ocean deep convection are the northwestern Mediterranean, the Labrador Sea and the Greenland Sea. In all of them, the water column may lose sufficient buoyancy through air-sea interaction (and ice-processes in the Greenland Sea), to allow erosion of the (weak) stratification and convective homogenization to great depth. In the Labrador Sea, a cyclonic gyre is found between the West Greenland Current to the northeast and the Labrador Current to the southwest, which has been known to be the site of deep convection since the early observations by Lazier (1973) and Clarke and Gascard (1983), and time series from ocean weather ship Bravo (Lazier, 1980). Deep convection in this gyre to depths of typically 1500-2000 m generates the Labrador Sea Water, a major component of the North Atlantic Deep Water circulating globally. The

E. P. Chassignet and J. Verron (eds.), Ocean Modeling and Parameterization, 191–214.

convection activity there has shown pronounced interannual variability, in terms of depth and water mass properties (including its virtual shutdown during the "great salinity anomaly"), which is now associated with the North Atlantic Oscillation (NAO) (Dickson et al., 1996). Currently the area is again the focus of convection studies on a more massive scale, involving groups of investigators from several countries using a wide range of modern techniques (see Lab Sea Group (1998) for a summary of the experiment).

In the central Greenland Sea, we again find a cyclonic gyre which is the site of deep convection processes. Between the West Spitsbergen Current to the east, the East Greenland Current to the west and the Jan Mayen Current to the south, homogenization of the water column to depths of 1500 m has been observed in some winters (Schott et al., 1993), but not at all in other years. In the Greenland Sea, ice formation and export seem to play an important role in removing sufficient buoyancy from the water column for deep mixing (Visbeck et al., 1995).

An intensely studied site of convection is the Gulf of Lions in the northwestern Mediterranean. This area of well-known intense atmospheric forcing via the Mistral and Tramontane winds blowing from the continent shows deep mixing in most winters at a rather well-defined location, probably related to the local cyclonic circulation and the topography of the Rhone fan. The first direct observations of deep convection took place here starting with the classic "Medoc" experiments begun in 1969 (Anati and Stommel, 1970; Stommel et al., 1971). More recent experiments with modern technology date from 1987 and 1991/92 (Schott and Leaman, 1991; Send et al., 1995; Schott et al., 1996). Figure 1 shows a density section in the Gulf of Lions during active convection. A sharply localized region of deep homogenization of approximately 50 km size is seen in the middle of the section. The central CTD station actually shows a statically unstable situation, consistent with ongoing buoyancy loss during the measurements. Schott et al. (1996), in their Figure 5, demonstrate the links between various variables (wind events, heat flux, stratification of the water column, and vertical velocity in the ocean) from during the cooling period (December/January) until after the convection (March/April).

2. General Aspects of Convection

2.1. SEQUENCE OF EVENTS

Typically, three different phases are distinguished in a deep convection scenario, which are termed "preconditioning", "deep or violent mixing", and "spreading". The "preconditioning" stage refers to the period during which the surface stratification that typically remains from the summer is gradually eroded by the action of cooling and evaporation during the winter

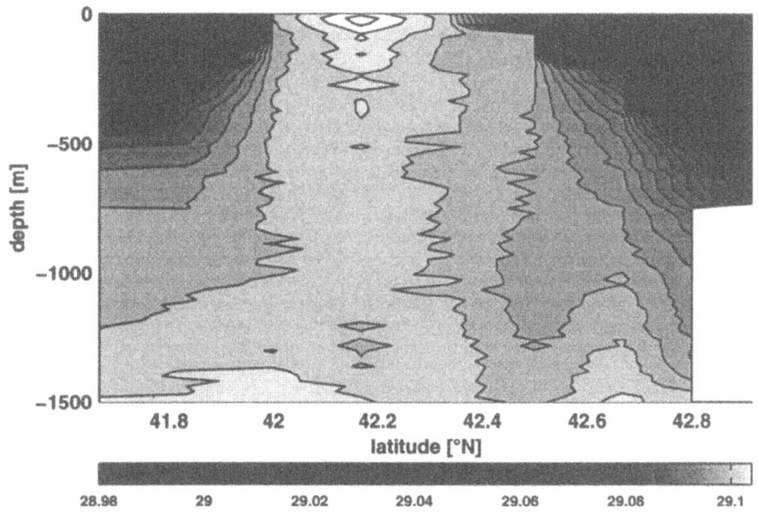

Figure 1. Potential density section in the Gulf of Lions taken during ongoing convection. Note the nonlinear gray scale, with light shades representing denser water.

months (fluxes on the order of $100-200\,W/m^2$, $0.2-1\times10^{-7}m^2/s^3$). Most of the buoyancy is contained in the upper 100-200 m, and once this layer has been homogenized in the course of the winter, little additional forcing is required to overcome the small remaining stratification and cause mixing to large depth. This is illustrated with the temperature and density profiles in Figure 2.

Frequently a single strong meteorological event is sufficient after this to trigger the "deep mixing" phase. Large heat and buoyancy fluxes (over 500 W/m^2, $0.5-3\times10^{-7}m^2/s^3$ depending on region) over only a few days are sufficient then to overturn the water column to depths of 1000-2000 m or more. The result is strikingly uniform conditions in the water mass properties over much of the water column (Figure 2).

During the "spreading" phase, the dense water which was generated during convection spreads and disperses under the action of eddy activity (this will be the topic of subsequent sections). The result of the removal of dense water from the convection site through this process is a gradual "restratification" in the area, via the horizontal advection of more buoyant water from the surroundings. This effect is also shown in Figure 2.

2.2. SPATIAL SCALES

In a schematic view of a convection regime, three scales can be distinguished, which lend themselves to different observational techniques. The

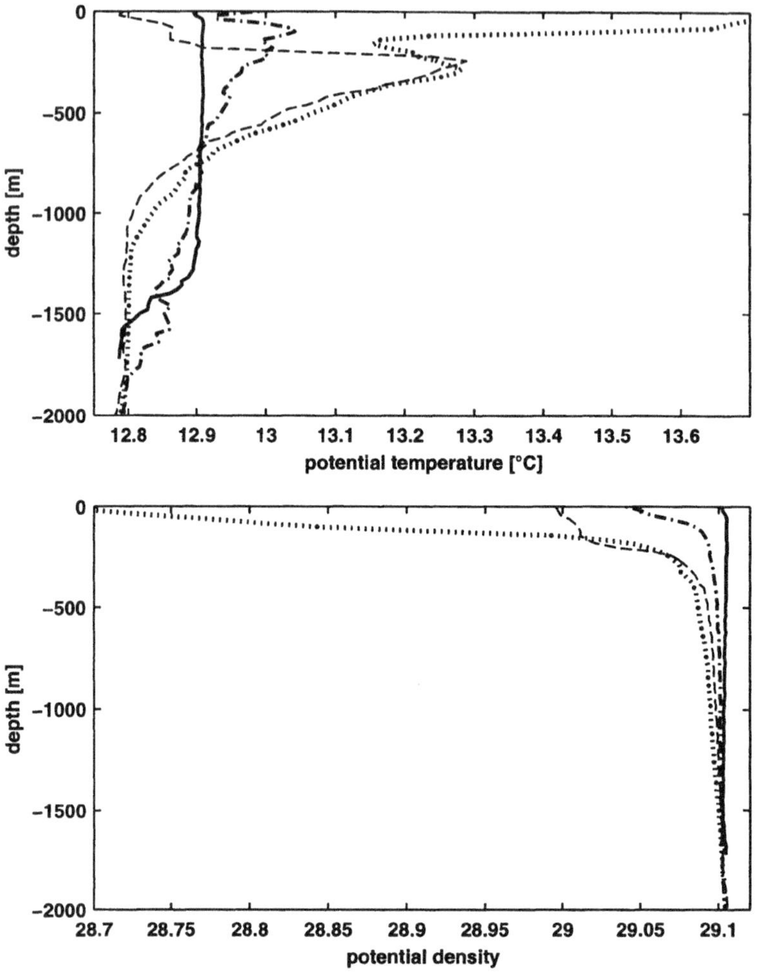

Figure 2. Typical profiles of potential temperature *(top)* and potential density *(bottom)*, showing the evolution of the stratification before and after convection, during the winter of 1991/92 in the Gulf of Lions. The dotted profile shows a warm surface layer as found in November/December. The dashed profile is typical of January, where much of the stratification has been cooled away. Below 200 m, there is little density stratification, thus after mixing to this depth it is easy to break through into the deep layers. A homogenized situation (late February) is shown in solid, and the beginning restratification (early March) in dash-dot.

largest scale is that of the homogenized region or "patch", i.e. the entire area that is subject to deep convection activity. This scale is typically 50-100 km in size, but could be as large as 200 km in some instances. An attractive method for observing the mean properties and evolution on this large scale is acoustic tomography, which senses horizontal averages over tens to hundreds of kilometers (Worcester et al., 1993; Send et al., 1995).

The next smaller scale corresponds to the scale of the eddies or instabilities (pre-existing or developing at the edge of the homogenized region) which break up the patch and disperse the water. Their scale is typically 5-10 km, and relates to the small Rossby radius under these conditions. Deep floats, especially if acoustically tracked, are ideally suited for detecting and following these eddies, as has been documented as early as 1975 (Gascard, 1978).

The smallest scale in a convection regime corresponds to the individual cells or "plumes" in which the turbulent downward motions actually take place. Observations with moored acoustic Doppler (ADCP) techniques have shown that these are characterized by large vertical motions (5-10 cm/s) vertically coherent over many hundred meters of the water column [e.g., Schott et al. (1996)]. This view is supported by the alternative perspective provided with the new technology of 3-D floats (E. D'Asaro, pers. comm.), which have performed vertical excursions of similar extent and speeds in the recent Labrador Sea experiments. The horizontal scale of plumes has been inferred to be 500-1000 m (from the assumption that the features are advected past the mooring with the observed horizontal flow velocity). The plumes are believed to be the actual "mixing agents". They can be modelled explicitly with non-hydrostatic models, as was documented by Jones and Marshall (1993), who obtained velocity and space scales for the plumes in reasonable agreement with the observations.

2.3. REQUIREMENTS FOR DEEP CONVECTION

Several ingredients seem to be necessary for the existence of a deep convection regime, based on observations and quantitative considerations. These include

- Weak stratification: The convection sites are located in oceanic regions of low vertical stratification. In all three sites, the stratification is further reduced by the upward doming of isopycnals, resulting from the cyclonic circulation in the area. This brings high densities closer to the surface, with the consequence that less buoyancy needs to be extracted for penetration to large depth.
- Strong buoyancy loss: Substantial buoyancy fluxes should be present for the winter months in order to effect the preconditioning, explained above. A few strong meteorological events may help to give the final "kick" for the deep mixing phase, even though one would expect a steady continued cooling to also do the job (note however the potentially equilibrating effect of horizontal eddy fluxes in some cases). The

buoyancy flux is calculated according to

$$B = \frac{g}{\rho}[\frac{\alpha}{c_p}Q + \beta S(E - P)] \tag{1}$$

where Q is the net heat flux and $E - P$ the evaporation minus precipitation in $kg\, s^{-1} m^{-2}$

– Residence time: In order to be subject to the high buoyancy fluxes in the region for a sufficient time, the water needs to have a residence time there which allows for the required buoyancy loss to take place. Exchanges with the surroundings due to mean flows, boundary currents, or eddy fluxes must not be too large. For example, estimation of area-averaged 1-D heat budgets in Send et al. (1995) led to exchanges with the surroundings corresponding to a residence time of order 2 months. There are indications that the residence time results from the wind-forced circulation (Madec et al., 1996).

It is worth noting that an intermediate temperature and salinity maximum does not seem to be a strict requirement for generating deep convection, even though this is found in all convection regions. In principle, the heat and buoyancy loss does increase somewhat, when a warmer intermediate water mass is mixed up, due to the resulting elevated air-sea temperature difference. Quantitatively this effect is rather small: simulations calculating the heat loss from bulk formulae for stratifications with and without an intermediate T/S maximum (but identical density profiles) changed convection depths from, say, 1400 m to 1700 m.

Of the ingredients listed above, it seems to be mainly the stratification (doming) that is responsible for the sharp localization of the deep-mixed regions. Reasonable predictions of the convection depth are therefore possible by applying typical or region-average buoyancy fluxes to the observed stratification in the area of weakest stratification. One-dimensional buoyancy budgets are sufficient in many cases, calculating the amount of buoyancy (buoyant potential energy) that needs to be extracted for mixing to depth h as

$$E_b = \int_{-h}^{0} [b(z) - b(h)]dz = \int_{0}^{t} Bdt \ , \tag{2}$$

where $b = -g/\rho_0\rho$. This quantity can be calculated and plotted as a function of depth for any given CTD profile, and the time integral of the surface buoyancy flux $\int_0^t Bdt$ then immediately gives an estimate of the mixing depth.

Evaluation of the integral in (2) for a constant $N^2 = \partial b/\partial z$ gives the well-known time-dependence for mixed-layer deepening of

$$h = \sqrt{2Bt}/N \tag{3}$$

which will be useful below for scaling relations.

3. Effect of the Plumes

3.1. MEAN VERTICAL TRANSPORT

It is not a priori clear whether there should be a net downward transport associated with the convective plume activity. If between the intense downward flow of the plumes there was completely compensating upward motion, there would be no horizontally averaged $\overline{w}$ in the interior of the convection patch. The only effect of the plumes would then be a vertical mixing. Some "slumping" of the dense water still is expected in this scenario at the edge of the region due to the convective adjustment ("cylinder collapse"), giving rise to a narrow "rim current" around the periphery of the patch (Dewar and Killworth, 1990). However, this $w < 0$ is concentrated at the edge as well, and just outside of it there is compensating $w > 0$. The other possibility is incomplete compensation of the downward motion in the plumes. The net result would be a $\overline{w} < 0$, implying a stretching ($\partial w/\partial z > 0$) in the upper water column and a squashing ($\partial w/\partial z < 0$) below. Applying a linear vorticity balance $\partial\zeta/\partial t = f\partial w/\partial z$ [the validity of which was confirmed by model results analyzed in Send and Marshall (1995)], the $\overline{w}$ would generate a mean vorticity $\overline{\zeta}$, i.e. a circulation around the convection regime. Send and Marshall (1995) show that this leads to unreasonable consequences, thus supporting the mixing view of convection ($\overline{w} = 0$, $\overline{\zeta} = 0$), also confirmed by their analyses of non-hydrostatic numerical model results. Further, modern ADCP observations of $\overline{w}$ and tomographic estimates of the mean vorticity are consistent with this.

One consequence of this result is the way one would calculate deep water mass formation rates. Since there is no mean w, the formation rate is not given by the time and area integral of the vertical velocity during the convection activity. Instead, the quantity to consider is the additional volume of dense water that is found as a result of the convection process, as compared to the typical climatological stratification. This is the shaded area sketched in Figure 3. Assuming that some fraction of this water eventually settles into the deep water pool (by processes discussed below), this amount gives the formation rate of new volume of deep water. This quantity then is available e.g. for export via the deep circulation.

3.2. SCALING AND PARAMETERIZATION

Scaling relations exist for the plume properties which are in approximate agreement with model and laboratory results and observations [see e.g. Jones and Marshall (1993), Fernando et al. (1991)]. Starting from scalings

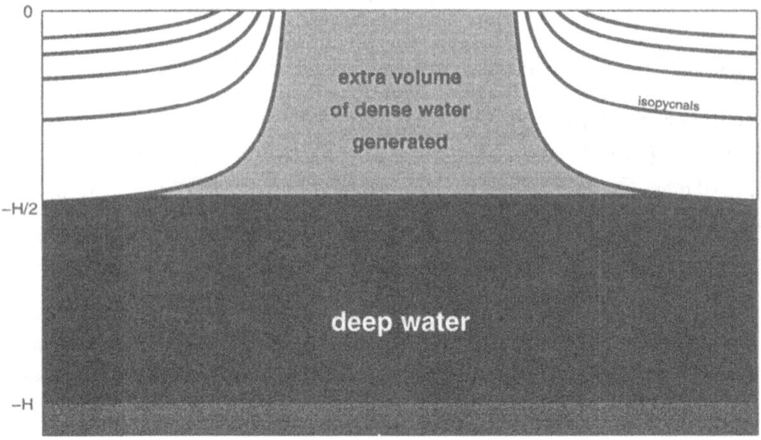

Figure 3. Cartoon of the density field after a deep convection event. The gray shaded area highlights the water which, according to its density, would be expected to settle into deep water pool (black).

for growing isotropic turbulent motions, and assuming that at $t \sim f^{-1}$ the growth of the cells is constrained by the earth's rotation, a length and velocity scale for convective plumes under rotational control can be obtained as

$$l_{rot} \sim (B_0/f^3)^{1/2}\ , \qquad u_{rot} \sim (B_0/f)^{1/2}\ , \tag{4}$$

and a measure for the importance of rotation is the 'natural Rossby number'

$$R_0^* = [B_0/(f^3H^2)]^{1/2}\ . \tag{5}$$

In these relations B_0 is the surface buoyancy flux, f the rotation rate, and H the depth of the ocean or of the convective layer. For typical oceanic conditions (H $= 2\,km$, $f = 10^{-4}s^{-1}$, $Q = 500\,W/m^2$), these scalings give $R_0^* = 0.24$, $l_{rot} = 0.47\,km$, $u_{rot} = 5\,cm/s$, i.e. rotation would start to be important already at a depth of $500\,m$.

Since the downward velocity w_{plume} of a plume is given by u_{rot} and it is the plumes which mix and carry density anomalies downward, a vertical mixing time scale can be defined as

$$T_{mix} = H/w_{plume} \sim H(f/B)^{1/2}\ . \tag{6}$$

This is the timescale for vertical mixing within the unstratified water column during ongoing convection. It is not the same as the timescale for the vertical advance of the homogenized layer, which is usually more limited by stratification (even in the deep ocean) according to Eq.(3). Nonetheless, T_{mix} should govern the maintenance of unstable density profiles and is the time scale a parameterization for convective mixing should have. If a vertical diffusivity K_v was chosen for this purpose, the mixing time scale would be obtained with a value of

$$K_v \sim H^2/(\pi T_{mix}) = \pi^{-1} H (B/f)^{1/2} \; . \tag{7}$$

This gives values of order $50\,m^2/s$ for typical conditions. Such a diffusivity should then be "switched on" in a model, when instabilities occur due to a buoyancy loss B at the surface.

Klinger et al. (1996) tested a variety of mixing schemes in direct comparison with plume-resolving model runs. They confirmed the expectation that schemes with the appropriate mixing time scale can represent the plume effects. Furthermore, even instantaneous convective adjustment gave reasonable results, at least few days after the overturning, when water masses had slumped and started to form eddies.

4. Effect of the Eddies

4.1. SCALINGS FOR THE PATCH

It is known from laboratory and numerical experiments, as well as field observations, that eddies which pre-exist or develop after the convection will break up the homogeneous patch and cause an exchange of fluid with the surroundings. If the homogenized region as sketched in Figure 3 is a result of localized cooling with buoyancy flux B, then the density anomaly measured by g' between the interior and exterior develops according to $g' = N^2 H = 2Bt/H$. The Rossby radius $L_R = \sqrt{g'H}/f$ then also is a function of time $L_R \sim \sqrt{2Bt}/f$. The horizontal temperature gradient at the surface will be in balance with a cyclonic shear, yielding a rim current around the patch which can be scaled via the thermal wind relation as $u_{rim} \sim g'/L_R \cdot h/f \sim \sqrt{2Bt}$. If this flow becomes unstable, we can estimate the length scale of the instabilities according to Eady theory as $2L_R$ at the time of break-up, which would scale like the inverse Eady growth rate. Putting all this together yields a length scale for the eddies of $L_\rho \sim H^{1/2}(B/f^3)^{1/4}$.

One should note, however, that the horizontal density gradient between the interior and exterior of a convection region exists even before convection as a result of the preconditioned stratification, i.e. the doming isopycnals (in fact one can show that the surface density gradient should decrease as a result of convection, consistent with our field data). Therefore, the above time-dependent scalings for g' and L_R should probably be replaced with their pre-existing values, the eddy length scale then becoming simply $2L_R$ based on this pre-existing stratification.

4.2. EQUILIBRIUM DUE TO EDDY FLUXES

We now assume that horizontal eddy buoyancy fluxes $\overline{v'b'}$ can offset the surface buoyancy loss and halt the convective deepening at some equilibrium depth h_{eq}, i.e.

$$\int_{-h_{eq}}^{0} \oint \overline{b'v'} dl dz = \int \int B dA , \tag{8}$$

which gives a scaling like $\overline{v'b'}h2\pi r \sim B\pi r^2$ for a convective region of radius r. Visbeck et al. (1996) have applied the theory of Green (1970) to this setup, relating $\overline{v'b'}$ to the horizontal and vertical buoyancy gradients M^2 and N^2, via $\overline{v'b'} = \alpha^* f Ri^{-1/2} M^2 l^2$, where $Ri = N^2 f^2 / M^4$ is a Richardson number and l a lateral scale. Substituting the scalings, the equilibrium depth is obtained as

$$h_{eq} = 1/N(\frac{Br}{2\alpha^*})^{1/3} . \tag{9}$$

Visbeck et al. (1996) determined the value of α^* from numerical experiments to be $\alpha^* = 0.008 \pm 0.002$. An equilibrium time scale was also derived and turned out to be in the range of 30-60 days for some typical oceanic parameters, suggesting that equilibration might be important in some oceanic cases. The solution (9) for the depth of the mixing layer here differs from that in Eq. (3) because of the assumed equilibration by horizontal fluxes.

4.3. RESTRATIFICATION DUE TO EDDY FLUXES

The ongoing effect of the horizontal eddy buoyancy fluxes is to restratify the homogenized convection region. The time it takes this process after the deep mixing to flux in the buoyancy that had been extracted during the convection therefore provides a scaling for the restratification time τ from

$$\tau \int_{-h_{eq}}^{0} \oint \overline{b'v'} dl dz \sim \int \int B dt dA \sim \pi r^2 0.5 N^2 h^2 . \tag{10}$$

Using the same expressions for $\overline{v'b'}$ as above, Jones and Marshall (1998) obtain the result

$$\tau_{restrat} = 56 r (h \Delta b)^{1/2} \tag{11}$$

where the numerical constant was determined from model experiments. In these relations, h is the depth of the stratified layer, and Δb measures the density jump across it.

Field observations with tomographic techniques in the Mediterranean yielded a restratification time scale of 40 days for the layer of the warm intermediate water (150-500 m depth). Applying the above expression for typical Mediterranean parameters, Jones and Marshall (1998) obtained a time scale of 20-30 days, not dissimilar to the observations.

5. Fate of the Convected Water

Figure 3 showed the schematic state of the stratification after completion of the deep homogenization. Clearly, this state of high available potential energy due to the dense water reaching up to the surface will not remain forever, and the question is what happens to this dense water. A number of possibilities exist for the fate of the water highlighted in Figure 3:

1. Frictional spin-down of the geostrophic gyre could lead to a lowering of the isopycnals and thus to a settling of the dense water into the deep water pool. This is unlikely however, since bottom or interfacial frictions are generally much too small.
2. Energy conversion due to baroclinic instability will lead to a lowering of the isopycnals and an associated sinking of the dense water. This process is clearly seen in all numerical model simulations of the decay of a convection region, and can be parameterized by a "transformed Eulerian mean" (TEM) velocity field which adiabatically advects the isopycnals (and tracers) (Gent and McWilliams, 1990; Visbeck et al., 1997). The net effect of the eddy fluxes on the density field thus is an inward advection of buoyant water at the surface, sinking of isopycnals in the interior, and outward advection below. Other chapters in this volume address this process in detail and therefore the focus here will be on the other mechanisms.
3. It appears that part of the dense water generated during convection may be drained out of the region in a boundary current, if the convection site is located near a side wall or continental slope. This hypothesis had been tested with a simple mooring downstream of the Gulf of Lions in 1992, and indeed a temperature "pulse" was observed at 1500 m depth some three weeks after the presumed convection activity (Send et al., 1996). The temperature anomaly, its vertical structure, and the timing all agreed with this anomaly actually corresponding to convected water, and a volume estimate indicated that a significant part of the dense water could have been drained from the convection site this way.
4. A fraction of the dense water may remain in the area in the form of a stable vortex, possibly trapped by the local circulation or topographic effects. Numerical simulations (such as those presented below) and our field observations indicate that a core of dense fluid may persist for weeks or months after break-up. Tracer observations in the Gulf of Lions by Rhein (1995) showed elevated tracer concentrations in the

center of the site before deep convection began, which was was believed to be the remnant of convection in the previous year.

5. The eddies which form from the high potential energy pool of dense water may be stable and long-lived structures. We will show below that these are expected to trap density and tracer anomalies at different depths and carry them horizontally over long distances through propagation in the open ocean or along boundaries. There is observational evidence for the existence of such "convection lenses".
6. Isopycnal and diapycnal mixing by eddy stirring and intrusive processes must be important processes in the post-convection phase. The last subsection will give some ideas from numerical simulations and suggestions from observational data concerning these effects.

The last two mechanisms listed will be explored in somewhat more detail in the remainder of the present paper. In order to gain insight into the relevant processes and to guide in the interpretation of observations, a variety of numerical model simulations were performed. A short summary of the model features is first given, followed by some analyses concerning the relevant eddy and lens properties and the mixing processes.

5.1. NUMERICAL MODEL SIMULATIONS

The main objective of the numerical simulations was to study the spreading and propagation of density and tracer anomalies as a result of the decay of a convection regime. The setup and general evolution is similar to that in Jones and Marshall (1997). Particular emphasis was on the effect of transient and coherent eddy processes on water mass property distributions and mixing. For the simulation we adopted the sigma coordinate primitive equation model SPEM, version 5.1 (Haidvogel et al., 1991), which is basically a hydrostatic approximation solver of the hydrodynamic equations. The horizontal grid spacing in the runs described below was better than 1 km, in order to make sure the eddy structures and associated streaks that wrap up water properties were well resolved. Depth-dependent high vertical resolution allowed for adequate representation of the surface-intensified restratification process. The diffusivities (vertically Laplacian, horizontally biharmonic) were made as small as possible to allow near adiabatic dynamics and realistic long-term evolution on the energetic eddy scales ($K_v = 2 \times 10^{-4} m^2 s^{-1}$, $K_h = 5 \times 10^6 m^4 s^{-1}$). The lateral extent of the simulated domains was 60 km to 100 km, resolution was always better than 1 km, and runs with 11 to 51 layers in the vertical were carried out. We used both closed and periodic domains to cover eddy-topography and eddy-eddy interaction without wall effects. Some long-time integrations included the β-effect.

The initial condition in all runs was a near-cylindrical column of homogenized dense water, embedded in a background stratification that was taken from data from the Gulf of Lions. The density profile used for the interior was similar to the mixed profile shown in Figure 2, while the exterior one resembled the strongly stratified one there with about 30% less stratification. In the model, density depends only on temperature. The dense water was then marked by a tracer with a concentration of unity in the layers above the depth at which the water of this density is normally found (in the background). This represents the excess amount of dense water generated through the convective process, and the objective was to follow its evolution. Figure 4 shows the initial condition for density and tracer. The different runs cover the adjustment to geostrophy and the baroclinic instability breakup of the cylindrical convection region as well as its subsequent spreading.

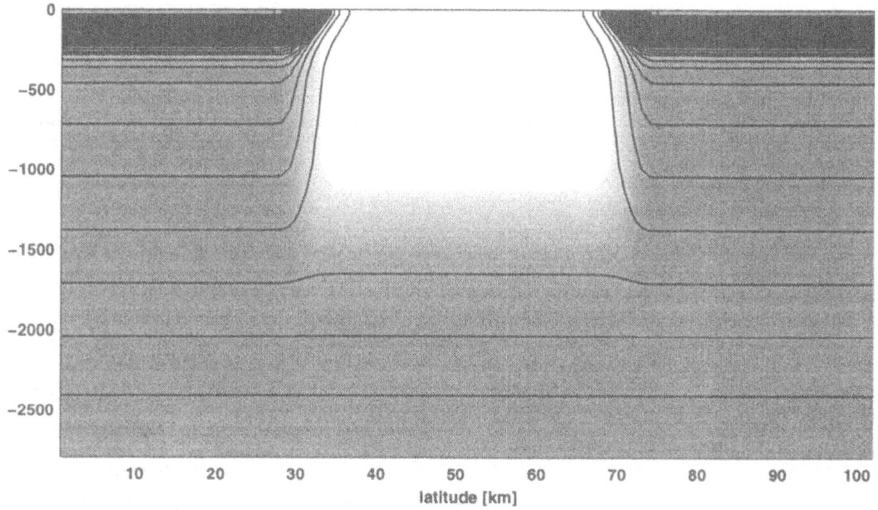

Figure 4. Initial density (lines) and tracer (shading) field for the numerical simulations presented here. White shades denote high tracer content.

5.2. WATER TRANSPORT IN COHERENT VORTICES

In the model simulations, initially a cylinder collapse geostrophic adjustment occurs (Dewar and Killworth, 1990; Send and Marshall, 1995), which corresponds to a limited slumping of the isopycnals, a surface inward displacement by a Rossby radius, and a corresponding outward movement below. Associated with this is a slight sinking of the dense tracer-marked water in the interior of the cylinder (see Figure 5 below). The cylinder collapse generates a sheared flow with cyclonic circulation at the top and anticyclonic flow at depth.

Subsequently instabilities develop with a cyclonic/anticyclonic ordering which is offset between the surface and bottom flow. At finite amplitude these form heton pairs (Hogg and Stommel, 1985; Helfrich and Send, 1988), i.e. surface cyclonic eddies locked with bottom anticyclonic ones offset sideways (and vice versa) such that a mutual advection and propagation results. The meander growth is such that, initially, the dense and tracer-rich water is contained in cyclonic eddies at the surface, while the dense water at depth is found in the anticyclonic ones.

The energy for the eddy growth is drawn from the state of high potential energy, with the net effect of lowering the isopycnals and bringing down dense, tracer-rich water. It is this sinking of "newly formed" dense water that we want to follow with the tracer. We can quantify the sinking now by plotting the fraction of total tracer content found in the upper and lower water column as a function of time (Figure 5). During the first day, about

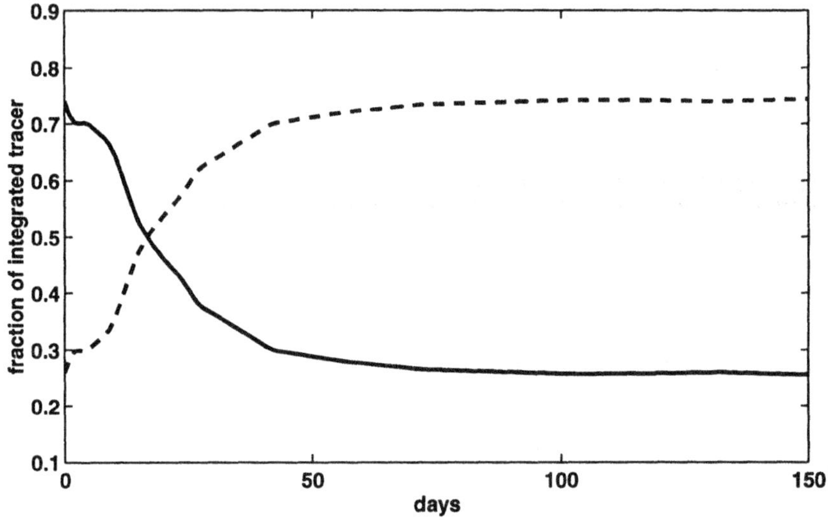

Figure 5. Fraction of total tracer amount above (solid) and below (dashed) 1000 m depth. The asymptotic values do not change after 150 days in runs as long as 500 days and are therefore not shown here.

5% of the tracer sinks due to the cylinder collapse. Then for a few days there is little activity while the instabilities develop. Starting at approximately day 5, much of the tracer sinks, as the eddy formation is in progress and extracts energy by lowering the elevated isopycnals. The timescale is 20-30 days. This is the process which is parameterized by the TEM scheme, and the sinking of dense, tracer-rich water is well described by it. An important distinction, however, is contained in the longer-term behaviour, seen in Figure 5. In the TEM scheme, all the dense (and tracer-rich) water eventually settles into its neutrally buoyant depth, i.e. a state with flat isopycnals. In contrast, the different runs we performed with a range of parameters all maintain 25-30% of tracer in the upper part of the water column as in Figure 5. This difference is further borne out by following water parcel trajectories that started out in the dense water near the surface. Figure 6 shows that, instead of settling into a layer corresponding to its density, these particles later occupy the entire water column. The typical vertical velocity for these particles during the first 10 days is $0.1\,cm/s$.

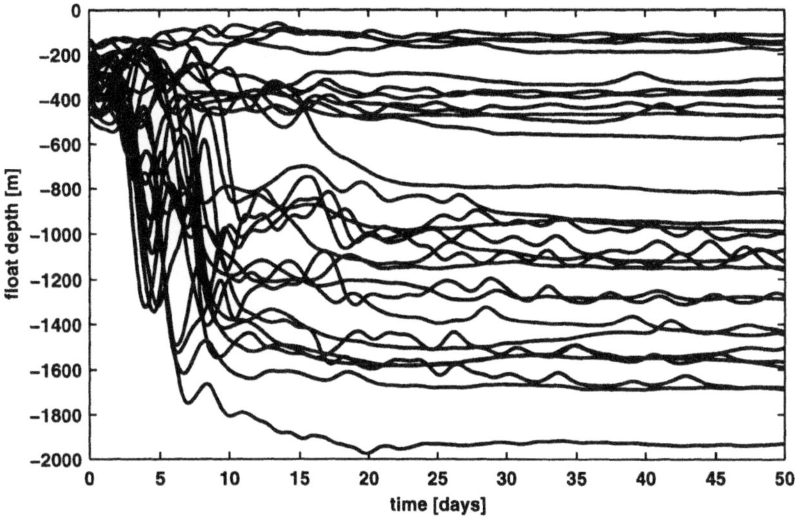

Figure 6. Depth of neutral floats released at 200-300 m depth inside the homogenized region.

How are these distributions achieved? It is instructive to investigate the kinematics and trajectories in the eddy regime. The first observation is that, due to the lateral offset of surface cyclones and deep anticyclones, there is a large barotropic component to the eddy flow, which develops from the purely baroclinic flow within the time span of the major instability (20-30 days). Thus the cyclones are surface-intensified but strongly cyclonic from top to bottom, while the anticyclones are bottom intensified with a large barotropic component. Since the vertical shear has the same sense in both cases, both types of eddies have a similar vertical density structure (a lens-like high-density anomaly), and thus the sense of an eddy cannot be recognized from CTD data alone. Trajectories show that tracer-rich water spirals down from the periphery of the cyclones into the anticyclones and accumulates in those in the lower layers. Similarly low-tracer water ascends in helical paths, and some particles can be found that cycle up and down in inclined orbits (along isopycnals). The end-state is that high tracer-content is found in the deeper layers inside the anticyclones and near the surface some tracer remains inside the cyclones, Figure 7, while the density is enhanced inside both cyclones and anticyclones. The flow field is essentially governed by 2-D turbulence dynamics (McWilliams, 1984), with equally-signed vorticity patches and streaks being merged to stabilize the coherent structures (if they have the chance to interact before leaving the region). Note that in the surface layer, capping with low-density water may mask the density and tracer signature of the eddies in vertical sections, even though in horizontal maps the anomalies are clearly visible (see later figures).

Not only the vertical distribution of parcels and tracer is different from the result of TEM type schemes. The parameterization schemes implicitly destroy the energy that was drawn from the initial state of high potential energy, while our simulations suggest that an important aspect is the non-local nature of the eddy processes and the energy still contained in them. Since the newly formed dense and tracer-rich water remains inside dynamically active eddies, the processes of eddy propagation, self-advection, interaction with boundaries or topography can significantly modify the spreading of this water in the ocean. For example, since the eddies eventually reach a highly stable size through instability/merger processes, they can live for long times and travel over larger distances before decaying and releasing the water elsewhere. At the same time, the core of tracer-rich water that remains in the convection area in some cases also is stabilized by eddy dynamics. As already pointed out by Tandon and Garrett (1996), the energy contained in the eddies is likely to be used in diapycnal mixing processes when ultimately the eddies decay through processes like internal wave breaking, interleaving, and double diffusion. This can lead to a horizontal release of the high-density water, instead of a sinking to its neutrally buoyant level.

Observational evidence for the "convection lenses" is sketchy. In the Mediterranean, cold anomalies at intermediate depths are found along the boundary downstream of the convection site after winter in historical data (Krahmann, 1997). It is not yet clear whether these also have a density anomaly, due to the sparser salinity data base. Pickart et al. (1996) have found eddies of convectively formed Labrador Sea water downstream of the formation region, with clear T/S, tracer, and density signatures. Exceptionally low temperatures are also found on moored time series in the Labrador Sea at intermediate depth occasionally, well before the deep convection phase, which suggests the presence of eddies of convectively generated water from the previous year (Lilly, 1998).

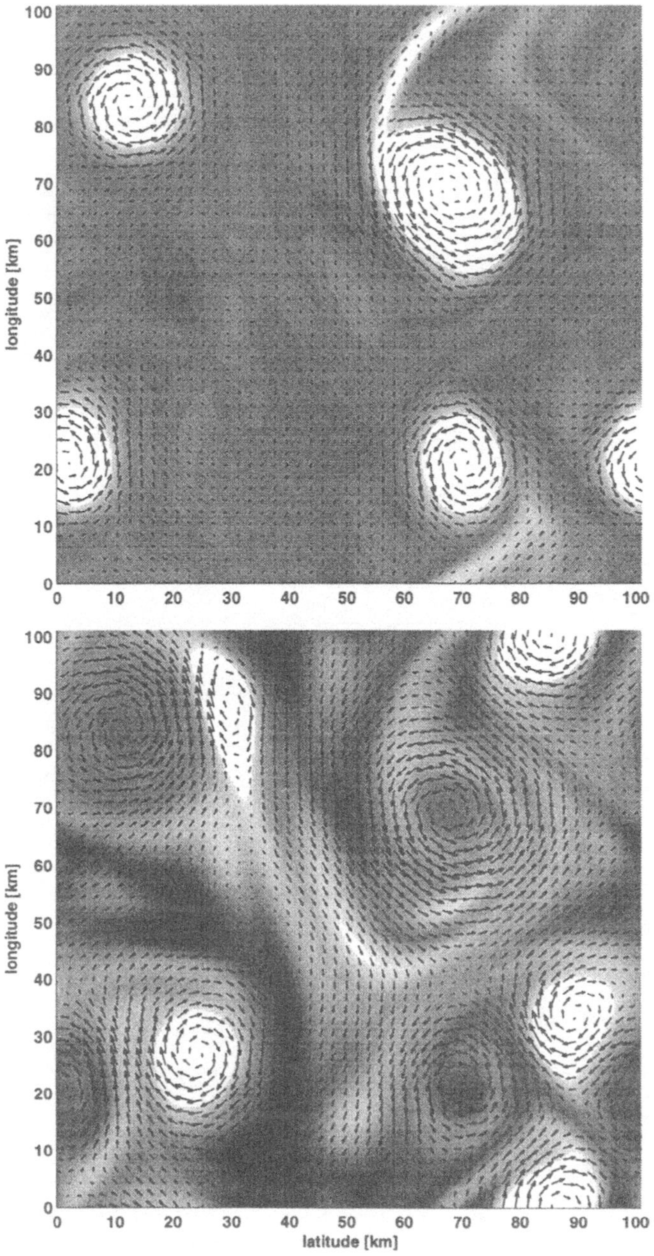

Figure 7. Horizontal flow field and tracer at the surface *(top)* and at 1560 *m* *(bottom)* showing the tracer content of cyclones and anticyclones. As before, white shades represent high tracer content.

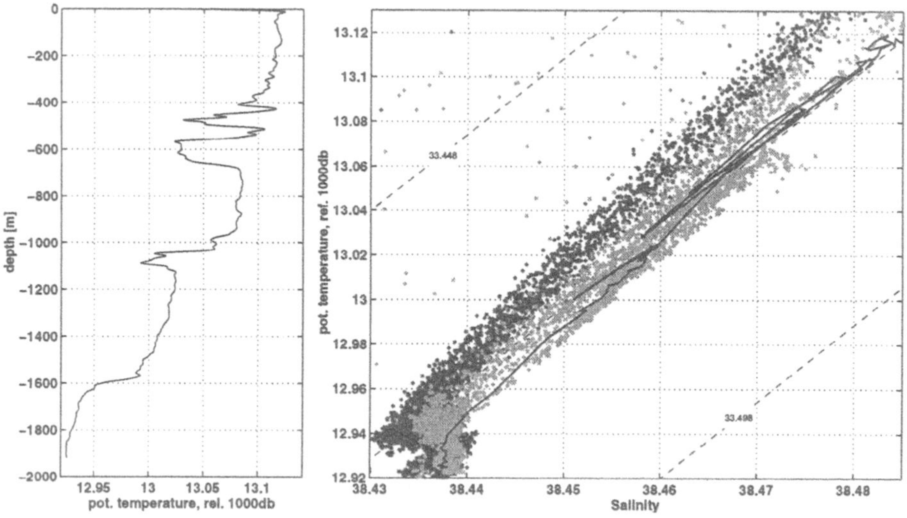

Figure 8. Isopycnal intrusions in a potential temperature profile and the corresponding θ-S diagram from the winter 1991/92 in the Gulf of Lions. Temperatures and densities are referenced to a pressure of 1000 *db*. The light dots in the diagram are data from all available profiles showing convective layers, while the dark dots show the normal θ-S relation from before convection. Later in the season, the observed data lie between the light and dark dots, the progression apparently being due to the isopycnal intrusive mixing.

5.3. MIXING AFTER CONVECTION

5.3.1. *Isopycnal*

Observations in the Mediterranean and the Labrador Sea show the appearance of strong intrusive activity during and after the convection phase. For the Mediterranean data, comparing various pre- and post-convection cruises, this can be demonstrated objectively for example by the mean square curvature in the CTD spiciness profiles. Figure 8 gives an example of a temperature profile from the Mediterranean.

The associated T-S diagram shows pronounced along-isopycnal zig-zagging, documenting the intense exchanges of different water types along isopycnals. This enhances the spiciness gradients in the water column which makes isopycnal mixing more efficient. The evolution in the T-S diagram seems to happen via this pathway. Initially, the profiles in the convection region show anomalously high densities over much of the water column (light dots in Figure 8). Over the weeks and months that follow the convection activity, the original lower densities are gradually re-established, but this seems to happen through the long intrusive excursions along isopycnals. Mixing along these excursions in the T-S diagram can generate the water properties found two months after convection, which are found to be a mixture of the convective properties and the pre-convection ones.

It is likely that the intrusions are a combination of double-diffusive interleaving (see Schmitt (1998), this volume), accounting at least for the smaller vertical scales, and of vertical helical advection in the eddy flow field. The latter mechanism is illustrated by our numerical simulations described above, which show particles moving in inclined vertical planes over hundreds of meters upwards and downwards. Extreme intrusions in CTD profiles support this mechanism. An intrusion of water not affected by convection (based on its T-S properties), which appears at shallow depths in a high-density convection profile, must originate from much larger depths. One such case is presented in the CTD section in Figure 9 taken one week after convection. The cold anomaly at 400-600 *m* must have come from at least 1100 *m* depth. The large vertical excursions and rearrangements of water on isopycnal surfaces due to the 3-D eddy flow field also enhance horizontal spiciness gradients. We speculate that these are the cause of the typical pronounced double-diffusive interleaving observed after convection.

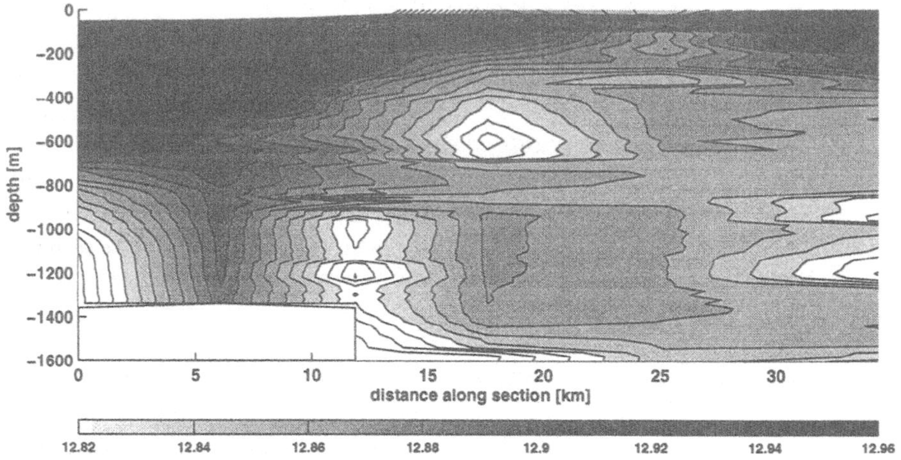

Figure 9. Potential temperature section in the Gulf of Lions one week after deep convection. Consistent with previous diagrams, light shades represent low temperatures. Pronounced intrusive activity is apparent in this section, and based on its water mass properties, the dense anomaly at 18 *km* must originate at least 500-600 *m* below.

5.3.2. *Diapycnal*

As mentioned in the previous section, diapycnal processes take on a special role in the post-convective regime, since they govern to what extent the convectively generated water retains its properties (density, salinity, tracer). The ultimate density in turn sets the level at which the (mixing) products of convection will be found in the water masses and the general circulation. In the model simulations, significant diapycnal mixing takes place, as can be seen from the density changes along particle trajectories, from dilution of the tracer concentrations found in the cyclones near the surface, and the fact that clear tracer and density anomalies are found at the surface even though the area has been capped by a low-density layer. This efficient diapycnal mixing is a consequence of the pronounced stirring by the eddy field, resulting in streaks and streamers of density and tracer both in surface and deep layers (Figure 10). This then leads to diapycnal mixing either by convective adjustment in unstable layers generated by the sheared flow or by numerical diffusion once the gradients have reached the grid scale. As a result, high-density and tracer-rich water is mixed into the upper layers - water that would otherwise be lost through sinking. At the same time, the near-surface density and tracer anomaly of the cyclonic eddies remains clearly visible in the long-term end state (e.g., Figure 7) but with diluted values that can only result from mixing.

There is some observational evidence that the streaking indeed has a diapycnal component, i.e. that in nature the streaks do not immediately adjust by a stable adiabatic re-layering in the vertical. Thermosalinograph observations of the surface temperature and salinity fields along the ship track in the Mediterranean convection regime can be compared during active convection and in the following weeks. Initially, the density is essentially bi-modal with remarkably homogeneous values in convecting areas near 29.1 or with much lower values in non-convecting regions (28.9 or less) (Figure 11a). One week later, there are many sharp alternations between high and low density water, as one would expect from cutting through a streaky field (Figure 11b). Another week later, the small-scale streaking appears qualitatively weaker, but larger isolated regions are found with elevated density values that lie between the original convecting and non-convecting ones (Figure 11c).

6. Conclusions

Some aspects of open-ocean deep convection are accessible to parameterization in models at this point. These include the ensemble effect of the plume vertical motions as a mixing process. Also the baroclinic instability dynamics and associated eddy fluxes require a parameterization in low-

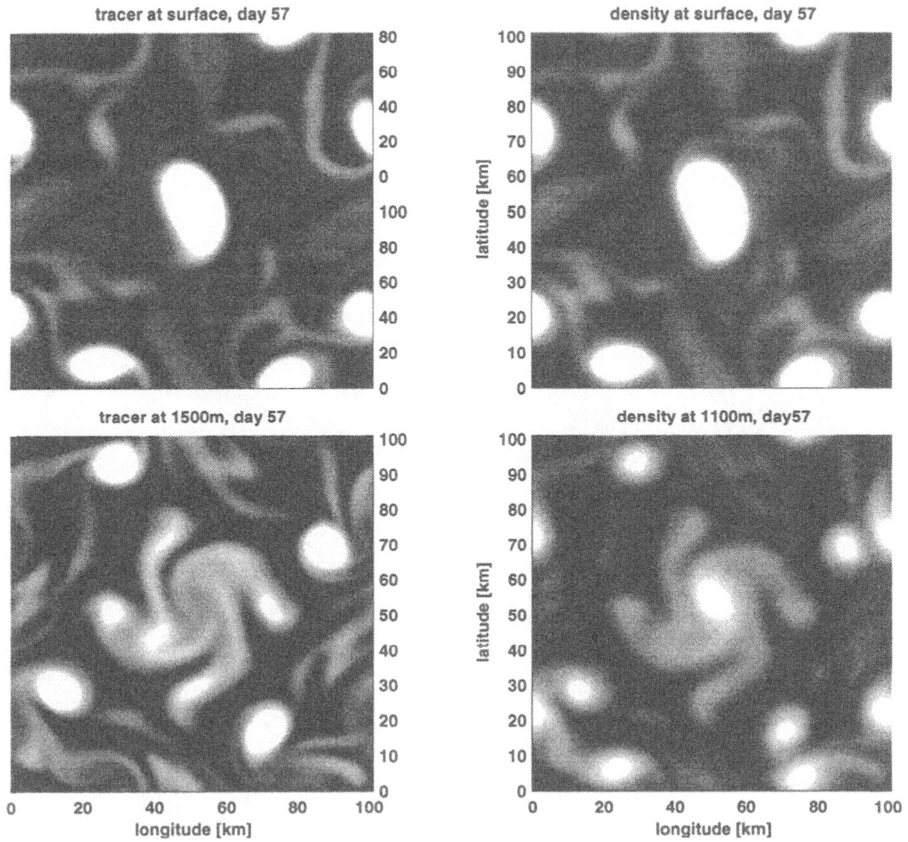

Figure 10. Surface and deep tracer and density field, showing streaking activity and mixing. Light shades correspond to high tracer content and high densities, but the scale is different in each panel. At the surface, tracer and density are strongly correlated (maximum inside cyclones), while at depth density is high in both cyclones and anticyclones.

resolution models. Gent and McWilliams (1990) suggest that its effect on the large scale circulation can be regarded as an equivalent thickness diffusion. A "bolus-velocity" (or TEM) parameterization rotates sloping isopycnals back into the horizontal plane. This scheme effectively extracts (all) available potential energy locally (the energy has vanished then) and thus allows all the dense and tracer-rich water to settle into the deep water. In contrast, we have shown that the kinetic energy produced results in both cyclonic and anticyclonic eddy motions (with large barotropic components), which are responsible for a variety of isopycnal, diapycnal, spreading, and propagation processes. These may be important for representing the effect of convection in models. One consequence is that not all the dense wa-

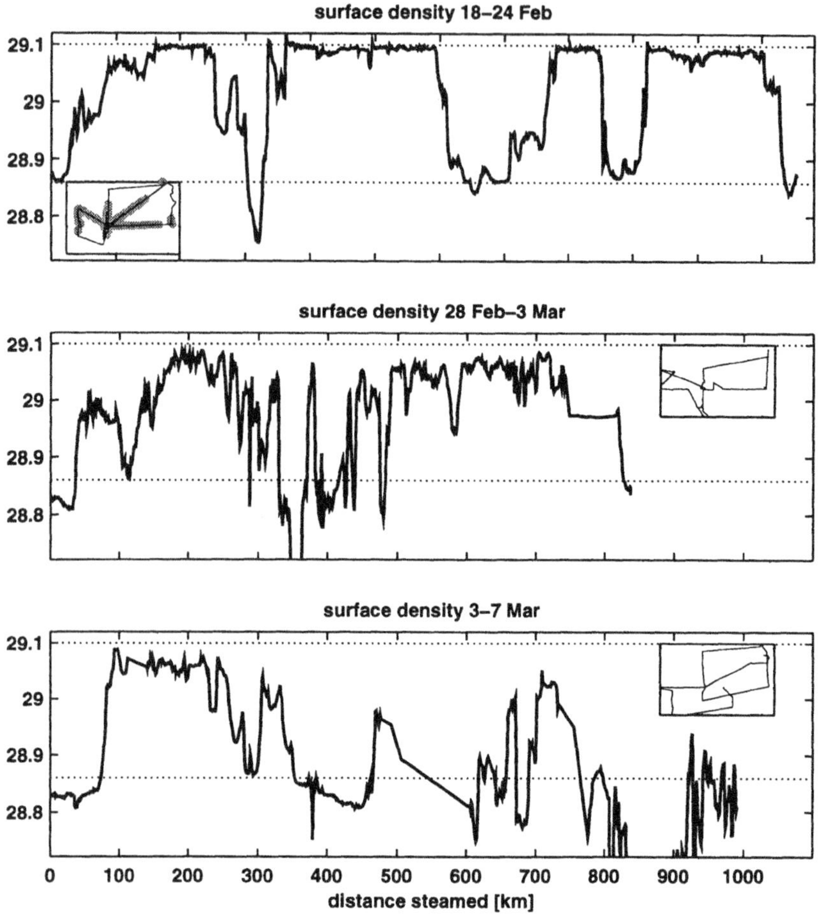

Figure 11. Surface density along the ship track through the convection region in the Gulf of Lions, during convection (top), and one (middle) and two (bottom) weeks after convection. The cruise track for each leg is sketched in the inserts, the first of which has areas of high density marked via shading. The typical densities found inside and outside the convecting region during the first leg are marked in all panels with dotted lines.

ter may sink to its neutrally buoyant levels. Instead, water mass, tracer, and density anomalies resulting from the convection process can remain as long-lived features and may be expected to propagate into the interior of the basins or along boundaries. These, together with the diapycnal mixing effect cited, may have profound effects on water mass properties (spiciness) and tracers. The processes are not understood well enough yet and dedicated field investigations might help to shed more light on them. It should also be noted that the ideas presented here are expected to apply not only to the isolated sites of deep convection, but might have similar relevance to subtropical and subpolar areas of mode water formation.

Acknowledgements

U.S. gratefully acknowledges the opportunity to attend the stimulating environment of the ASI and the support for this. We thank the two reviewers, as well as the ASI students, for their constructive comments.

References

Anati, D. and H. Stommel (1970). The initial phase of deep-water formation in the northwestern mediterranean, during medoc'69. *Cah. Oceanogr. 22*, 347–351.

Clarke, R. A. and J. C. Gascard (1983). The formation of Labrador Sea Water. Part i: Large-scale processes. *J. Phys. Oceanogr. 13*, 1764–1778.

Dewar, W. K. and P. D. Killworth (1990). On the cylinder collapse problem, mixing, and the merger of isolated eddies. *J. Phys. Oceanogr. 20*(10), 1563–1575.

Dickson, R., J. Lazier, P. Rhines, and J. Swift (1996). Long-term coordinated changes in the convective activity of the North Atlantic. *Prog. Oceanogr. 38*, 241–295.

Fernando, H. J. S., D. L. Boyer, and R. Chen (1991). Effects of rotation on convective turbulence. *J. Fluid Mech. 228*, 513–547.

Gascard, J. C. (1978). Mediterranean deep water formation, baroclinic instability and oceanic eddies. *Oceanol. Acta 1*, 315–330.

Gent, P. and J. McWilliams (1990). Isopycnal mixing in ocean circulation models. *J. Phys. Oceanogr. 20*, 150–155.

Green, J. A. (1970). Transfer properties of the large-scale eddies and the general circulation of the atmosphere. *Quat. J. Roy. Met. Soc. 96*, 157–185.

Haidvogel, D. B., L. Wilkin, and R. Young (1991). A semi-spectral primitive equation ocean circulation model using vertical sigma and orthogonal curvilinear horizontal coordinates. *J. Comput. Phys. 94*, 151–185.

Helfrich, K. R. and U. Send (1988). Finite-amplitude evolution of two-layer geostrophic vortices. *J. Fluid Mech. 197*, 331–348.

Hogg, N. G. and H. M. Stommel (1985). The heton, an elementary interaction between discrete baroclinic geostrophic vortices, and its implications concerning eddy heat-flow. *Proc. Roy. Soc. London A397*, 1–20.

Jones, H. and J. Marshall (1993). Convection with rotation in a neutral ocean; a study of open ocean deep convection. *J. Phys. Oceanogr. 23*, 1009–1039.

Jones, H. and J. Marshall (1998). Restratification after deep convection. *J. Phys. Oceanogr.*. in press.

Killworth, P. D. (1983). Deep convection in the world ocean. *Rev. Geophys. Space Phys. 21*(1), 1–26.

Klinger, B. A., J. Marshall, and U. Send (1996). Representation of convective plumes by vertical adjustment. *J. Geophys. Res. 101*, 18175–18182.

Krahmann, G. (1997). *Saisonale und zwischenjährliche Variabilität im westlichen Mittelmeer - Analyse historischer Daten.* Ph. D. thesis, Institut für Meereskunde, Kiel.

Lab Sea Group, T. (1998). The Labrador Sea deep convection experiment. *Bull. Amer. Met. Soc.*. in print.

Lazier, J. R. (1973). The renewal of Labrador Sea water. *Deep-Sea Res. 20*, 341–353.

Lazier, J. R. (1980). Oceanographic conditions at ocean weather ship "Bravo", 1964-1974. *Atmosph.-Ocean 18*(3), 227–238.

Lilly, J. M. (1998). The annual cycle of water properties in the Labrador Sea. In *1998 Ocean Sciences Meeting.* American Geophysical Union. Supplement to EOS Vol.79 (1).

Madec, G., F. Lott, P. Delecluse, and M. Crepon (1996). Large-scale preconditioning of deep-water formation in the Northwestern Mediterranean Sea. *J. Phys. Oceanogr. 26*, 1393–1408.

Marshall, J. and F. Schott (1998). Open-ocean convection: observations, theory and models. *Rev. Geophys.*. in print.

McWilliams, J. C. (1984). The emergence of isolated vortices in turbulent flow. *J. Fluid Mech. 164*, 21–43.

Pickart, R. S., W. M. Smethie, J. R. N. Lazier, E. P. Jones, and W. J. Jenkins (1996). Eddies of newly formed upper labrador sea water. *J. Geophys. Res. 101*(C90), 20711–20726.

Rhein, M. (1995). Deep water formation in the western mediterranean. *J. Geophys. Res. 100*, 6943–6959.

Schmitt, R. (1998). Double-diffusive convection: Its role in ocean mixing and parameterization schemes for large scale modeling. In *Ocean Modeling and Parameterization.* Kluwer Academic Publishers. E. P. Chassignet and J. Verron, Eds., 215–234.

Schott, F. and K. D. Leaman (1991). Observations with moored acoustic doppler current profilers in the convection regime in the Golfe du Lion. *J. Phys. Oceanogr. 21*, 558–574.

Schott, F., M. Visbeck, and J. Fischer (1993). Observations of vertical currents and convection in the central Greenland Sea during the winter of 1988/89. *J. Geophys. Res. 98*, 14401–14421.

Schott, F., M. Visbeck, U.Send, J. Fischer, L. Stramma, and Y. Desaubies (1996). Observations of deep convection in the gulf of lions, northern mediterranean, during the winter of 1991/92. *J. Phys. Oceanogr. 26*, 505–524.

Send, U., J. Font, and C. Mertens (1996). Recent observation indicates convection's role in deep circulation. *EOS 77*(7), 61–65.

Send, U. and J. Marshall (1995). Integral effects of deep convection. *J. Phys. Oceanogr. 25*, 855–872.

Send, U., F. Schott, F. Gaillard, and Y. Desaubies (1995). Observation of a deep convection regime with acoustic tomography. *J. Geophys. Res. 100*, 6927–6941.

Stommel, H., A. Voorhis, and D. Webb (1971). Submarine clouds in the deep ocean. *Am. Scientist 59*, 717–723.

Tandon, A. and C. Garrett (1996). On a recent parameterization of mesoscale eddies. *J. Phys. Oceanogr. 26*, 406–411.

Visbeck, M., J. Fischer, and F. Schott (1995). Preconditioning the Greenland Sea for deep convection: Ice formation and ice drift. *J. Geophys. Res. 100*, 18489–18502.

Visbeck, M., J. Marshall, T. Haine, and M. Spall (1997). Specification of eddy transfer coefficients in coarse-resolution ocean circulation models. *J. Phys. Oceanogr. 27*, 381–402.

Visbeck, M., J. Marshall, and H. Jones (1996). Dnamics of isolated convective regions in the ocean. *J. Phys. Oceanogr. 26*, 1721–1734.

Worcester, P. F., J. F. Lynch, W. M. L. Morawitz, R. Pawlowicz, P.J.Sutton, B. D.Cornuelle, O. M.Johannessen, W. H. Munk, W. B. Owens, R. Shuchman, and R. C. Spindel (1993). Evolution of the large-scale temperature field in the greenland sea during 1988-89 from tomographic measurements. *Geophys. Res. Letters 20*, 2211–2214.

DOUBLE-DIFFUSIVE CONVECTION

Its role in ocean mixing and parameterization schemes for large scale modeling

RAYMOND W. SCHMITT
Woods Hole Oceanographic Institution,
Woods Hole, MA, U.S.A.

1. Introduction

One of the distinguishing features of mixing in the ocean is that it is quite commonly unstable to double-diffusive instabilities. These arise because of the two orders of magnitude difference between the thermal conductivity of water and the diffusivity of dissolved salts, and the tendency of gradients of temperature and salinity to oppose one another in their effects on density. This leads to a number of counter-intuitive effects, such as up-gradient density fluxes, especially when heat and salt contrasts make nearly equal contributions to the vertical buoyancy structure. This situation contrasts with atmospheric mixing, where the two buoyancy-affecting elements of heat and humidity have nearly the same diffusivities.

While the basic double-diffusive phenomena in oceanography have been known for nearly four decades, they have received relatively little attention from large-scale modelers, perhaps because of a perception that turbulence would overwhelm the slower diffusive effects. It seems timely to correct this misconception because:

1. We now know that turbulence is actually extremely weak in most of the interior ocean;
2. Evidence has accumulated for the widespread presence of double diffusion throughout the oceans; and
3. Recent model results indicate significant effects of double-diffusive processes on water mass structure and the strength of the thermohaline circulation.

The double-diffusive processes are divided into three categories for the bulk of this paper: salt fingers, diffusive convection, and lateral intrusions. These are addressed in Sections 2-4, and Section 5 concludes with a discussion of the parameterization problem for general circulation models. The

E. P. Chassignet and J. Verron (eds.), Ocean Modeling and Parameterization, 215–234.

rest of the introduction gives an account of the basic physics. A recent review of the field has been provided in [1].

In most of the subtropical ocean, evaporation exceeds precipitation at the same time that heating exceeds cooling. This leads to a warm, salty water mass at the surface. It overlies cooler, fresher water from higher latitudes. The contributions of heat and salt to the vertical buoyancy gradient are often quite comparable, with the temperature overcompensating the unstable salt gradient by a factor of only two or less. If the temperature contrast could be removed, there would be a large-scale overturning of the water column, releasing the very substantial energy available in the salt distribution. However, this does not happen except on a small scale, where the greater diffusivity of heat can establish thermal equilibrium in adjacent water parcels that still have strong salt contrasts. The salt finger instability appears as a close packed array of up- and downward flowing convection cells which exchange heat laterally but diffuse little salt [2]. The result is an advective transport of salt and, to a lesser extent, heat, in the vertical. Typical cell widths in the ocean are 2-3 centimeters. The salt finger instability is "direct", in the sense that initial displacements are accelerated, and is accurately modeled with an exponential growth rate. When most intense, the fingers tend to exist on high gradient interfaces separating weak gradient or well mixed layers in the adjacent fluid.

In much of the polar oceans, the stratification is the reverse of that just described, with cold fresh water (arising from excess precipitation and ice melting) overlying warmer, saltier fluid from lower latitudes. Here it is the distribution of heat that drives instability. Since heat is also the faster diffusing component, the initial perturbation appears as an "overstability", where oscillatory modes have positive growth rate [2]. That is, parcels displaced upward from equilibrium experience loss of heat but not salt, and therefore become more dense. This accelerates them back toward their equilibrium positions with more momentum than when they left, so they overshoot in a growing oscillation [3]. This quickly becomes unstable, and a system of sharp interfaces separated by convecting layers appears. Heat diffuses rapidly across the interfaces, but not much salt, leading to an upward heat transport. The system is similar to Rayleigh-Benard convection, with the salt acting as a slight drag on the convection, though the free-boundary conditions may allow fluxes in some situations to exceed those in the analogous salt-free convection.

These two double-diffusive processes both cause redistributions of mass that tend to increase the vertical buoyancy gradient of the ocean, rather than decrease it as ordinary turbulence does. If there exist horizontal gradients in temperature and salinity, then it is easy for the double-diffusive mixing to set up horizontal pressure gradients that drive complex interleaving on finescales (5-100 m in the vertical) [4-7]. Because the finescale vertical shear acting on the isopycnal gradients of T and S is very effective at producing strong vertical gradients that favor the finger and diffusive instabilities mentioned above, double-diffusive lateral intrusions are self-driven. Amplitudes are quite large in some places [8], and appear to be an important mode of lateral mixing of the water mass anomalies due to eddies [9]. However, parameterization of their effects on larger scales is more challenging than the purely vertical processes of fingering and diffusive convection discussed above. In the following we review the evidence for each of these phenomena and conclude with a discussion of how they may be parameterized in models.

2. Salt Fingers

As one would expect, the propensity toward salt fingering is a strong function of the strength of the vertical salinity gradient. The instability can grow at extremely weak values of the salinity gradient, because the diffusivity of salt is two orders of magnitude less than the thermal conductivity. When expressed in terms of the effects on density, all that is required is a top-heavy density gradient due to salt that is only about one one-hundredth of the gradient due to temperature. That is, the density ratio, R_ρ, must be less than the diffusivity ratio:

$$R_\rho = \frac{\alpha\overline{\theta_z}}{\beta\overline{S_z}} < \frac{\kappa_T}{\kappa_S} \approx 100, \qquad (1)$$

where $\alpha = -\frac{1}{\rho}\frac{\partial\rho}{\partial\theta}$, $\beta = \frac{1}{\rho}\frac{\partial\rho}{\partial S}$, are the thermal expansion and haline contraction coefficients and κ_S, κ_T are the molecular diffusivity for salt and the thermal conductivity. This criterion is met over vast regions of the tropical and subtropical thermocline. However, while the required salt gradient is very small, the growth rate of salt fingers does not become "large" until R_ρ approaches one. The local buoyancy frequency, N, provides a convenient scaling. We find that, for finger perturbations with exponential growth in a region of constant gradients, the exponential growth rate λ can be expressed as [1,10]:

$$\frac{\lambda}{N} = \frac{G(R_\rho, \gamma, K)}{(1 - R_\rho^{-1})^{1/2}}, \qquad (2)$$

where G is the nondimensional growth rate, which is a function of R_ρ, the heat/salt flux ratio

$$\gamma = \frac{\overline{\alpha\theta' w'}}{\overline{\beta S' w'}}, \qquad (3)$$

and/or the horizontal wavenumber, K. The wavenumber of the fingers is closely related to the flux ratio, since very narrow fingers lose nearly all their thermal contrast and thus can advect little heat, and have low flux ratio, while wide fingers retain most of the temperature contrast, advect more heat, and have a flux ratio near one. An intermediate wavenumber is found to have the fastest growth; it has a flux ratio of 0.6-0.7. This indicates that most of the energy extracted from the salt field is used to mix the temperature field, and rather little is dissipated. The relation (2) above is plotted in Figure 1 for the fastest growing finger in the heat-salt system. For $R_\rho = 2$, a very common value in the subtropical thermocline, the fingers take about one buoyancy period to e-fold in amplitude, and fluxes would e-fold in half this time.

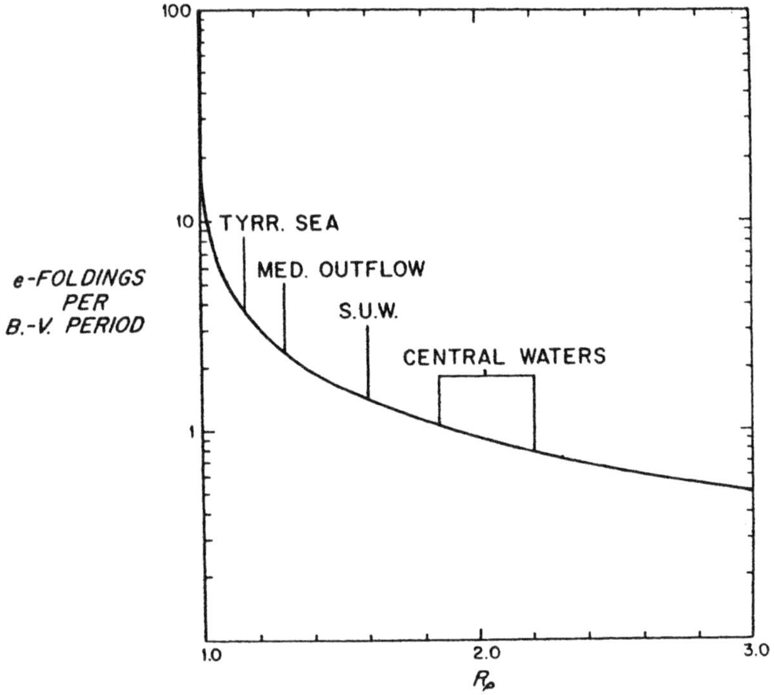

Figure 1. The exponential growth rate of salt fingers as a function of density ratio (R_ρ). The growth rate has been scaled with the buoyancy frequency, N, and multiplied by 2π for expression in "e-foldings per buoyancy period". Most of the main thermocline of the subtropical gyres has $R_\rho \sim 2$, while the regions with strong salt finger staircases have R_ρ closer to one.

This strong dependence of finger intensity on the density ratio is thought to explain one of the primary observable features of double-diffusive fine-structure. That is, when fingers become active they transport more salt than heat in the vertical and have a net up-gradient buoyancy flux. Since the growth rate and fluxes increase with the strength of the stratification, high gradient regions will harbor greater fluxes than adjacent weak gradient intervals. This leads to a flux convergence that can cause the weaker gradient region to overturn and mix. The resulting structure has thin interfaces separating thicker, well mixed, layers. The layers are continuously mixed by the downward salt flux, and the convective turbulence of the layers serves to keep the interface thin and limits the length of the fingers. Observations of the "thermohaline staircase" have been reported from many sites with strong salinity gradients. A necessary condition for an organized salt finger staircase seems to be that the density ratio is less than 1.7 (Figure 2). Such conditions are found occasionally near the surface, where evaporation produces the unstable salinity gradient [11], but more often at depth, where the presence of isopycnal gradients of temperature and salinity can lead to a minimum in R_ρ, provided there is a component of differential advection (shear) parallel to the isopycnal gradients of T and S. Examples are the Mediterranean Outflow in the eastern Atlantic [12, 13], within the Mediterranean [14] and Tyrrhenian Seas [15], and the western tropical Atlantic [16, 17].

However, while staircases are prominent and apparently permanent features of certain regions, they are not that common in the subtropical gyres, even though the density ratio is only marginally higher. The distribution of density ratio in the thermocline has a remarkable tendency to be just below 2 in many subtropical gyres. Figure 3 displays a histogram of R_ρ for the upper kilometer of the Atlantic at 24^oN. Here we find that 96% of this portion of the water column is salt finger-favorable with a smaller percentage very strongly so. The mode of $R_\rho \sim 1.9$ is a striking feature which begs for explanation. Schmitt [18] suggested that if fingering becomes more intense as R_ρ drops below 1.9 (due to staircase formation), then there will be a tendency to remove the low R_ρ sites due to the vertical flux convergences and the fact that salt is transported at a greater rate than heat.

In 1985, a detailed examination of one particular staircase system was made in the C-SALT (Caribbean Sheets And Layers Transects) program [16, 19]. Over a large area in the western tropical North Atlantic ($\sim$ 1 million square kilometers), a sequence of $\sim$ 10 mixed layers, 5-40 m thick, can be observed. Data over several decades indicates that the layers are a permanent feature of the region, despite layer splitting and merging. One of the most remarkable features was the observed change in layer properties across the region. Layers got colder, fresher and lighter from north to south (Figure 4). This can only be due to the flux convergence of salt fingers, which transport more salt than heat; turbulence transports the two components equally and isopycnal mixing, by definition, transports them in density-compensating amounts [20]. The specific value of the lateral

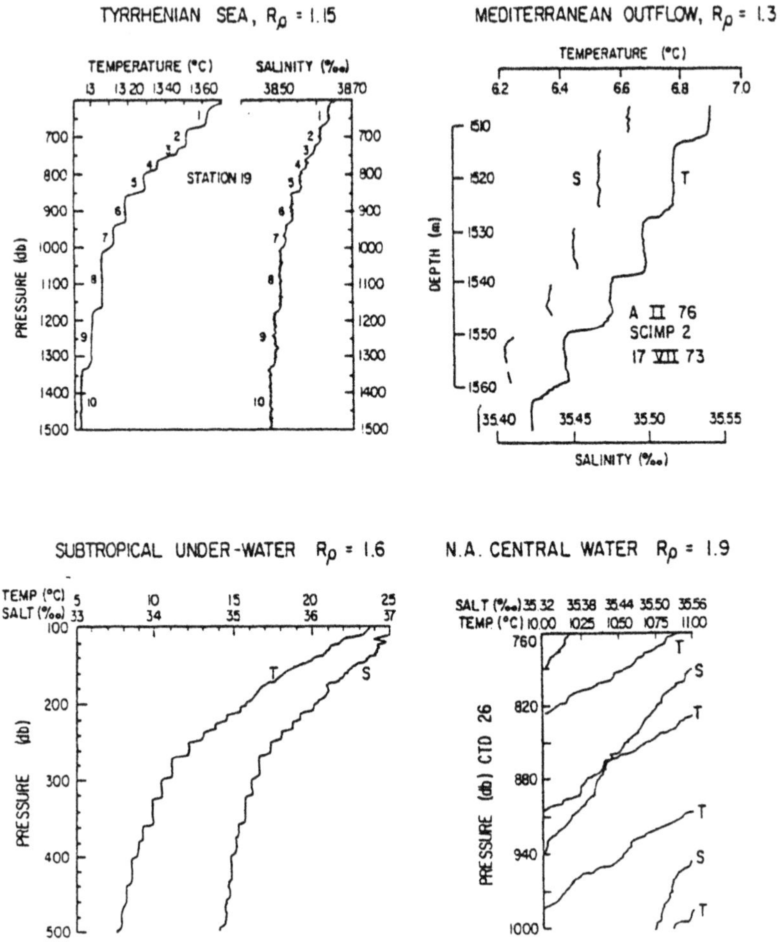

Figure 2. The occurrence of thermohaline staircases as a function of density ratio. All of the strong staircases have $R_\rho < 1.7$. In contrast, irregular steppiness characterizes the Central Waters of the subtropical gyres, where $R_\rho \sim 2$.

density ratio of ~ 0.85 is slightly above that expected by salt fingers alone (0.7-0.8). This elevation is readily understood as the effect of nonlinearities in the equation of state on the vertical flux convergence [21]. But it is also consistent with an interpretation of the staircases as fundamentally intrusive structures [22].

Microstructure measurements taken in the staircase revealed limited amplitude, narrow band temperature structure within the interfaces [23,

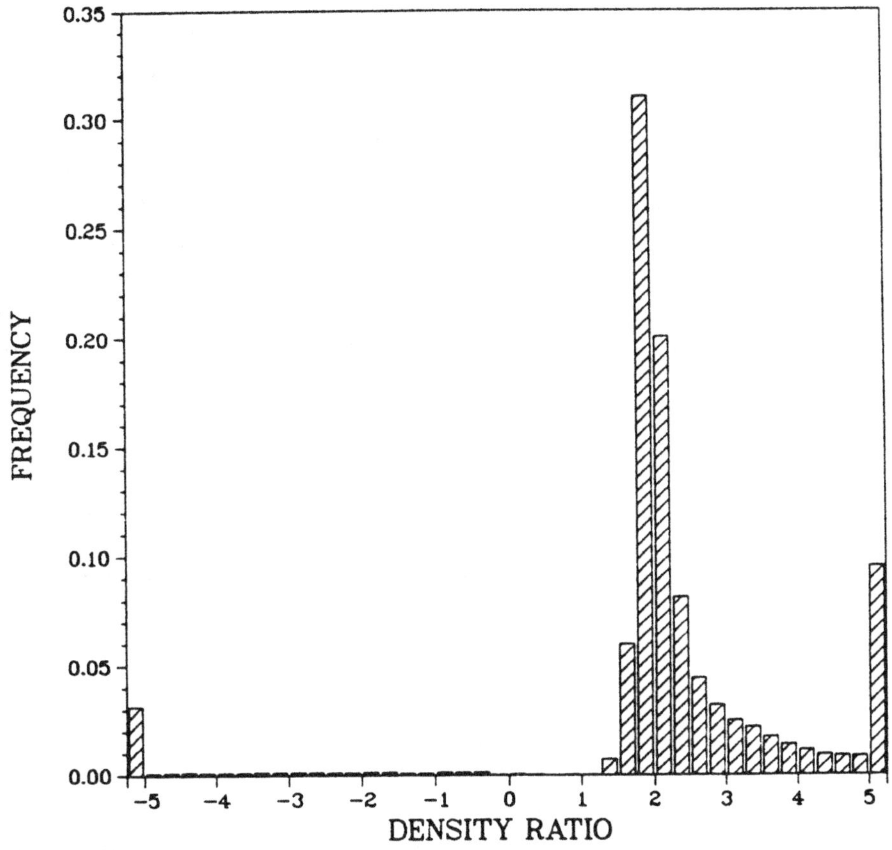

Figure 3. Histogram showing the frequency of occurrence of density ratio in the upper 1000 m along $24^o N$ in the Atlantic. R_ρ is computed over a 50 m vertical window. Over 65% of the observations have $1.5 < R_\rho < 2.5$, where fingering should be active.

24]. The dominant wavelength was $\sim$ 5 cm, in excellent agreement with the theoretical finger scale. Recently, a model of salt finger spectra at an interface has been introduced which is able to explain the +2 slope of the temperature gradient spectra observed in the salt finger regions [25]. This is a useful discriminator from turbulence, which has a +1 slope for temperature gradients between the Kolmogorov and Batchelor wavenumbers [26].

The magnitude of the fluxes in the C-SALT staircase is an issue of some importance. Whereas the turbulent dissipation estimates were smaller than had been expected from simple extrapolation of laboratory flux laws [24,

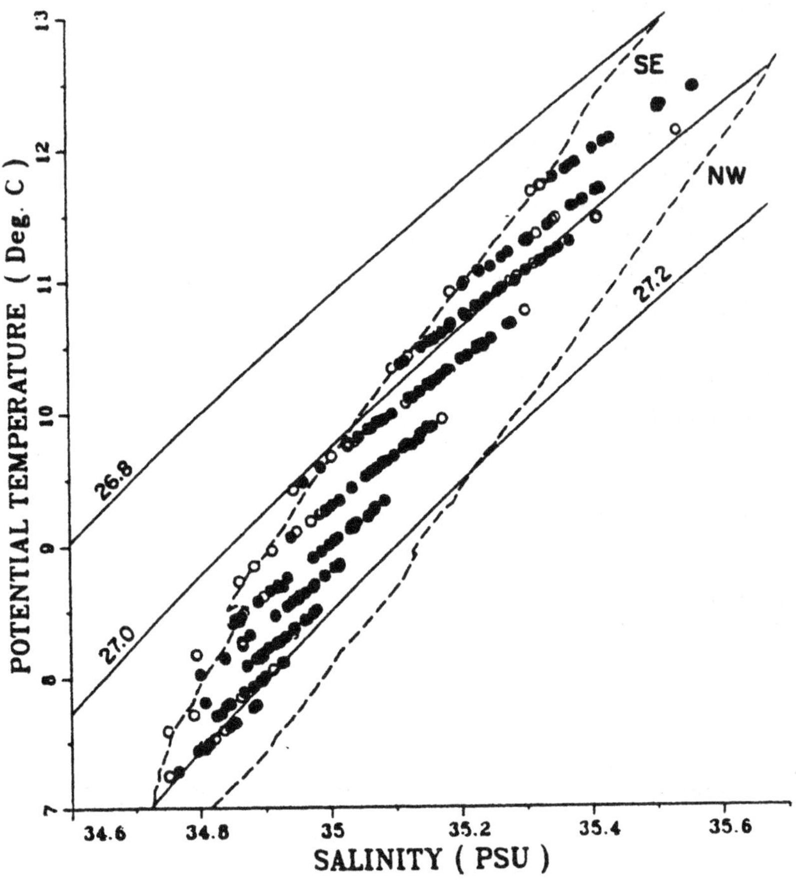

Figure 4. Potential temperature - salinity values of the mixed layers observed during C-SALT. The solid circles are from mixed layers more than 10 m thick; the open circles are from layers 5-10 m thick. Temperature-salinity relationships from the northwest and southeast corners of the survey are also shown (dashed line). The evolution of layer properties across the region is such that layers become warmer, saltier, and denser from southeast to northwest, as would be expected from the vertical convergence of salt finger fluxes. The layer properties cross isopycnals (the 26.8, 27.0, and 27.2 potential density surfaces are shown) with an apparent heat/salt density flux convergence of 0.85.

27], the rate of dissipation of thermal variance was substantial. Schmitt [20] and Kunze [28] estimate an eddy diffusivity for salinity of $1-2\times10^{-4}$ m^2/s within the staircase, based on the observed microstructure levels. Since the staircase occupies about 1/4 of the area of the Atlantic between $10-15^oN$, the vertical salt flux in this high gradient area is predicted to be 3-4 times as large as the flux in the remaining area of this latitude band, which is expected to have a diffusivity ten times smaller as well as a weaker salinity gradient. Thus, we cannot ignore the substantial vertical transports predicted for the staircase areas; they appear to be very significant sites of enhanced vertical transport. An important way to confirm the microstructure-based estimates will be by the use of a deliberate tracer release experiment in a staircase, which has quite different physics from those of internal wave-dominated mixing. The most recent evidence for mixing due to salt fingers comes from analysis of microstructure data taken during the North Atlantic Tracer Release Experiment [29]. A 400×400 km grid of fine- and microstructure profiles was obtained prior to the injection of a passive tracer. These showed an excess of thermal dissipation relative to the dissipation of turbulent kinetic energy. When expressed in terms of the "dissipation ratio" as defined by Oakey [30], a distinct signature of salt fingers is observed in the "low density ratio, high Richardson number" part of parameter space. This can be explained by the greater efficiency of salt fingers ($\sim$ 70%); for a given heat flux they dissipate relatively little kinetic energy. In contrast, turbulence dissipates most of its energy, using only about 15% to change the density stratification. Applying the finger formula for the vertical fluxes where appropriate leads to a significant increase in the diapycnal diffusivity estimate, bringing it into excellent agreement with the value deduced from the vertical dispersion of tracer. In addition, the salt finger-based estimate can explain the observed downward advection of the tracer, whereas the conventional turbulence model predicts upward diapycnal advection [29].

All of these data provide strong evidence that salt fingers are a significant ocean mixing process when the density ratio is below 2. There is some evidence for active salt fingers at higher R_ρ as well [31, 32], but these seem to be infrequently observed, perhaps occurring only when internal waves or turbulence are particularly weak. Thus, it appears that the differential transports due to salt fingers are best modeled as a process that only "turns on" when the density ratio becomes sufficiently close to one. The effects of a density ratio-dependent mixing rate have been studied in a one-dimensional numerical model [18]. Since salt is transported in the vertical at a greater rate than heat, the density ratio tends to increase with time, unless boundary fluxes (or shears in the presence of isopycnal gradients [33]) maintain the excess salt supply. Also, there is a rapid removal of density ratio anomalies, since they tend to be the sites of flux convergences that act to remove the perturbations. This was offered as a explanation for the remarkable constancy of R_ρ in much of the Central Waters of the subtropical gyre. The role of fingers in maintaining the tightness and structure of the T-S relation may be one of their most important effects. Other effects are discussed in section 5.

3. Diffusive Convection

The "diffusive" type of double-diffusive convection is the term used to describe the inverse of the salt finger stratification, that is, a stable salt gradient heated from below. As noted in the introduction, the initial instability appears as an overstable oscillation. However, this is hard to observe, as it breaks down into convective cells within mixed layers separated by very thin interfaces. Heat conduction across the interfaces drives Rayleigh-type convection which maintains the uniformity of the mixed layers, a much weaker salt diffusion acts as a slight brake on the overturning. Thermohaline staircases are readily formed in laboratory experiments with bottom heating of a stable salt gradient. In addition, large areas of the polar oceans harbor such staircase profiles in T and S, especially under the ice where the internal wave field is particularly weak.

The diffusive convection system is more profitably analyzed in terms of classic Rayleigh-Benard convection than in terms of salt fingers, which define their own internal length scale. In diffusive convection the limiting effect of diffusion through the boundaries is an important part of the physics, with the adjacent convection cells being of the order of the layer depth. Veronis [34, 35] studied the finite amplitude behavior using numerical integration of a truncated spectral model. Relative to the non-double-diffusive problem, the presence of the solute delays the onset of convection. However, at high Rayleigh number, fluxes are nearly as large. Also of interest is the salt/heat buoyancy flux ratio, which Veronis suggested tends toward $(\frac{\kappa_S}{\kappa_T})^{\frac{1}{2}}$. This ratio has been roughly confirmed in experiments for density ratios not too close to one.

The low value of the flux ratio compared to fingers can be understood by considering the different roles of the two components in the two systems. That is, the energy in the salt field is released in salt fingers by diffusion of heat, and advection is the vertical transport mechanism. In diffusive convection the temperature field has both the energy supply and the means of its release, with conduction across the interface providing the vertical flux. We expect that the diffusion of heat and salt across the interface yields penetration distances into the adjacent mixed layer proportional to the one-half power of the diffusivity. Since heat penetrates further, an unstable region forms which convects away from the interface when sufficiently thick. The relative portions of heat and salt carried away will reflect their different penetration scales, giving a salt/heat flux ratio as specified above. However, when the stability is low (when the density ratio approaches one), the flux ratio increases toward unity, apparently due to penetration of the interface by buoyant elements in the mixed layers [36]. The laboratory experiments support a flux proportional to the 4/3 power of the temperature difference across the interface. However, [37] questions the validity of the 4/3 power law, suggesting that an exponent between 4/3 and 5/4 (depending on Rayleigh number) may be more accurate.

Strong but localized diffusive staircases are found at the hot, salty brines in ocean deeps found at spreading centers [38]. There the separation of heat and salt could be contributing to pooling of the brines in topographic depressions [39] and possibly ore formation as well. The diffusive process also plays a role in intrusions and on the upper side of warm, salty water masses such as the Mediterranean water in the Atlantic. It may be a factor in the evolution of fresh mixed layers laid down by rain or eddy processes [40]. However, the most extensive regions of importance are found in the polar and subpolar oceans. Steps under the Arctic ice were first reported by [41]. More recent observations have been provided by Padman and Dillon [42-44] in other parts of the Arctic. A "diffusive" thermohaline staircase between about 200 and 400 m depth appears to be a ubiquitous feature under most of the Arctic ice field away from the boundaries. It supports a heat flux from the intruding warm, salty Atlantic water to the cooler, fresher Arctic surface waters. The extensiveness of the staircase may be due to the especially weak internal wave field under the ice. This is likely due to the rigid lid, but also possibly due to an enhanced wave decay within the convectively mixed staircase [42]. Areas near topography with stronger internal waves and more frequent turbulent mixing events are less likely to harbor a staircase. The down-gradient buoyancy flux from the turbulence (an up-gradient buoyancy flux is necessary to maintain a staircase), and the destruction of the small scale property gradients by isotropic turbulence, are competing factors to the double diffusion.

The waters around Antarctica also display prominent diffusive staircases [45-47]. Muench *et al.* [47] find staircases to be a common feature over much of the Weddell Sea. The layers are much thicker (10-100 m) than those found in the Arctic and may support a much larger flux. They apply laboratory flux laws to obtain an upward heat flux of 15 W/m^2 in open waters. This is sufficient to be important in upper ocean heat budgets and may help to maintain ice-free conditions in the summer. Lower fluxes are estimated for the Arctic steps [44]. However, recent observations of a much stronger incursion of Atlantic water into the Arctic are characterized by very extensive double-diffusive intrusions [48]. This suggests that the lateral processes (next section) may be dominant over the vertical in the halocline of the Arctic Ocean, and that the significant climatic changes currently underway there may be mediated by small-scale double-diffusive processes.

The importance of diffusive convection in polar regions lies in its ability to produce a cold, salty and dense water mass without air-sea interaction.

That is, heat can be extracted from a subsurface water mass without much change in salinity. The resulting water may be dense enough to become a bottom water. This idea has been applied to the formation of Antarctic Bottom Water [49] and Greenland Sea Bottom Water [50]. The T-S characteristics are in agreement with the model predictions but quantification of the rates of mixing remains uncertain.

Kelley [37] has provided a parameterization for the eddy diffusivities resulting from diffusive staircases. This work is based on his finding of a general scaling law for the thickness of convective layers [51]. The estimated eddy diffusivities for T and S both increase as the density ratio approaches one, much like the proposed salt finger parameterization, though they are generally about an order of magnitude smaller (Figure 5). However, an alternative treatment would predict larger fluxes [52].

4. Double-Diffusive Intrusions

The realization that double diffusion has consequences for motion in two and three dimensions arose from the work of Stern [4], who predicted that finescale intrusive motions could grow in the presence of horizontal gradients of T and S. The mechanism relied on the buoyancy flux convergences due to salt fingers, which generated lateral pressure gradients that drove the motion. Thorpe, Hutt, and Soulsby [53] showed that horizontal intrusive layering arose in a stable salt gradient heated from the side, with sites for fingering and diffusive layering. Laboratory experiments by Turner [5] confirmed the theoretical prediction that warm salty (cold fresh) intrusions should rise (sink) across density surfaces when salt fingers dominate the fluxes. He also found that when diffusive layering dominated, the obverse was true. Further laboratory studies by Ruddick and Turner of intrusions in the presence of sharp lateral property changes suggested that the vertical scale was set by the potential energy available across the front [6].

Important supporting oceanic observations were reported by Joyce *et al.* [54] and Gregg and McKenzie [55], who showed that the intrusions did indeed slope as expected. Joyce *et al.* [54] showed further that the amplitude of the intrusions was far greater than could be explained by internal/inertial wave displacements acting on the background gradients. A continuous displacement process such as that predicted by the double-diffusive mechanism was called for. Similar conclusions were reached by Schmitt and Georgi [8], who found remarkably strong intrusions in the North Atlantic Current, where waters of the subtropical and subpolar gyres are brought into close proximity. They also found evidence for double-diffusive mixing at the boundaries of the intrusions in optical microstructure profiles.

The most complete description and analysis of oceanic intrusions has been contributed by Ruddick [9]. Intrusions around a Mediterranean salt lens (Meddy) in the Atlantic showed slopes indicating diffusive-dominated intrusions in the upper parts of the lens and finger-dominated intrusions around the deeper portions of the lens. The observed slopes were inconsistent with the McIntyre [56] mechanism involving a double diffusion of momentum and mass and provide strong evidence for the importance of the double-diffusive intrusion in Meddy decay. This Meddy was tracked over two years and mixing by intrusions was found to be more important than the purely vertical mixing caused by fingering below the central core of the eddy.

The original theory of Stern [4] had no viscosity, and consequently the modes with the smallest vertical scale had the largest growth rate. Toole and Georgi [57] developed a viscous model of the process which showed that an intermediate wavelength instability grew fastest. They estimated growth rates of $O(days^{-1})$ and vertical length scales of $5 - 40\ m$, in reasonable agreement with observations at oceanic fronts. However, the model depends on rather uncertain eddy diffusivities and viscosities, and is very sensitive to the strength of the horizontal salinity gradient, which is difficult to estimate from sparse oceanic data sets. Theoretical work of McDougall [22, 58] found that the "along leaf" density ratio is closer to unity than to the salt finger flux ratio, and that even in a rotating system the fastest growing intrusion goes across the front and not along it. Further work on the theory of double-diffusive intrusions has been provided in [59] and [60]. Also, Richards [7] has extended the theory to the equatorial beta-plane. The model shows promise in explaining the thin intrusions with remarkably long horizontal coherence that are found in the equatorial Pacific. These are estimated to be a significant factor in the equatorial heat budget [72].

While we know that double-diffusive intrusions are an important lateral mixing mechanism, attempts to estimate their effects on larger scales have been limited to empirical studies, and no numerical models have incorporated any of the possible parameterizations. Joyce [61] suggested a formalism to infer the effective lateral diffusivity due to intrusive interleaving. Based on a "production-dissipation" balance for the finescale thermal variance, he obtained the following expression relating the lateral diffusivity to the vertical diffusivity and the ratio of the vertical temperature gradient variance to the mean square horizontal gradient. That is:

$$K_H = K_V \frac{\overline{(\tilde{T}_Z^2)}}{(\overline{T_Y})^2}. \tag{4}$$

Since intrusion regions harbor large vertical gradient variance, this approach yields lateral diffusivities much in excess of the modest diffusivities estimated from the shear dispersion of internal waves. However, it relies on empirical observations and a still uncertain vertical diffusivity, so its use as a parameterization may be limited.

Garrett [62] considered the effect of intrusions on diapycnal mixing, since they represent a mechanism for the strong intensification of vertical gradients. He relied on the Ruddick and Turner estimate of the vertical scale of the intrusion [6], and on simple models of the frontal width and eddy diameter, to derive the formula:

$$K_S = 10^{-3} D^2 N \left(\frac{g\beta S_X}{N_2} \right)^3 . \tag{5}$$

Here K_S is the effective diapycnal diffusivity, D is the eddy diameter, N the buoyancy frequency and $g\beta S_X$ the horizontal (isopycnal) buoyancy gradient due to salt. This suggests that the intrusion mechanism yields especially enhanced vertical diffusivities in weakly stratified frontal regions, a situation particularly common in higher latitudes.

A somewhat different approach was taken by McDougall [58]. He develops an explicit model of steady state intrusions, with the vertical scale set by the linear stability analysis (after [57]). Explicit expressions are given for the along-intrusion changes in T and S which are in good agreement with observations. He finds the diapycnal fluxes of heat, salt, and density to be upgradient, giving negative diapycnal diffusivities for all three quantities. His formula for a basin scale diffusivity (after [62]) is:

$$K_S = -0.004 \frac{g\beta S_X}{N} h^2 \tag{6}$$

where h is the intrusion thickness. He obtains a negative diffusivity of order $10^{-5}\ m^2/s$ for typical parameters; this diffusivity represents the combined effects of salt fingers, diffusive interfaces, and the advection of individual intrusions across isopycnals. This is of comparable magnitude, but opposite sign, to the general diffusivity in the open ocean, so the mechanism certainly deserves further study.

As mentioned above, no attempts have been made to use a parameterization of intrusions in a large-scale GCM. Much depends on still uncertain diffusivities and eddy viscosities for the purely vertical fluxes, so caution is reasonable at this juncture. However, the intrusion process appears to be extremely important in the decay of water mass anomalies at fronts and eddies, and may play a role in equatorial heat balances and recent changes in Arctic Ocean temperature structure [48]. It may be the most important means of fluxing variance to small scales; it certainly seems a stronger mechanism than shear dispersion by internal waves. The negative diffusivities implied could lead to the spin-up of mesoscale features [63]. Such dynamics, as well as the dependence of vertical fluxes on horizontal gradients, are features which will require a great deal of work to incorporate into models. There is a clear need for intermediate scale models that directly address the intrusion growth and dynamics, but these must rely on a firm understanding of the microscale physics, which is just emerging. An additional problem is the serious observational challenge posed by intrusions; they are difficult to map in the dynamic environment around fronts and eddies. However, these are problems which must be attacked, as no other mechanism appears to approach the intrusions in importance for the destruction of variances produced by eddy stirring, and the upgradient fluxes have a number of dynamical consequences.

5. Large-scale Modeling of Double-diffusive Processes

Only a few studies have examined the effect of salt fingers on the general circulation. Gargett and Holloway [64] applied uniformly different diffusivities for T and S to a general circulation model. They found strong sensitivity of the steady-state characteristics of low resolution GCMs used in climate studies to the ratio of the vertical eddy diffusivities for salinity and temperature. Large differences in meridional transports resulted due to the upgradient buoyancy flux, which caused different advective-diffusive balances to be realized. They also found that the intermediate water salinity minimum was more readily produced with unequal heat and salt diffusivities.

Simple box models have also been examined for the effects of unequal mixing rates. Ruddick and Zhang [65] have found that the salt oscillator mechanism of Welander [66] in a one-dimensional box model could be completely stabilized by incorporating salt fingers into the model. More recently, Gargett and Ferron [67] have found that multi-box models with non-equal heat and salt diffusivities exhibited extended ranges of multiple equilibria, a different mode transition near present-day values of freshwater forcing, and the possibility of quasi-periodic oscillatory states, compared to an equal diffusivity run.

The most recent examination of the large-scale effects of the double diffusion has been provided by Zhang *et al.* [68]. They used a more realistic parameterization scheme in which the diffusivities varied with density ratio in a manner similar to that in Schmitt [18] for fingers and in Kelley [37] for diffusive convection (Figure 5). Thus, the double-diffusive mixing is applied only where the appropriate stratifications call for it. They obtained modest changes in water mass structure and in the strength of the meridional overturning in the GFDL Modular Ocean Model with relaxation boundary conditions. About a 20% reduction in the meridional overturning was observed along with improved representation of temperature and salinity structure. Experiments with flux boundary conditions are currently underway. The intimate connection between diapycnal mixing and the strength of the meridional overturning circulation has been further explored in a scale analysis by Zhang *et al.* [69]. They find that the stability of the "conveyor belt" is a sensitive function of the diffusivity and the hydrologic forcing, with the transition point between thermal and haline modes lying close to the conventional metrics of 1×10^{-4} m^2/s for diffusivity and 1 m/yr for evaporation minus precipitation. In a fundamental sense, the thermohaline circulation can be thought of as being driven by interior mixing, and thus any process that affects the strength and distribution of diapycnal mixing will affect the thermohaline circulation. Knowing the interior patterns of vertical mixing is as important for our understanding of the thermohaline circulation as knowledge of the wind stress is for understanding the wind-driven circulation. As "greenhouse warming" models project that the collapse of the thermohaline circulation is the most dramatic consequence of potential climate change [70, 71], it is important that we develop much more confidence in our parameterizations of vertical mixing.

In addition to modification of the thermohaline circulation, double-diffusive processes appear to play an important role in maintaining the general shape and tightness of the T-S relation in the Central Waters [18]. These water masses retain a tight and consistent T-S relation, despite climatic variability in the properties of subducted waters driven by atmospheric fluxes. Double diffusion is thought to be an effective means of destroying "spiciness" (salinity variance) on isopycnal surfaces, whereas the small diffusivity now ascribed to internal wave breaking appears to be too weak. The double-diffusive processes also seem likely to be the main cause of the thin water mass features with extremely long horizontal coherence scales found in both high- and low-latitude oceans. They provide a "zero-frequency" baroclinic driving agent for such structures, which are otherwise difficult to explain.

In summary, many different aspects of oceanic microstructure, finestructure, and large-scale T-S correlations pass the consistency tests for double-diffusive convection, while the traditional assumption that ocean mixing is fully turbulent can no longer be maintained. Much more work remains to be done to develop quantitative estimates of the rates of these processes, especially for thermohaline intrusions. Efforts along these lines will pay off in a more accurate representation of temperature and salinity structure in numerical simulations of the oceanic circulation. Ultimately, the proper representation of subgrid-scale mixing processes should help to rid coupled ocean-atmosphere models of the need for "flux corrections" invoked to maintain T-S fidelity in long-term runs, and to provide a more secure basis for predicting future climatic states.

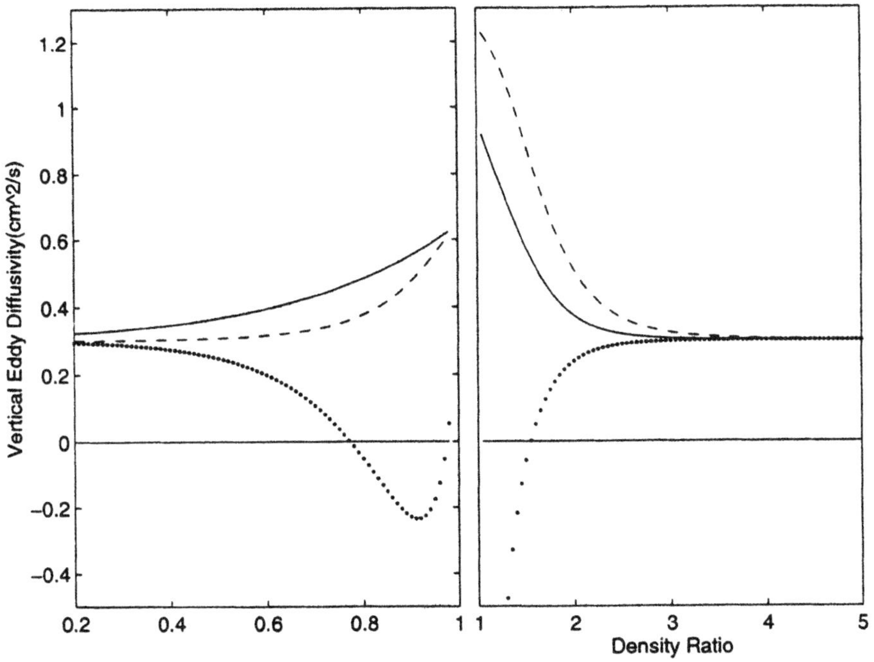

Figure 5. The dependence of diapycnal eddy diffusivities on density ratio, for temperature (solid line), salinity (dashed line), and density (dotted line), as applied in the model runs of Zhang *et al.* [68]. The left panel, with $R_\rho < 1$, is where diffusive layering occurs, and the prescription of Kelley [37] is applied. The right panel, with $R_\rho > 1$, is the salt finger regime, and a modified version of the mixing rules of Schmitt [18] is used. In both cases a background turbulent diffusivity of 0.3 cm^2/s is invoked as well. Note the gap as $R_\rho = 1$ represents a statistically unstable regime.

6. Acknowledgements

Thanks to Trevor McDougall, Kelvin Richards, and John Toole for reviews of the manuscript. The Office of Naval Research and the National Science Foundation have generously supported my research. The conference organizers, Eric Chassignet and Jacques Verron, are complimented on their management of an excellent conference in an exceptional venue.

References

1. Schmitt, R.W. (1994) Double diffusion in oceanography, *Annual Review of Fluid Mechanics*, **26**, 255-285.
2. Stern, M.E. (1960) The 'salt fountain' and thermohaline convection, *Tellus*, **12**, 172-175.

3. Shirtcliffe, T. (1967) Thermosolutal convection: observation of an overstable mode, *Nature*, **213**, 489-490.
4. Stern, M.E. (1967) Lateral mixing of water masses, *Deep-Sea Research*, **14**, 747-753.
5. Turner, J.S. (1978) Double-diffusive intrusions into a density gradient, *Journal of Geophysical Research*, **83**, 2887-2901.
6. Ruddick, B.R. and Turner, J.S. (1979) The vertical length scale of double-diffusive intrusions, *Deep-Sea Research*, **26A**, 903-913.
7. Richards, K.J. (1991) Double-diffusive interleaving at the Equator, *Journal of Physical Oceanography*, **21**(7), 933-938.
8. Schmitt, R.W. and Georgi, D.T. (1982) Finestructure and microstructure in the North Atlantic Current, *Journal of Marine Research*, **40**(Supplement), 659-705.
9. Ruddick, B. (1992) Intrusive mixing in a Mediterranean salt lens-intrusion slopes and dynamical mechanisms, *Journal of Physical Oceanography*, **22**, 1274-1285.
10. Schmitt, R.W. (1979) The growth rate of super-critical salt fingers, *Deep-Sea Research*, **26**A, 23-40.
11. Gordon, A.L. (1981) South Atlantic thermocline ventilation, *Deep-Sea Research*, **28**, 1236-1264.
12. Elliot, A.J. and Tait, R.I. (1977) On the steady-state nature of the Mediterranean Outflow step structure, in M.V. Angel (Ed.), A Voyage of Discovery, Pergamon Press: New York. pp. 213-220.
13. Williams, A.J. (1974) Salt fingers observed in the Mediterranean outflow, *Science*, **185**, 941-943.
14. Krahman, G. (1998) Horizontal variability of thermohaline staircases in the Western Mediterranean, (Personal Communication).
15. Molcard, R. and Tait, R.I., (1977) The steady state of the step structure in the Tyrrhenian Sea, in M.V. Angel (Ed.), A Voyage of Discovery, Pergamon Press: New York. pp. 221-233.
16. Schmitt, R.W., *et al.* (1987) C-SALT: an investigation of the thermohaline staircase in the western tropical North Atlantic, *Deep-Sea Research*, **34**(10), 1697-1704.
17. Boyd, J.D. (1989) Properties of thermal staircase off the northeast coast of South America, *Journal of Geophysical Research* **94**, 8303-8312.
18. Schmitt, R.W. (1981) Form of the temperature-salinity relationship in the Central Water: evidence for double-diffusive mixing, *Journal of Physical Oceanography*, **11**, 1015-1026.
19. Schmitt, R.W. (1987) The Caribbean Sheets And Layers Transects (C-SALT) Program, *EOS*, Transactions of the American Geophysical Union, **68**(5), 57-60.
20. Schmitt, R.W., (1988) Mixing in a thermohaline staircase, in J. Nihoul and B. Jamart (Eds.), Small-Scale Turbulence and Mixing in the Ocean, Elsevier: New York. pp. 435-452.
21. McDougall, T.J. (1991) Interfacial advection in the thermohaline staircase east of Barbados, *Deep-Sea Research*, **38**(3), 367-370.
22. McDougall, T.J. (1985) Double-diffusive interleaving. Part I: Linear stability analysis, *Journal of Physical Oceanography*, **15**, 1532-1541.
23. Marmorino, G.O., Brown, W.K., and Morris, W.D. (1987) Two-dimensional temperature structure in the C-SALT thermohaline staircase, *Deep-Sea Research*, **23**(10), 1667-1675.
24. Lueck, R. (1987) Microstructure measurements in a thermohaline staircase, *Deep-Sea Research*, **34**(10), 1677-1688.
25. Shen, C. and Schmitt, R.W. (1996) The wavenumber spectrum of salt fingers, in A. Brandt and H. Fernando (Eds.), Double-Diffusive Convection, AGU Geophysical Monograph 94. pp.305-312.
26. Mack, S.A. and Schoberlein, H.C. (1993) Discriminating salt fingering from turbulence-induced microstructure: analysis of towed temperature-conductivity chain data., *Journal of Physical Oceanography*, **23**(9), 2073-2106.
27. Gregg, M.C. and Sanford, T. (1987) Shear and turbulence in a thermohaline stair-

case, *Deep-Sea Research*, **34**(10), 1689-1696.
28. Kunze, E. (1990) The evolution of salt fingers in inertial wave shear, *Journal of Marine Research*, **48**, 471-504.
29. St. Laurent, L. and Schmitt, R.W. (1998) The contribution of salt fingers to vertical mixing in the North Atlantic Tracer Release Experiment, *Journal of Physical Oceanography*, submitted.
30. Oakey, N.S. (1985) Statistics of mixing parameters in the upper ocean during JASIN phase 2, *Journal of Physical Oceanography*, **10**(1), 83-98.
31. Gargett, A.E. and Schmitt, R.W. (1982) Observations of salt fingers in the central waters of the eastern North Pacific, *Journal of Geophysical Research*, **87**(C10), 8017-8029.
32. Osborn, T.R. (1991) Observations of the Salt Fountain, *Atmosphere-Oceans*, **29**(2), 340-356.
33. Schmitt, R.W. (1990) On the density ratio balance in the Central Water, *Journal of Physical Oceanography*, **20**(6), 900-906.
34. Veronis, G. (1965) On finite amplitude instability in thermohaline convection, *Journal of Marine Research*, **23**, 1-17.
35. Veronis, G. (1968) Effect of a stabilizing gradient of solute on thermal convection., *Journal of Fluid Mechanics*, **34**, 315-336.
36. Marmorino, G.O. and Caldwell, D.R. (1976) Heat and salt transport through a diffusive thermohaline interface, *Deep-Sea Research*, **23**, 59-67.
37. Kelley, D.E. (1990) Fluxes through diffusive staircases: A new formulation, *Journal of Geophysical Research*, **95**, 3365-3371.
38. Swallow, J.C. and Crease , J. (1965) Hot salty water at the bottom of the Red Sea, *Nature*, **205**, 165-166.
39. McDougall, T.J. (1985) Convective processes caused by a dense, hot saline source flowing into a submarine depression from above, *Deep-Sea Research*, **29**, 953-965.
40. Schmitt, R.W. and Olson, D.B. (1985) Wintertime convection in warm core rings: thermocline ventilation and the formation of mesoscale lenses, *Journal of Geophysical Research*, **90**, 8823-8837.
41. Neal, V.T., Neshyba, S., and Denner, W. (1969) Thermal stratification in the Arctic Ocean, *Science*, **166**, 373-374.
42. Padman, L. and Dillon, T.M. (1989) Thermal microstructure and internal waves in the Canada Basin diffusive staircase, *Deep-Sea Research*, **36**, 531-542.
43. Padman, L. and Dillon, T.J. (1988) On the horizontal extent of the Canada Basin thermohaline steps, *Journal of Physical Oceanography*, **18**, 1458-1462.
44. Padman, L. and Dillon, T.M. (1987) Vertical fluxes through the Beaufort sea thermohaline staircase, *Journal of Geophysical Research*, **92**, 10799-10806.
45. Foster, T.D. and Carmack, E.C. (1976) Temperature and salinity structure in the Weddell Sea, *Journal of Physical Oceanography*, **6**, 36-44.
46. Middleton, J.H. and Foster, T.D. (1980) Finestructure measurements in a temperature-compensated halocline, *Journal of Geophysical Research*, **85**, 1107-1122.
47. Muench, R.D., Fernando, H.J.S., and Stegan, G.R. (1990) Temperature and salinity staircases in the northwestern Weddell Sea, *Journal of Physical Oceanography*, **20**, 295-306.
48. Carmack, E.C., *et al.* (1997) Changes in temperature and tracer distributions within the Arctic Ocean: results from the 1994 Arctic Ocean section, *Deep-Sea Research*, **44**(8), 1487-1502.
49. Gill, A.E. and Turner, J.S. (1969) Some new ideas about the formation of Antarctic bottom water, *Nature*, **224**, 1287-1288.
50. McDougall, T.J. (1983) Greenland Sea Bottom Water formation: a balance between advection and double-diffusion, *Deep-Sea Research*, **30**, 1109-1118.
51. Kelley, D.E. (1984) Effective diffusivities within ocean thermohaline staircases, *Journal of Geophysical Research*, **89**, 10484-10488.

52. Fernando, H.J.S. (1987) The formation of layered structure when a stable salinity gradient is heated from below, *Journal of Fluid Mechanics*, **182**, 425-442.
53. Thorpe, S.A., Hutt, P.K., and Soulsby, R. (1969) The effect of horizontal gradients on thermohaline convection, *Journal of Fluid Mechanics*, **38**, 375-400.
54. Joyce, T.M., Zenk, W., and Toole, J.M. (1978) The anatomy of the Antarctic Polar Front in Drake Passage, *Journal of Geophysical Research*, **83**, 6093-6113.
55. Gregg, M.C. and McKenzie, J.H. (1979) Thermohaline intrusions lie across isopycnals, *Nature*, **280**, 310-311.
56. McIntyre, M.E. (1970) Diffusive destabilization of the baroclinic vortex, *Geophysical Fluid Dynamics*, **1**, 19-57.
57. Toole, J.M. and Georgi, D.T. (1981) On the dynamics and effects of double-diffusively driven intrusions, *Progress in Oceanography*, **10**, 123-145.
58. McDougall, T.J. (1985) Double-diffusive interleaving. Part II: Finite amplitude steady state interleaving, *Journal of Physical Oceanography*, **15**, 1542-1556.
59. Niino, H. (1986) A linear stability theory of double-diffusive horizontal intrusions in a temperature-salinity front, *Journal of Fluid Mechanics*, **171**, 71-100.
60. Yoshida, J., Nagashima, H., and Niino, H. (1989) The behavior of double-diffusive intrusion in a rotating system, *Journal of Geophysical Research*, **94**(C4), 4923-4938.
61. Joyce, T.M. (1977) A note on the lateral mixing of water masses, *Journal of Physical Oceanography*, **7**, 626-629.
62. Garrett, C. (1982) On the parameterization of diapycnal fluxes due to double-diffusive intrusions, *Journal of Physical Oceanography*, **12**, 952-959.
63. Posmentier, E.S. and Kirwan, A.D. (1985) The role of double diffusive interleaving in mesoscale dynamics: an hypothesis, *Journal of Marine Research*, **43**, 541-552.
64. Gargett, A.E. and Holloway, G. (1992) Sensitivity of the GFDL Ocean Model to different diffusivities for heat and salt. *Journal of Physical Oceanography*, **22**, 1158-1177.
65. Ruddick, B. and Zhang, L. (1989) The mythical thermohaline oscillator?, *Journal of Marine Research*, **47**(4), 717-746.
66. Welander, P. (1982) A simple heat-salt oscillator, *Dynamics of Atmospheres and Oceans*, **6**, 233-242.
67. Gargett, A.E. and Ferron, B. (1996) The effects of differential vertical diffusion of T and S in a box model of thermohaline circulation, *Journal of Marine Research*, **54**, 827-866.
68. Zhang, J., R. W. Schmitt and R. X. Huang (1998) Sensitivity of GFDL Modular Ocean Model to the parameterization of double-diffusive processes, *Journal of Physical Oceanography*, **28**, in press.
69. Zhang, J., Schmitt, R.W., and Huang, R.X. (1998) The relative influence of diapycnal mixing and hydrologic forcing on the stability of the thermohaline circulation, *Journal of Physical Oceanography*, submitted.
70. Broecker, W.S. (1997) Thermohaline circulation, the Achilles heel of our climate system: will man-made CO2 upset the current balance?, *Science*, **278**, 1582-1588.
71. Manabe, S. and Stouffer, R.J. (1994) Multiple-century response of a coupled ocean-atmosphere model to an increase of atmospheric carbon dioxide, *Journal of Climate*, **7**, 5-23.
72. Richards, K.J. (1998) Interleaving at the Equator, in *Ocean Modeling and Parameterization*, E.P. Chassignet and J. Verron (Eds.), Kluwer Academic Publishers, 235-252.

INTERLEAVING AT THE EQUATOR

Its parameterization and effect on the large scale dynamics

KELVIN J. RICHARDS
Southampton Oceanography Centre,
Southampton, United Kingdom

1. Introduction

The robustness of climate predictions depends on the effectiveness of atmosphere and ocean models to capture the essential physics of the problem. Recent studies have shown that for the ENSO phenomenom the predictability of coupled models is very dependent on the initial state of the ocean model (*e.g.* Latif and Graham, [15]). What dictates the state of the ocean, its density and current structure, is still a matter for debate. Neelin *et al.* [20] suggest that the lack of ability of coupled models to reproduce major features of the tropical climatology may be attributable to a sensitivity of models to parameters used in subgrid scale processes. McPhaden [19] notes the importance of improving model physics, through parameterizing small scale mixing processes, to accurately model the equatorial ocean and predict El Niño.

Studies such as Chen *et al.* [6] and Blanke and Delecluse [3] focus on the role of vertical mixing in establishing the vertical structure of the equatorial currents. However a recent model intercomparison [31] has shown the important role that lateral mixing plays in determining the SST in the equatorial Pacific. Maes *et al.* [16] in a study to assess the relative roles of lateral and vertical mixing find that there is a nonlinear interplay between horizontal and vertical diffusion. The nature of the Equatorial Undercurrent is found to be particularly sensitive to the relative importance of the components of diffusion. Lateral mixing conditions affect the meridional heat transport and surface heat budget as well as the zonal momentum budget. In their low lateral mixing simulation Maes *et al.* [16] find that the level of vertical mixing required to keep the current system in balance is excessively high. At a value of the lateral diffusion coefficient of 10^3 m^2 s^{-1} they find that the model produces more acceptable rates of vertical mixing and that lateral mixing starts to take on a dominant role.

E. P. Chassignet and J. Verron (eds.), Ocean Modeling and Parameterization, 235–251.

Unlike vertical mixing where various forms of mixing scheme have been tested (*e.g.* Pacanowski and Philander [21], Chen *et al.* [6]) the appropriate form for the horizontal mixing used in OGCMs has received scant attention. In most cases a crude Laplacian mixing term with a constant coefficient of diffusion is employed in the momentum and heat equations. Horizontal mixing is often thought of as being produced by the action of mesoscale eddies. However horizontal mixing brought about by the interleaving of water masses may be equally or more important. It has been suggested that interleaving may be the primary cause of the destruction of such features as meddies [27]. At the equator where the interleaving is considerably more pronounced the action of the interleaving on the dynamics requires consideration and evaluation.

The role of interleaving in the dynamics of the equatorial ocean needs to be understood and quantified. Here we investigate the properties of interleaving at the equator and study the impact that interleaving has on the large scale dynamics of the the equatorial ocean. The ultimate aim is to provide an improved parameterisation of lateral mixing processes for future climate models.

2. Observations of Interleaving

The salinity structure within the thermocline in the equatorial Pacific is characterised by strong meridional gradients produced by the convergence of the relatively salty South Pacific and fresher North Pacific waters within the thermocline (see figure 1a). This convergence of water masses is compensating the surface divergence caused by the predominantely easterly winds.

In the vertical the salinity has a layered structure seen most clearly in the spice (or salinity) anomaly and Turner angle. The spice anomaly is given by $S(\rho) - \overline{S}(\rho)$, where the overbar represents a low–pass filter of the data. The Turner angle, Tu, defined as

$$\mathrm{Tu} = \arctan\left(\alpha\frac{\partial\theta}{\partial z} - \beta\frac{\partial S}{\partial z}, \alpha\frac{\partial\theta}{\partial z} + \beta\frac{\partial S}{\partial z}\right),$$

is related to the stability ratio, $R_\rho = \alpha\theta_z/\beta S_z$ (where α and δ are the coefficients of thermal expansion and saline contraction, respectively). The stability ratio gives the relative importance of vertical gradients of temperature, θ, and salinity, S, in determining the gradient of density. The Turner angle can be used to identify regions of the water column susceptible to double–diffusive processes. For values of $45^o < \mathrm{Tu} < 90^o$ salt fingering can occur whilst for $-90^o < \mathrm{Tu} < -45^o$ diffusive convection can occur. Figure 1 shows a large region favourable to salt fingering with interfaces between layers alternating between fingering and diffusive regimes.

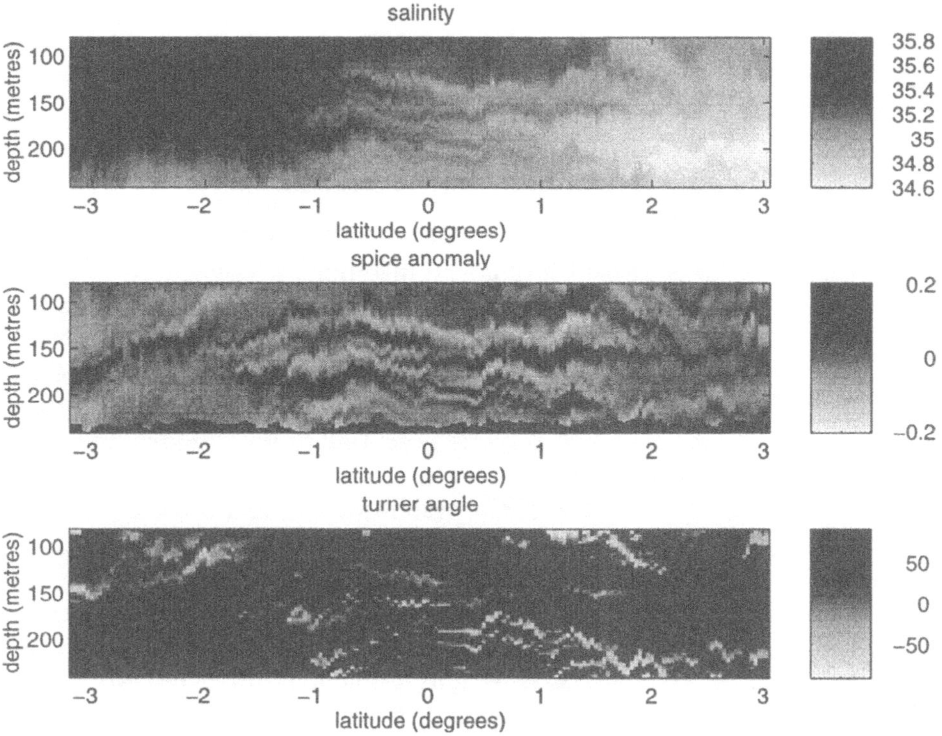

Figure 1. The salinity, spice (salinity) anomaly and Turner angle along 156°E in the Western Equatorial Pacific from 3°S to 3°N and from 80 m to 250 m depth (see text for definitions). The figure shows layers a few metres thick of alternating anomalously warm/salty water (dark) and cooler/fresh water (light) extending a few hundred kilometres in the north/south direction.

A layered structure in the equatorial Pacific was first noted by Toole [32] in the eastern Pacific and later by McPhaden [19] in the central Pacific. Observations made by Richards and Pollard [26] first showed the large meridional coherence of the layering, much greater than that observed at mid–latitudes (see *e.g.* Joyce [14]).

Banks [1] (see also Banks and Richards [2]) analysed a large number of high resolution meridional sections. She concluded that the layers, or intrusions, have typical vertical scales O(20 m), meridional scales O($>30\,km$) and zonal scales O(100 km). Individual layers slope both upward and downward across isopycnic surfaces. By considering the spice and density anomalies, Turner angle and temperature–salinity ratio she finds that at least 50% of the layers show evidence that double–diffusion (both fingering and diffusive) is playing a role in the fluxes of temperature and salinity between the layers. There is also a tendency for strong layering (in terms of spice energy) to be in regions of strong meridional salinity gradient.

Based on the observations of interleaving we can make estimates of lateral fluxes and suggest a possible parameterization scheme. But first we consider how the interleaving is formed.

3. Formation Mechanism: Diffusive or Shear Instability?

The often regular nature of the interleaving with a definite vertical wavenumber suggests that it is formed through an instability process. A word of caution is appropriate here. Although double diffusion is seen to be important for the mature layers it is not necessarily the mechanism for their formation. As well as double diffusive interleaving, inertial instability is a possible candidate. The current structure is highly sheared in both the horizontal and vertical (see *e.g.* Eldin *et al.* [11]) potentially providing conditions for instability. We shall consider both.

To investigate the stability of the system we consider a basic state consisting of a zonal current, $\overline{u}(y, z)$, which varies in the meridional and vertical and prescribed distributions of temperature, $\overline{\theta}(y, z)$, and salinity, $\overline{S}(y, z)$. The density field, $\rho = -\rho_o(\alpha\theta - \delta S)$ is such that the flow is geostrophically balanced.

The linearized conservation equations for momentum, mass, salinity and potential temperature for a perturbation to the basic state on an equatorial β–plane, where $f = \beta y$, are then

$$u'_t + \overline{u}u'_x + v'\overline{u}_y + w'\overline{u}_z - \beta y v' = -\rho_o^{-1}p'_x - \nabla \cdot \boldsymbol{F^x} \tag{1}$$

$$v'_t + \overline{u}v'_x + \beta y u' = -\rho_o^{-1}p'_y - \nabla \cdot \boldsymbol{F^y} \tag{2}$$

$$w'_t + \overline{u}w'_x = -\rho_o^{-1}p'_z - g(\delta S' - \alpha\theta') - \nabla \cdot \boldsymbol{F^z} \tag{3}$$

$$u'_x + v'_y + w'_z = 0 \tag{4}$$

$$S'_t + \overline{u}S'_x + v'\overline{S}_y + w'\overline{S}_z = -\nabla \cdot \boldsymbol{F^S} \tag{5}$$

$$\theta'_t + \overline{u}\theta'_x + v'\overline{\theta}_y + w'\overline{\theta}_z = -\nabla \cdot \boldsymbol{F^\theta} \tag{6}$$

where $\boldsymbol{F^x}$, $\boldsymbol{F^y}$ and $\boldsymbol{F^z}$ are the diffusive fluxes of x, y, z momentum and $\boldsymbol{F^S}$ and $\boldsymbol{F^\theta}$ are the diffusive fluxes of salt and potential temperature.

The perturbation is assumed to be of the form

$$u' = U_0(y)\exp[\hat{\lambda}t + i(\hat{k}x + \hat{m}z)].$$

The growth rate of the perturbation, $\hat{\lambda}$ then depends on the nature of the basic state.

3.1. DOUBLE–DIFFUSIVE INTERLEAVING

Double–diffusive interleaving can arise when the vertical fluxes of salinity and temperature are dominated by double–diffusive processes. The mismatch of salinity and temperature fluxes means that there is an associated flux of buoyancy in the vertical, and it is the divergence of these buoyancy fluxes that can drive the interleaving.

To isolate the process we assume the basic state is at rest ($\overline{u} = 0$) and that density surfaces are initially horizontal. However along the density surfaces the temperature and salinity varies in the northward direction such that

$$\delta\overline{S}_y = \alpha\overline{\theta}_y.$$

This situation has been considered without and with rotation on an f–plane by Stern [29], Toole and Georgi [33] and McDougall [17] and on an equatorial β–plane by Richards [24].

Here we assume the double–diffusive fluxes of momentum and salinity can themselves be modelled by a diffusive process such that

$$-\nabla \cdot \boldsymbol{F}^{x} = K_v u'_{zz} \tag{7}$$

and

$$-\nabla \cdot \boldsymbol{F}^{S} = K_S S'_{zz}. \tag{8}$$

The diffusive flux of temperature is assumed to be proportional to that of salinity,

$$-\nabla \cdot \boldsymbol{F}^{T} = \gamma(\delta/\alpha) K_S S'_{zz}, \tag{9}$$

where the buoyancy flux ratio, γ, is less than one for fingering and greater than one for diffusive layering. We require it to be different from one for the interleaving instability to occur. As pointed out by McDougall [17], this form of parameterization for the vertical flux divergence may not be the most appropriate for fluxes across sharp interfaces. The functional form of the results will change using the scheme suggested by McDougall, but the numerical values are similar – see [24])

The meridional structure of the perturbation meridional velocity takes the form

$$V_0(y) \sim e^{iay} H_n\left(\frac{y-y_0}{\Psi}\right) \exp\left(-\frac{(y-y_0)^2}{2\Psi^2}\right). \tag{10}$$

H_n is the n*th*-order Hermite polynomial. There are three length scales, a^{-1}, an inverse wavenumber, y_0, a northward displacement, and Ψ, an equatorial trapping scale.

The inverse growth rate of the fastest growing perturbation is

$$\hat{\lambda}_r^{-1} \simeq \frac{1}{N}\frac{4\sigma^{1/2}}{\varepsilon_y} \simeq 23\ days. \tag{11}$$

The corresponding vertical wavenumber

$$\hat{m} \simeq \left(\frac{N}{K_s}\frac{|\varepsilon_y|}{4\sigma^{1/2}}\right)^{1/2} \simeq 0.13\ m^{-1}, \tag{12}$$

and the latitudinal extent of the layering is

$$\Psi = \left(\frac{K_v N \sigma^{1/2}}{|\varepsilon_y|\beta^2}\right)^{1/4} \simeq 60\ km. \tag{13}$$

where N is the buoyancy frequency, $\sigma = K_v/K_S$, the Prandtl number and

$$\varepsilon_y = \frac{(1-\gamma_f)g\delta\overline{S}_y}{N^2}, \tag{14}$$

a measure of the strength of the meridional salinity gradient. The values of parameters have been obtained using $N = 0.02\ s^{-1}$, $\varepsilon_y = -1 \times 10^{-4}$, $K_s = 3 \times 10^{-5}\ m^2 s^{-1}$ and $\sigma = 1$. (See Richards [24] for details.)

The vertical and meridional length scales are very similar to those of the observed interleaving. The timescale is rather long but the growth rate is increased by a factor of 5 if we take into account the fact that the flux ratio, γ, is expected to vary with the stability ratio, R_ρ (Walsh and Ruddick [34]).

3.2. INERTIAL INSTABILTY

When the basic state has a sheared flow the flow may be susceptible to inertial instability (Rayleigh [23]). In our case with both horizontal and vertical shear the condition for instability is

$$1 - \overline{u}_z^2/N^2 - \overline{u}_y/f < 0\,. \tag{15}$$

Inertial instability on the equator has been studied by Dunkerton [9], Stevens [30], Clark and Haynes [7] and Hua *et al.* [12]. The simplest case is for a laterally linearily sheared inviscid flow, in which case the meridional structure of the perturbation meridional velocity takes the form

$$V_o(y) \sim \exp\left(-\frac{1}{2}y^{*2}\right) H_n(y^*) \tag{16}$$

where

$$y^* = (y - y_o)/\Psi$$

and y_0, the northward displacement, and Ψ, the equatorial trapping scale are given by

$$y_o = \overline{u}_y/(2\beta), \;\; \Psi^2 = N/(m\beta) \tag{17}$$

For the inviscid case the flow is unstable for any value of the shear and the growth rate of the perturbation increases with vertical wavenumber. There is however an upper bound on the growth rate. Its inverse is given by

$$\hat{\lambda}_r^{-1} = \frac{\overline{u}_y}{2}. \tag{18}$$

With friction Dunkerton [9] has shown that there is a viscous cut-off at the vertical wavenumber

$$m_c = \left(\frac{N\beta}{4K_v^2}\right)^{1/5} \simeq 0.1\ m^{-1} \tag{19}$$

and a critical shear

$$\overline{u}_y = \sqrt{5}\left(2K_v N^2 \beta^2\right)^{1/5} \simeq 4 \times 10^{-6}\ s^{-1}. \tag{20}$$

(taking $N = 0.02\ s^{-1}$ and $K_v = 10^{-4}\ m^2 s^{-1}$). The latitudinal scale is then

$$\Psi \simeq 100\ km. \tag{21}$$

Again we find the length scales are comparable to those observed. The critical shear is equivalent to a change of 0.4 $m\,s^{-1}$ over 1° of latitude, large but not a totally unrealistic value. With this value for the meridional shear the maximum inviscid growth $\hat{\lambda}_r = 2 \times 10^{-6}\ s^{-1}$, the inverse being 6 days. This timescale is somewhat shorter than the diffusive case but it is for the inviscid case and the shear is large.

3.3. COMBINED DIFFUSIVE AND INERTIAL INSTABILTY

The similaritity of scales for the double diffusive and inertial instabilities requires us to look carefully at the combined case. This is work underway

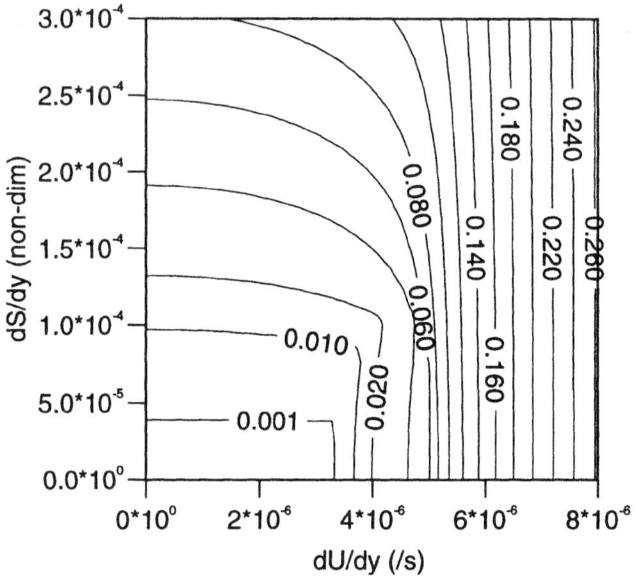

Figure 2. Growth rate of the fastest growing perturbation for combined diffusive and inertial instability, plotted as a function of the meridional gradients of the non-dimensional salinity gradient ($-\varepsilon_y$) and zonal velocity shear. $K_v = K_S = 3 \times 10^{-5}$. Units: day^{-1}.

and will be reported in Edwards and Richards [10]. Here initial results will be presented.

The simplest case to consider is one in which there is a linear meridional shear together with a temperature compensated linear meridional salinity gradient. Further simplifications include application of the hydrostatic approximation and no zonal variation (i.e. $\hat{k} = 0$). The growth rate and associated vertical wavenumber of the fastest growing perturbation as a function of the meridional gradients of salinity, ε_y, and zonal velocity, $\overline{u}_y$, is shown in figures 2 and 3, respectively. Increasing these gradients makes the flow more unstable with respect to double–diffusive and inertial instability respectively.

For values of $|\varepsilon_y| < 10^{-4}$, the diffusive instability is unaffected by the presence of shear until the shear reaches its critical value. For larger values of ε_y, the shear has a significant effect on the instability even for low values of the shear. Clearly the combined instability has some interesting prop-

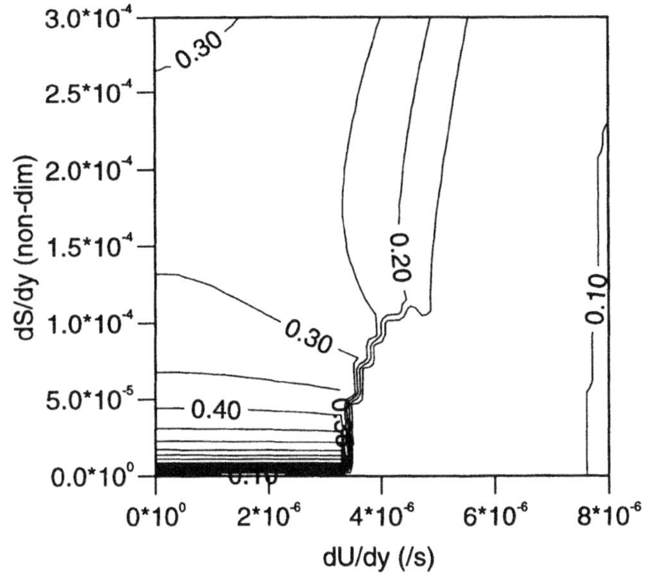

Figure 3. Vertical wavenumber of the fastest growing perturbation for combined diffusive and inertial instability, plotted as a function of the meridional gradients of the non-dimensional salinity gradient ($-\varepsilon_y$) and zonal velocity shear. $K_v = K_S = 3 \times 10^{-5}$. Units: m^{-1}.

erties. It should be noted that the fastest growing mode for the diffusive instability with values of $|\varepsilon_y| < 3 \times 10^{-4}$ occurs for perturbations with zonal variation [24] so the time and length scales given by (11) and (12) do not apply. The assumptions used in the present analysis need to be relaxed but the results presented should be indicative of the more general problem.

4. Finite Amplitude Balance: Proposed Parameterization Scheme

When the layers have reached a quasi–steady state we expect the changes along a layer to be brought about by a divergence of the vertical fluxes across the layer interfaces. Following the approach of Joyce [13] and using the parameterization of vertical fluxes used in Section 3.1 we therefore assume the following balance for salinity

$$v_L \frac{\partial S}{\partial y} \simeq K_S \frac{\partial^2 S}{\partial z^2} \tag{22}$$

where v_L is the along layer advective velocity caused by the interleaving and K_S the vertical diffusion coefficient for salt. In a fingering situation K_S is estimated to $3 \times 10^{-5}\, m^2\, s^{-1}$ by Ruddick and Hebert [28]. Using the observed vertical and along layer gradients of salinity then (22) gives a value for the along layer velocity of

$$v_L = 0.02\, m\, s^{-1}. \tag{23}$$

This is a crude estimate of the advective velocity, but it does demonstrate that the velocity has to be reasonably large. (McDougall [18] shows that a steady state exists only if the interfaces alternate between fingering and diffusive regimes. In this case the details of the balance in (22) will be different although its general nature will be similar.)

Using the velocity given by (23), we estimate the salt flux due to interleaving to be

$$(\overline{v'S'})_{\text{inter}} \sim v_L \Delta S \simeq 0.004\, \text{psu}\, m\, s^{-1} \tag{24}$$

where the overbar represents some areal average over the region of interleaving and ΔS the salinity difference between alternate layers ($\simeq 0.2$ psu in our case). Here we have assumed a strong correlation between the sign of v_L and ΔS, a reasonable assumption for interleaving.

The associated temperature flux is

$$(\overline{v'\theta'})_{\text{inter}} \simeq \frac{\delta}{\alpha} (\overline{v'S'})_{\text{inter}} \simeq 0.012^o\, \text{C}\, m\, s^{-1} \tag{25}$$

This value is comparable to estimates of eddy fluxes, $(\overline{v'\theta'})_{\text{eddy}}$ made from measurements taken from moorings (see *e.g.* Bryden and Brady [5], Banks [1]). The eddy fluxes estimated from mooring measurements are produced by relatively low vertical wavenumber motions having a timescale in excess of O(10 days), what is considered to constitute the more classical lateral eddy flux. Here we suggest that motions with considerably higher vertical wavenumber (and therefore unresolved by conventional ocean models) make a significant contribution to the lateral flux of tracers.

To extend the point further we can estimate an effective diffusion coefficient for the interleaving. Interleaving is acting to laterally mix water properties across the equator. Assuming (and this is a big assumption) that the mixing caused by interleaving can be modelled as a diffusive process on vertical scales large compared to the individual layer thickness and operating over a lateral distance comparable to the meridional extent of the layering then we can write

$$(\overline{v'S'})_{\text{inter}} \simeq K_{\text{inter}} \frac{\partial \overline{S}}{\partial y} \tag{26}$$

Using the above estimate of $(\overline{v'S'})_{\text{inter}}$ and a value for the large scale salinity gradient $\frac{\partial \overline{S}}{\partial y} = 2 \times 10^{-6}\,\text{psu}\,m^{-1}$ then we find

$$K_{\text{inter}} \simeq 2 \times 10^{3}\, m^{2}\, s^{-1} \tag{27}$$

This value for the effective lateral diffusion coefficent is such that we would expect the dynamics of the largescale flow to be affected (see Maes *et al.* [16]).

With the above in mind it is therefore important that the effects of interleaving on the dynamics of the equatorial ocean together with its heat and salinity budgets be considered. The vertical scale of the interleaving is such that its effect needs to be parameterized in ocean GCMs. To proceed we need to know the relationship between the fluxes caused by interleaving and the largescale flow properties. Our knowledge of the interleaving at the equator is still rather rudimentary. Here we suggest a parameterization scheme for interleaving, but it should be borne in mind that the approach is our first attempt and we expect the scheme to be refined as we learn more about the interleaving process.

We base the scheme on the diffusive parameterization given by (26). We then need to determine the diffusion coefficient K_{inter} which can be written as

$$K_{\text{inter}} = v''\mathcal{L} \tag{28}$$

where v'' is an appropriate velocity scale and $\mathcal{L}$ a mixing length.

Assuming the finite amplitude balance is controlled by double–diffusive fluxes (22), equating v'' with v_L, and assuming that the salinity difference ΔS is related to the lateral salinity difference (a reasonable assumption for interleaving) then

$$v''\mathcal{L} \sim K_S\, m^2 \mathcal{L}^2 \tag{29}$$

The vertical wavenumber, m, and the lateral length scale, $\mathcal{L}$, will depend on the mechanism setting these scales. For double–diffusive interleaving they are given by (12) and (13) respectively. The lateral diffusion coefficient is then given by

$$K_{\text{inter}} \sim \left(\frac{N^3 K_v \varepsilon_y}{16\beta^2 \sigma^{1/2}} \right)^{1/2} \tag{30}$$

From the results of inertial instability theory (eqs (19) and (17)) we get

$$K_{\text{inter}} \sim \frac{1}{\sigma} \left(\frac{N^6 K_v^3}{4\beta^4} \right)^{1/5} \tag{31}$$

The functional form for the two different estimates of the diffusion coefficient is remarkable similar, although the double–diffusive case does have a slightly greater dependency on stratification. The double–diffusive case is also dependent on the meridional gradient of salinity.

5. Effect of Lateral Mixing on the Equatorial Current System

The large effective lateral diffusion coefficient for interleaving implies that it will have an impact on the largescale dynamics of the equatorial Pacific. How the dynamics depends on the representation of interleaving needs to be considered. To this end Banks [1] and Richards and Banks [25] have investigated the effect of interleaving in a simplified ocean model using the diffusive representation (30) of the lateral diffusion coefficent.

The analysis in Section 4 applies to the lateral flux of tracers such as salinity and temperature. What about momentum? The relationship between the transfer of momentum and heat/salt in doubly diffusively driven flows is unclear in either the vertical or the horizontal. However two points should be borne in mind. Firstly a study by Brentnall [4] using the generalised thermal wind equation of Colin de Verdière and Schopp [8] and data from TOGA COARE finds that the equatorial current system was in a geostrophic balance even at the equator. Adjustments to the density structure will therefore have an impact on the current structure. Secondly the existence of the layers themselves suggests that mixing of momentum through shear instability is unimportant and can only be intermittent at most. For the results presented below the effective diffusion coefficient for momentum, A_{inter}, is set to a constant value.

The approach is similar to that of Yin and Sarachik [35]. We consider a two–dimensional model (in the vertical–meridional plane) with no zonal variation. The model equations are

$$\frac{\partial u}{\partial t} + v\frac{\partial u}{\partial y} + w\frac{\partial u}{\partial z} - \beta y v = -\frac{1}{\rho}\frac{dP}{dx} + \frac{\partial}{\partial y}A_h\frac{\partial u}{\partial y} + \frac{\partial}{\partial z}A_z\frac{\partial u}{\partial z} \tag{32}$$

$$\frac{\partial v}{\partial t} + v\frac{\partial v}{\partial y} + w\frac{\partial v}{\partial z} + \beta y u = -\frac{1}{\rho}\frac{\partial p}{\partial y} + \frac{\partial}{\partial y}A_h\frac{\partial v}{\partial y} + \frac{\partial}{\partial z}A_z\frac{\partial v}{\partial z} \tag{33}$$

$$\frac{\partial v}{\partial y} + \frac{\partial w}{\partial z} = 0 \tag{34}$$

$$\frac{\partial \theta}{\partial t} + v\frac{\partial \theta}{\partial y} + w\frac{\partial \theta}{\partial z} = \frac{\partial}{\partial y}K_h\frac{\partial \theta}{\partial y} + \frac{\partial}{\partial z}K_z\frac{\partial \theta}{\partial z} \tag{35}$$

$$\frac{\partial S}{\partial t} + v\frac{\partial S}{\partial y} + w\frac{\partial S}{\partial z} = \frac{\partial}{\partial y}K_h\frac{\partial S}{\partial y} + \frac{\partial}{\partial z}K_z\frac{\partial S}{\partial z} \tag{36}$$

$$\frac{\partial p}{\partial z} = -\frac{g}{\rho_o}\rho \tag{37}$$

$$\rho = f(\theta, S) \tag{38}$$

The flow is driven by a surface easterly wind stress balanced by a prescribed zonal pressure gradient, dP/dx. It can be shown that these equations are mathematically inconsistent, in the sense that $\partial/\partial x$ cannot be set to be identically zero, if the vertical variation of the prescribed pressure gradient is non–linear (a necessary condition to obtain a well-defined undercurrent). However this fact should not compromise the results presented here.

Here we consider cases with a constant lateral tracer diffusion coefficient, K_h and one that varies according to (30). The lateral diffusion coefficient for momentum, A_h is kept constant and equal to 5000 $m^2\,s^{-1}$. The vertical diffusion coefficients, A_z and K_z vary according to the Richardson number based scheme of Pacanowski and Philander [21].

The results using a constant value for $K_h = 5000\ m^2\,s^{-1}$ are shown in figure (4). The initial temperature and salinity approximate those during TOGA COARE. The model has been spun–up to a quasi–steady state. The current structure shows a well developed westward surface flow, equivalent to the south equatorial current and an eastward undercurrent. There is reasonably strong upwelling, cooling the SST at the equator. Lateral mixing controls both the momentum and heat budgets in the thermocline, maintaining the strength of the undercurrent. Within the upper 50 m of the ocean vertical mixing plays a dominant role.

Various properties of the flow are given in table 1. Results are also presented for $K_h = 1000\ m^2\,s^{-1}$ and K_h varying as (30), the parameterization for double–diffusive interleaving. In the latter case the maximum value of the coefficient is taken to be 5000 $m^2\,s^{-1}$. A lower bound on K_{inter} of 1000 $m^2\,s^{-1}$ had to be set for numerical stability reasons. Lowering the lateral

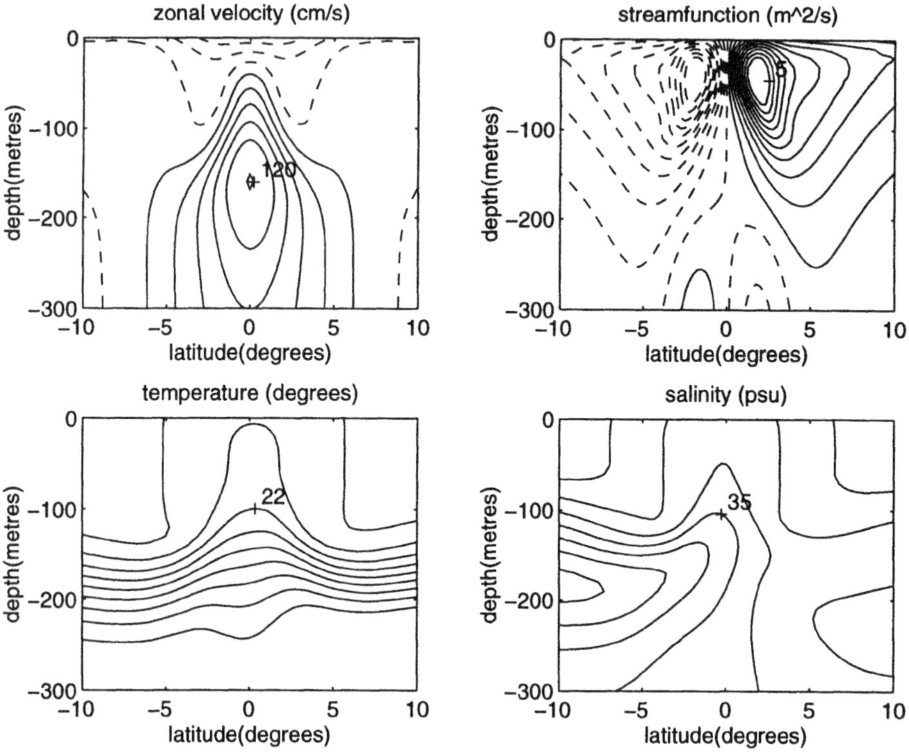

Figure 4. The zonal velocity, streamfunction, temperature and salinity fields for K_h = const. = 5000 $m^2\ s^{-1}$. The contour intervals are 20 $cm\ s^{-1}$, 0.5 $m^2\ s^{-1}$, 2°C and 0.2 psu, respectively.

diffusion coefficient increases the strength of the undercurrent and a further reduction in the SST by over a degree C. The width of the circulation cell also broadens, affecting a larger area around the equator. The variable diffusion coefficient case fits approximately halfway between the two constant cases. Compared with the high constant K_h case, the lowering of the lateral diffusion coefficient brings about a deeper penetration of the effect of vertical mixing.

6. Discussion

The interleaving of water masses at the equator is a persistent feature of the Pacific ocean. Our estimates suggest that the fluxes of heat and salt produced by the interleaving are at least as large as lateral eddy fluxes in the western equatorial Pacific. As noted by Maes *et al.* [16] there is an interplay between vertical and lateral mixing in the equatorial ocean; reducing one increases the effect of the other with an associated change in the structure

TABLE 1. Results from the 2D model for different values of the lateral tracer diffusion coefficient K_S (K_{inter} refers to the use of (30)). U_{EUC} is the maximum speed of the equatorial undercurrent, w_{max} the maximum speed of upwelling and SST_{eq} the sea surface temperature at the equator.

K_S $m^2\,s^{-1}$	U_{EUC} $cm\,s^{-1}$	w_{max} $cm\,s^{-1}$	SST_{eq} ^oC
5000	120	0.0046	24.1
1000	144	0.0031	23.0
K_{inter}	130	0.0048	23.4

and dynamics of the ocean. There will also be an interplay between the processes producing lateral mixing. The presence of interleaving will affect the low frequency eddying motions and vice versa.

Because of the small vertical scale of the interleaving its effect will have to be parameterized in ocean GCMs designed for seasonal to decadal studies. In order to do this we need to know the properties of the interleaving and how it interacts with other scales of motion. We have presented two plausible mechanisms for the formation of the layering and have noted the strong interaction between the two. We have also put forward a parameterization scheme based on the characteristics of both.

The numerical experiments demonstrate that the functional form of the lateral diffusion coefficient impacts on the dynamics and temperature and salinity fields of the equatorial ocean. Changing the direction of the lateral mixing to be along density surfaces or at some angle to them also has a significant effect [1]. The atmosphere over the western equatorial Pacific is very sensitive to the SST. Moderate changes of SST of the size found in the numerical experiments reported here can have a large effect on the convection occuring in the atmosphere which influences the largescale atmospheric circulation [22].

Our knowledge of interleaving is far from complete and there are probable shortcomings in some of the assumptions we have been forced to make. We know little about the eventual breakdown of the layering and subsequent mixing (preliminary results from numerical experiments using a non-hydrostatic model that explicitly resolves the layering suggest that this is accomplished by gravitational overturning as dense fluid is advected over lighter fluid). Is a diffusive parameterization appropriate? There are large zonal variations in the structure and dynamics of the ocean which will influence the development of the layers. The results from the 2D numerical model are only indicative of the effects of lateral mixing and experiments are required with a 3D GCM. However I leave you with the thought that we have here a mechanism by which double–diffusive processes at the small scale may influence the circulation of the equatorial Pacific and hence ENSO events.

Acknowledgments

I am greatly indebted to Helene Banks and Neil Edwards who have worked with me on much of the material presented in these lecture notes. Helpful comments from students and lecturers of the winter school on an earlier draft of these lecture notes are also very much appreciated.

References

1. Banks, H. T., 1997: Intrusions and mixing in the western equatorial Pacific, Ph.D. thesis, University of Southampton.
2. Banks, H. T., and K. J. Richards, 1998: Intrusions in the western equatorial Pacific, To be submitted to J. Geophys. Res.
3. Blanke, B., and P. Delecluse, 1993: Variability of the tropical Alantic ocean simlulated by a general circulation model with two different mixed layer physics, *Journal of Physical Oceanography*, **23**, 1363–1388.
4. Brentnall, S. J., 1995: Geostrophic calculations near the equator, Master's thesis, University of Southampton, Department of Oceanography.
5. Bryden, H. L., and E. C. Brady, 1985: Dianostic model of the three-dimensional circulation in the upper equatorial Pacific Ocean, *Journal of Physical Oceanography*, **15**, 1255–1273.
6. Chen, D., L. M. Rothstein and A. J. Busalacchi, 1994: A hybrid vertical mixing scheme and its application to tropical ocean models, *Journal of Physical Oceanography*, **24**, 2156–2179.
7. Clark, P. D., and P. H. Haynes, 1996: Inertial instability on an asymmetric low–latitude flow, *Quarterly Journal of the Royal Meteorological Society*, **122**, 151–182.
8. Colin de Verdière, A., and R. Schopp, 1994: Flows in a rotating shell: the equatorial case, *Journal of Fluid Mechanics*, **276**, 233–260.
9. Dunkerton, T. J., 1981: On the inertial stability of the equatorial middle atmosphere, *Journal of Atmospheric Sciences*, **38**, 2354–2364.
10. Edwards, N., and K. J. Richards, 1998: Combined double–diffisive and inertial instability on an equatorial β–plane, to be submitted to Journal of Physical Oceanography.
11. Eldin, G., T. Delcroix, C. Henin, K. J. Richards, Y. du Penhoat, J. Picaut and P. Rual, 1994: Large-scale current and thermohaline structures along 156° E during the COARE IOP, *Geophysical Research Letters*, **21**, 2681–2684.
12. Hua, B. L., D. W. Moore and S. Le Gentil, 1997: Inertial nonlinear equilibration of equatorial flows, *Journal of Fluid Mechanics*, **331**, 345–371.
13. Joyce, T. M., 1977: A note on the lateral mixing of water masses, *Journal of Physical Oceanography*, **7**, 626–629.
14. Joyce, T. M., W. Zenk and J. M. Toole, 1978: The anatomy of the Antarctic Polar front in the Drake Passage, *Journal of Geophysical Research*, **83**, 6093–6113.

15. Latif, M., and N. E. Graham, 1992: How much predictive skill is contained in the thermal structure of an oceanic GCM?, *Journal of Physical Oceanography*, **28**, 951–962.
16. Maes, C., G. Madec and P. Delecluse, 1997: Sensitivity of an equatorial Pacific OGCM to the lateral diffusion, *Monthly Weather Review*, **125**, 958–971.
17. McDougall, T. J., 1985a: Double–diffuve interleaving. Part I: Linear stability analysis, *Journal of Physical Oceanography*, **15**, 1532–1541.
18. McDougall, T. J., 1985b: Double–diffuve interleaving. Part II: Finite amplitude. steady state interleaving, *Journal of Physical Oceanography*, **15**, 1542–1556.
19. McPhaden, M. J., 1985: Fine-structure variability observed in CTD measurements from the central equatorial Pacific, *Journal of Geophysical Research*, **90**, 11726–11740.
20. Neelin, J. D., et al., 1992: Tropical air-sea interaction in general circulation models, *Climate Dynamics*, **7**, 73–104.
21. Pacanowski, R., and S. G. H. Philander, 1981: Parameterization of vertical mixing in numerical models of tropical oceans, *Journal of Physical Oceanography*, **11**, 1443–1451.
22. Palmer, T. N., and D. A. Mansfield, 1986: A study of wintertime circulation anomalies during past El Niño events using a high resolution general circulation model. ii. variability of the seasonal mean response, *Quarterly Journal of the Royal Meteorological Society*, **112**, 639–660.
23. Rayleigh, Lord, 1916: On the dynamics of revolving fluids, *Proc. R. Soc. Lond. A*, **93**, 148–154.
24. Richards, K. J., 1991: Double-diffusive interleaving at the equator, *Journal of Physical Oceanography*, **21**, 933–938.
25. Richards, K. J., and H. T. Banks, 1998: the effect of interleaving on the dynamics of the equatorial Pacific, To be submitted to J. Phys. Ocean.
26. Richards, K. J., and R. T. Pollard, 1991: Structure of the upper ocean in the western equatorial Paific, *Nature*, **350**, 48.
27. Ruddick, B. R., 1992: Intrusive mixing in a Mediterranean salt lens – intrusion slopes and dynamical mechanisms, *Journal of Physical Oceanography*, **22**, 1274–1285.
28. Ruddick, B. R., and D. Hebert, 1988: The mixing of meddy 'sharon', in *Small scale turbulence and mixing in the ocean*, edited by J. Nihoul, and B. Jamart, pp. 249–261, Elsevier, Amsterdam.
29. Stern, M. E., 1967: Lateral mixing of water masses, *Deep-Sea Research*, **14**, 747–753.
30. Stevens, D., 1983: On symmetric stability of horizontally sheared flow away from the equator, *Journal of Atmospheric Sciences*, **40**, 882–893.
31. Stockdale, T. D., D. Anderson, M. Davey, P. Delecluse, A. Kattenberg, Y. Kitamura, M. Latif and T. Yamagata, 1993: Intercomparison of tropical Pacific ocean GCM's, *WMO/TD-545 43pp*, WCRP79.
32. Toole, J. M., 1981: Anomalous characteristics of equatorial thermocline finestructure, *Journal of Physical Oceanography*, **11**, 871–876.
33. Toole, J. M., and D. T. Georgi, 1981: On the dynamics and effects of double–diffusively driven intrusions, *Progress in Oceanography*, **10**, 123–145.
34. Walsh, D., and B. Ruddick, 1995: Double–diffusive interleaving: the influence of nonconstant diffusivities, *Journal of Physical Oceanography*, **25**, 348–358.
35. Yin, F. L., and E. S. Sarachik, 1993: Dynamics and heat balance of steady equatorial undercurrents, *Journal of Physical Oceanography*, **24**, 1647–1669.

EDDY PARAMETERISATION IN LARGE SCALE FLOW

PETER D. KILLWORTH
Southampton Oceanography Centre,
Southampton, United Kingdom

1. Background

The best-known difficulty in numerical ocean modelling has been present since the earliest runs of eddy-permitting models: the inherent conflict between the need for simulations of centuries to examine climate issues, and the enormous computer resources necessary to undertake such integrations while retaining at least some degree of eddy activity in the model. For the foreseeable future, climate models will have to continue to include parameterisations of the eddy field from a purely pragmatic perspective. It is also frustrating to need a model delineating every eddy in order to understand broader-scale climate issues, so that a physically meaningful parameterisation is useful for comprehension also.

Until the last five years, oceanographic representations of eddy effects were largely crude, save for early work by Welander (1973) and Marshall (1981), and homogeneous turbulence studies by, *e.g.*, Haidvogel and Held (1980) and Panetta and Held (1988). The representation used simple down-gradient diffusion terms for momentum and tracers. With the centred differences used in numerical ocean models, such terms were necessary to maintain numerical stability, rather than as representations of eddy effects per se. Solomon (1971) and Redi (1982) had shown how to rotate the mixing tensor to account for predominantly isopycnic mixing, but horizontal lateral mixing terms remained for stability; Griffies *et al.* (1997) have shown how to avoid the lateral terms and maintain stability.

Advances in the meteorological literature (*e.g.*, Plumb and Mahlman, 1987) proceeded independently of the two main – and distinct – oceanographic approaches. (Only more recently, in discussion of transformed Eulerian means, are the two approaches converging.) One approach, by Gent and McWilliams (1990), sought to parameterise the effect on tracers by

E. P. Chassignet and J. Verron (eds.), Ocean Modeling and Parameterization, 253–268.

time- and space-varying eddy fields using a mixing of layer thickness. The other, by Eby and Holloway (1994), parameterised the effects on momentum of the presence of a statistical equilibrium field of eddies. They argue for a relaxation of momentum towards a value dominated by a barotropic flow around topography in the Kelvin wave sense. (cf. Sommeria (1998) and Alvarez and Tintoré (1998) in this volume for more details.)

These two approaches both appear to yield improved large-scale oceanographic representations, despite their differing approaches. (Although appealing to different equations, terms in a tracer equation can be manoeuvred into the momentum equation and vice versa; cf. Lee and Leach, 1996, for example.) We shall restrict attention here to conserved tracers such as potential temperature and salinity. These tracers are moved both by time- and space-mean flow, but also by eddy transport terms which can be thought of as an additional 'bolus' advective velocity (the effect has many names and a long history; cf. Gent *et al.*, 1995 for a discussion). Attempts stemming from Gent and McWilliams' (1990) work have largely concentrated on prescriptions for the bolus velocity. They assume that eddy motions represent a release of mean energy through instability mechanisms. Since most parts of the mean ocean circulation have length scales large compared with a deformation radius, much of this release is assumed to occur through a baroclinic, rather than a barotropic, mechanism, following the early suggestions of Green (1970). Underlying these attempts, then, is the concept of reduction of mean available potential energy by eddy fluxes, as well as motions preferentially on density surfaces (we follow other authors here and do not consider difficulties relating to neutral surfaces, etc.; cf. McDougall, 1987 for details). Such concepts cannot, by their nature, deal with long-distance advection of tracers within sub-mesoscale coherent vortices like Meddies (Armi et al, 1989), or many other aspects of ocean flow: barotropic instability, spatial growth and decay, etc. Thus we foresee a suite of parameterisations being necessary for a full representation of nonlinear effects. Some – *e.g.* for barotropic instability – may be local, while others – *e.g.* for Meddies – may be nonlocal.

Gent *et al.* (1995) give an excellent summary of the requirements of a parameterisation based on tracers, including suggestions for the treatment of horizontal boundaries, and discuss two versions of their parameterisation (their equations (18) and (22)), involving down-gradient mixing of isopycnic layer thickness in a manner similar, but not identical, to that in use in isopycnic co-ordinate models (Bleck *et al.*, 1992). Tréguier *et al.* (1997) investigated, inter alia, constraints on such parameterisations caused by horizontal boundaries; they also noted that potential vorticity, rather than layer thickness, was the relevant conserved quantity, and discussed some of the changes using this quantity would produce. Visbeck *et al.* (1997)

examined four potential parameterisations in the light of eddy-resolving initial value problems. Lee *et al.* (1997) have examined how tracers are spread in a statistically steady eddy-resolving channel model, also drawing attention to the importance of potential vorticity conservation.

2. Parameterisation

As a preliminary guide to possible eddy effects, we consider linear perturbation theory applied to a slowly varying mean flow. Linear theory is not a good descriptor of fully developed geostrophic turbulence, particularly for momentum fluxes (*e.g.* Simmons and Hoskins, 1978). Nonetheless, linear theory has the virtue that its motions are solutions of the relevant equations, and fluxes of buoyancy – which are better described by linear theory – do behave in a physically reasonable way.

We assume the background (mean) flow is slowly varying, in the sense that its length scales are large compared with a deformation radius. For coarse ocean climate models this holds automatically on resolution grounds. This slow variation thus precludes barotropic instability, and mitigates against areas of the ocean which have smaller length scales, such as western boundary layers (which are poorly represented in coarse models anyway). It is possible that eddy interactions can remove this convenient scale separation.

Density co-ordinates are used, to facilitate the expansion. Accordingly density is assumed a linear function of temperature and salinity in what follows, and the effects of more complicated equations of state (cf. McDougall, 1987) are ignored. The β-effect is included; it can act to stabilise or destabilise the flow. Conservation of momentum, mass, and the geostrophic relation become

$$u_t + \mathbf{u} \cdot \nabla u - fv = -\frac{B_x}{\rho_0} \tag{1}$$

$$v_t + \mathbf{u} \cdot \nabla v + fu = -\frac{B_y}{\rho_0} \tag{2}$$

$$z_{\rho t} + \nabla \cdot (\mathbf{u} z_\rho) = 0 \tag{3}$$

$$B_\rho = gz \tag{4}$$

where the axes are (x, y) oriented east and north, z is directed upwards, with zero at the surface, ρ is the density (ρ_0 is a reference density), B represents the linear Bernoulli or Montgomery function $p + \rho g z$, g is the acceleration due to gravity, f the Coriolis parameter (with northward variation β), and t the time. The gradient and divergences are taken in the horizontal directions only.

From these equations we can derive conservation of potential vorticity $q = (f + v_x - u_y)/z_\rho$,

$$q_t + \mathbf{u} \cdot \nabla q = 0 \tag{5}$$

Small perturbations are taken against a mean background (denoted by an overbar) which is in steady (or, more formally, slowly varying) geostrophic and mass balance. The expansion procedure will be terminated at the zeroth order, representing a local vertical problem. The first use of such a slowly-varying expansion appears to be Simmons (1974) for a quasi-geostrophic parallel jet shear flow; Robinson and McWilliams (1974) wrote a description in terms of fast and slow space and time variables, but terminated this expression at the leading term. Killworth (1980) extended Simmons' (1974) approach to a general quasi-geostrophic parallel shear flow, and recently Killworth *et al.* (1997) have used a similar theory for primitive equation ring instability.

The perturbations, denoted by primes, satisfy

$$u'_t + \bar{u}u'_x + u'\bar{u}_x + \bar{v}u'_y + \bar{u}_y v' - fv' = -\frac{B'_x}{\rho_0} \tag{6}$$

$$v'_t + \bar{u}v'_x + u'\bar{v}_x + \bar{v}v'_y + v'\bar{v}_y + fu' = -\frac{B'_y}{\rho_0} \tag{7}$$

$$z'_{\rho t} + \nabla \cdot (\bar{\mathbf{u}} z'_\rho + \mathbf{u}'\bar{z}_\rho) = 0 \tag{8}$$

$$B'_\rho = gz' . \tag{9}$$

As in Robinson and McWilliams (1974), we suppose that the mean flow is spatially slowly varying, i.e., that its length scale L is much larger than a, where a is the local deformation radius. Here $a \approx (gH\Delta\rho/\rho_0)^{1/2}/f$, where H is a depth scale, $\Delta\rho$ a density scale, so that the velocity scales on $gH(\Delta\rho/\rho_0)/fL$. We set

$$\frac{a}{L} \equiv \varepsilon \tag{10}$$

which is assumed to be a small quantity. We also assume the mean flow varies temporally on a long time scale $T = \varepsilon^{-2} f^{-1}$.

The perturbation quantities vary on more rapid scales: spatially, of order a, and temporally of order $\varepsilon^{-1} f^{-1}$. Thus

$$B = \overline{B}(X, Y, \rho, T) + B'(x, y, \rho, t, X, Y, T) \tag{11}$$

is assumed, where (X, Y, T) represent the slow variables and (x, y, t) the fast. We now pose

$$B' = \mathrm{Re}\,[B(\rho) \exp ik(x \cos\theta + y \sin\theta - ct)] \tag{12}$$

with similar expressions for (u', v', z'), plus smaller ageostrophic terms, where (k, θ) are the local wavenumber and direction of the perturbation and are functions of the slow variables. The phase velocity c will be complex in general. Choices for k and θ will be made later.

From these equations, we can derive a single equation for B:

$$(\tilde{u} - c)\left(B_{\rho\rho} + \frac{g}{f^2 \rho_0} k^2 \bar{z}_\rho B\right) - B\left(\frac{g}{f^2 \rho_0} \beta \bar{z}_\rho \cos\theta + \tilde{u}_{\rho\rho}\right) = 0 \tag{13}$$

where

$$\tilde{u} = \bar{u}\cos\theta + \bar{v}\sin\theta \tag{14}$$

is the projection of the mean flow in direction θ. This is merely the standard quasi-geostrophic instability problem written in density co-ordinates. To leading order the perturbations are geostrophic:

$$v' = \frac{ik\cos\theta}{\rho_0 f} B', \; u' = -\frac{ik\sin\theta}{\rho_0 f} B' \tag{15}$$

Equation (13) has boundary conditions (for flat surface and floor):

$$(\tilde{u} - c)B_\rho = \tilde{u}_\rho B, \; \rho = \rho_b, \rho_t \, . \tag{16}$$

where ρ_b, ρ_t are the bottom and top densities respectively. We now examine fluxes of perturbed quantities.

2.1. THICKNESS FLUXES

Perturbation thickness fluxes, averaged over many eddy space scales, can be estimated in the usual way to give

$$\overline{u' z'_\rho} = \frac{1}{2}\mathrm{Re}(u z^*_\rho) \, , \; \overline{v' z'_\rho} = \frac{1}{2}\mathrm{Re}(v z^*_\rho) \tag{17}$$

where an asterisk denotes a complex conjugate.

We find

$$\begin{pmatrix} \overline{u' z'_\rho} \\ \overline{v' z'_\rho} \end{pmatrix} = -\frac{k c_i}{2 f^2 \rho_0^2}\left|\frac{B}{\tilde{u} - c}\right|^2 \mathbf{A}\begin{pmatrix} \bar{z}_{\rho x} \\ \bar{z}_{\rho y} \end{pmatrix} + \frac{k c_i \beta \cos\theta}{2 f^3 \rho_0^2}\left|\frac{B}{\tilde{u} - c}\right|^2 \begin{pmatrix} -\sin\theta \\ \cos\theta \end{pmatrix} \bar{z}_\rho$$

$$\equiv -\kappa\left[\mathbf{A}\cdot\nabla\bar{z}_\rho - \frac{\beta}{f}\bar{z}_\rho \mathbf{A}_2\right] = -\kappa f \mathbf{A}\cdot\nabla\left(\frac{\bar{z}_\rho}{f}\right) \tag{18}$$

where

$$\mathbf{A} = \begin{pmatrix} \sin^2\theta & -\sin\theta\cos\theta \\ -\sin\theta\cos\theta & \cos^2\theta \end{pmatrix} \tag{19}$$

is a non-negative matrix (its eigenvalues are 1 and 0), $\mathbf{A}_2$ is its second column, and

$$\kappa(x, y, \rho) = \frac{kc_i}{2f^2\rho_0^2}\left|\frac{B}{\tilde{u} - c}\right|^2 \tag{20}$$

is a diffusivity, proportional to the mean-square particle excursions, and would normally be largest in the interior of the fluid. The diffusivity is everywhere non-negative, from (20), and varies in all three dimensions. It decreases towards zero as horizontal gradients of density decrease. The overall amplitude will be discussed later.

2.2. POTENTIAL VORTICITY FLUXES

The relative vorticity term is uncorrelated with $\mathbf{u}'$, because the Laplacian of B is out of phase with the geostrophic velocity, so that

$$\begin{pmatrix} \overline{u'q'} \\ \overline{v'q'} \end{pmatrix} = -\kappa \mathbf{A} \cdot \nabla \bar{q} \tag{21}$$

with no additional terms (because q, and not z_ρ, is a conserved quantity). Here $\bar{q} = f/\bar{h}$ is the large-scale potential vorticity only. In other words, a parameterisation of thickness fluxes implies one for q, and vice versa. (It is unclear at this time whether, in a global context, mixing of full q, or just the planetary part, as here, would be preferred.)

2.3. DISCUSSION

For the rest of this work we will explore parameterisations of fluxes of quantities, where the shape of the parameterisation is that of (21) or (18), without reference to linear perturbation theory. In other words, (21) is used as a *guide*. (Since instability involves fully-developed eddies, this is all it can be.) The diffusivity κ is permitted to depend on all three spatial coordinates (and time) while the matrix $\mathbf{A}$ depends only on (x, y, t). If $\mathbf{A}$ is the identity matrix, (21) would correspond to purely downgradient (Fickian) transfer. Another example would be the special case of a channel flow (discussed in Section 8) in which there is no mean along-channel gradient ($\partial/\partial x = 0$), where the problem would reduce to two dimensions and $\mathbf{A}$ would become unity; (21) would then be completely downgradient.

On an f-plane, layer thickness and large-scale potential vorticity are equivalent; the form (21) would apply in both cases. On a β-plane, this is not the case: only q is fluxed pseudo-downgradient – the 'pseudo' indicating the inclusion of the matrix $\mathbf{A}$ – while thickness has an extra term. For a channel flow, this extra term in (18) is $(\beta/f)\kappa\bar{z}_\rho$, and is of uniform sign; converted to actual layer thickness this yields a poleward flux which would be present in the absence of mean thickness gradients. Such a flux is clearly visible in Lee et al.'s (1997, Fig. 4b) eddy-resolving channel experiments, where their middle layer is essentially of uniform thickness 500 m, and yet there is a uniform northward flux of about 0.2 $m^2\,s^{-1}$. Using the diffusivity they estimate (about 2000 $m^2\,s^{-1}$) implies a flux $(\beta/f)\kappa h$ of 0.1 $m^2\,s^{-1}$, which is clearly of the right order. Put another way, the term is equivalent to an advective velocity of $\beta\kappa/f$ which is typically of order $10^{-4}\,m\,s^{-1}$. This is a very small velocity; but in the channel case is of uniform direction (poleward) and, being persistent, can exert a strong influence in the steady state. It also tends to advect across property fronts, and thus could be active in their modification.

This β-term raises an apparent difficulty. Lee *et al.*'s (1997) example demonstrates that eddy fluxes of q, rather than z_ρ, lend themselves naturally to parameterisation; indeed, Tréguier *et al.* (1997) and Lee *et al.* (1997) call for such a parameterisation. Greatbatch and Lamb (1990) also demonstrate that downgradient mixing of z_ρ leads to mixing of f/q rather than of q itself. However, most numerical models are not posed with q as a prognostic variable, so that unless some inversion procedure is undertaken – which would slow down the computation of a climate model drastically – knowledge of $\nabla \cdot (\overline{\mathbf{u}'q'})$ needs to be used indirectly.

Two further points should be made. First, this scheme does not fully flux potential vorticity: at this stage, no momentum effects have been included. Second, all schemes based on baroclinic instability have at least conceptual problems near the equator; this scheme, like others, would need modification, which is under consideration at present.

3. The Form of the Solution

It is more convenient to work in z-co-ordinates. The conversion is not straightforward. In particular, the velocity to be produced here is more correctly relevant for the temporal residual mean circulation rather than the divergent 'bolus' velocity (McDougall and McIntosh, 1996; also McDougall, 1998, this volume). The differences between the two are O(amplitude^2), which is relevant here.

If we define the depth-weighted velocity

$$\mathbf{u}^* = \frac{\overline{\mathbf{u}'z'_\rho}}{\bar{z}_\rho}; \quad \mathbf{U} = \bar{\mathbf{u}} + \mathbf{u}^* \tag{22}$$

then

$$\bar{z}_{\rho t} + \nabla \cdot (\mathbf{U}\bar{z}_\rho) = 0 \tag{23}$$

so that

$$\bar{z}_{\rho t}+\nabla\cdot(\bar{\mathbf{u}}\bar{z}_\rho) = \nabla\cdot(\kappa\mathbf{A}\cdot\nabla\bar{z}_\rho)-\beta\nabla\cdot\left(\frac{\kappa\bar{z}_\rho}{f}\mathbf{A}_2\right) = \nabla\cdot\left\{\kappa f\mathbf{A}\cdot\nabla\left(\frac{\bar{z}_\rho}{f}\right)\right\}. \quad (24)$$

On an f-plane, since $\mathbf{A}$ is non-negative definite, solutions of (24) conserve integrals of layer thickness (and hence density itself), and the r.h.s. acts to decrease layer thickness variance. These are desirable properties, as noted by Gent and McWilliams (1990). If β is included, the additional effects have no such properties.

Conversion to (x, y, z) co-ordinates gives the horizontal part of $\mathbf{u}^*$ as

$$\mathbf{u}_H{}^* = \kappa\mathbf{A}\cdot\frac{\partial}{\partial z}\left(\frac{\nabla_H\bar{\rho}}{\bar{\rho}_z}\right) + \frac{\beta}{f}\kappa\mathbf{A}_2. \quad (25)$$

(The vertical component is obtained from the requirement of zero divergence.) Note that the diffusivity κ is outside the derivative in (25), unlike the original shape proposed by Gent and McWilliams (1990)[1]. Gent *et al.* (1995) suggested that this form, without the β-term, permits a steady solution in which the isopycnals are flat (another desirable feature). The addition of the extra term appears to invalidate this; but recall that the diffusivity is that for a local instability problem and would vanish for flat isotherms. The large-scale density satisfies, in z-co-ordinates (Gent et al., 1995)

$$\bar{\rho}_t + \mathbf{U}\cdot\nabla\bar{\rho} = 0 \quad (26)$$

since *any* formulation of TRM velocity implies (26), the differences occurring within the parameterisations. It is straightforward to convert the velocity to a vector streamfunction form following Gent *et al.* (1995), i.e. $\mathbf{u}^* = (\psi_{1z}, \psi_{2z}, -(\psi_{1x} + \psi_{2y}))$.

4. Boundary Conditions

At all boundaries, there must be no normal flow of bolus velocity. In a streamfunction formulation this is easily achieved by requiring the vector streamfunction to vanish at all boundaries, although – as here – its prescription is not in closed form and so would cause difficulties.

In the density co-ordinate form here, other features come into play which prove enlightening. In density co-ordinates it is usual to think of outcropping lines as areas where density layers are flat and have no thickness (indeed, numerical isopycnic models adopt precisely this formulation). Thus

[1]The (here arbitrary) adjustment of placing κ *inside* the horizontal derivative would restore the variance reduction above.

the existence of nonzero thickness fluxes immediately adjacent to surface and floor implies the existence of δ-function changes, so that we must have

$$\mathbf{u}^*\bar{z}_\rho = \kappa\left[-\mathbf{A}\cdot\nabla\bar{z}_\rho + \beta\bar{z}_\rho\mathbf{A}_2/f\right] - \kappa(\rho_s)\mathbf{A}\cdot\nabla\bar{z}\,\delta(\rho-\rho_s) + \kappa(\rho_b)\mathbf{A}\cdot\nabla\bar{z}\,\delta(\rho-\rho_b) \quad (27)$$

where the signs are consistent with the downward increase of density, and we have assumed an immediate change from non-zero to zero slope in $\bar{z}$. The terms in $\nabla\bar{z}$ are evaluated just in the fluid interior.

The two δ-functions in (27) are responsible for the intrusion at the surface of light water into the denser domain, and the intrusion at the bottom of dense water into the lighter domain, when baroclinic instability occurs. In the Eady problem, for example, density gradients are uniform across the channel, and there is identically no thickness flux in the fluid interior. The entire thickness flux occurs in the two δ-functions at top and bottom. Converted into z-co-ordinates, it is clear that the δ-functions have the correct physical behaviour:

$$\mathbf{u}^* = \kappa\left(\mathbf{A}\cdot\frac{\partial}{\partial z}\left(\frac{\nabla_H\bar{\rho}}{\bar{\rho}_z}\right) + \frac{\beta}{f}\mathbf{A}_2\right) - \kappa(\rho_s)\mathbf{A}\cdot\frac{\nabla_H\bar{\rho}}{\bar{\rho}_z}\,\delta(z) + \kappa(\rho_b)\mathbf{A}\cdot\frac{\nabla_H\bar{\rho}}{\bar{\rho}_z}\,\delta(z+H). \quad (28)$$

In the Gent and McWilliams (1990) formulation, either the diffusivity κ was required to vanish at top and bottom or, equivalently, the streamfunction was set to zero to ensure no normal flux. Here it becomes important that κ not vanish, so that it may permit the correct additional fluxes of thickness near the boundaries.

The δ-functions serve two functions: first, they become part of a necessary condition which any parameterisation of form (21) must satisfy; second, they permit an elegant numerical treatment which will not be discussed here further[2].

5. A Necessary Condition

By integrating (21) vertically, including the δ-functions, it is possible to show that parameterisations which mix potential vorticity and which – as here – have the diffusivity outside the horizontal gradient operator, must satisfy

$$\int_{-H}^{0}\frac{\kappa_z}{\rho_z}\mathbf{A}\cdot\nabla\bar{\rho}\,dz = \frac{\beta}{f}\int_{-H}^{0}\kappa\mathbf{A}_2\,dz\,.$$

[2]In fully developed eddies, McDougall (1998, this volume) suggests that the δ-functions could be replaced by a tapering of the interior streamfunction over a height scale $a\,|\nabla\rho/\rho_z|$.

This is trivially satisfied by the linear theory above. It can be shown to be directly equivalent to requiring the mixing terms to yield no net momentum flux. In particular, the simple solution with κ independent of height, is *not* a solution satisfying the consistency condition, except on an f-plane.

6. Approximate Forms for the Parameterisation

With sufficient computer time, it would be straightforward to pose the slowly-varying linear problem at each vertical column and each time step, find the fastest growing mode together with wavenumber and orientation, and compute the *shape* of the diffusivity in the vertical. All that would remain would be an estimate of the overall *size* of the diffusivity.

This is of course not possible under most circumstances. Instead, approximate forms can be created which are both physically motivated and quick to calculate. Neither are derived here: one uses a small wavenumber approach, the other an iterative procedure; cf. Killworth (1997) for details. Both approaches need only $O(n)$ computations for n gridpoints in the vertical, and yield estimates of wavenumber and orientation as required[3].

The overall scaling, with units of velocity times time, partly follows Visbeck et al (1997), and we choose the final scaling as $A \max(a, \Delta) \cdot c_i$, where A is some unknown scaling coefficient of order unity, to be tuned by numerical experiment (see below), a is an estimate of the local deformation radius, and Δ is the grid spacing. The max is suggested as a partial method to handle poor resolution, which would by itself have underestimated density gradients and hence geostrophic flow[4]. Note that the use of c_i for a velocity scale means that stable flow does not mix, as required. It should also be noted that this parameterisation predicts a diffusivity which varies both horizontally and vertically.

Calculations of fastest growth rate in simple problems show that these forms give reasonable answers in such cases. A consistent feature is a mid-depth maximum of diffusivity, as reported by Tréguier (1998) using an eddy-resolving model to estimate fluxes and diffusivities directly.

7. Comparison with an Eddy-resolving Model

Parameterisations need testing in as wide a variety of scenarios as possible. Here we report briefly on an eddy-resolving channel calculation which was then emulated using both the Gent and McWilliams (1990) formulation and the parameterisation discussed here.

[3]It is possible that undesirable numerical effects could occur, *e.g.* grid point noise involving the angle θ. The limited tests below (which did not involve θ) have shown no such problem.

[4]To maintain invariance to grid size, retaining a might be preferable.

The channel was oriented east-west over a longitude range of 2.6°, which is sufficiently wide to contain three wavelengths of the fastest growing linear mode of the initial configuration. The channel occupied a latitudinal extent of 5.2°, centred on 30° N. Its depth was 300 m; this shallow depth was chosen because the configuration had previously been used for other purposes. The model used a straightforward z-co-ordinate code in a fully spherical configuration, with grid spacings 5 m vertically and 0.2° east-west, 0.18° north-south (giving a spacing of 2 km horizontally); the deformation radius, about 10 km, was well resolved by this choice. However, the relatively narrow channel meant that the β-effect was small. All walls are insulating; there is no slip on the north-south walls, and no stress conditions at surface and floor. Numerical eddy coefficients were fairly normal for such resolution: viscous terms were 50 $m^2\, s^{-1}$ horizontally and 5 vertically; diffusion terms were 10 $m^2\, s^{-1}$ horizontally and 1 vertically. This is fairly viscous, but has the virtue of equilibrating with realistic currents.

Initially the channel was occupied by a temperature front (salinity was uniform, with a linear equation of state); the channel is much wider either than the front or than the width the front would slump to following initial baroclinic instability. There was a small seeding of the maximum linear disturbance added at time zero. Eddies build up rapidly on this front, producing significant distortions within 24 days. Left to itself, the release of baroclinic energy would run the system down, eventually producing the flat isopycnals seen as the end result in Gent *et al.* (1995). To prevent this, energy was re-introduced by a strong relaxation of the temperature field back to its initial distribution in 30 grid points at the north and south boundaries.

During the initial slump of the front the available potential energy falls to almost the value relevant for a temperature field linearly interpolated between the two vertical boundaries. Over a period of several years, the available potential energy increases slowly as the front is re-created by the boundary forcing. After eight years, the front begins a second collapse. Computer restrictions prevented following of the cycle further, but there is no reason to doubt that another collapse and recreation of potential energy will occur. However, there is no direct proof that the second cycle will parallel the first since it starts from a different configuration. There is, simply, no good time to halt the experiment; were additional computer time available, a gyre-scale calculation would probably be preferred to an extension of the experiment here.

The channel model can be described by zonal averages (along channel) and time averages over the time period 300 to 2950 days (i.e., over one complete recovery cycle, about 7.25 years). Such a description is hardly statistically reliable, but gives a good guide to the behaviour of the system.

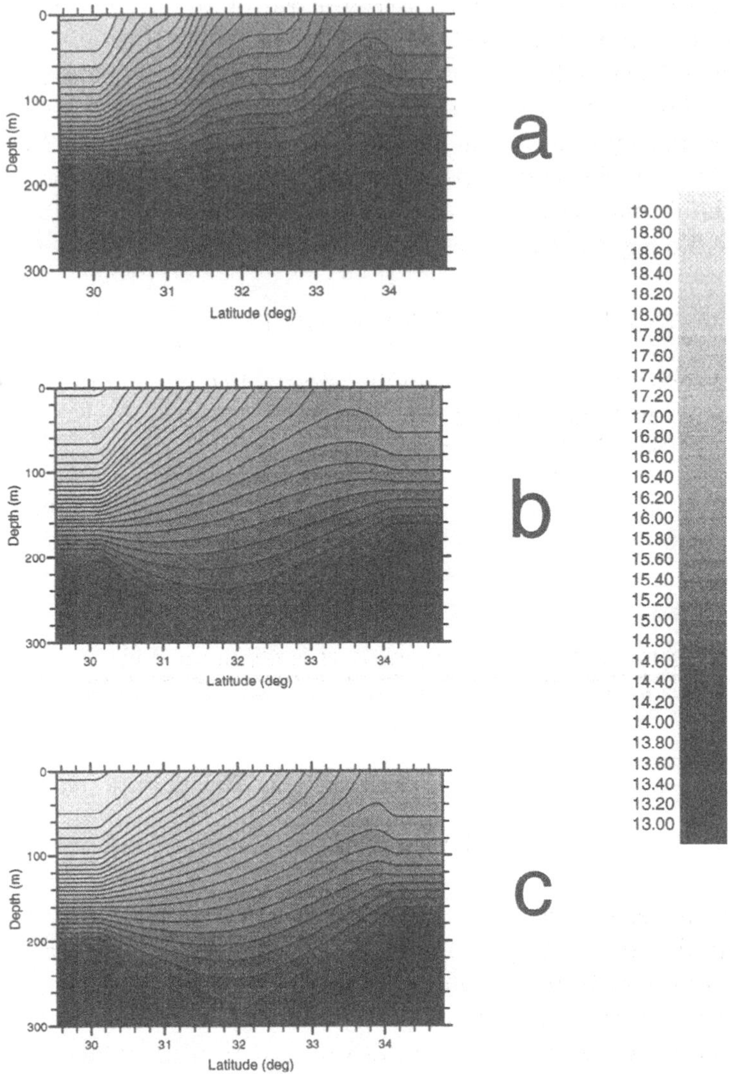

Figure 1. Fig. 1. (a) shows the time- and longitude-mean of the temperature field from the eddy-resolving channel model. (b) shows the best simulation of this using a tuned Gent and McWilliams (1990) scheme in a two-dimensional model, and (c) the equivalent using a tuned version of the new parameterisation.

Space restrictions limit the discussion; the reader is referred to Killworth (1998a) for fuller details, in particular the statistical tests performed. The use of a channel removes the turning matrix and any angular dependence; these features of the parameterisation scheme are not yet tested.

Fig. 1a shows the average temperature field. It displays two fronts, with a region of flatter isopycnals between them. There is evidence of upwelling against the northern region of relaxation, which has led to isopycnals raised above their relaxed level.

This field was simulated by running the equivalent two-dimensional (N-S and vertical) problem using parameterisations until a steady state was achieved. This necessitated tuning of both the Gent and McWilliams diffusivity (to 200 $m^2\, s^{-1}$) and the nondimensional A parameter in the new parameterisation (to 3.3). The tuning was such that both models attained states with potential energy equal to the average over the time interval for the nonlinear problem.

Remarkably, both parameterisations performed essentially well, as Figs. 1b and 1c demonstrate, under all measures. However, the new parameterisation was also able to simulate the initial-value problem used to start the channel model, as well as the long-time average.

Further tests will be necessary using global models to determine whether refinements are necessary in the formulation.

8. Other Aspects

It is possible to use the small-amplitude theory to examine the mixing of tracers. This derivation (Killworth 1998b) finds that tracers are advected by the bolus velocity, and mixed along density surfaces using the same diffusivity as used for thickness and potential vorticity.

Small-amplitude theory can also be used to answer the awkward question of how to handle bolus velocity when it cannot usefully be defined, i.e. in regions of vertical overturn – mixed layers, and regions of deep convection (Killworth, 1998b). Linear instability has to be reposed (since normal quasi-geostrophic theory has broken down) and depth has to be used in place of density as a vertical co-ordinate. It is found that diffusivity does not vary vertically within overturned regions, and connects smoothly with the value at neighbouring stratified areas. The effect of mixing again appears in the equation for density, which is both quasi-diffused horizontally ('quasi' because the Coriolis parameter again enters), as does the turning matrix, and an extra advective term appears which acts to slump the overturned fluid towards stability, as one would expect from instability theory. This latter term is the only one to release available potential energy. This mixing scheme is applied in the overturned layer while the original theory is applied in the stratified region.

This theory is as yet untested.

9. Conclusions

Eddy parameterisation is an active field, with most schemes concentrating, as here, on the effects of predominantly baroclinic eddies. The scheme discussed here predicts a mixing of potential vorticity rather than layer thickness (though these are identical on an f-plane). Simple eddy-resolving model runs have suggested that potential vorticity is mixed rather than thickness; on the other hand, thickness mixing can be shown to reduce the net energy of a closed system, while potential vorticity mixing cannot so be shown (although in almost all circumstances it will reduce energy). The additional β-term, while small, is typically oriented across property fronts, which may have interesting long-term effects.

One important offshoot from the theoretical approach here is that for any variety of potential vorticity mixing, eddy diffusivity cannot be uniform in the vertical if momentum is not to be fed to the system. (The limited observations (Jenkins, private communication) support the view that diffusivity varies strongly in the vertical, although this may be due to many possible effects.) Is this variation of order one importance, or a detail? Until more robust comparisons between the parameterisations discussed here – and others yet to come – can be made, this question, like many others related to parameterisation, cannot be satisfactorily answered.

Acknowledgments

My thanks to the lecturers and students at the school who provided detailed critiques of this manuscript, and to Eric Chassignet and Jacques Verron for an excellent meeting.

References

Alvarez, A. and J. Tintoré. (1998) Topographic stress: Importance and parameterization. In *Ocean Modeling and Parameterization*, E.P. Chassignet and J. Verron (Eds.), Kluwer Academic Publishers, 327-350.

Armi, L., D. Hebert, N. Oakey *et al.* (1989) Two years in the life of a Mediterranean salt lens. *J. Phys. Oceanogr.*, **19**, 354-370.

Bleck, R., C. Rooth, D. Hu, and L.T. Smith. (1992) Salinity-driven thermocline transients in a wind- and thermohaline-forced isopycnic coordinate model of the North Atlantic. *J. Phys. Oceanogr.*, **22**, 1486-1505.

Eby, M. and G. Holloway. (1994) Sensitivity of a large scale ocean model to a parameterization of topographic stress. *J. Phys. Oceanogr.*, **24**, 2577-2587.

Gent, P.R. and J.C. McWilliams. (1990) Isopycnal Mixing in Ocean Circulation Models. *J. Phys. Oceanogr.*, **20**, 150-155.

Gent, P.R., J. Willebrand, T.J. McDougall and J.C. McWilliams. (1995) Parameterizing Eddy-Induced Transports in Ocean Circulation Models. *J. Phys. Oceanogr.*, **25**, 463-474.

Green, J.S.A. (1970) Transfer properties of large scale eddies and the general circulation of the atmosphere. *Quart. J. R. Met. Soc.*, **96**, 157-185.

Griffies, S.M., A. Gnanadeskina, R.C. Pacanowski, V.D. Larichev, J.K. Dukowicz and R.D. Smith. (1997) Isoneutral diffusion in a z-coordinate ocean model. *J. Phys. Oceanogr.*, in press.

Haidvogel, D.B. and I.M. Held. (1980) Homogeneous Quasi-Geostrophic Turbulence Driven by a Uniform Temperature Gradient. *J. Atmos. Sci.*, **37**, 2644-2660.

Killworth, P.D. (1980) Barotropic and baroclinic instability in rotating stratified fluids. *Dyn. Atmos. Ocean.*, **4**, 143-184.

Killworth, P.D. (1981) Eddy fluxes and mean flow tendencies in open ocean baroclinic instability. *Dyn. Atmos. Ocean.*, **5**, 175-186.

Killworth, P.D. (1997) On the parameterisation of eddy transfer. Part I: Theory. *J. Mar. Res.*, 1171-1197.

Killworth, P.D. (1998a) On the parameterisation of eddy transfer. Part II: Tests with a channel model. *J. Mar. Res.*, in press.

Killworth, P.D. (1998b) On the parameterisation of eddy transfer. Part III: Treatment of tracers and areas with no vertical stratification. Submitted to *J. Mar. Res.*

Killworth, P.D., J.R. Blundell and W.K. Dewar. (1997) Primitive-equation instability of large oceanic rings. I. Linear theory. *J. Phys. Oceanogr.*, **27**, 941-962.

Lee, M.-M., D.P. Marshall and R.G. Williams. (1997) On the Eddy Transfer of Tracers: Advective or Diffusive? *J. Mar. Res.*, **55**, 483-505.

Lee, M.-M. and H. Leach. (1996) Eliassen-Palm Flux and Eddy Potential Vorticity Flux for a Nonquasigeostrophic Time-Mean Flow. *J. Phys. Oceanogr.*, **26**, 1304-1319.

Marshall, J.C. (1981) On the parameterization of geostrophic eddies in the ocean. *J. Phys. Oceanogr.*, **11**, 257-271.

McDougall, T.J. (1987) Neutral surfaces. *J. Phys. Oceanogr.*, **17**, 1950-1964.

Dougall, T.J. (1998) Three-dimensional residual-mean theory. In *Ocean Modeling and Parameterization*, E.P. Chassignet and J. Verron (Eds.), Kluwer Academic Publishers, 269-302.

McDougall, T.J. and P.C. McIntosh (1996). The temporal-residual-mean velocity. Part I: derivation and the scalar conservation equations. *J. Phys. Oceanogr.*, **26**, 2653-2665.

Panetta, R.L. and I.M. Held. (1988) Baroclinic Eddy Fluxes in a One-Dimensional Model of Quasi-Geostrophic Turbulence. *J. Atmos. Sci.*, **45**, 3354-3365.

Plumb, R.A. and J.D. Mahlman. (1987) The Zonally Averaged Transport Characteristics of the GFDL General Circulation/Transport Model. *J. Atmos. Sci.*, **44**, 298-327.

Redi, M. (1982) Oceanic isopycnal mixing by coordinate rotation. *J. Phys. Oceanogr.*, **12**, 1154-1157.

Robinson, A.R. and J.C. McWilliams. (1974) The Baroclinic Instability of the Open Ocean. *J. Phys. Oceanogr.*, **4**, 281-294.

Simmons, A.J. (1974) The meridional scale of baroclinic waves. *J. Atmos. Sci.*, **31**, 1515-1525.

Simmons, A.J. an B.J. Hoskins. (1978) The life cycle of some nonlinear baroclinic waves. *J. Atmos. Sci.*, **35**, 414-432.

Solomon, H. (1971) On the Representation of Isentropic Mixing in Ocean Circulation. *J. Phys. Oceanogr.*, **1**, 233-234.

Sommeria J. (1998) Statistical mechanics of potential vorticity for parameterizing mesoscale eddies. In *Ocean Modeling and Parameterization*, E.P. Chassignet and J. Verron (Eds.), Kluwer Academic Publishers, 303-326.

Tréguier, A.M. (1998) Evaluating eddy mixing coefficients from eddy resolving ocean models: a case study. Submitted to J. Phys. Oceanogr.

Tréguier, A.M., I.M. Held, and V.D. Larichev. (1997) On the parameterization of quasi-geostrophic eddies in primitive equation ocean models. *J. Phys. Oceanogr.*, **27**, 567-580..

Visbeck, M., J. Marshall, T. Haine and M. Spall. (1997) On the specification of eddy transfer coefficients in coarse resolution ocean circulation models. *J. Phys. Oceanogr.*, **27**, 381-402.

Welander, P. (1973) Lateral friction in the ocean as an effect of potential vorticity mixing. *Geophys. Fl. Dyn.*, **5**, 173-189.

THREE-DIMENSIONAL RESIDUAL-MEAN THEORY

TREVOR J. MCDOUGALL
CSIRO Marine Research,
Hobart, Tasmania, Australia

Abstract

Mesoscale eddies act to mix fluid parcels in a way that is highly constrained by the stratified nature of the fluid. The residual-mean theory provides the link between the different views that are apparent from averaging these turbulent flow fields in height coordinates and in density coordinates. It reduces the parameterization problem from three dimensions to two dimensions and it shows how the eddy fluxes are skew-symmetric in height coordinates so that the total advection velocity can be adiabatic.

1. Introduction

The residual-mean theory of Andrews and McIntyre (1976) was developed for the zonal averaging operator. It was realized that the Eulerian-mean northward velocity was not a reliable indicator of the mean velocity at which heat and tracers were advected meridionally. The zonal-residual-mean theory involves a revised total velocity field, called the residual-mean velocity field, which better represents the velocity at which heat and tracers were advected by the "mean" atmospheric circulation. This zonal-residual-mean theory has been extended to the oceanic case (where continents interrupt the zonal integrals) by McIntosh and McDougall (1996).

The task of parameterizing unresolved mesoscale eddy motions in a forward ocean model is significantly more complicated than forming an appropriate zonally-averaged velocity field. There are two principal reasons for the added complexity. First, the mean flow field in a forward ocean model is three dimensional and is of zeroth order in perturbation amplitude (in contrast to the two-dimensional velocity field of the zonally-averaged problem where the Eulerian-mean flow is of second order in perturbation amplitude), and second, one needs to account for unresolved fields in both

E. P. Chassignet and J. Verron (eds.), Ocean Modeling and Parameterization, 269–302.

space and time. The temporal-residual-mean (TRM) theory addresses the first issue, while there is as yet no published solution to the second issue (a potential solution to this problem is sketched below).

We begin by summarizing the TRM theory, first emphasizing its advective nature, then examining the vector decomposition of the eddy density flux that is inherent in the TRM approach. This is followed by a physical interpretation of the TRM conservation equations as being equivalent to the corresponding equations averaged in density coordinates. The TRM theory provides the physical interpretation of the quasi-Stokes (or eddy-induced) streamfunction of a method such as that of Gent *et al.* (1995). This streamfunction is the contribution of eddies to the horizontal transport of water of a certain density class, and this physical interpretation assists in applying boundary conditions to the quasi-Stokes streamfunction.

This article concentrates on theoretical features of residual-mean theory and we argue that the Gent and McWilliams (1990) scheme for advecting tracers in models is not equivalent to down-gradient diffusion of thickness despite this being the original justification of the scheme. The construction of the vertical component of the eddy-induced velocity assumes that the eddy-induced velocity is three-dimensionally non-divergent and this means that the eddy-induced velocity cannot be proportional to the down-gradient flux of thickness. This also is apparent in the diapycnal nature of the resolved-scale velocity in such ocean models, and we argue that the resolved-scale velocity in these models is the Eulerian-mean velocity, that is, it is the velocity averaged at constant height, not at constant density as we have previously stated (Gent *et al.*, 1995). In this way, it is argued that the Gent *et al.* (1995) scheme is really a parameterization of the quasi-Stokes velocity of the TRM theory. This realization points to the route for improved parameterizations, and it also has implications for how we must interpret the model variables in such coarse-resolution models.

Coarse resolution ocean models suffer from two separate problems because of their lack of horizontal resolution. The first problem is that they do not exhibit mesoscale eddy activity and the temporal-residual-mean (TRM) theory addresses this issue. In this way we can imagine a model that has fine horizontal resolution but, for whatever reason, does not have mesoscale eddies (perhaps because the model is too viscous). The addition of the skew diffusion of the TRM theory can correct for the effects of the missing mesoscale eddies. The second problem is that these models have no knowledge of the spatial correlations of properties on scales smaller than their resolution. This second problem would arise whether mesoscale eddies and temporal perturbations were present or not. In this article, a tentative solution to this lack of horizontal resolution is advanced. The end result of this procedure is to suggest more terms that should be added to the quasi-Stokes streamfunction. These extra contributions effectively put into a coarse resolution model the missing horizontal spatial correlations. The extra contributions do not require any parameterization since they are already written in terms of the variables carried by the coarse resolution model.

This article concentrates on the theory of three-dimensional residual-mean circulations and it does not discuss the state of the art in actually parameterizing the quasi-Stokes streamfunction of the TRM circulation, although a few remarks are made along these lines in the concluding section. Nor does this article concentrate on the significant advantages that have been identified to date when eddy-induced advection is included in coarse-resolution ocean models.

2. Review of the TRM Mean Density Equation

In what follows the averaging operator (the overbar) is defined to be a low-pass temporal average and primed quantities are the deviations from this low-passed value. In developing the theory it is convenient to deal with the density conservation equation and to ignore any non-linearity in the equation of state. In practice ocean models do not carry a conservation equation for density but rather for conservative tracers such as salinity and potential temperature. Diapycnal mixing in the ocean is relatively weak and it proves very useful to develop the residual-mean theory for the density conservation equation and to treat density as a conservative variable for this purpose. The usual conservation equations for Eulerian-averaged density and for half the density variance are

$$\bar{D}_t\bar{\gamma} = \bar{Q} - \nabla \cdot \left(\overline{\boldsymbol{U}'\gamma'}\right) \tag{1}$$

and

$$\bar{D}_t\bar{\phi} = \overline{Q'\gamma'} - \overline{\boldsymbol{U}'\gamma'} \cdot \nabla\bar{\gamma} + O\left(\alpha^3\right) \tag{2}$$

Here $\bar{\phi} \equiv \frac{1}{2}\overline{\gamma'^2}$ is half the density variance measured at a fixed point in space and the terminology $O\left(\alpha^3\right)$ indicates terms that are of cubic or higher order in perturbation amplitude.

In order to develop residual-mean conservation equations that apply to unsteady flows we need to admit the possibility that the Eulerian-mean density may not be the most appropriate mean density to appear in the mean density conservation equation. For example, the Eulerian-mean density, $\bar{\gamma}(x, y, z, t)$, describes a density surface whose average height is not that of the original Eulerian averaging, z. The appropriate mean density is the one whose surface is, on average, at the height of the averaging. This density can be expressed in terms of $\bar{\gamma}$ and $\bar{\phi}$ by

$$\tilde{\gamma} = \bar{\gamma} - \left(\frac{\bar{\phi}}{\bar{\gamma}_z}\right)_z + O\left(\alpha^3\right) \tag{3}$$

The distinction here is between averaging density at a given height, z, and averaging the height of a given density surface $\gamma = \tilde{\gamma}$. McDougall and McIntosh (1998a) used the density variance equation, (2), to rewrite the mean density conservation equation, (1), in terms of this new mean density, $\tilde{\gamma}$, the TRM velocity, $\bar{\boldsymbol{U}}^{\#}$ and a new "diabatic" term, $\bar{Q}^{\#}$ as

$$\bar{D}_t^{\#}\tilde{\gamma} \equiv \tilde{\gamma}_t + \bar{\boldsymbol{U}}^{\#} \cdot \nabla\tilde{\gamma} = \bar{Q}^{\#} + O\left(\alpha^3\right) \tag{4}$$

where

$$\bar{\boldsymbol{U}}^{\#} \equiv \bar{\boldsymbol{U}} + \boldsymbol{U}^{+}; \quad \boldsymbol{U}^{+} \equiv \nabla \times (\boldsymbol{\Psi} \times \mathbf{k}) = \boldsymbol{\Psi}_z - \mathbf{k}\left(\nabla_H \cdot \boldsymbol{\Psi}\right) \tag{5}$$

$$\boldsymbol{\Psi} \equiv -\frac{\overline{\boldsymbol{V}'\gamma'}}{\bar{\gamma}_z} + \frac{\bar{\boldsymbol{V}}_z}{\bar{\gamma}_z}\left(\frac{\bar{\phi}}{\bar{\gamma}_z}\right) \tag{6}$$

and

$$\bar{Q}^{\#} = \bar{Q} + \left[-\frac{\overline{Q'\gamma'}}{\bar{\gamma}_z} + \frac{\bar{Q}_z}{\bar{\gamma}_z}\left(\frac{\bar{\phi}}{\bar{\gamma}_z}\right)\right]_z \tag{7}$$

We will return to a detailed discussion of these expressions, but first the TRM density conservation, (4), will be derived a little differently; this time examining the nature of the eddy density flux itself rather than simply its divergence.

3. Vector Decomposition of the Eddy Density Flux

Combining (3) and (1), the conservation equation for $\tilde{\gamma}$ can be written as

$$\tilde{\gamma}_t + \nabla \cdot \left(\bar{\boldsymbol{U}}\tilde{\gamma}\right) = \bar{Q}^{\#} - \nabla \cdot \boldsymbol{F}^{M} + O\left(\alpha^3\right) \tag{8}$$

where the modified density flux, $\boldsymbol{F}^{M}$, is

$$\boldsymbol{F}^{M} = \overline{\boldsymbol{U}'\gamma'} + \bar{\boldsymbol{U}}\left(\frac{\bar{\phi}}{\bar{\gamma}_z}\right)_z + \mathbf{k}\left\{\left(\frac{\bar{\phi}}{\bar{\gamma}_z}\right)_t - \frac{\overline{Q'\gamma'}}{\bar{\gamma}_z} + \frac{\bar{Q}_z}{\bar{\gamma}_z}\left(\frac{\bar{\phi}}{\bar{\gamma}_z}\right)\right\} \tag{9}$$

Now the density variance equation, (2), is used to eliminate $\overline{w'\gamma'}$ from the expression (9), and guided by the results above, the modified density flux, can be expressed as

$$\boldsymbol{F}^{M} = -\boldsymbol{A}\nabla\tilde{\gamma} + \boldsymbol{N} + O\left(\alpha^3\right) \tag{10}$$

where the asymmetric matrix $\boldsymbol{A}$ is defined in terms of the two components of the quasi-Stokes streamfunction, $\boldsymbol{\Psi} = (\Psi^x, \Psi^y)$, as

$$\boldsymbol{A} \equiv \begin{bmatrix} 0 & 0 & \Psi^x \\ 0 & 0 & \Psi^y \\ -\Psi^x & -\Psi^y & 0 \end{bmatrix} \tag{11}$$

and the non-divergent density flux, $\boldsymbol{N}$, is given by

$$\boldsymbol{N} \equiv \left(\frac{\bar{\boldsymbol{V}}\bar{\phi}}{\bar{\gamma}_z}\right)_z - \mathbf{k}\nabla_H \cdot \left(\frac{\bar{\boldsymbol{V}}\bar{\phi}}{\bar{\gamma}_z}\right) = \nabla \times \left[\left(\frac{\bar{\boldsymbol{V}}\bar{\phi}}{\bar{\gamma}_z}\right) \times \mathbf{k}\right] \tag{12}$$

The density variance equation, (2), can be written in terms of the modified density flux, $\boldsymbol{F}^M$, as (subscripts denote differentiation)

$$\bar{D}_t\bar{\phi} = -\boldsymbol{F}^M \cdot \nabla\tilde{\gamma} + \left\{\bar{\gamma}_z\left(\frac{\bar{\phi}}{\bar{\gamma}_z}\right)_t - \bar{\gamma}_t\left(\frac{\bar{\phi}}{\bar{\gamma}_z}\right)_z + \left(\frac{\bar{Q}\bar{\phi}}{\bar{\gamma}_z}\right)_z\right\} + O\left(\alpha^3\right) \tag{13}$$

Apart from the three terms in curly brackets (the two unsteady terms and the diabatic term), it is apparent that the advection of density variance by the mean flow is balanced by the diapycnal component of the modified density flux. Furthermore, the vector decomposition, (10), shows that the diapycnal component of $\boldsymbol{F}^M$ is due only to the non-divergent flux, $\boldsymbol{N}$, since the skew flux is directed in the density surface (see Figure 1c). That is, apart from the diabatic and unsteady terms, we have $\bar{D}_t\bar{\phi} = -\boldsymbol{N}\cdot\nabla\tilde{\gamma} + O\left(\alpha^3\right)$. In the same three-dimensional turbulent situation as we are considering, Marshall and Shutts (1981) have found that the horizontal advection of variance is balanced by a non-divergent two-dimensional flux. This finding was subject to the assumption that the mean flow in the horizontal plane follows the mean density contours. Without having to invoke such an assumption, the present result generalizes Marshall and Shutt's finding to three dimensions, involving the advection of variance by the three-dimensional mean flow, $\bar{D}_t\bar{\phi}$, and the non-divergent part, $\boldsymbol{N}$, of the three-dimensional modified density flux. However, this result appears to be of no real use but rather is in the nature of a truism because neither side of $\bar{D}_t\bar{\phi} = -\boldsymbol{N} \cdot \nabla\tilde{\gamma} + O\left(\alpha^3\right)$ involves the eddy density flux.

4. Discussion of the TRM Density Equation

It is apparent that the modified density flux, $\boldsymbol{F}^M$, represents the eddy forcing of the mean density equation, (8). The TRM theory (McDougall and McIntosh, 1998a) shows

- that $\boldsymbol{F}^M$ is the sum of a skew flux, $-\boldsymbol{A}\nabla\tilde{\gamma}$ plus some non-gradient-terms,

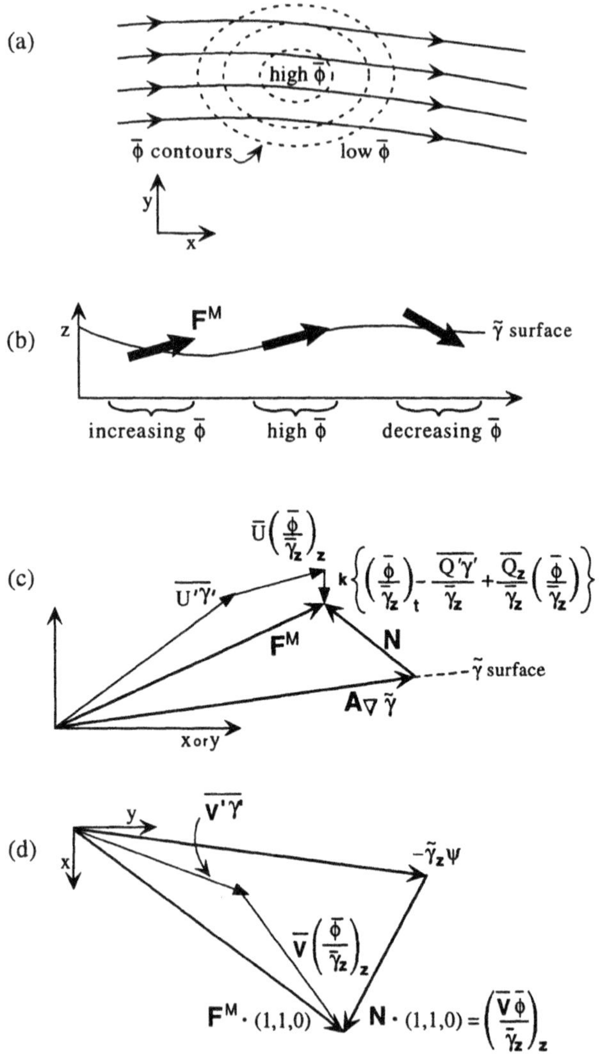

Figure 1. As the mean flow moves through a region of increasing density variance (panel (a)) the modified density flux, $\boldsymbol{F}^M$, must have an upwards directed diapycnal component (panel (b), where it is assumed that the unsteady and diabatic terms in (13) do not dominate). This diapycnal component of $\boldsymbol{F}^M$ is supplied by only the non-divergent flux, $\boldsymbol{N}$, since the skew flux is directed in the density surface [panel (c)]. Panel (c) also indicates the construction of the modified density flux according to its definition, (9). The TRM decomposition of the modified density flux into the two parts $\boldsymbol{A}\nabla\tilde{\gamma}$ and $\boldsymbol{N}$ is fascinating as it is only the skew flux that affects the mean density equation, while it is only the non-divergent flux that affects the density variance equation. Panel (d) shows just the horizontal components of the full three-dimensional vectors that are sketched in panel (c).

- that these non-gradient-terms are non-divergent and so do not affect the mean density equation,
- that the diapycnal component of $\boldsymbol{F}^M$ is due only to the non-divergent flux, $\boldsymbol{N}$,
- that the skew flux, $-\boldsymbol{A}\boldsymbol{\nabla}\tilde{\gamma}$, does not contribute to the density variance equation,
- that the symmetric diffusion tensor, $\boldsymbol{S}$, (which we anticipate exists and causes diffusion of passive tracers along density surfaces) must be such that it does not cause a density flux, that is, $\boldsymbol{S}\boldsymbol{\nabla}\tilde{\gamma} = \boldsymbol{0}$, consistent with current practice in ocean GCMs (beginning with Redi, 1982), and
- that the divergence of the skew flux is the advection of density by the extra (quasi-Stokes) velocity, that is, $\boldsymbol{\nabla} \cdot (\boldsymbol{A}\boldsymbol{\nabla}\tilde{\gamma}) = -\boldsymbol{U}^+ \cdot \boldsymbol{\nabla}\tilde{\gamma}$.

These points are illustrated in Figure 1.

The above has involved various manipulations of (i) the mean density equation, (ii) the density variance equation, and (iii), the continuity equation. Apart from simply writing these equations in different forms, what has been achieved? One achievement can be described as a major simplification. In the original Reynolds-averaged mean density equation, (1), the task of parameterizing the eddy density flux is daunting and has never been done successfully because each of the three components of the eddy flux, $\overline{\boldsymbol{U}'\gamma'}$, needs to be parameterized and one has no intuition about how much of this flux should be divergent, how much should be diapycnal or what form the non-gradient terms might take. An indication of how difficult it would be to parameterize $\overline{\boldsymbol{U}'\gamma'}$ directly can be obtained by combining (9) and (10) to write $\overline{\boldsymbol{U}'\gamma'}$ as

$$\overline{\boldsymbol{U}'\gamma'} = -\boldsymbol{A}\boldsymbol{\nabla}\tilde{\gamma} + \boldsymbol{N} - \bar{\boldsymbol{U}}\left(\frac{\bar{\phi}}{\bar{\gamma}_z}\right)_z - \mathbf{k}\left\{\left(\frac{\bar{\phi}}{\bar{\gamma}_z}\right)_t - \frac{\overline{Q'\gamma'}}{\bar{\gamma}_z} + \frac{\bar{Q}_z}{\bar{\gamma}_z}\left(\frac{\bar{\phi}}{\bar{\gamma}_z}\right)\right\} + O\left(\alpha^3\right) \tag{14}$$

All terms except the first are non-gradient terms; that is, they are not of the form of a diffusion tensor operating on the appropriate mean density gradient, $\boldsymbol{\nabla}\tilde{\gamma}$. How would one have guessed the nature of these non-gradient terms without the TRM theory? In contrast, all that needs to be parameterized in the TRM density conservation (4) or (8) is the quasi-Stokes streamfunction, $\boldsymbol{\Psi}$. It is expected that it will be substantially easier to parameterize this two-dimensional vector rather than having to parameterize the three-dimensional vector $\overline{\boldsymbol{U}'\gamma'}$.

Another major achievement of TRM theory (and equally of the Gent and McWilliams (1990) eddy parameterization scheme) is that even if the parameterization of $\boldsymbol{\Psi}$ is imperfect, because the parameterized term enters the density conservation equation as a skew flux which is equivalent to an extra advection of density, the total velocity, the TRM velocity,

$\bar{\boldsymbol{U}}^{\#} = \bar{\boldsymbol{U}} + \boldsymbol{\Psi}_z - \mathbf{k}\,(\nabla_H \cdot \boldsymbol{\Psi})$, will have a diapycnal component only if the diabatic source term, $\bar{Q}^{\#}$, is non-zero. That is, uncertainty in the parameterization of the quasi-Stokes streamfunction will not cause unwanted changes in water masses. This feature occurs because the TRM theory shows that the symmetric diffusion tensor does not cause any flux of density, which provides belated proof of this assumption in the rotated symmetric diffusion tensor of Redi (1982) and Griffies *et al.* (1998).

Hand-in-hand with this progress in understanding the mean density equation has come the realization that the conservation equations should be written in terms of the modified density variable, $\tilde{\gamma}$, or equivalently, that this is the correct interpretation for the mean density that is carried by an eddyless general circulation model. Also, just as the mean density field and the total advection velocity are re-interpreted in the TRM approach, so too the appropriate diabatic source term is changed from $\bar{Q}$ to $\bar{Q}^{\#}$. The need for this change is best understood by averaging the conservation equation in density coordinates, and this is discussed in the next section.

There are other approaches to developing mean conservation equations in turbulent stratified fluids. The generalized-Lagrangian-mean approach (Andrews and McIntyre, 1978) is an exact theory but it has not proved of practical use because the Lagrangian-mean velocity and the Stokes velocity are both divergent at leading order (second order) in perturbation quantities. This is the reason why we call our additional velocity, $\boldsymbol{U}^{+}$, the "quasi-Stokes" velocity because not only is it an approximation to the true Stokes velocity, but also the divergence of the quasi-Stokes velocity is identically zero and so differs from the divergence of the true Stokes velocity at leading order. Plumb and Mahlman (1987) have pursued a Lagrangian perturbation analysis of the tracer conservation equations under the zonal averaging operator and we have extended this technique to the density equation under temporal averaging in three dimensions (this is not discussed here), but to date we have not been successful in applying Plumb and Mahlman's "effective transport" method to the tracer equations under temporal averaging in three dimensions. The "effective" density conservation equation has much in common with the TRM density conservation equation in that the density variable is different to $\bar{\gamma}$ and, like the TRM velocity, the effective velocity is non-divergent. The "effective" approach is however less attractive that the TRM approach because of the difficulty in evaluating the correlations between Lagrangian perturbation quantities as compared to the TRM approach where the correlations are between perturbation quantities evaluated at fixed points in space.

5. The Isopycnal Interpretation of TRM

With errors that are cubic in perturbation quantities, it is possible to show that the TRM approach of forming averaged quantities for use in height coordinates corresponds to temporally averaging the instantaneous conservation equations in density coordinates. McDougall and McIntosh (1998b) have proven that the quasi-Stokes streamfunction, $\boldsymbol{\Psi}$, is the contribution of perturbations to the horizontal transport of fluid that is denser than $\tilde{\gamma}(z)$, the density of the density surface whose average height is z. Their approach consisted of writing a vertical Taylor series expression for the horizontal transport of fluid of this density class, and is illustrated in Figure 2a and by the equation

$$\boldsymbol{\Psi} \equiv -\frac{\overline{\boldsymbol{V}'\gamma'}}{\bar{\gamma}_z} + \frac{\bar{\boldsymbol{V}}_z}{\bar{\gamma}_z}\left(\frac{\bar{\phi}}{\bar{\gamma}_z}\right) = \overline{\left(\boldsymbol{V} + \frac{1}{2}\boldsymbol{V}_z\eta'\right)\eta'} + O\left(\alpha^3\right) \tag{15}$$

where η' is the perturbation height of the density surface, so that $\overline{\eta'} \equiv 0$ and $\eta' = -\gamma'|_z / \bar{\gamma}_z + O(\alpha^3)$. The velocity $\left(\boldsymbol{V} + \frac{1}{2}\boldsymbol{V}_z\eta'\right)$ is the instantaneous velocity at the height mid way between the fixed height, z, and the instantaneous height of the density surface, $z + \eta'$, so that the quasi-Stokes streamfunction is the average horizontal transport of fluid between these same two heights as indicated in Figure 2a. It is this horizontal transport of fluid that must be added to the volume transport found by using the Eulerian-averaged velocity in order to correctly represent the transport of shaded fluid in Figure 2a.

The Taylor series approach also shows that the thickness-weighted value, $\hat{A}$ of any quantity, A, obtained by averaging A between a pair of closely-spaced $\tilde{\gamma}$ surfaces is

$$\hat{A} \equiv \bar{\gamma}_z \overline{\left(\frac{A}{\gamma_z}\right)\Big|_{\tilde{\gamma}}} \equiv \tilde{A} + \overline{A'|_{\tilde{\gamma}}\eta'_z} = \bar{A} + \left(-\frac{\overline{A'\gamma'}}{\bar{\gamma}_z} + \frac{\bar{A}_z}{\bar{\gamma}_z}\left(\frac{\bar{\phi}}{\bar{\gamma}_z}\right)\right)_z + O\left(\alpha^3\right) \tag{16}$$

where $\tilde{A}$ is the value averaged on the density surface (not thickness-weighted) and $\overline{A'|_{\tilde{\gamma}}\eta'_z}$ is due to the correlations between perturbations evaluated on a density surface and the thickness between two closely-spaced surfaces. With A equal to the horizontal velocity vector, $\boldsymbol{V}$, (16) shows that

$$\hat{\boldsymbol{V}} \equiv \tilde{\boldsymbol{V}} + \boldsymbol{V}^B = \bar{\boldsymbol{V}} + \boldsymbol{\Psi}_z + O\left(\alpha^3\right) = \bar{\boldsymbol{V}}^{\#} + O\left(\alpha^3\right) \tag{17}$$

where the bolus velocity, $\boldsymbol{V}^B = \overline{\boldsymbol{V}'|_{\tilde{\gamma}}\eta'_z}$, is due to correlations between the horizontal velocity evaluated on the density surface and the thickness between density surfaces. Hence the thickness-weighted horizontal velocity

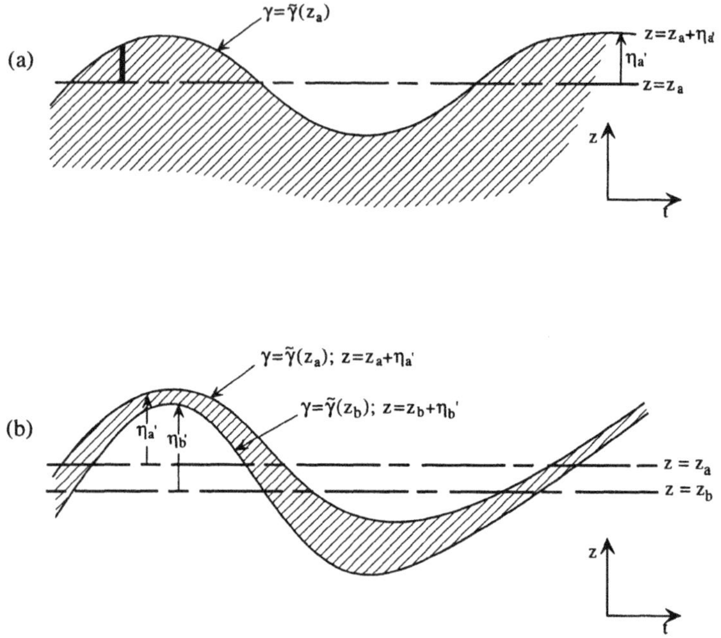

Figure 2. (a) Sketch of the height of a $\tilde{\gamma}$ density surface as a function of time at a given latitude and longitude. The quasi-Stokes streamfunction, $\boldsymbol{\Psi}$, is equal to the horizontal volume transport between the fixed height, $z = z_a$, and the instantaneous height of this density surface. This volume flux is equal to the temporal average of the perturbation height, η', and the velocity appropriate to the column of fluid shown, namely, $\left(\boldsymbol{V} + \frac{1}{2}\boldsymbol{V}_z\eta'\right)$ where both $\boldsymbol{V}$ and $\boldsymbol{V}_z$ are evaluated at the fixed height, $z = z_a$. It is this volume transport that must be added to the volume transport found by using the horizontal Eulerian-mean velocity in order to correctly estimate the transport of the shaded fluid that is more dense than $\tilde{\gamma}(z)$. (b) The thickness-weighted horizontal velocity of density coordinates, $\hat{\boldsymbol{V}}$, is defined to be the average horizontal velocity of the the shaded fluid, and this is equal to the horizontal TRM velocity, $\bar{\boldsymbol{V}}^{\#} \equiv \bar{\boldsymbol{V}} + \boldsymbol{\Psi}_z$, with an error that is cubic in the amplitude of the temporal perturbations.

obtained by averaging in density coordinates is equal to the horizontal TRM velocity, $\bar{\boldsymbol{V}}^{\#}$, up to error terms of cubic order in perturbation quantities.

When the instantaneous density conservation equation, $D_t\gamma = Q$, is averaged in density coordinates, one finds that the time-averaged diapycnal velocity, $\tilde{e} \equiv \overline{e|_{\tilde{\gamma}}}$, is given by

$$\tilde{e} \equiv \overline{\left(\frac{Q}{\gamma_z}\right)\Big|_{\tilde{\gamma}}} \equiv \frac{\hat{Q}}{\tilde{\gamma}_z} = \frac{\bar{Q}^{\#}}{\tilde{\gamma}_z} + O\left(\alpha^3\right) \tag{18}$$

where the last part of this expression has used the general thickness-weighted relationship, (16), and it is noted (from (4)) that $\bar{Q}^{\#}/\tilde{\gamma}_z$ is also the diapycnal component, $\tilde{e}^{\#}$, of the TRM velocity, $\bar{\boldsymbol{U}}^{\#}$. That is, $\bar{D}_t^{\#}\tilde{\gamma} = \tilde{e}^{\#}\tilde{\gamma}_z = \bar{Q}^{\#} + O\left(\alpha^3\right)$.

These results, (17) and (18), derived by McDougall and McIntosh (1998b) from a Taylor series approach, show that both the horizontal and the diapycnal components of the TRM velocity, $\bar{\boldsymbol{U}}^{\#}$, are the same as those obtained by averaging the instantaneous flow with respect to $\tilde{\gamma}$ surfaces (within cubic errors in perturbation amplitude). In this way, it is clear that the TRM theory is the way of representing in height coordinates the relevant terms that arise from averaging in density coordinates. The reason why the TRM approach is needed in height coordinates is that unless significant changes are made to the horizontal momentum equations, the resolved-scale velocity in height-coordinate models is the Eulerian-mean velocity.

The above results have errors that are cubic in the amplitude of perturbation quantities and so the question arises whether these errors may be significant. On the basis of the corresponding zonal-residual-mean theory, we have reason to believe that the cubic terms will be quite small. McIntosh and McDougall (1996) extended the zonal-residual-mean theory of Andrews and McIntyre (1976) to the oceanic situation where the continents interrupt the zonal integrals, and they showed that the zonally-thickness-weighted circulation (evaluated by zonally averaging in density coordinates) was well approximated by the zonal-residual-mean circulation which was evaluated in height coordinates. This close correspondence occurred despite the fact that at a constant latitude in the Southern ocean, density surfaces vary in height by about 1,500 m. In the TRM case of temporal averaging at a given latitude and longitude, density surfaces undulate much less than 1,500 m, with root-mean-square vertical excursions of 150 m being more typical. Hence we expect that ignoring the cubic terms in perturbation amplitude will be quite adequate for our purposes in the TRM theory.

6. The TRM Tracer Conservation Equation

The divergence form of the instantaneous tracer conservation equation is $(\tau/\gamma_z)_t + \nabla_\gamma \cdot (\boldsymbol{V}\tau/\gamma_z) + (e\tau)_\gamma = X/\gamma_z$, where the temporal derivative is taken following a density surface, e is the diapycnal velocity and X is the source term which represents local production or consumption, as well as the flux divergence of unresolved mixing processes. Temporally averaging this tracer equation yields

$$\left(\frac{\hat{\tau}}{\tilde{\gamma}_z}\right)_t + \nabla_{\tilde{\gamma}} \cdot \left(\frac{\hat{\boldsymbol{V}}\hat{\tau}}{\tilde{\gamma}_z}\right) + (\tilde{e}\hat{\tau})_{\tilde{\gamma}} = \frac{\hat{X}}{\tilde{\gamma}_z} - \overline{\left(\frac{Q''\tau''}{\gamma_z}\right)_{\tilde{\gamma}}} - \nabla_{\tilde{\gamma}} \cdot \overline{\left(\frac{\boldsymbol{V}''\tau''}{\gamma_z}\right)} \qquad (19)$$

where both the overbars on the right represent quantities averaged at constant density and subscripts denote differentiation. Here the mean value of the tracer is the thickness-weighted tracer value, with the double primed variables being the deviation of the instantaneous variables from their thickness-weighted values. These variables obey

$$\hat{\tau} \equiv \tilde{\gamma}_z \left(\overline{\frac{\tau}{\gamma_z}\Big|_{\tilde{\gamma}}} \right) \quad \text{and} \quad \overline{\frac{\tau''}{\gamma_z}\Big|_{\tilde{\gamma}}} \equiv 0 \tag{20}$$

To make progress it is necessary to assume that the eddy fluxes of tracer in (19) are representable as a diffusion tensor operating on the gradient of thickness-weighted tracer. McDougall and McIntosh (1998b) assume that the first two terms on the right-hand side of (19) amount to $(1/\tilde{\gamma}_z)$ times the divergence of a diapycnal diffusivity, D, operating on the diapycnal gradient of $\hat{\tau}$. That is, the sum of these two terms are regarded as $(1/\tilde{\gamma}_z)\,(D\hat{\tau}_z)_z$. The epipycnal flux (*i.e.* the flux directed along a density surface) in (19) is also assumed to take a Fickian form, namely, $\overline{\left(\frac{\boldsymbol{V}''\tau''}{\gamma_z}\right)} = -\frac{K}{\tilde{\gamma}_z}\nabla_{\tilde{\gamma}}\hat{\tau}$. These assumptions mean that the right-hand side of (19) can be written in height coordinates as $(1/\tilde{\gamma}_z)\,\nabla\cdot(\boldsymbol{S}\nabla\hat{\tau})$, where $\boldsymbol{S}$ is the symmetric diffusion tensor of Redi (1982). It is worth noting that these Fickian assumptions have not yet been proven for temporal averaging of three-dimensional flow. Under the simpler zonal averaging operator, Plumb and Mahlman (1987) expanded perturbation quantities in terms of linearized Lagrangian displacements and they were able to establish the diffusive nature of the fluxes for the tracer equation in that case. To date such an approach has not been successful for temporal averaging in three dimensions in either Cartesian or density coordinates.

The left-hand side of (19) is now transformed into height coordinates, using (17) to equate $\hat{\boldsymbol{V}}$ to $\bar{\boldsymbol{V}}^{\#}$ and (18) to equate $\tilde{e}$ to $\bar{Q}^{\#}/\tilde{\gamma}_z$ which from (4) is seen to be the diapycnal component, $\bar{e}^{\#}$, of the TRM velocity. The result of this transformation is

$$\hat{\tau}_t|_z + \nabla_H \cdot \left(\bar{\boldsymbol{V}}^{\#}\hat{\tau}\right) + \left(\bar{w}^{\#}\hat{\tau}\right)_z = \nabla\cdot(\boldsymbol{S}\nabla\hat{\tau}) + O\left(\alpha^3\right) \tag{21}$$

or

$$\hat{\tau}_t|_z + \nabla_H \cdot \left(\bar{\boldsymbol{V}}\hat{\tau}\right) + \left(\bar{w}\hat{\tau}\right)_z = \nabla\cdot([\boldsymbol{S}+\boldsymbol{A}]\nabla\hat{\tau}) + O\left(\alpha^3\right) \tag{22}$$

This conservation equation is of exactly the same form as is solved in modern GCMs of the ocean. The form (21) applies when the quasi-Stokes streamfunction is used to form the quasi-Stokes velocity whereas the form (22) uses the quasi-Stokes streamfunction in the antisymmetric diffusion tensor, $\boldsymbol{A}$. Griffies (1998) has pointed out that adopting the skew-diffusion approach, (22), is preferable numerically because the quasi-Stokes streamfunction is spatially differentiated one less time in comparison with (21). It is perhaps worth noting that while the skew flux, $-\boldsymbol{A}\nabla\hat{\tau} = -\hat{\tau}_z\boldsymbol{\Psi} + \mathbf{k}\left[\boldsymbol{\Psi}\cdot\nabla_H\hat{\tau}\right]$ has no component in the direction of $\nabla\hat{\tau}$ the horizontal and vertical components of the skew flux are often down-gradient and up-gradient respectively (see Plumb, 1979).

There are two major conclusions to take away from the TRM tracer conservation equation, (22). First, one needs to not only diffuse the tracer with a symmetric diffusion tensor, but also with the asymmetric diffusion tensor, $\boldsymbol{A}$, defined by (11) and (6). This ensures that the tracer is transported with the TRM velocity. The second point is that the mean tracer is not the Eulerian-averaged tracer value. If one wanted to insist that the model's tracer value was the Eulerian-mean tracer value, then one would need to impose an additional flux divergence in the tracer equation as follows,

$$\bar{\tau}_t|_z + \nabla_H\cdot\left(\bar{\boldsymbol{V}}\bar{\tau}\right) + \left(\bar{w}\bar{\tau}\right)_z = \nabla\cdot\left(\left[\boldsymbol{S}+\boldsymbol{A}\right]\nabla\bar{\tau}\right) - \nabla\cdot\boldsymbol{E} + O\left(\alpha^3\right) \qquad (23)$$

where the additional flux, $\boldsymbol{E}$, is (using (16)),

$$\boldsymbol{E} = \mathbf{k}\left(-\frac{\overline{\tau'\gamma'}}{\bar{\gamma}_z} + \frac{\bar{\tau}_z}{\bar{\gamma}_z}\left(\frac{\bar{\phi}}{\bar{\gamma}_z}\right)\right)_t + \bar{\boldsymbol{U}}\left(-\frac{\overline{\tau'\gamma'}}{\bar{\gamma}_z} + \frac{\bar{\tau}_z}{\bar{\gamma}_z}\left(\frac{\bar{\phi}}{\bar{\gamma}_z}\right)\right)_z \qquad (24)$$

Such non-gradient terms would be very difficult to parameterize.

It is perhaps surprising, but also fortunate, that the tracer that is naturally carried by an eddyless height-coordinate model should be interpreted in exactly the same way as the tracer carried by an eddyless density-coordinate model, namely, as the thickness-weighted tracer value, $\hat{\tau}$. Lozier *et al.* (1994) have drawn attention to the damage that can be done to water masses by averaging salinity and potential temperature data at fixed height. The theoretical result presented in this section dictates that tracer data should be averaged between density surfaces not only for the purposes of forming atlases of hydrographic data, but also for comparing with the output of ocean models, for the assimilation of observed data into ocean models and before using data in inverse models.

While the tracer variables in height-coordinate and density-coordinate models should be interpreted identically, the same cannot be said of the velocity vectors. The resolved-scale horizontal velocity in a height-coordinate model is the Eulerian-mean velocity, $\bar{\boldsymbol{V}}$, whereas the resolved-scale horizontal velocity in a density coordinate model is the horizontal velocity averaged on a density surface, $\tilde{\boldsymbol{V}}$. This interpretation of the horizontal velocities in the two different coordinates is consistent (through geostrophy) with the manner in which the horizontal pressure gradient is calculated in height-coordinates and the isopycnal gradient of the Montgomery potential is evaluated in density coordinates.

7. The Diapycnal Nature of the Eulerian-mean Velocity

The advection of $\tilde{\gamma}$ by the Eulerian-mean flow is given by (from (8) using (10) and (11))

$$\begin{aligned} \tilde{\gamma}_t + \bar{\boldsymbol{U}} \cdot \nabla \tilde{\gamma} &= \bar{Q}^{\#} + \nabla \cdot (\boldsymbol{A} \nabla \tilde{\gamma}) + O\left(\alpha^3\right) \\ &= \bar{Q}^{\#} - \boldsymbol{U}^{+} \cdot \nabla \tilde{\gamma} + O\left(\alpha^3\right) \\ &= \bar{Q}^{\#} + \tilde{\gamma}_z \nabla_{\tilde{\gamma}} \cdot \boldsymbol{\Psi} + O\left(\alpha^3\right) \end{aligned} \tag{25}$$

where $\nabla_{\tilde{\gamma}}$ is the spatial gradient operator in a $\tilde{\gamma}$ surface. Hence the diapycnal component of the Eulerian-mean velocity, $\bar{e}$ is

$$\bar{e} = \bar{Q}^{\#}/\tilde{\gamma}_z + \nabla_{\tilde{\gamma}} \cdot \boldsymbol{\Psi} + O\left(\alpha^3\right) \tag{26}$$

This is a compact way of expressing how the eddy forcing affects the Eulerian-mean density equation:- namely by the epipycnal divergence of the quasi-Stokes streamfunction, $\nabla_{\tilde{\gamma}} \cdot \boldsymbol{\Psi}$. Since the physical interpretation of the quasi-Stokes streamfunction is the contribution of eddy motions to the transport of water of a certain density class, it is perhaps not surprising to see the epineutral divergence of this same physical quantity appear as a diapycnal advection. (Note that we use the word "diapycnal" to mean the part of the vertical velocity that flows through the $\tilde{\gamma}$ surface, not through the $\bar{\gamma}$ surface.)

In contrast to the Eulerian-mean flow, the diapycnal component of the TRM velocity, $\bar{e}^{\#}$ is

$$\bar{e}^{\#} = \bar{Q}^{\#}/\tilde{\gamma}_z + O\left(\alpha^3\right) \tag{27}$$

Gent and McWilliams (1990) realized that their suggested parameterization of eddies caused the resolved-scale velocity to have a diapycnal component and for several years this feature was considered a weakness of their scheme. This was considered a sufficiently negative feature that it was emphasised that the areal average of this diapycnal velocity component was zero. Now, of course, we realise that the Eulerian-mean velocity should have a diapycnal component caused by mesoscale eddies. Gent *et al.* (1995) pointed out that the relevant velocity at which tracers are advected was the total velocity (which we are now interpreting as the TRM velocity, $\bar{\boldsymbol{U}}^{\#}$) and they showed that this TRM velocity was diapycnal only to the extent of explicit diapycnal mixing processes, as is evident in (27). A good way of illustrating the diapycnal nature of the Eulerian-mean flow is to consider the relaxation of the density front of section 6 of Gent *et al.* (1995). There the Eulerian-mean flow is zero and so the advection of density following the Eulerian-mean flow is simply the temporal change of density at a point in space. As the front relaxes, the density at a fixed point certainly changes and this occurs [see (25)] due to the variation of the quasi-Stokes streamfunction along density surfaces. If one were to interpret these changes of density following the Eulerian-mean flow as a symmetric diffusion of density then it would be as often up-gradient as down-gradient. The key is to realize that these density changes are not due to diffusion in the normal sense of the word, but rather they are due to the divergence of a skew-diffusive density flux. This realization is totally equivalent to viewing the process with respect to the TRM velocity, when one finds that the front relaxes adiabatically; this being the explanation advanced by Gent *et al.* (1995).

The vector relationship between the velocity vectors $\bar{\boldsymbol{U}}$, $\boldsymbol{U}^{+}$ and $\bar{\boldsymbol{U}}^{\#}$ is illustrated in Figure 3. The positions of a given density surface are shown at an initial time and at a later time and the labelled velocity vectors are actually the displacements achieved by those velocities in this time interval. The vertical component of the Eulerian-mean velocity is due to the sum of (i) the sliding along the density surface due to its slope, $-\bar{\boldsymbol{V}} \cdot \nabla_H \tilde{\gamma}/\tilde{\gamma}_z$, (ii), the vertical advection of the density surface, $-\tilde{\gamma}_t/\tilde{\gamma}_z$, (iii), diabatic mixing processes, $\bar{Q}^{\#}/\tilde{\gamma}_z$, and (iv), the diapycnal eddy forcing of the Eulerian-mean flow, $\nabla_{\tilde{\gamma}} \cdot \boldsymbol{\Psi}$. The vertical component of the TRM velocity has no diapycnal eddy forcing but rather is the sum of just three terms, namely (i) the sliding along the density surface due to its slope, $-\bar{\boldsymbol{V}}^{\#} \cdot \nabla_H \tilde{\gamma}/\tilde{\gamma}_z$, (ii), the vertical advection of the density surface, $-\tilde{\gamma}_t/\tilde{\gamma}_z$ and (iii), the diabatic mixing processes, $\bar{Q}^{\#}/\tilde{\gamma}_z$. The cubic terms in perturbation quantities are not included on this figure as they are of higher order and are believed to be unimportant.

The eddy forcing of the mean density equation can be expressed as $-\boldsymbol{U}^{+} \cdot \nabla\tilde{\gamma}$ [see the middle line of (25)] and the relative contribution of the vertical quasi-Stokes velocity to this is

$$\frac{-w^{+}\tilde{\gamma}_z}{-\boldsymbol{U}^{+} \cdot \nabla\tilde{\gamma}} = \frac{\nabla_H \cdot \boldsymbol{\Psi}}{\nabla_{\tilde{\gamma}} \cdot \boldsymbol{\Psi}} \tag{28}$$

In the quasi-geostrophic (QG) limit this ratio is assumed to be unity because only the vertical component of the extra velocity appears as eddy forcing in the density equation (see Tréguier *et al.*, 1997). This equation shows that the QG limit is equivalent to assuming that the quasi-Stokes streamfunction varies much more strongly with horizontal position than with height. To the extent that QG theory is applicable to the real ocean,

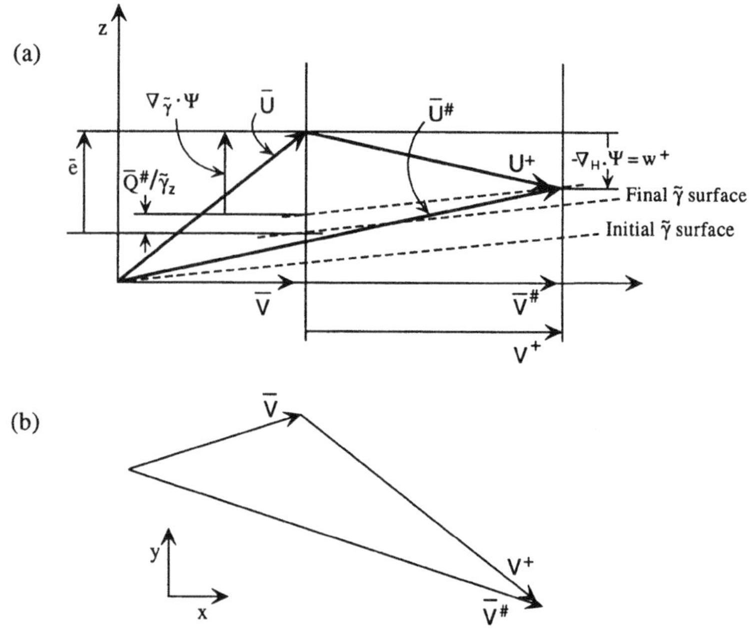

Figure 3. (a) Sketch of the Eulerian-mean velocity, $\bar{\boldsymbol{U}}$, the quasi-Stokes velocity, $\boldsymbol{U}^+$, and the TRM velocity, $\bar{\boldsymbol{U}}^{\#}$. (b) The horizontal components of the same three velocities.

the implication would be that the horizontal quasi-Stokes velocity is unimportant. The following argument suggests that this implication of QG theory is inappropriate to the real ocean. McDougall (1995) pointed out that while the eddy-induced horizontal velocity, $\boldsymbol{V}^+$, might be small (rarely larger than 1 mm/s), the fact that it tends to point in a direction across the mean epipycnal property gradients means that it can be the dominant term in a conservation equation. Also, GCM experiments with the Gent and McWilliams (1990) scheme show that the quasi-Stokes circulation accounts for up to 0.5 PW of meridional heat flux in the Southern Ocean (Danabasoglu and McWilliams, 1995). Hence it seems that this aspect of QG theory is inappropriate to the problem of parameterizing eddies in the ocean.

Ocean GCM simulations using the Gent and McWilliams (1990) scheme have found large diapycnal transports in the Southern Ocean for both the Eulerian-mean flow and the quasi-Stokes circulation. Hirst and McDougall (1998) specifically plotted the zonally-averaged streamfunctions of the Eulerian-mean flow and the quasi-Stokes flow in density coordinates to illustrate the diapycnal nature of both circulations (their figure 6), and they found about 14Sv of zonally-averaged diapycnal transport in the Southern

Ocean in both the Eulerian-mean and the quasi-Stokes circulations. The sum of these circulations was much smaller. The Gent and McWilliams (1990) scheme is often described as being a parameterization of "bolus transport", but if this were the case one would expect the extra advection of the Gent *et al.* (1995) scheme to have no diapycnal transport. In Gent *et al.* (1995) we also assumed that the resolved-scale velocity of an eddy-less ocean model could be interpreted as the velocity averaged on density surfaces, but the large diapycnal transport of the resolved-scale velocity illustrates that this is not the case. Rather the resolved-scale velocity in such models is the Eulerian-mean velocity. The observation that the quasi-Stokes advection has substantial diapycnal transport raises the important question:- "in what sense can we regard the quasi-Stokes velocity as bolus transport?" This question sets the scene for section 10 of this paper.

8. The Quasi-Stokes Velocity is not Due to the Down-gradient Thickness Flux

The aim of Gent and McWilliams (1990) and of Gent *et al.* (1995) was to provide a parameterization for the bolus velocity and it was argued that this should be directed down the isopycnal gradient of thickness so that the extra horizontal velocity would be $\kappa\left(\nabla_H\tilde{\gamma}/\tilde{\gamma}_z\right)_z$. However, in order to make the streamfunction a locally-determined quantity, the diffusivity was moved inside the vertical derivative so that the streamfunction of Gent *et al.* (1995) became $\kappa\nabla_H\tilde{\gamma}/\tilde{\gamma}_z$. This form also allowed the boundary conditions at the top and bottom of the ocean to be easily satisfied by having the diffusivity go to zero.

An immediate consequence of the TRM theory is that it justifies the appearance of κ inside the vertical derivative and so provides theoretical support for a closure of the Gent and McWilliams (1990) form. This is seen by making the most elementary assumption about the horizontal density flux in the expression, (6), for the quasi-Stokes streamfunction. If one assumes that the horizontal density flux is directed down the horizontal gradient of $\tilde{\gamma}$ with diffusivity κ (that is, $\overline{\mathbf{V}'\gamma'} = -\kappa\nabla_H\tilde{\gamma}$) and if one ignores the second term in the quasi-Stokes streamfunction, then (6) becomes $\mathbf{\Psi} = \kappa\nabla_H\tilde{\gamma}/\tilde{\gamma}_z$ exactly as in Gent *et al.* (1995) and as used by many other authors.

It was realized by Gent *et al.* (1995) that in many situations it was crucial that the diffusivity be located inside the vertical derivative in the expression for the eddy-induced velocity. For example, in the Eady problem (and to a good approximation also in the Southern Ocean), the slopes of density surfaces are independent of height so that the original idea of down-gradient thickness flux would mean that the eddy-induced horizontal velocity was zero at all heights except for delta functions of eddy-induced horizontal velocity at the top and bottom. With the diffusivity inside the vertical derivative, and with the streamfunction set to zero on all boundaries, the Gent and McWilliams (1990) parameterization quite naturally gives boundary layers of eddy-induced advection at the top and bottom of the ocean just as in the Eady problem, and the thickness of these boundary layers can be readily controlled by the height over which the diffusivity is tapered to zero. It is in fact these boundary layers that carry most of the meridional circulation of the quasi-Stokes velocity in GCM simulations of the Southern Ocean because in this region the northward slope of density surfaces tends to be almost independent of depth.

Quite apart from the issue of having the diffusivity inside the vertical derivative, there are two other compelling reasons why the eddy-induced velocity of Gent *et al.* (1995) (which we are interpreting as the quasi-Stokes velocity) is not the down-gradient thickness flux. The first of these is that the quasi-Stokes velocity is three-dimensionally non-divergent whereas the epipycnal flux of thickness is divergent. The second (which has been discussed above) is that the epipycnal thickness flux has no diapycnal component whereas the quasi-Stokes velocity has a large diapycnal component. These differences arise through the use of the streamfunction (that is, the continuity equation) to construct the vertical component of the quasi-Stokes velocity.

9. Boundary Conditions on the Quasi-Stokes Streamfunction

The physical interpretation of the quasi-Stokes streamfunction, as described in section 5 above, provides guidance on the boundary conditions that should be imposed at the top and bottom of the ocean. Figure 4 displays the temporal variations in the heights of three different $\tilde{\gamma}$ surfaces (panels (a) to (c)) when the ocean's density field displays harmonic temporal variations. When a given density is less than any in the ocean at that time, the height of that density surface is taken to be at the sea surface. The modified density, $\tilde{\gamma}$, appropriate to each height has the property (by definition) that the height of this $\tilde{\gamma}$ surface averages to zero, as is indicated by the shading in Figure 4 (the shaded fluid appearing below the mean height is equal to the shaded fluid above the mean height). As the sea surface (or the ocean floor) is approached, the shaded area reduces to zero and so the correlation of velocity and thickness in this shaded region also must tend to zero. That is, the contribution of eddies to the transport of water that is more dense than $\tilde{\gamma}$ reduces to zero as the sea surface (or ocean floor) is approached. By definition, this transport is the quasi-Stokes streamfunction and so it must approach zero smoothly at the top and bottom boundary.

In the absence of a horizontal boundary one would estimate that a fluid

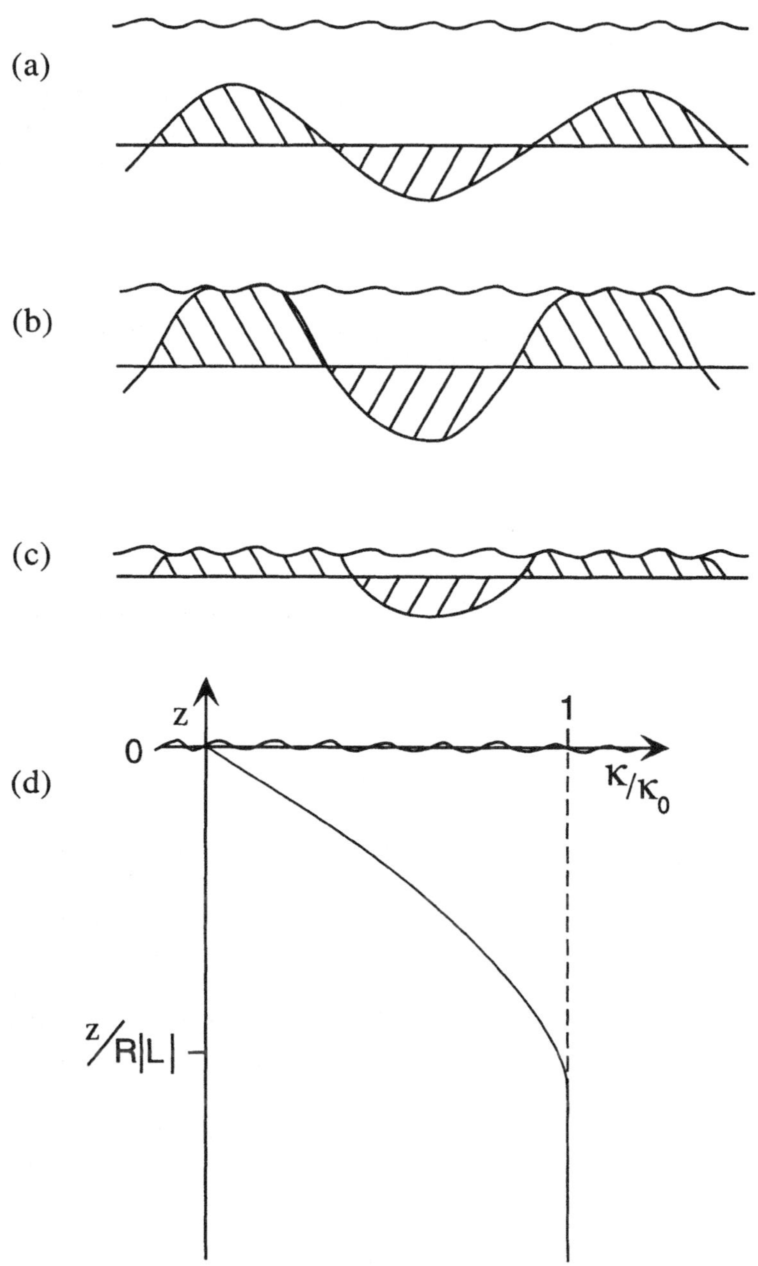

Figure 4. Sketch of the temporal variation of the heights of three different $\tilde{\gamma}$ surfaces as the sea surface is approached. Panel (d) sketches the implication for the vertical tapering of the diffusivity.

parcel in a mesoscale eddy would undergo vertical excursions (of height) that would scale with the Rossby radius, R, multiplied by the magnitude of the slope of the density surface, $|\boldsymbol{L}| = |\nabla_H \tilde{\gamma}/\tilde{\gamma}_z|$. Both R and $|\boldsymbol{L}|$ can be estimated locally at every point in a GCM. One would estimate that if $R|\boldsymbol{L}|$ were greater than the depth of a certain grid-point, then the density surface, $\tilde{\gamma}$, appropriate to that point would have begun to clip the sea surface as indicated in Figure 4b and 4c. This suggests that the quasi-Stokes streamfunction (or a diffusivity that is used to parameterize the quasi-Stokes streamfunction) should be tapered to zero according to the scaled height, $z/(R|\boldsymbol{L}|)$ (as sketched in Figure 4d) and such a procedure is already in use (see appendix B of Large *et al.*, 1997). The physical interpretation of the quasi-Stokes streamfunction makes it possible to justify this procedure.

Edmon *et al.* (1980) have contoured the divergence of the Eliassen-Palm flux of the zonally-averaged flow and have shown that for a finite amplitude situation (their figure 3d), the contours are not all bunched into surface boundary layers as one might expect from the linear solution to the Eady problem. These several lines of evidence lead us to conclude that in addition to numerical convenience, there are good physical justifications for tapering of the quasi-Stokes streamfunction over several hundred metres when approaching a horizontal boundary. Tréguier *et al.* (1997) have also suggested that the explicit horizontal diffusion should be increased as the diffusivity of the quasi-Stokes streamfunction is decreased in these boundary layers, and this appears to be a suggestion worth trying.

With no-slip boundary conditions it is clear that the quasi-Stokes streamfunction should be put equal to zero on all vertical boundaries. Mesoscale eddies would actually feel the presence of the vertical boundary when they were within a few Rossby radii of the boundary, so this would be a good choice of distance for the horizontal tapering of the quasi-Stokes streamfunction of the TRM circulation.

10. The Quasi-Stokes Velocity is not the Bolus Velocity

The horizontal quasi-Stokes velocity is the vertical derivative of the quasi-Stokes streamfunction, and we vertically differentiate the second expression in (15) to find that

$$\begin{aligned} \boldsymbol{V}^{+} = \boldsymbol{\Psi}_z &= \overline{(\boldsymbol{V}' + \bar{\boldsymbol{V}}_z \eta')\,\eta'_z} + \overline{\left(\boldsymbol{V}_z + \frac{1}{2}\boldsymbol{V}_{zz}\eta'\right)\eta'} + O\left(\alpha^3\right) \\ &= \quad \boldsymbol{V}^B \quad + \quad \left(\tilde{\boldsymbol{V}} - \bar{\boldsymbol{V}}\right) \quad + O\left(\alpha^3\right) \end{aligned} \tag{29}$$

The expression $(\boldsymbol{V}' + \bar{\boldsymbol{V}}_z \eta')$ is the first-order Taylor series expansion for the velocity perturbation at constant density, $\boldsymbol{V}'|_{\tilde{\gamma}}$ so that the first term of (29) is the bolus velocity, $\boldsymbol{V}^B$ to within an error that is third order in perturbation quantities. At leading order, the expression $\left(\boldsymbol{V}_z + \frac{1}{2}\boldsymbol{V}_{zz}\eta'\right)$ is the instantaneous shear half way between the fixed height and the instantaneous height of the density surface so that the second term in (29) is equal to the difference between the horizontal velocities averaged at constant density and at constant height, $\left(\tilde{\boldsymbol{V}} - \bar{\boldsymbol{V}}\right)$.

It is apparent from (29) that the task of the quasi-Stokes streamfunction is not only to parameterize the bolus velocity, but also to take account of $\left(\tilde{\boldsymbol{V}} - \bar{\boldsymbol{V}}\right)$. The reason why this dual role exists for the quasi-Stokes streamfunction is that it is to be used in a model that has a vertical coordinate whose height does not vary in time. That is, the TRM theory applies to height-coordinate models and sigma-coordinate models such as SPEM. The momentum equations in this type of model dictate that the model's horizontal velocity in the Eulerian-mean velocity, $\bar{\boldsymbol{V}}$.

The task of parameterizing the effects of mesoscale eddies is very different for isopycnal models than for height-coordinate models. In an isopycnal model the extra horizontal velocity that is needed is indeed the bolus velocity. The sum of the bolus velocity and the resolved-scale horizontal velocity of an isopycnal model, $\tilde{\boldsymbol{V}}$, gives the thickness-weighted velocity of isopycnal coordinates, $\hat{\boldsymbol{V}}$. In stark contrast to the large diapycnal component of the quasi-Stokes velocity of height-coordinate models, the bolus advection that is needed in isopycnal models has zero diapycnal component. Also, the resolved scale velocity, $\tilde{\boldsymbol{V}}$, has no diapycnal component except for explicitly imposed small-scale mixing processes. A further important contrast is that whereas both $\bar{\boldsymbol{U}}$ and $\boldsymbol{U}^+$ of height-coordinate models are non-divergent, both $\tilde{\boldsymbol{V}}$ and $\boldsymbol{V}^B$ of isopycnal models are divergent (while their sum is not). It is this divergent nature of the two velocities in density coordinates that resolves the "curious point" raised by Tréguier, Held and Larichev (1997) (on their page 571). For the above three reasons it is clear that the task of parameterizing the effects of eddies on the transport of tracers in height-coordinate models is different to the corresponding task in isopycnal coordinates.

The following approximate expression can be found for $\left(\tilde{\boldsymbol{V}} - \bar{\boldsymbol{V}}\right)$ using the thermal wind relation, $\boldsymbol{V}_z = -\left(g/f\rho_o\right)\mathbf{k} \times \nabla_H \gamma$, which applies to both the mean and the perturbation fields.

$$\left(\tilde{\boldsymbol{V}} - \bar{\boldsymbol{V}}\right) = \left(\frac{g}{f\rho_o}\right)\mathbf{k} \times \nabla_H \left(\frac{\bar{\phi}}{\tilde{\gamma}_z}\right) + O\left(\alpha^3\right) \tag{30}$$

Taking a maximum expected value of half the density variance of 10^{-2} kg^2m^{-6} (equivalent to an rms vertical heaving by meso-scale eddies of about 150 m) and assuming the epineutral gradient of $(\bar{\phi}/\bar{\gamma}_z)$ to vary by its own magnitude in a horizontal distance of 10^6 m, we find from (30) that $\left(\tilde{\boldsymbol{V}} - \bar{\boldsymbol{V}}\right)$ is about 10^{-3} m/s. One assumes that it is sufficiently accurate to use the geostrophic approximation (thermal wind) in this estimation. While this estimate of $\left(\tilde{\boldsymbol{V}} - \bar{\boldsymbol{V}}\right)$ is quite large, it must be noticed that this velocity is directed along the (usually closed) contours of $(\bar{\phi}/\bar{\gamma}_z)$ in the horizontal plane. Because of this, when the normal component of $\left(\tilde{\boldsymbol{V}} - \bar{\boldsymbol{V}}\right)$ is integrated along a closed path (for example, a path that encircles the globe passing through Drake Passage) then the average of the normal component of $\left(\tilde{\boldsymbol{V}} - \bar{\boldsymbol{V}}\right)$ is zero. This term can contribute to zonally-averaged tracer budgets if there is a correlation between the tracer values and the locations where $\left(\tilde{\boldsymbol{V}} - \bar{\boldsymbol{V}}\right)$ is northward or southward, but this is unlikely to be of leading order importance.

Another consequence of the form (30) is that the horizontal divergence of $\left(\tilde{\boldsymbol{V}} - \bar{\boldsymbol{V}}\right)$ is zero except for a small term due to the beta effect (as pointed out by Tréguier *et al.* (1997) for the QG case). This implies that if the quasi-Stokes streamfunction was made to satisfy $\boldsymbol{\Psi}_z \approx \boldsymbol{V}^B$ rather than (29), this approximation would not affect the determination of the correct vertical quasi-Stokes velocity, w^+.

Hence we conclude that while it may well be sufficiently accurate for many purposes to regard the horizontal quasi-Stokes velocity of the TRM circulation as the bolus velocity, the eddy forcing of the mean density equation, (25), is caused by the part of the quasi-Stokes velocity that is directed normal to the density surfaces, $-\boldsymbol{U}^+ \cdot \nabla\tilde{\gamma}$ and this is just the part that would be zero if the three-dimensional quasi-Stokes velocity were in fact the bolus velocity. It is the construction of the vertical component of the quasi-Stokes velocity that causes the full three-dimensional quasi-Stokes velocity to have a component normal to density surfaces. This construction uses the non-divergent nature of the quasi-Stokes velocity, and this is the key assumption that makes it different to the bolus velocity which is three-dimensionally divergent.

It is concluded that the eddy-induced velocities that are needed in eddy-less models are quite different in isopycnal models than in height-coordinate models. In isopycnal models the extra velocity that is needed is the bolus velocity which has no diapycnal component and is three-dimensionally divergent. In direct contrast, in height coordinates the eddy-induced velocity is non-divergent and has a substantial diapycnal component. Despite these differences, it seems that if one achieves a close correspondence between the horizontal bolus velocity and the vertical derivative of the quasi-Stokes streamfunction, the very nature of the residual-mean approach will ensure a good three-dimensional flow.

11. How Important is the Second Part of the Expression for Ψ?

Here we investigate the contribution to the mean density equation of the second term in the expression, (6), for the quasi-Stokes streamfunction, $(\bar{\boldsymbol{V}}_z/\bar{\gamma}_z)(\bar{\phi}/\bar{\gamma}_z)$. From (25), the forcing of the density equation due to the presence of mesoscale eddy motions is $\tilde{\gamma}_z \nabla_{\tilde{\gamma}} \cdot \Psi$ so that the second term of the quasi-Stokes streamfunction contributes to the eddy forcing of the mean density equation the amount

$$\tilde{\gamma}_z \nabla_{\tilde{\gamma}} \cdot \left(\frac{\bar{\boldsymbol{V}}_z}{\bar{\gamma}_z} \frac{\bar{\phi}}{\bar{\gamma}_z} \right) \tag{31}$$

The geostrophic relationship in density coordinates implies that $\nabla_{\tilde{\gamma}} \cdot (f\tilde{\boldsymbol{V}}) = 0$ so that we have $\nabla_{\tilde{\gamma}} \cdot (f\tilde{\boldsymbol{V}}) = O\left(\alpha^2\right)$, and taking the derivative with respect to $\tilde{\gamma}$ of this expression we have

$$\nabla_{\tilde{\gamma}} \cdot (f\bar{\boldsymbol{V}}_{\tilde{\gamma}}) = \nabla_{\tilde{\gamma}} \cdot \left(\frac{f\bar{\boldsymbol{V}}_z}{\bar{\gamma}_z} \right) + O\left(\alpha^2\right) = O\left(\alpha^2\right) \tag{32}$$

This equation is used in (31) to show that

$$\tilde{\gamma}_z \nabla_{\tilde{\gamma}} \cdot \left(\frac{\bar{\boldsymbol{V}}_z}{\bar{\gamma}_z} \frac{\bar{\phi}}{\bar{\gamma}_z} \right) = f\bar{\boldsymbol{V}}_z \cdot \nabla_{\tilde{\gamma}} \left(\frac{\bar{\phi}}{f\bar{\gamma}_z} \right) + O\left(\alpha^4\right) \tag{33}$$

and the mean thermal wind equation, $\bar{\boldsymbol{V}}_z = -\left(g/f\rho_o\right)\mathbf{k} \times \nabla_H \bar{\gamma}$, is used to show that

$$\tilde{\gamma}_z \nabla_{\tilde{\gamma}} \cdot \left(\frac{\bar{\boldsymbol{V}}_z}{\bar{\gamma}_z} \frac{\bar{\phi}}{\bar{\gamma}_z} \right) = \left(\frac{g}{f\rho_o} \right) \mathbf{k} \times \nabla_H \left(\frac{\bar{\phi}}{\bar{\gamma}_z} \right) \cdot \nabla_H \bar{\gamma} + O\left(\alpha^4\right) \tag{34}$$

plus a smaller term that is proportional to the variation of the Coriolis parameter with latitude. This expression is unlikely to make a significant contribution to the mean density equation. One way of seeing this is to write (34) as $\left(g/f\rho_o\right)\mathbf{k} \cdot \nabla_H \times \left(\left[\bar{\phi}/\bar{\gamma}_z\right] \nabla_H \bar{\gamma}\right) + O\left(\alpha^4\right)$ and Stokes' theorem can be used to relate the contribution of this term in a closed region in a horizontal plane to the values of $\left[\bar{\phi}/\bar{\gamma}_z\right]$ and of $\bar{\gamma}$ around the periphery of the closed area. This leads to the conclusion that the expression (34) will oscillate in sign in neighbouring spatial regions so that its areal average will be small.

Interestingly, this contribution, (34), of the second term in the streamfunction to $\tilde{\gamma}_z \nabla_{\tilde{\gamma}} \cdot \Psi$ is the same expression but of opposite sign to the contribution of $\left(\tilde{\boldsymbol{V}} - \bar{\boldsymbol{V}}\right)$ to the eddy forcing of the mean density equation, namely $-\left(\tilde{\boldsymbol{V}} - \bar{\boldsymbol{V}}\right) \cdot \nabla_H \tilde{\gamma}$ [see (25)]. That is, these two different parts of the eddy forcing of the mean density equation are not simply two different ways of viewing the same process, rather, one is the negative of the other. We would not have expected the two contributions to the density equation to be the same since, from (29), the second term in the streamfunction is seen to contribute to both $\boldsymbol{V}^B$ and $\left(\tilde{\boldsymbol{V}} - \bar{\boldsymbol{V}}\right)$. However, to find that the terms are of the same magnitude but of opposite sign seems coincidental.

Taking the result of this section and the previous section together, it is concluded that as far as the eddy forcing of the mean density equation is concerned, one can ignore the second term in the expression, (6), for the streamfunction, so that it is sufficiently accurate to seek a parameterization of $-\overline{\boldsymbol{V}'\gamma'}/\bar{\gamma}_z$ and to regard this as the quasi-Stokes streamfunction, $\boldsymbol{\Psi}$. Alternatively, if a good parameterization can be found for the horizontal bolus velocity, $\boldsymbol{V}^B$ (perhaps by way of the potential vorticity equation) then it is sufficiently accurate to equate this to $\boldsymbol{\Psi}_z$ and, together with suitable boundary conditions, the quasi-Stokes streamfunction can be estimated in this manner. Hence there appear to be at least these two equally good routes to obtain approximate parameterizations of the quasi-Stokes streamfunction.

12. The Horizontal-&-Temporal-Residual-Mean Circulation

So far the averaging operator has been a low-pass temporal average and the aim of the temporal-residual-mean theory has been to incorporate in eddyless models the effects of the missing temporal correlations of variables. Such temporal correlations can be thought of as being caused by mesoscale eddies. Coarse resolution models also lack knowledge of spatial correlations between variables on scales smaller than the grid scale. The best-known example of this is in western boundary currents, where, as pointed out by Rintoul and Wunsch (1991), the spatial correlations of northward velocity and temperature (at a given height) in the Gulf Stream is an important contributor to the northward heat flux. This section summarises a preliminary attack on this issue. Note that the parameterization of these spatial correlations would be necessary even for a stratified fluid that contained no eddies, that is, for a completely steady and laminar ocean.

The physical motivation for the total quasi-Stokes streamfunction of the horizontal and temporal residual-mean (H&TRM) circulation remains the same as for the quasi-Stokes streamfunction of the temporal-residual-mean circulation. That is, we seek a two-dimensional streamfunction so that the sum of it and the vertical integral of the Eulerian mean velocity gives the horizontal transport, per unit horizontal distance, of water that is denser than $\tilde{\gamma}$. Equivalently, we seek the horizontal velocity at which tracers are advected given both the lack of mesoscale eddies and the limited horizontal resolution, and we want this horizontal velocity to be given by the sum of the Eulerian mean horizontal velocity plus the vertical derivative of the total quasi-Stokes streamfunction.

Figure 5a shows the height of a particular density surface, $\tilde{\gamma}\,(< z_a >)$ as a function of x at various times and we call the fluid that is denser than $\tilde{\gamma}$ "marked fluid". This particular density surface is chosen so that its temporal and x-average height is at the known height, $< z_a >$. At each location, x, the temporal mean height of the same density surface is at $z_a(x)$ and the instantaneous height of the density surface is $z = z_a(x) + \eta'_a(x,t)$ so that $\overline{\eta'_a(x,t)} = 0$. The spatial average of $z_a(x)$ in the x direction in the range $-\frac{1}{2}\Delta x < x < \frac{1}{2}\Delta x$ is labelled $< z_a >$. The horizontal distance over which the spatial averaging is being performed is Δx which corresponds to the horizontal grid size of an eddyless height-coordinate model in the east-west direction. The temporal average of the northward transport of "marked fluid" integrated over the width, Δx of this box is

$$\text{Mean Northward Transport of Marked Fluid} = \overline{\int_{-\Delta x/2}^{\Delta x/2} \int_{-H}^{z_a(x)+\eta'_a(x,t)} v\, dz dx} \tag{35}$$

where the bottom of the ocean is at $z = -H$. The overbar continues to represents a low-pass temporal average and the angle brackets denotes a box-car x-average over the distance Δx. The inner vertical integral in (35) is split into three separate integrals so that (35) becomes

$$\int_{-\Delta x/2}^{\Delta x/2} \int_{-H}^{<z_a>} \bar{v}\, dz dx + \int_{-\Delta x/2}^{\Delta x/2} \int_{<z_a>}^{z_a(x)} \bar{v}\, dz dx + \overline{\int_{-\Delta x/2}^{\Delta x/2} \int_{<z_a>}^{z_a(x)+\eta'_a(x,t)} v\, dz dx} \tag{36}$$

Because the limits of the vertical integration in the first two terms are independent of time, the temporal averaging operator has been applied to the integrand so that the northward velocity that enters the first two terms is the Eulerian mean northward velocity, $\bar{v}$.

Figure 5a shows several snapshots of the height of the density surface as a function of x across this face of a box and when the height of this density surface is temporally averaged, the resulting average height, $z = z_a(x)$ is sketched as a function of x in Figure 5b. The figure indicates that this mean height is a linear function of x but in the subsequent analysis this is not a necessary requirement. If a higher order spatial function of x is included in the function $z_a(x)$, the resulting expression for the volume transport merely contains higher order terms that we subsequently ignore. The important feature is that the spatially and temporally averaged height of this density

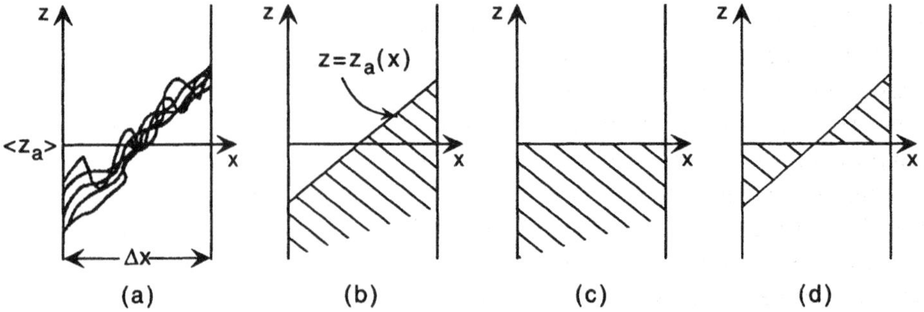

Figure 5. (a) Snapshots at various times of the height of a particular density surface are sketched as a function of the distance, x, across the face of a box of a coarse-resolution model. When the height of this density surface is temporally averaged, the smooth curve in panel (b) is found, and the northward transport of shaded water in (b) can be calculated as the sum of the transports of shaded water in panels (c) and (d).

surface is the known height, $< z_a >$. Figures 5c and 5d illustrate how the vertical integral up to the height $z = z_a(x)$ in Figure 5b can be decomposed into the first two terms in (36), namely the first integral whose integration limits are not a function of x and the second whose integration limits are a function of x. We now discuss each of the three terms in (36) in turn.

The order of integration of the first term in (36) is reversed showing that it is

$$\Delta x \int_{-H}^{<z_a>} < \bar{v} > dz \tag{37}$$

and the model's resolved-scale northward velocity is interpreted as the box-car x-average of the temporal mean velocity at fixed height, $< \bar{v} >$. Note that the geostrophic relationship inherently yields the spatially averaged velocity rather than say the velocity at the centre of the face, so the interpretation of the model's resolved-scale northward velocity as $< \bar{v} >$ is the only interpretation of the model's velocity that is consistent with geostrophy. We conclude that the first term in (36) is the northward flux of marked fluid that can be calculated using the model's resolved-scale velocity.

In order to evaluate the second term in (36), the Eulerian-mean northward velocity, $\bar{v}$ is written as a spatial Taylor series as

$$\bar{v} = \bar{v}\,(x, z) = \bar{v_o} + \bar{v_x}x + \bar{v_z}[z- < z_a >] + \ldots \tag{38}$$

where $\bar{v_x}$ and $\bar{v_z}$ are understood to be evaluated at $z =< z_a >$ and at the centre of the box in the x direction, that is, at $x = 0$, while $\bar{v_o}$ is independent of both x and z. Substituting this Taylor series into the middle term in (36) it can be shown to be

$$\int_{-\Delta x/2}^{\Delta x/2}\int_{<z_a>}^{z_a(x)} \bar{v}\, dzdx = \frac{1}{12} < \bar{v} >_x L^x[\Delta x]^3 + \frac{1}{24} < \bar{v} >_z (L^x)^2 [\Delta x]^3 + O\left([\Delta x]^4\right) \quad (39)$$

where L^x is the mean slope of the density surface in the x direction, $L^x = -\tilde{\gamma}_x/\tilde{\gamma}_z$.

The inner integral of the third term in (36) is what appears in the Taylor series analysis of McDougall and McIntosh (1998b) so that we can write this integral as

$$\overline{\int_{z_a(x)}^{z_a(x)+\eta_a'(x,t)} vdz} = \left(-\frac{\overline{v'\gamma'}}{\bar{\gamma}_z} + \frac{\bar{v}_z}{\bar{\gamma}_z}\frac{\bar{\phi}}{\bar{\gamma}_z}\right)\Bigg|_{z=z_a(x)} + O\left(\alpha^3\right) \quad (40)$$

which is the northward component of the quasi-Stokes streamfunction of the purely temporal-residual-mean circulation, evaluated at the height $z = z_a(x)$. Hence the last term in (36) is Δx times the spatial average of the streamfunction, (40), along the sloping density surface, $z = z_a(x)$, plus higher order error terms. It is readily shown that evaluating the streamfunction, (40), at the fixed height, $z =< z_a >$ incurs only higher order error terms.

Dividing (36) by Δx and subtracting the vertical integral of the model's resolved-scale velocity, $< \bar{v} >$ gives the expression for the northward component of the new quasi-Stokes streamfunction:- the quasi-Stokes streamfunction that now includes the effects of both temporal and spatial averaging, namely

$$\frac{1}{12} < \bar{v} >_x L^x[\Delta x]^2 + \frac{1}{24} < \bar{v} >_z (L^x)^2 [\Delta x]^2 + \left(-\frac{\overline{v'\gamma'}}{\bar{\gamma}_z} + \frac{< \bar{v} >_z}{\bar{\gamma}_z}\frac{\bar{\phi}}{\bar{\gamma}_z}\right) + O\left(\alpha^3, [\Delta x]^3\right) \quad (41)$$

and the eastward component of the new quasi-Stokes streamfunction, where the spatial box-car averaging is over the northward length scale Δy is

$$\frac{1}{12} < \bar{u} >_x L^y[\Delta y]^2 + \frac{1}{24} < \bar{u} >_z (L^y)^2 [\Delta y]^2 + \left(-\frac{\overline{u'\gamma'}}{\bar{\gamma}_z} + \frac{< \bar{u} >_z}{\bar{\gamma}_z}\frac{\bar{\phi}}{\bar{\gamma}_z}\right) + O\left(\alpha^3, [\Delta y]^3\right) \quad (42)$$

Having defined the total quasi-Stokes streamfunction so that it has the quasi-Lagrangian property (transport of fluid denser than $\tilde{\gamma}$), it is clear that the difference between two such streamfunctions in the vertical gives the contributions of both temporal and horizontal box-car averaging to the appropriate thickness-weighted mean velocity. It is this horizontal velocity that is required in the layered continuity equation and in the layered tracer conservation equation. The total quasi-Stokes streamfunction whose northward and eastward components are (41) and (42) must be made to satisfy the same boundary conditions as discussed earlier for the quasi-Stokes streamfunction of the TRM circulation.

We can make some preliminary remarks about the direction of the extra quasi-Stokes streamfunction whose role is to overcome the inherent horizontal box-car averaging of coarse resolution models. Consider the situation where $L^x = 0$ and $\partial L^x/\partial z = 0$ so that the density surface slopes to the north or south and, from (41), there is no contribution from the "spatial averaging terms" to the total northward velocity. Even though the density surface does not slope to the east, there is a contribution from the horizontal averaging part of (42) to the eastward velocity. This example implies that, contrary to temporal averaging, the terms that correct for the box-car horizontal averaging of coarse resolution models do not necessarily cause a contribution to the total quasi-Stokes streamfunction in the down-gradient direction of the horizontal density gradient, but can cause a contribution normal to this direction.

In the Gulf Stream and the East Australian Current, I expect that it will be the first term in (41) that will dominate. The first two terms in (41) correspond to the northward heat flux due to the unresolved spatial correlations of northward velocity and potential temperature, as discussed by Rintoul and Wunsch (1991). The sign of the first term in (41) is such as to cause an increase in the northward flux of heat, and when the expression, (41), is used in a coarse resolution model, this flux of heat will be represented by an additional northward flux of "marked" water, just as it would appear in an eddy-resolving isopycnal model of the same region. This is a case where the extra advection velocity that must be added to a coarse-resolution model is not down the horizontal density gradient and so one would not normally think that it is important because the extra advection will not tend to contribute to the local heat or salt conservation equation. Despite this, the transport of water of certain density classes clearly is important for the section-averaged heat and fresh-water fluxes. The mixing that is required to transform one water mass into another can occur at locations very distant to the section across which the poleward heat flux is being budgeted.

At face value it appears that the terms in (41) and (42) that correct for the horizontal box-car averaging should not be important because the numerical discretization involved in writing conservation equations on a grid of boxes has errors that are quadratic in the grid size of the model. The first two terms in (41) and (42) are also quadratic in the grid size and so it seems inconsistent to be including terms at this order. However, in western boundary currents, the horizontal shear of the northward velocity actually scales as the reciprocal of the grid resolution so that the first term in (41) is really one order of lower than it appears (Bill Dewar, personal communication, 1997).

Notice that in the quasi-Stokes streamfunction of TRM we need to parameterize the horizontal flux of density, $\overline{\mathbf{V}'\gamma'}$ whereas in the horizontal box-car averaging parts of the H&TRM quasi-Stokes streamfunction, no parameterization is required. That is, the quantities that appear in the first two terms in (41) and (42) are already known to the coarse-resolution model. This is because we have chosen to do the temporal averaging first, followed by the box-car horizontal averaging, so that the variables that are being considered at the stage of box-car horizontal averaging are assumed to be smooth in space. They key assumption appears to be that the temporally-averaged velocity field is sufficiently smooth that the Taylor series, (38), converges and can be approximated by the leading linear terms.

13. Concluding Remarks

This article has concentrated on theoretical aspects of residual-mean theory as applied to scalar conservation equations in three-dimensions. Little has been said about the important issue of obtaining a practical parameterization of the quasi-Stokes streamfunction of the TRM circulation, and the few remarks below will not be comprehensive in this regard.

Tréguier (1998) has analysed a primitive equation model for the zonally-averaged bolus velocity and has found a small diffusivity for use in the Gent *et al.* (1995) scheme. If it can be shown that it is appropriate to use a such a small diffusivity for the quasi-Stokes streamfunction then this should have some benefits so long as the models remain stable with these smaller diffusivities. The two benefits that come to mind are avoiding the slowing of the horizontal circulation of the sub-tropical gyres that occurs with the larger values of the quasi-Stokes diffusivity, and reducing the intrusion of Antarctic Bottom Water into the North Atlantic which is too strong with present values of the quasi-Stokes diffusivity (Hirst and McDougall, 1998).

To date coarse resolution models have mainly been run with diffusivities that are independent of space, but the work of Visbeck *et al.* (1997) suggests that the diffusivity should depend on the slope of the density surfaces and so should be a rather strong function of space. It is hoped that by having the diffusivity be an increasing function of the slope of density sufaces, it may be possible to avoid numerical instabilities while having a much smaller diffusivity in most of the ocean. In this way it may be possible to avoid numerical instabilities, which tend to occur on boundaries, while having the strength of the skew diffusion of the H&TRM circulation set at realistically small values in the ocean interior such as in the Antarctic Circumpolar Current.

As discussed in section 8 above, the Gent and McWilliams (1990) parameterization scheme is essentially equivalent to the TRM theory under the assumption that the horizontal flux of density is directed down the horizontal density gradient. While this is a common assumption, several authors have recently made the point that there is probably better theoretical support for the flux of potential vorticity along neutral density surfaces being directed down the epineutral gradient of potential vorticity. This naturally leads to a parameterization for the bolus velocity and then one has to address the questions of applying boundary conditions at the top and bottom of the ocean and of avoiding a singularity at the equator. Killworth (1998) has made great strides with the issue of boundary conditions, but so far only for one horizontal dimension. When the extra horizontal dimension is added, another constraint arises that such a scheme must satisfy. These constraints are automatically satisfied by the robust Gent *et al.* (1995) implementation of the TRM circulation.

The theory of Killworth (1998) and parameterization schemes based on the down-epineutral-gradient of potential vorticity are, by construction, applicable to the parameterization of eddies in eddyless density-coordinate models. In such density-coordinate models, the additional velocity that is required is the bolus velocity and this velocity is three-dimensionally divergent and adiabatic. It is not obvious a priori that such a parameterization scheme should be applied to a z-coordinate model where the use of the continuity equation ensures that whatever choice is made for its horizontal components, the extra velocity is three-dimensionally non-divergent and most importantly, diabatic at leading order. One may indeed be able to get away with using the horizontal velocity of one theory in the three-dimensional parameterization machinery of another, but one must be aware of the fact that one is mixing apples and oranges in the process.

It seems that the topic of parameterization is in its infancy. We have a robust scheme that has been used to date with a diffusivity that may be too large in many places. We also have some ideas about how to develop a parameterization based on the assumption that the epineutral flux of potential vorticity is down-gradient, but this has not yet matured into a robust scheme in three dimensions. In addition to the issues of temporal averaging, there is the issue of the lack of horizontal resolution, and in this article it has been argued that this should be treated separately from the

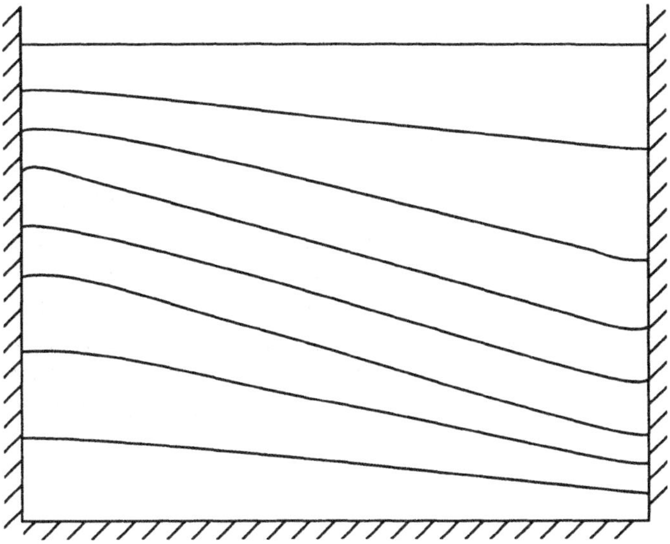

Figure 6. Sketch of a vertical section showing several density contours.

issue of parameterizing the effects of the missing mesoscale eddies.

The actual scheme that is used for down-gradient thickness diffusion in the MICOM isopycnic model seems to have very different characteristics to the Gent and McWilliams (1990) scheme. Consider an ocean section as sketched in Figure 6 where the slope of a given density surface is constant in the interior. The Gent and McWilliams (1990) scheme will act on this structure to destroy the available potential energy and to relax the isopycnals towards a final (flat) state of rest. This is achieved by a large-scale overturning motion as described in Gent *et al.* (1995). However the MICOM scheme (as described by Bleck *et al.*, 1992) acts to smooth the interface heights and as these are already smooth in the ocean interior, the scheme has no effect there, but will lead to some motion only near the edges of Figure 6. The reasons for this different behaviour are not entirely clear at this time, but it seems that the Gent and McWilliams (1990) method, by supplying an extra velocity field, has the ability of achieving a far field effect whereas the MICOM scheme that is implemented only in the thickness equation seems to respond only to the small-scale variations in layer thickness. A valid scheme in isopycnal coordinates would provide an estimate of the bolus velocity and this extra velocity would be inserted into all the conservation statements.

Another issue concerns the mechanical (sum of potential and kinetic) energy equation. Work in progress with Professor Jürgen Willebrand shows that the expected extraction of energy by baroclinic instability is guaranteed if $\mathbf{\Psi} \cdot \nabla_H \tilde{\gamma} < 0$. The Gent *et al.* (1995) scheme for $\mathbf{\Psi}$ satisfies this requirement at every point in space, and indeed this ability of mesoscale eddies to deplete available potential energy was a key motivation of the original work of Gent and McWilliams (1990). In contrast, a parameterization scheme based on the down-epineutral-gradient flux of potential vorticity does not satisfy this requirement at every point. One then will have to rely on other sensible features of the scheme to achieve a global extraction of energy by baroclinic instability. The terms in the quasi-Stokes streamfunction of the H&TRM circulation that correct for the inherent box-car horizontal averaging of coarse resolution models also do not satisfy this energy requirement at every point in space.

Finally, there is the issue of whether the present implementation of the TRM circulation in ocean models is displaying improvements for the correct reasons. Several authors have described substantial improvements including greatly improved deep water masses, less unwanted deep convection and less drift in coupled atmosphere-ocean models (see Hirst *et al.*, 1996). An important common element of these improvements is that the bottom water of the world's oceans has been able to sink from the surface to the ocean bottom with very little dilution. In fact, in the work of Hirst and McDougall (1996) it was found that there was insufficient diapycnal mixing occurring in the overflow regions. Previously such a result had only been possible using a density-coordinate model. In this way, a height-coordinate model has been shown to be sufficiently "adiabatic" for the purposes of climate modelling.

In practice this vertical motion of the Deep and Bottom Water occurs in canyons and across sills which are not part of the coarse resolution models. The TRM advection scheme achieves this "adiabatic" sinking motion because of two almost equal effects, as demonstrated by Hirst and McDougall (1996). First, the unwanted horizontal diffusion is eliminated, and second, an extra advection (or skew diffusion) is added which assists in the transport of water from the surface to the deep. The elimination of horizontal diffusion is thought to be physically required, but the extra advection at the bottom of the ocean seems to be more an artefact of the bottom boundary condition on the quasi-Stokes streamfunction than a representation of the actual boundary current mechanisms that achieve the transport of bottom water. In this way it may be that half of the benefits that we are seeing to date have been obtained by stealth:- obtaining the right results for the wrong reasons. If so, what fraction of this half will respond incorrectly to changing boundary conditions associated with, for example, climate change?

In summary, this article has shown how the three-dimensional residual-mean theory allows coarse-resolution models to incorporate the effects of temporal eddies and of limited spatial resolution. By seeking a parameterization for the quasi-Stokes streamfunction, (6), and adding this to the components shown in (41) and (42) that account for the limited horizontal resolution, an extra skew diffusive flux, (see (10) and (11)), can be added to a coarse-resolution ocean model. The equivalence of this procedure to averaging the instantaneous flow in density coordinates has been demonstrated, and this equivalence provides the physical motivation for the H&TRM theory.

References

Andrews, D. G. and M. E. McIntyre, 1976: Planetary waves in horizontal and vertical shear: The generalized Eliassen-Palm relation and the zonal mean acceleration. *J. Atmos. Sci.*, **33**, 2031-2048.

Andrews, D. G. and M. E. McIntyre, 1978: An exact theory of nonlinear waves on a Lagrangian-mean flow. *J. Fluid Mech.*, **89**, 609-646.

Bleck, R., C. Rooth, D. Hu and L. Smith, 1992: Salinity-driven thermocline transients in a wind- and thermohaline-forced isopycnic coordinate model of the North Atlantic. *J. Phys. Oceanogr.*, **22**, 1486-1505.

Danabasoglu, G. and J. C. McWilliams, 1995: Sensitivity of the global ocean circulation to parameterizations of mesoscale tracer transports. *J. Climate*, **8**, 2967-2987.

Gent, P. R., and J. C. McWilliams, 1990: Isopycnal mixing in ocean circulation models. *J. Phys. Oceanogr.*, **20**, 150-155.

Gent, P. R., J. Willebrand, T. J. McDougall, and J. C. McWilliams, 1995: Parameterizing eddy-induced tracer transports in ocean circulation models. *J. Phys. Oceanogr.*, **25**, 463-474.

Griffies, S. M., 1998: The Gent-McWilliams skew-flux. Submitted to *J. Phys. Oceanogr.*

Griffies, S. M., A. Gnanadesikan, R. C. Pacanowski, V. Larichev, J. K. Dukowicz, and R. D. Smith, 1998: Isoneutral diffusion in a z-coordinate ocean model. Submitted to *J. Phys. Oceanogr.*

Hirst, A. C., H. B. Gordon and S. P. O'Farrell, 1996: Global warming in a coupled climate model including oceanic eddy-induced advection. *Geophys. Res. Lett.*, **23**, 3361-3364.

Hirst, A. C. and T. J. McDougall, 1996: Deep water properties and surface buoyancy flux as simulated by a z-coordinate model including eddy-induced advection. *J. Phys. Oceanogr.*, **26**, 1320-1343.

Hirst, A. C. and T. J. McDougall, 1998: Meridional overturning and dianeutral motion in a z-coordinate ocean model including eddy-induced advection. *J. Phys. Oceanogr.*, in press.

Killworth, P. D., 1998: On the parameterization of eddy transfer. Part I: Theory. *J. Mar. Res.*, **55**, 1171-1197.

Large, W. G., G. Danabasoglu, S. C. Doney, and J. C. McWilliams, 1997: Sensitivity to surface forcing and boundary layer mixing in a global ocean model: Annual-mean climatology. *J. Phys. Oceanogr.*, **27**, 2418-2447.

Lozier, M. S., M. S. McCartney, and W. B. Owens, 1994: Anomalous anomalies in averaged hydrographic data. *J. Phys. Oceanogr.*, **24**, 2624-2638.

Marshall, J. and G. Shutts, 1981: A note on rotational and divergent eddy fluxes. *J. Phys. Oceanogr.*, **11**, 1677-1680.

McDougall, T. J., 1995: The influence of ocean mixing on the absolute velocity vector. *J. Phys. Oceanogr.*, **25**, 705-725.

McDougall, T. J. and P. C. McIntosh, 1996: The temporal-residual-mean velocity. Part I: Derivation and the scalar conservation equations. *J. Phys. Oceanogr.*, **26**, 2653-2665.

McDougall, T. J. and P. C. McIntosh, 1998a: A residual-mean circulation suitable for unsteady flows. Submitted to *J. Phys. Oceanogr.*

McDougall, T. J. and P. C. McIntosh, 1998b: The temporal-residual-mean velocity. Part II: Isopycnal interpretation. Submitted to *J. Phys. Oceanogr.*

McIntosh, P. C. and T. J. McDougall, 1996: Isopycnal averaging and the residual mean circulation. *J. Phys. Oceanogr.*, **26**, 1655-1660.

Plumb, R. A, 1979: Eddy fluxes of conserved quantities by small-amplitude waves. *J. Atmos. Sci.*, **36**, 1699-1704.

Plumb, R. A. and J. D. Mahlman, 1987: The zonally-average transport characteristics of the GFDL general circulation/transport model. *J. Atmos. Sci.*, **44**, 298-327.

Redi, M. H., 1982: Oceanic isopycnal mixing by coordinate rotation. *J. Phys. Oceanogr.*, **12**, 1154-1158.

Rintoul, S. R., and C. Wunsch, 1991: Mass, heat, oxygen and nutrient fluxes and budgets in the North Atlantic Ocean. *Deep-Sea Res.*, **38**(Suppl. 1), S355 - S377.

Tréguier, A. M., 1998: Evaluating eddy mixing coefficients from eddy resolving ocean models: A case study. Submitted to *J. Marine. Res.*

Tréguier, A. M., I. M. Held, and V. D. Larichev, 1997: Parameterization of quasi-geostrophic eddies in primitive equation ocean models. *J. Phys. Oceanogr.*, **27**, 567-580.

Visbeck, M., J. Marshall, T. Haine and M. Spall, 1997: Specification of eddy transfer coefficients in coarse-resolution ocean circulation models. *J. Phys. Oceanogr.*, **27**, 381-402.

STATISTICAL MECHANICS OF POTENTIAL VORTICITY FOR PARAMETERIZING MESOSCALE EDDIES

JOËL SOMMERIA
Laboratoire de Physique,
Ecole Normale Supérieure de Lyon,
Lyon, France

Abstract

We justify the use a Maximum Entropy Production (MEP) principle (*i.e.* the production of a mixing entropy is assumed maximum with the constraint of a bounded diffusion current) in parameterizing the diffusion of a conserved quantity. In the case of potential vorticity, complete mixing is forbidden by energy conservation, and the MEP leads to a subgrid-scale parameterization with a diffusion term, but also with a drift term. The end result of the free inertial evolution is an organized flow (the statistical equilibrium) in which these two terms cancel each other. In the limit of strong β-effect, the drift term corresponds to the Neptune effect of Holloway (1992). Application to a wind-driven barotropic ocean basin is presented, and extensions to baroclinic cases are sketched. A more mechanistic approach, using kinetic models, confirms the existence of the drift term and estimates the diffusivity.

1. Introduction

A fundamental difficulty in modeling turbulent systems at high Reynolds number, like oceanic currents, is the influence of motion at scales smaller than the numerical grid mesh, requiring some statistical averaging. Subgrid-scale motion transports dynamical quantities in a complex way, often modeled as an eddy diffusion. In physics, diffusion is a process of relaxation toward a thermodynamical equilibrium, characterized by uniform solutant concentration and temperature. This state maximizes entropy with the constraint of given global conserved quantities, such as the total mass of solutant or energy. Entropy is defined in usual statistical mechanics as a measure of the number of microscopic configurations (atomic positions and

E. P. Chassignet and J. Verron (eds.), Ocean Modeling and Parameterization, 303–326.

velocities) corresponding to a given macroscopic state, which is characterized by locally averaged quantities of interest (such as concentration or internal energy). The state of maximum entropy is then by far the most likely outcome of a complex evolution.

It is tempting to justify turbulent mixing as a similar relaxation toward a statistical equilibrium, for which fluid particle motion replaces atomic motion. Models of eddy diffusion have been initially devised with this analogy in mind, but have never received a precise justification. The situation is different in two-dimensional (2D) turbulence, or in quasi-geostrophic (QG) systems, as discussed in this paper. This provides a rationale for developing a parameterization for mesoscale eddies, for which geostrophic balance maintains a quasi-2D dynamics. In this regime, the motion is fully determined by the potential vorticity (PV) field. This is the fundamental quantity that is transported and mixed by eddy transport. However, flow energy is related to PV through the induced velocity field, resulting in a long range interaction leading to non-trivial statistical equilibria with organized vortices and jets. This is formally like the non-uniform concentration obtained as statistical equilibrium in a gravity field, under an overall balance between diffusion and sedimentation.

Similarly, it will appear that the relaxation toward equilibrium involves a drift force in addition to eddy diffusion. The "Neptune" effect of Holloway [1], discussed in this volume by Alvarez and Tintoré [2], appears as a particular limit. The present approach has a similar philosophy, but is based on a more general and systematic statistical theory. We shall provide a rapid overview of this theory, stressing applications to the oceanic context. More complete derivations and discussions can be found in [3-7].

2. Statistical Mechanics of Two-dimensional Turbulence

We consider a divergenceless 2D flow, satisfying the barotropic vorticity equation. The velocity $\mathbf{u}$ derives from a streamfunction ψ, $\mathbf{u} = -\nabla \times \psi \mathbf{i_z}$ ($\mathbf{i_z}$ is the upward vertical unit vector), and the vorticity is vertical, with value $\omega = \Delta\psi$. In the absence of forcing and viscous friction, the potential vorticity q is conserved by fluid particles:

$$\frac{dq}{dt} \equiv \frac{\partial q}{\partial t} + J(\psi, q) = 0, \qquad \text{with } q = \omega + h(x, y), \tag{1}$$

where J is the Jacobian operator, and the given function $h(\mathbf{r})$ of the coordinates $\mathbf{r} = (x, y)$ represents a topographic or planetary β-effect. For a bottom topography with elevation $\eta(\mathbf{r})$, small with respect to the mean water depth H_0, $h \equiv f\eta/H_0$, where f is the Coriolis parameter. The planetary β-effect corresponds to $h(x, y) = \beta y$. We shall also consider the case of the Euler equation, for which $h = 0$. Although we restrict ourselves to a barotropic case, the statistical approach can be extended in principle to any balanced dynamics, for which the flow evolution is fully determined by the PV field. This should be a good assumption for mesoscale eddies, but does not hold for internal waves or three-dimensional turbulence.

Three approaches of statistical mechanics have been applied to the Euler or barotropic equation. In a first approach, due to Onsager [8], the system is modeled by a set of point vortices with singular vorticity. The dynamics has a Hamiltonian structure, allowing the standard procedure of statistical mechanics. Onsager noticed that "negative temperature states" are possible, for which elementary vortices of the same sign tend to clump together, forming large coherent vortices. This was a remarkable statement at that time, as there was no real observational evidence for such coherent vortices. Onsager also correctly remarked upon the strong difference from three-dimensional turbulence, for which the process of energy cascade toward small scales prevails. This process is forbidden in 2D turbulence, due to conservation of vorticity, so that energy is preserved, justifying an approach in terms of statistical equilibrium. This program has been pursued by Montgomery and Joyce [9], making an explicit prediction with a mean field approximation.

Modeling the continuous vorticity field in terms of point vortices leads, however, to fundamental difficulties. This can be a good approximation for short term predictions, as used for instance in numerical schemes for the Euler equation, but discrepancies appear on long time scales, relevant for statistical mechanics. Then the prediction is not unique, but depends on the method of approaching the continuous field by point vortices. These difficulties are particularly obvious in the presence of the β-effect, and this point vortex approach was apparently ignored for geophysical applications.

Another statistical approach was proposed by Kraichnan [10,11], based on a spectrally truncated model of the barotropic equation. Spectral truncation leads to spurious behavior in comparison to the initial continuous system. In particular it suppresses conservation laws, like the integral of vorticity. In the absence of the β-effect, an equilibrium energy spectrum is obtained, without any mean flow, so it is not appropriate for studying spatial organization. However, with the β-effect, a non-zero mean velocity is obtained, correlated with topography [12, 13]. Comparisons with numerical computations indicate reasonable agreement [14,15], which has prompted geophysical applications, such as those developed in Alvarez's paper [2].

We use here a theory that removes limitations of both previous approaches, as developed by Robert [16] and Robert and Sommeria [3], and independently by Miller [17]. The predicted statistical equilibrium appears to be a particular steady solution of the Euler equations, superposed with fine scale vorticity fluctuations, which will be referred to as subgrid-scale fluctuations in the present context. This equilibrium state can be calculated by maximizing a mixing entropy, with the constraints brought by the conserved quantities, which are the energy and the global probability distribution of vorticity values. The extension of the theory to QG systems is straightforward [18, 19], replacing vorticity by PV.

The statistical equilibrium is a steady flow (superposed with random local fluctuations), characterized by a given relationship between PV and streamfunction. We find that in the oceanic case, this relationship is nearly linear, so that the inertial solution of Fofonoff [20] is obtained. The relevance of this steady solution to the strongly turbulent oceanic circulation was recognized and discussed for various conditions [21, 24]. This fact has already been explained in terms of statistical mechanics, using truncated spectral models (Salmon et al. [12]). An alternative justification of the Fofonoff flow, as a state of minimum potential enstrophy, has been also proposed by Bretherton and Haidvogel [25]. It coincides with the maximum entropy state for the truncated system, in the natural limit of a truncation at an infinite wavenumber (Carnevale and Frederiksen [26]), for which the fluctuations accumulate at very small scale, with vanishing energy. Our different statistical approach yields again a similar result for this particular problem, but places it in a wider context of self-organization in 2D turbulence. For instance, our approach still provides a good prediction when the β-effect is removed [27, 4].

Although parts of the mid-latitude ocean show tendencies to organize into an equilibrium Fofonoff state, this is clearly not a complete description: wind forcing and dissipation permanently drive the system out of equilibrium. However, mesoscale oceanic turbulence tends to restore statistical equilibrium, and this provides a natural guideline for turbulence parameterization, as emphasized by Holloway [1]. A similar idea has been proposed by Robert and Sommeria [28] for the Euler equation, and adapted to the oceanic case by Kazantsev et al. [6]. It relies on a principle of Maximum Entropy Production (MEP), with appropriate dynamical constraints, in particular keeping constant all the known conserved quantities of the inviscid system. In other words, it most efficiently mixes PV, but generally does not reach the state of uniform PV forbidden by energy conservation.

3. Determination of the Statistical Equilibrium

The subgrid-scale eddies are characterized (at a given time) by a probability distribution $\rho(\sigma, \mathbf{r})$ of PV levels σ at point $\mathbf{r}$. This probability (normalized such that $\int \rho(\sigma, \mathbf{r}) d\sigma = 1$) can be thought of as the area fraction occupied by the PV value σ in a neighborhood of the position $\mathbf{r}$. We are mainly interested in the local PV average

$$\overline{q}(\mathbf{r}) \equiv \int \rho(\sigma, \mathbf{r})\, \sigma \, d\sigma, \tag{2}$$

describing the explicitly resolved scales (in the following, the overbar designates a local average over the subgrid-scales). The streamfunction $\overline{\psi}$ and the associated velocity are obtained from this field as the solution of the PV inversion equation

$$\Delta\overline{\psi} = \overline{\omega} = \overline{q} - h(x, y), \tag{3}$$

with the impermeability boundary condition $\overline{\psi} \mid_{\partial\mathcal{D}} = 0$. We shall also need to consider the local PV variance

$$q_2(\mathbf{r}) \equiv \int \rho(\sigma, \mathbf{r})(\sigma - \overline{q})^2 d\sigma. \tag{4}$$

Many fine scale (i.e., microscopic) PV fields correspond to a given macroscopic field ρ. The number of such microscopic states can be evaluated by introducing a given (arbitrary small) resolution mesh (see [3]). Its logarithm is proportional to a mixing entropy

$$S = -\int \rho(\sigma, \mathbf{r}) \;\log \rho(\sigma, \mathbf{r})\; d^2\mathbf{r}\; d\sigma. \tag{5}$$

A randomly chosen microscopic state very likely corresponds to the macroscopic state that maximizes the mixing entropy, with the constraints due to conservation laws. A first constraint is the conservation of the spatial integral of ρ, equal to the global area proportion $\gamma(\sigma)$ of PV level σ. This area proportion is preserved by inviscid evolution, which involves area-preserving rearrangements of fluid parcels, without change of their PV level. Entropy maximization with this constraint yields a uniform field ρ for each PV level σ. Each PV level then behaves like a chemical species and is expected to mix over all the available space.

However, energy conservation generally prevents such complete PV mixing. For our barotropic model, the energy $\mathcal{E}$ is purely kinetic, $\mathcal{E} = \frac{1}{2}\int \mathbf{u}^2\, d^2\mathbf{r}$, and is classically expressed in terms of vorticity, using an integration by parts,

$$\mathcal{E} = -\frac{1}{2}\int \psi\, \omega\, d^2\mathbf{r} \simeq -\frac{1}{2}\int \overline{\psi}\, \overline{\omega}\, d^2\mathbf{r}, \tag{6}$$

where we have neglected the energy of the subgrid-scale fluctuations (this point will be justified in section 5). Energy is therefore related, using (3), to the local average $\overline{q}$, bringing a constraint of the macroscopic field ρ by (2).

With this energy constraint, non-trivial states of statistical equilibrium are obtained, defined by a given relationship between the potential vorticity $\overline{q}$ and the streamfunction $\overline{\psi}$. Such a relationship defines a steady solution of the barotropic equation, expected to be reached by complex PV stirring. It depends on the conserved quantities, which are the energy $\mathcal{E}$, and the global distribution $\gamma(\sigma)$ of PV levels, fully determined by the initial condition. The set of statistical equilibria can involve bifurcations depending on the control parameters in a fairly complex way, as computed for instance by Juttner et al. [27].

However, the result simplifies in the limit of strong β-effect, defined by

$$E_\beta \equiv \frac{(2\mathcal{E})^{1/2}}{\beta L^3} \ll 1, \tag{7}$$

where β is the gradient of planetary vorticity (or a typical value of ∇h in the topographic case), and L a horizontal length scale. This condition is generally well-satisfied in an oceanic basin, with typical value $E_\beta = 10^{-2}$. Then Kazantsev et al. [6] showed that the statistical equilibrium relationship between PV and streamfunction is close to linear, yielding the classical Fofonoff inertial flow, represented in Fig. 1. This flow is therefore expected as the most likely outcome of strong eddy stirring.

In this limit (7), the statistical equilibrium depends only on energy, circulation $\Gamma = \int q \, d^2\mathbf{r}$, and total enstrophy $\Gamma_2 = \int q^2 \, d^2\mathbf{r}$. The higher order moments of PV are also conserved for inviscid evolution, but it has been shown [6] that the corresponding constraints do not influence the statistical equilibrium. Then entropy maximization can be performed in two steps. We first maximize entropy locally, with the normalization condition $\int \rho d\sigma = 1$, a given local average $\overline{q}$, and local variance q_2. In a second step, we determine the fields $\overline{q}$ and q_2, which globally maximize entropy.

The local maximization yields a Gaussian probability distribution. This result is easily shown by introducing Lagrange multipliers, and by stating that the first variations must vanish, $\delta S - \lambda_0 \, \delta(\int \rho d\sigma) - \lambda_1 \, \delta\overline{q} - \lambda_2 \, \delta q_2 = 0$, so that $\int [(log \; \rho + 1) - \lambda_0 - \lambda_1 \sigma - \lambda_2 (\sigma - \overline{q})^2] \, \delta\rho \, d\sigma = 0$ for any variation $\delta\rho(\sigma)$. This implies that the integrand vanishes, yielding a Gaussian probability distribution $\rho = \frac{1}{\sqrt{2\pi q_2}} \exp\left[-\frac{(\sigma-\overline{q})^2}{2q_2}\right]$. The corresponding local entropy, given by (5), is proportional to $log \; q_2$.

We need then to globally maximize the entropy

$$S = \int log \; q_2 \, d^2\mathbf{r} \tag{8}$$

with a given energy $\mathcal{E}$, circulation Γ, and potential enstrophy $\Gamma_2 = \int (\overline{q}^2 + q_2) \, d^2\mathbf{r}$. Introducing the corresponding Lagrange multipliers, we obtain $\delta S -$

$b\ \delta\mathcal{E} - c\ \delta\Gamma - d\ \delta\Gamma_2 = 0$. The energy variation is $\delta\mathcal{E} = (1/2)\int[\overline{\psi}\ \delta\overline{q} + \overline{q}\ \delta\overline{\psi}]\ d^2\mathbf{r}$. The second integral in this expression can be made equal to the first one by integrating twice by part. Note also that $\delta\overline{\omega} = \delta\overline{q}$ (since the topographic term $h(\mathbf{r})$ is given), so that $\delta\mathcal{E} = \int\overline{\psi}\ \delta\overline{q}\ d^2\mathbf{r}$ and $\int[(1/q_2 - d)\delta q_2 - (b\overline{\psi}+c+2d\overline{q})\delta\overline{q}]\ d^2\mathbf{r} = 0$ for any (independent) variations $\delta\overline{q}(\mathbf{r})$ and $\delta q_2(\mathbf{r})$. The statistical equilibrium has therefore a uniform variance q_2 and a linear relationship between PV and streamfunction. The two coefficients of this relationship are set by the total energy $\mathcal{E}$ and circulation Γ.

Note that since the explicit (or coarse-grained) potential enstrophy $\Gamma_2^{c.g.} = \int\overline{q}^2\ d^2\mathbf{r}$ can be expressed as $\Gamma_2^{c.g.} = \Gamma_2 - \int q_2\ d^2\mathbf{r}$, and q_2 is uniform at statistical equilibrium, entropy maximization becomes equivalent to minimization of the potential enstrophy $\Gamma_2^{c.g.}$ with given energy and circulation.

4. The Maximum Entropy Production (MEP) Parameterization

4.1. GENERAL FORMALISM

Out of statistical equilibrium, we still characterize the subgrid-scale fluctuations by a local probability distribution, now evolving with time. The subgrid-scale PV patches are transported and conserved by the explicit velocity field, with area preservation. This results in an advection equation for the probability density ρ. In addition, the subgrid-scale velocity fluctuations carry an eddy flux $\mathcal{J}(\sigma,\mathbf{r},t)$, so that the general conservation law for the probabilities is written as (for each value σ)

$$\frac{\partial\rho}{\partial t} + J(\overline{\psi},\rho) = -\nabla\cdot\mathcal{J}, \tag{9}$$

where $\overline{\psi}$ is still related to the ρ field by (2) and (3). At this stage, we have written an exact probabilistic translation of the PV conservation law. A corresponding conservation equation for the local PV average $\overline{q}$ is deduced by integration of (9) over σ:

$$\frac{\partial\overline{q}}{\partial t} + J(\overline{\psi},\overline{q}) = -\nabla\cdot\mathcal{J}_{\mathbf{q}}, \tag{10}$$

involving the PV eddy flux

$$\mathcal{J}_{\mathbf{q}} = \int\sigma\mathcal{J}\,d\sigma. \tag{11}$$

We expect that the random eddy stirring parameterized by the eddy fluxes $\mathcal{J}$ drives the system toward a statistical equilibrium, with a monotonic entropy increase. We can express the rate of entropy increase by time

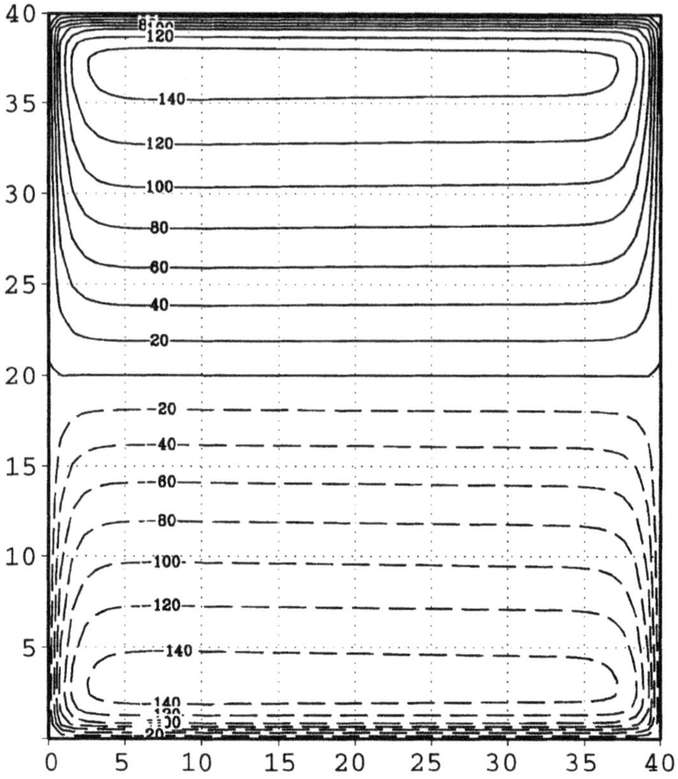

Figure 1. Fofonoff flow (streamlines) obtained as statistical equilibrium in a square basin, characterized by a linear relationship between PV and streamfunction. A slow westward current in the interior is closing along intense parietal jets. The flow is determined by the energy, or non-dimensional parameter E_β, defined by (7), and by the total circulation Γ. The ratio of the jet width to the basin width increases with energy, scaling like $E_\beta^{2/3}$. The symmetric configuration shown here corresponds to a zero circulation. Non-zero circulation results in two gyres of different size (as in Fig.2).

differentiating (5), expressing $\partial\rho/\partial t$ by (9), and noting that $\rho \, log\rho$ is conserved by the advection term,

$$\dot{S} = -\int \mathcal{J} \cdot \nabla(log\rho) \, d^2\mathbf{r} d\sigma \geq 0. \tag{12}$$

In usual non-equilibrium thermodynamics, the fluxes are restoring forces linearly related to the gradients of thermodynamic potentials, which become uniform at equilibrium. Following Robert and Sommeria [28], Robert and Rosier [29] and Kazantsev et al. [6], we use here a related, but more systematic method, the Maximum Entropy Production (MEP) principle. The idea is to maximize (at each time) the rate of entropy production (12) with the appropriate dynamical constraints, which are:

- the conservation of the local normalization $\int \rho d\sigma = 1$, implying

$$\int \mathcal{J} d\sigma = 0 \tag{13}$$

- the energy conservation, expressed from (6) and (9) as

$$\dot{\mathcal{E}} = -\int \mathcal{J}_{\mathbf{q}} \cdot \nabla\overline{\psi}\, d^2\mathbf{r} = 0 \tag{14}$$

- a limitation on the eddy flux, characterized by a bound $C(\mathbf{r})$ (which exists but is not specified)

$$\int \frac{\mathcal{J}^2}{2\rho} d\sigma \leq C(\mathbf{r}). \tag{15}$$

(The term $\frac{\mathcal{J}^2}{2\rho}$ can be considered as the square of a diffusion velocity, a natural quantity to bound.)

This variational problem is treated by introducing (at each time t) Lagrange multipliers, denoted $\zeta(\mathbf{r})$, b, $1/A_E(\mathbf{r})$ for the three respective constraints, so that the condition

$$\delta\dot{S} - \int \zeta(\mathbf{r})\delta\mathcal{J}\, d^2\mathbf{r}d\sigma \; - b\,\delta\dot{\mathcal{E}} + \int \frac{1}{A_E(\mathbf{r})}\frac{\delta\mathcal{J}^2}{2\rho}\, d^2\mathbf{r}d\sigma = 0 \tag{16}$$

must be satisfied for any variations $\delta\mathcal{J}(\sigma, \mathbf{r})$. (It can be shown from a convexity argument that the bound (15) is reached, so that this constraint can be replaced by an equality.) This yields

$$\mathcal{J} = -A_E(\mathbf{r}, t)\,[\nabla\rho - b\,\rho\,(\sigma - \overline{q})\,\nabla\overline{\psi}]. \tag{17}$$

The Lagrange multiplier $\zeta(\mathbf{r})$ has been eliminated, using the condition (13) of local normalization conservation. The eddy flux (17) involves a usual diffusion term in $\nabla\rho$, plus a drift term proportional to the Lagrange multiplier b. At statistical equilibrium, this term balances diffusion, leading to vanishing eddy fluxes and stabilization into a statistical equilibrium.

The parameter b can be identified as the inverse of a temperature (here a purely dynamical quantity appearing in the statistical theory). It is determined by introducing the expression (17) in the condition (14) of energy conservation:

$$b = \frac{\int A_E\,(\nabla\overline{q}) \cdot (\nabla\overline{\psi})\, d^2\mathbf{r}}{\int A_E\,(\nabla\overline{\psi})^2\, q_2\, d^2\mathbf{r}}. \tag{18}$$

The integrals in (18) should be computed in principle over the whole domain, according to the *global* energy constraint (14). In practice, the integration can be limited to an active region, with large velocity $\nabla\overline{\psi}$. It would be more satisfactory to replace (14) by a *local* energy conservation law, introducing an energy diffusion flux, as discussed in section 7.

As the boundaries are impermeable, there is no eddy flux, so that

$$\overline{\psi}\mid_{\partial\mathcal{D}} = 0 \ , \ \ \mathcal{J}.\mathbf{n}\mid_{\partial\mathcal{D}} = 0. \tag{19}$$

The complete MEP formulation for the free inertial evolution is then completely defined, once the eddy diffusivity $A_E(\mathbf{r}, t)$ is determined.

This diffusivity is related to the unknown bound $C(\mathbf{r}, t)$ on the eddy fluxes and is not determined by the MEP, except for its sign, which must be positive to insure entropy increase. The entropy production (12) can indeed be expressed in the form

$$\dot{S} = \int \frac{\mathcal{J}^2}{A_E\,\rho}\, d^2\mathbf{r}\, d\sigma, \tag{20}$$

so the positivity of $\dot{S}$ is only warranted when $A_E > 0$. The determination of the diffusivity A_E by kinetic models will be discussed in section 5.

4.2. THE TWO-MOMENT APPROXIMATION

Instead of working with the probability density ρ of PV levels, it is convenient to work with the successive moments $\overline{q^k} \equiv \int \rho(\sigma)\sigma^k d\sigma$. The equation for each of these moments is easily obtained by taking the corresponding moment of (9) and (17), but it involves the moment $\overline{q^{k+1}}$. Robert and Rosier [29] propose closing the hierarchy by assuming that the density distribution ρ at any position maximizes the entropy $-\int \rho\ log\rho\ d\sigma$ with the constraint of the known first n moments.

Restricting this expansion to $n = 2$ (the simplest case), we find that ρ is a Gaussian, as shown at the end of section 3, and $\overline{q^3} = \overline{q}\overline{q^2} = 2q_2$. This system is then conveniently rearranged in terms of $\overline{q}$ and q_2:

$$\frac{\partial \overline{q}}{\partial t} + J(\overline{\psi}, \overline{q}) \ = \ \nabla \cdot [A_E\, (\nabla\overline{q} \ - \ bq_2\nabla\overline{\psi})] \tag{21}$$

$$\frac{\partial q_2}{\partial t} + J(\overline{\psi}, q_2) \ = \ \nabla \cdot [A_E \nabla q_2] + 2\, A_E\, [\nabla\overline{q} \ - \ bq_2\nabla\overline{\psi}] \cdot \nabla\overline{q}. \tag{22}$$

The coefficient b is still given by (18). For an oceanic basin, Kazantsev et al. [6] showed that this two-moment approximation provides results close to the complete MEP formulation. It evolves toward the simplified statistical equilibrium of minimum potential enstrophy $\Gamma_2^{c.g.}$ discussed in the previous section.

The impermeability boundary condition imposes a zero normal probability flux (19), so that the first two moments of these fluxes vanish. This yields the following boundary conditions:

$$\overline{\psi}\,|_{\partial\mathcal{D}} = 0, \tag{23}$$

$$\mathbf{n}\cdot\nabla\overline{q}\,|_{\partial\mathcal{D}} = \; b\, q_2\, \mathbf{n}\cdot\nabla\overline{\psi}\,|_{\partial\mathcal{D}}, \tag{24}$$

$$\mathbf{n}\cdot\nabla q_2\,|_{\partial\mathcal{D}} = 0. \tag{25}$$

Note that condition (24) is not standard: the normal PV gradient is proportional to the tangential velocity.

The physical interpretation of the different terms in (21) is the following. The Jacobian expresses as usual the advection by the explicit velocity and the β-effect. The eddy parameterization on the right hand side contains three terms, taking into account that $\overline{q} = \overline{\omega} + h(\mathbf{r})$. The first term is an ordinary diffusion of vorticity. The second term $\nabla\cdot[A_E\nabla h]$ is the curl of a drift force due to PV mixing, corresponding to current rectification. The third term corresponds to a kind of friction (positive or negative), adjusted to assure energy conservation. Equation (22) expresses the transport of the PV variance q_2. An eddy diffusion is described by the first term of the right hand side, while the second term is a production for q_2, of the form $2\mathcal{J}_q.\nabla\overline{q}$, associated with the PV flux $\mathcal{J}_q$.

This two-moment formulation, which we call MEPm2, exactly conserves the total energy and the PV integral over the domain (i.e. the circulation), as well as the total PV variance (enstrophy) $\int(q_2 + \overline{q}^2)\, d^2\mathbf{r}$, but the conservation of the higher moments is forgotten. The entropy always increases (if $A_E > 0$), at an optimal rate, under the simplified form (8).

4.3. EXPRESSION IN A VELOCITY FORMULATION AND LINK WITH THE "NEPTUNE" EFFECT

We show here that, in the presence of a strong topographic β-effect, our two-moment formulation corresponds to the "Neptune" effect [1]. Computations show that q_2 and the related diffusivity A_E vary smoothly. By contrast, the topographic slope varies more sharply, as well as the velocity $\nabla\psi$, forming jets. Therefore we can neglect the gradients of A_E and q_2 in (21), and write

$$\frac{\partial\overline{\omega}}{\partial t} + J(\overline{\psi},\overline{q}) \;= A_E\,[\nabla^2\overline{\omega} + \nabla^2 h \;-\; bq_2\nabla^2\overline{\psi}]. \tag{26}$$

Defining a local parameter $\mu \equiv (bq_2)^{-1}$, and a vorticity field $\omega_* \equiv \mu\nabla^2 h$, we obtain

$$\frac{\partial\overline{\omega}}{\partial t} + J(\overline{\psi},\overline{q}) \;= A_E\,\nabla^2\overline{\omega} \;+\; (A_E/\mu)(\omega_* - \overline{\omega}). \tag{27}$$

This is an ordinary viscosity plus a term driving the system toward the state with vorticity ω_*.

Assuming again that q_2 and A_E vary smoothly, we can write (27) in term of the velocity $\mathbf{u}$, by inversion of the curl operator for each term. The eddy parameterization then reduces to an ordinary diffusion $A_E\nabla^2\mathbf{u}$, plus a topographic stress $(A_E/\mu)(\mathbf{u}_* - \mathbf{u})$, where $\mathbf{u}_*$ is obtained from the stream function $\psi_* \equiv \mu\ h$. Since this parameterization is expressed in terms of velocity, it can be used in a model with primitive equations, although its derivation is strictly valid only in the frame of the QG model.

Holloway [1] introduced a topographic stress of the form $A_m\nabla^m(u_* - u)$, so we recover this model, with the choice $m = 0$. In Holloway's approach, the parameters m, A_m, and μ are free, and independent from the smoothing operator. By contrast, we find that the topographic stress and the eddy diffusion are two aspects of the *same PV mixing effect*. This specifies the exponent $m = 0$, and relates the parameter A_0 to the eddy diffusivity, by $A_E = \mu A_0$. Moreover, the coefficient μ is specified by (18), expressing energy conservation. This is not easy to handle for actual oceanographic modeling, as it depends on the q_2 field, obtained by solving the transport equation (22), and the assumed uniformity of b is not to be expected in a large ocean basin. Therefore we may limit the integrals in (18) to some neighborhood, in which q_2 is assumed uniform. Assuming furthermore that topography dominates the potential vorticity gradient in (18), we can propose the practical relation

$$\mu^{-1}(t) \equiv bq_2 \simeq -\frac{\int \mathbf{i_z} \cdot (\nabla h \times \mathbf{u})\, d^2\mathbf{r}}{\int \mathbf{u}^2\, d^2\mathbf{r}}, \tag{28}$$

where the integrals can be viewed as low pass filters.

5. Kinetic Approaches and Determination of the Eddy Diffusivity

The *thermodynamical* approach of the previous section provides the form of the parameterization, but does not specify the coefficient A_E of eddy diffusivity. It must be evaluated by more specific *kinetic* models. The simplest hypothesis is that vorticity is passively transported by subgrid-scale random velocity fluctuations $\tilde{\mathbf{u}}$. The diffusivity is then classically obtained as in Brownian motion (Taylor [30]),

$$A_E = \frac{1}{4}\tau_c\overline{\tilde{\mathbf{u}}^2}, \tag{29}$$

where τ_c is the (Lagrangian) correlation time of $\tilde{\mathbf{u}}$.

We can express the local velocity fluctuation $\tilde{u}$ in terms of the vorticity fluctuation $\tilde{\omega}$ by the integral formula $\tilde{\mathbf{u}}(\mathbf{r}) = \int \tilde{\omega}(\mathbf{r}')\mathbf{K}(\mathbf{r}-\mathbf{r}')d^2\mathbf{r}'$, where the kernel $K(\mathbf{r}-\mathbf{r}')$ expresses the velocity induced by a unit singular vortex. For a barotropic flow far from boundaries, its expression is $\mathbf{K}(\mathbf{r}-\mathbf{r}') = (1/2\pi)\mathbf{i_z} \times (\mathbf{r}-\mathbf{r}')/|\mathbf{r}-\mathbf{r}'|^2$ (and the effect of image vortices must be added to take boundaries into account). The energy $\overline{\tilde{\mathbf{u}}^2}$ is then expressed in terms of the spatial correlation function for vorticity as

$$\overline{\tilde{\mathbf{u}}^2} = \int \overline{\tilde{\omega}(\mathbf{r}')\tilde{\omega}(\mathbf{r}'')}\mathbf{K}(\mathbf{r}-\mathbf{r}')\mathbf{K}(\mathbf{r}-\mathbf{r}'')\, d^2\mathbf{r}' d^2\mathbf{r}''. \tag{30}$$

The subgrid-scale vorticity fluctuations $\tilde{\omega}$ are by definition at scales smaller than the grid mesh d, so the vorticity correlation drops for $|\mathbf{r}'-\mathbf{r}''| > d$. We can distinguish in (30) the contribution $\overline{\tilde{\mathbf{u}}^2}_{loc}$ of local vorticity fluctuations ($|\mathbf{r}'-\mathbf{r}''| < d$) and the non-local contribution $\overline{\tilde{\mathbf{u}}^2}_{non-loc}$ (with $|\mathbf{r}'-\mathbf{r}''| > d$). Dimensional arguments give a local contribution of order $d^2\overline{\tilde{\omega}^2}$. For the non-local contribution, one can assume that the vorticity correlation is a Dirac distribution, $\overline{\tilde{\omega}(\mathbf{r}')\tilde{\omega}(\mathbf{r}'')} = d^2\overline{\tilde{\omega}^2}(\mathbf{r}')\delta(\mathbf{r}'-\mathbf{r}'')$, so that $\overline{\tilde{\mathbf{u}}^2}_{non-loc} = d^2\overline{\tilde{\omega}^2} \int_d^\infty K^2(r')2\pi r' dr' \simeq (d^2/2\pi)\ log(L/d)\ \overline{\tilde{\omega}^2}$. The integral logarithmically diverges, which means that all scales contribute equally to $\overline{\tilde{\mathbf{u}}^2}_{non-loc}$. We have chosen the size of the domain L as the upper bound (the result is only weakly sensitive to this choice because of the logarithmic dependency). Adding the two contributions yields

$$\overline{\tilde{\mathbf{u}}^2} \simeq d^2(1 + \frac{1}{2\pi} log\frac{L}{d})\ \overline{\tilde{\omega}^2}. \tag{31}$$

We notice that, since $\overline{\tilde{\omega}^2}$ is of the same order as $\overline{\omega}$, the energy of the velocity fluctuations is of order $d^2\omega^2$, much smaller than the total energy, of order $L^2\omega^2$. This estimation provides a justification of the approximation made in (6). However the velocity fluctuations control the eddy diffusivity, obtained by introducing the expression (31) in (29).

Assuming the correlation time is of the order $q_2^{-1/2}$, this yields a diffusivity of the form

$$A_E \simeq\ k\, d^2\, q_2^{1/2}, \tag{32}$$

where $k = \frac{1}{4}(1 + \frac{1}{2\pi} log\frac{L}{d})$. In practice this non-dimensional parameter k will be adjusted as the minimum value required to smooth out the numerical noise (at a value $k \simeq 10^{-2}$ in the computations of section 6). We notice that the relaxation toward statistical equilibrium is faster in active regions, with strong vorticity fluctuations q_2. This can result in a flow organization limited to a "bubble" [5, 29], surrounded by an irrotational background. Notice that the β-effect should influence the eddy diffusivity, and this point needs to be investigated.

Furthermore, this derivation is far from rigorous, and it does not yield the drift term. A more precise approach has been recently proposed by

Chavanis [7], using a point vortex model. The vortices are randomly positioned, according to the mean field statistical equilibrium, and they follow the mean field velocity to a first approximation. Close vortex encounters result in a diffusivity. Furthermore any test vortex perturbs the other vortices (like a polarization effect in electrostatics), and it receives a response resulting in the systematic drift of the test vortex. Therefore the MEP result is further justified by this more mechanistic approach.

6. Application to a Wind-driven Barotropic Oceanic Basin

We have tested the MEP parameterization in the standard case of a wind-driven barotropic oceanic circulation. We consider a rectangular basin and introduce a forcing stress τ (whose curl is vertical and considered as a scalar denoted *curl* τ) and a bottom friction term with decay time T_F,

$$\frac{\partial \omega}{\partial t} + J(\psi, \omega + \beta y) \;=\; curl\; \tau - \frac{\omega}{T_F}. \tag{33}$$

We assume that the subgrid-scale dynamics is still inertial, with a negligible direct influence of forcing and friction, so we simply add the forcing and friction terms to the vorticity equation (21). We assume the forcing is fully resolved, so the variance q_2 is not directly forced, but decays by friction with the characteristic time $T_F/2$ (since it is quadratic in vorticity).

In most of our computations, the ocean is forced, in a square domain, by a steady zonal wind with a classical two gyre antisymmetric pattern:

$$curl\; \tau \;=\; \mathcal{F}_0 \sin \frac{2\pi y}{L}, \tag{34}$$

where L is the basin side, and $y = 0$ at the mid-basin latitude. This is seen as a schematic pattern for the mean wind stress over the mid-latitude North Atlantic ocean. To test the case of non-symmetric forcing, we have also considered the same forcing (33) in a half-basin, with y in the interval $[-L/2, 0]$ (keeping the same longitudinal extent).

Somewhat standard values have been chosen for the various model parameters: $L = 4000$ km, $\beta = 2 \times 10^{-11}$ m^{-1}s^{-1}, $T_F = 2 \times 10^7$s ~ 200 days. In the square basin, the physical wind stress is assumed to vary from its minimum $(-1/2\pi)$ N/m^2 at the southern and northern boundaries to a maximum of $(1/2\pi)$ N/m^2 at mid-latitude. The corresponding acceleration $\tau_m = (10/\pi) \times 10^{-7}$ms^{-2} is obtained after dividing this stress by the water density and by the depth of the active wind-driven current, chosen as 500 m. The reduced amplitude of the stress curl is therefore $\mathcal{F}_0 = 2\pi\tau_m/L = 5 \times 10^{-13} s^{-2}$. These parameters provide a long and active turbulent jet, with realistic velocities, appropriate to test the parameterization of barotropic eddies.

A first approach of the wind-driven ocean is provided by the Sverdrup balance between β-effect and wind forcing,

$$\beta \frac{\partial \psi}{\partial x} \simeq curl\ \tau, \tag{35}$$

giving a maximum y velocity component $v = \mathcal{F}_0/\beta = 2.5$ cm/s with our numerical parameters. Neglecting inertial effects, this flow must return along the Stommel boundary jets, characterized by a balance between β-effect and bottom friction. These jets would be very narrow and rapid (with width $1/(T_F\beta) = 1.6$ km and velocity $\mathcal{F}_0 T_F L = 40$ m/s), leading to strong instability.

Therefore inertial effects are necessarily important, and turbulence must arise. A lateral friction term is usually added to (33) to parameterize the horizontal eddy effects at the grid size. From a numerical point of view, such parameterization is actually required to dissipate the enstrophy cascade toward small scales. Otherwise, the development of finer and finer scales should forbid any explicit description with a reasonable resolution. In practice, the most commonly used parameterizations are the Laplacian and the biharmonic lateral friction, written respectively as $\nabla^4\psi$ and $\nabla^6\psi$ with some constant coefficient A_H. The rationale for using a biharmonic is to provide a weaker energy dissipation than a Laplacian. Energy conservation is indeed a genuine property of two-dimensional turbulence that the parameterization should preserve. However, a biharmonic lateral friction does not provide a good representation of PV conservation and mixing, another important property that the parameterization should preserve. Here we *derive* a term of eddy diffusivity from the statistical mechanics of an inviscid theory, so we do not add any a priori lateral viscosity term to (32). Note that, consequently, the boundary conditions are completely specified by the impermeability constraint. For real oceanic modeling, coastal diabatic (dissipative) effects and internal waves could introduce some horizontal viscosity and modify the boundary conditions, but this problem is not addressed by the present study.

We now apply the MEPm2 parameterization to a coarse resolution ($d =$ 80 km) computation of the model (32), and compare it to a high resolution (d = 20 km) reference computation (with a weak ordinary viscosity of the Laplacian type). We mainly consider the familiar double-gyre wind-forced box, which mimics the open ocean intense jet arising in mid-latitudes at the convergence of the subpolar and the subtropical counterrotating cells. Normally, this Gulf Stream-like jet is strongly instable with mixed barotropic/baroclinic instabilities. In the present model, only barotropic instability would be acting subject to the constraint that the resolution is fine enough. For each run, the oceanic flow is spun up from rest under the influence of the wind forcing, up to a permanent regime, reached after about 5 years. After that, the model is integrated for 30 more years, in order to provide for an adequate time sequence for computing the time average flow fields and the eddy statistics.

We notice that this standard double-gyre forcing cannot maintain the statistical equilibrium of Fig. 1: the wind forcing is, in most of the domain, opposite to the current. In order to make a closer connection with the statistical equilibrium, we first introduce a frontier at the middle of the basin, considering a single gyre forcing limited to the lower half-basin ($-L/2 < y < 0$). With our nominal forcing $\tau_m = 10/\pi \times 10^{-7}$ ms^{-2}, we observe the flow organization into a single gyre (Fig. 2). After the spinup time of about 5 years, a nearly steady state is reached. The jet is organized in agreement with the statistical equilibrium, with an intense jet along the northern boundary. As expected, this flow is well-reproduced by the MEPm2 computation (Fig. 2b).

We now consider the more realistic double-gyre case in the square basin. The velocity is globally much weaker than in the half-basin, illustration the contradiction between the statistical equilibrium and the forcing. Streamlines of a snapshot and time averaged mean flow are shown in Fig. 3. We observe the western boundary jets, detaching into a free jet (the "Gulf Stream"), surrounded by inertial gyres. The remainder of the mean flow is controlled by the Sverdrup balance, while variability is due to Rossby waves emitted by the instabilities of the detached jet. All these features are well-reproduced with the coarse resolution by the MEPm2 parameterization. The mean energy is $\mathcal{E}/L^2 = 0.127 \pm 0.05$ m^{2}s^{-2} (about 15 % higher than for the reference computation). Furthermore, properties of the statistical equilibrium help to understand the behavior of the system. Unlike in the half-basin, the forcing indeed tends to drive an eastward current in the interior, clearly in contradiction with a global statistical equilibrium, for which eastward currents are confined to boundaries. In reality, a partial statistical equilibrium is established on each side of the jet, where the gyres are sufficiently rapid for inertial stirring, but the system is globally out of equilibrium, with strong PV fluxes across the jet, maintaining the instability.

A good evaluation of the parameterization is obtained by comparing the velocity along the jet maximum, along a line $y = 0$. An excellent agreement with the reference model is indeed observed in Fig. 4. Other comparisons, involving fluctuation energy and meridional PV fluxes, confirm this good agreement. By contrast, other parameterizations, tested by Kazantsev et al. [6], behave poorly with the same coarse resolution.

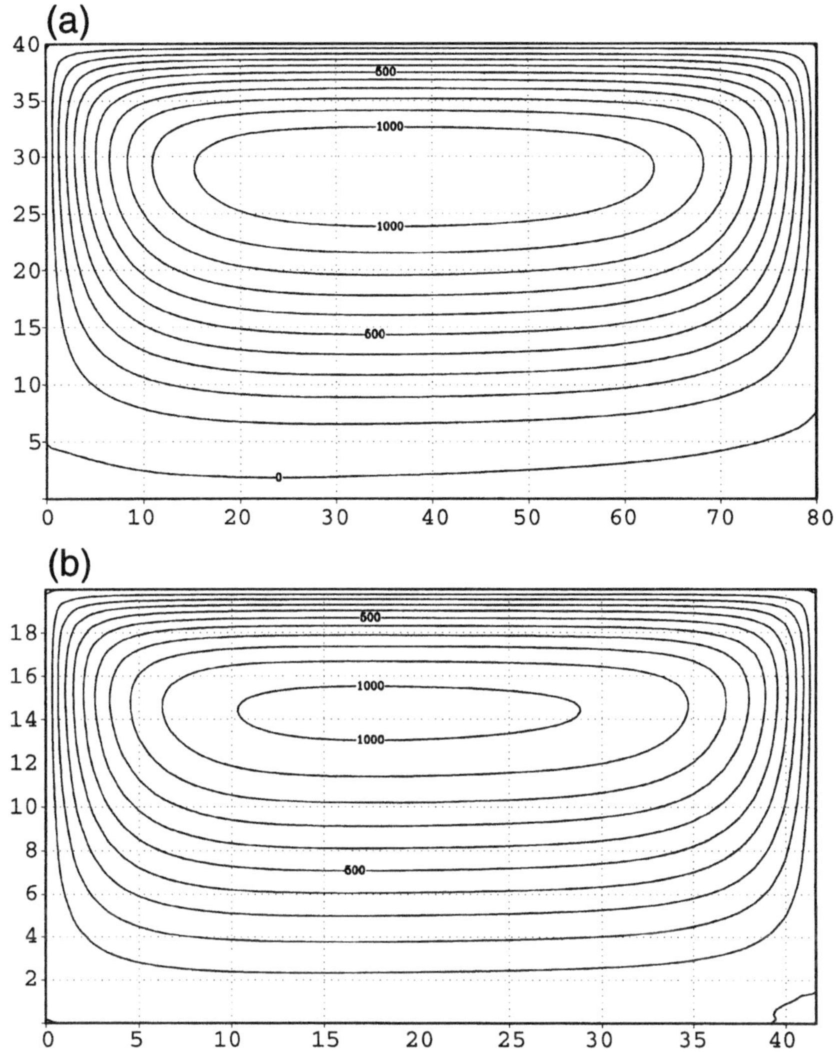

Figure 2. Single-gyre wind forcing in the half-basin: comparison between (a) a reference high resolution run, and (b) the coarse resolution with MEPm2 parameterization. The single-gyre forcing allows the flow to remain close to statistical equilibrium.

7. Extensions to a Stratified Fluid

The next step is to consider a fluid layer stratified in density, which can be modeled as a QG shallow water system, with the potential vorticity

$$q = \omega + h(\mathbf{r}) - \psi/R^2, \tag{36}$$

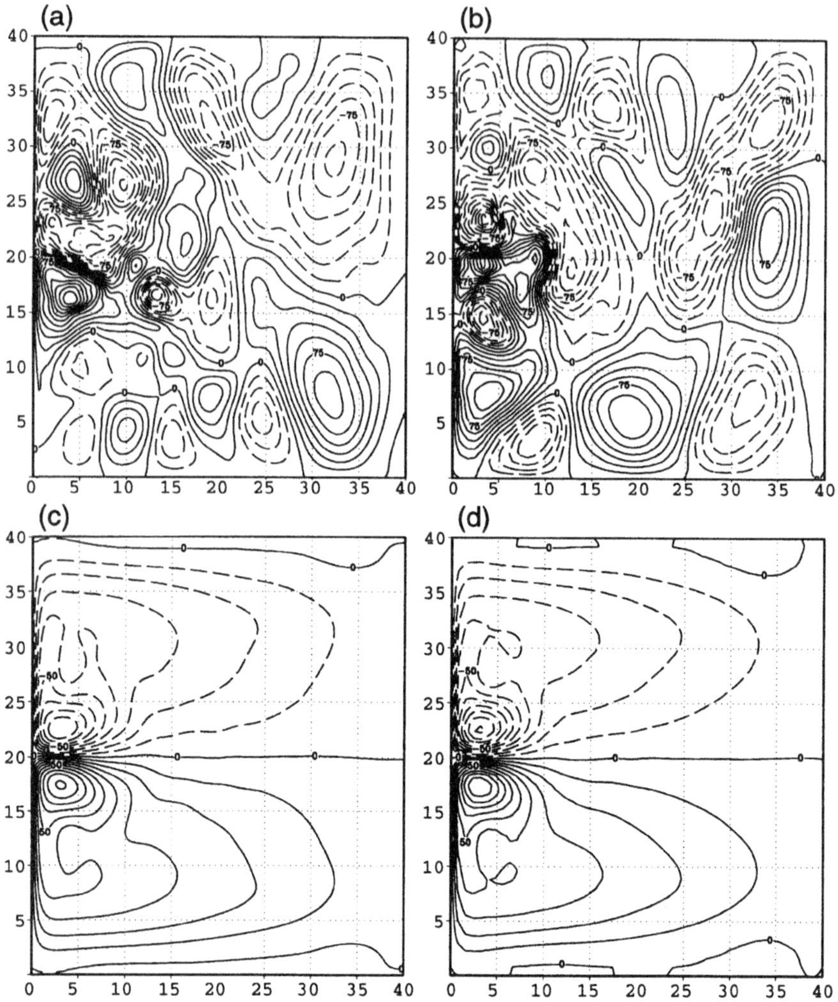

Figure 3. The double-gyre wind forcing: high resolution reference (REF) computations (left) compared to the MEP computation with coarse resolution (right). (a,b) Examples of instant streamlines, contour interval 15 Sv (1 Sv= 10^6 m^{3}s^{-1}); (b,c) streamlines of the 30 yr averaged velocity, contour int. 10 Sv, from -80 to 90 Sv (REF) and from -100 to 90 Sv (MEP), respectively.

introducing the radius of deformation R. When the grid mesh d is smaller than R, the modification is straightforward, just changing the expression of q in the results. However, when the mesh is on the order of or larger than R (the most interesting case for oceanic applications), the energy of the local fluctuations is no longer negligible. The energy now includes a contribution of potential energy, $\mathcal{E} = \frac{1}{2}\int(\mathbf{u}^2 + \frac{\psi^2}{R^2})\, d^2\mathbf{r}$. Distinguishing the explicit and

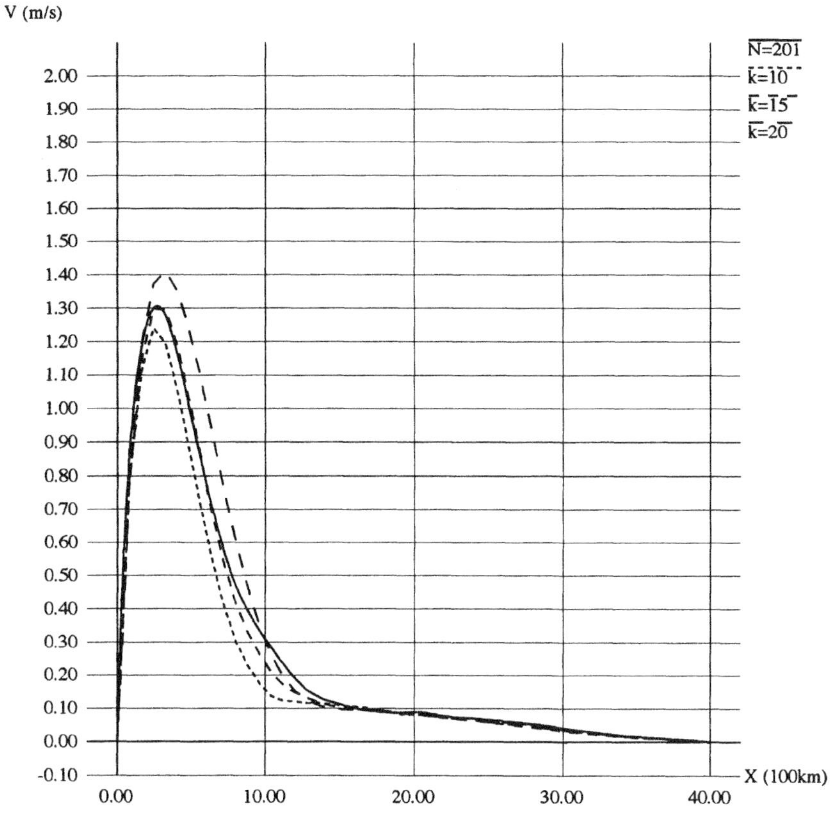

Figure 4. Mean velocity along the jet center (y=0) vs. longitude, with the double-gyre forcing. The MEP results with different diffusivity coefficients k are compared to the reference result with solid line.

subgrid scale contributions of $\mathbf{u}$ and ψ, and integrating by parts the explicit contribution, yields

$$\mathcal{E} = -\frac{1}{2}\int \overline{\psi}\,(\overline{q} - h(\mathbf{r}))\, d^2\mathbf{r} + \int \epsilon\, d^2\mathbf{r}, \tag{37}$$

where $\epsilon = \frac{1}{2}(\overline{\tilde{\mathbf{u}}^2} + \frac{\overline{\tilde{\psi}^2}}{R^2})$ is the energy of the local fluctuations.

We now need to express the local energy ϵ in terms of the moment q_2 of the probability distribution ρ, in order to apply our formalism. If the grid mesh d is much smaller than R, the effect of the stratification term will be negligible at the scale of the mesh, so we can still use (31) and estimate $\epsilon \sim (d^2/2)q_2$. In the opposite case $d \gg R$, the velocity and streamfunction fluctuations can be expressed from the correlation of the PV fluctuations by a formula like (30), but the kernel $\mathbf{K}$ expressing the velocity induced by a singular delta function PV is modified. It is now a Bessel function which exponentially decays beyond the distance R, so the relationship between PV fluctuations and ϵ is dominated by local terms. For a Fourier mode with wavenumber k, we have $\tilde{\psi} = -(k^2+1/R^2)^{-1}\tilde{q}$, and $\tilde{\mathbf{u}^2} = k^2(k^2+1/R^2)^{-2}\tilde{q}^2$. The corresponding contribution to ϵ is $(\tilde{q}^2/2)(k^2+1/R^2)^{-2}$. It reduces to $k^{-2}\tilde{q}^2$ for $kR \gg 1$ and tends to $\tilde{q}^2R^2/2$ for $kR \ll 1$. In the case $d \ll R$, all subgrid scales satisfy the latter condition, so that $\epsilon \sim q_2R^2/2$. As a summary, we can make the simple estimate

$$\epsilon = \frac{1}{2}B^2 q_2, \quad \text{with } B = d \text{ for } d \ll R, \quad B = R \text{ for } d \gg R. \tag{38}$$

The evolution equation for the local variance q_2 is obtained by introducing (4) in (9, 10),

$$dq_2/dt = -\nabla \cdot \mathcal{J}_2 - 2\mathcal{J}_q \cdot \nabla\overline{q}, \tag{39}$$

involving a diffusion flux $\mathcal{J}_2 \equiv \int \sigma(\sigma - 2\overline{q})\mathcal{J}(\sigma)d\sigma$ and a source term $-2\mathcal{J}_q \cdot \nabla\overline{q}$. The change of the explicit energy is still given by (14), but now it must be balanced by the change of $\int \epsilon d^2\mathbf{r}$, expressed from (38) and (39). This yields the global energy equation

$$\dot{\mathcal{E}} = -\int \mathcal{J}_\mathbf{q} \cdot \nabla\overline{\psi'}\, d^2\mathbf{r} = 0 \tag{40}$$

where $\psi' \equiv \psi + B^2\overline{q}$.

We can now use the MEP as in section 4, replacing the rate of change of energy (14) by (40), which results in replacing ψ by ψ' in the expression of the eddy flux (17) and in the corresponding PV flux $\mathcal{J}_\mathbf{q}$:

$$\mathcal{J} = -A_E(\mathbf{r},t)\, [\nabla\rho - b\,\rho\,(\sigma - \bar{q})\, \nabla(\overline{\psi} + B^2\overline{q})], \tag{41}$$

$$\mathcal{J}_\mathbf{q} = -A_E\, [(1 - bq_2B^2)\nabla\overline{q} - bq_2\nabla\overline{\psi}]. \tag{42}$$

The temperature b can be obtained by the global conservation of energy, leading to (18), replacing ψ by ψ'. However, in an attempt to improve the model, we shall replace instead the global constraint on energy by a local one [32], introducing an energy flux $\mathcal{J}_E$, writing

$$\mathcal{J}_\mathbf{q} \cdot \nabla\overline{\psi'} = \nabla \cdot \mathcal{J}_E. \tag{43}$$

We assume the boundary condition $\mathbf{n} \cdot \mathcal{J}_E = 0$, so the integral of the divergence vanishes, and this local expression is equivalent to (40). However we shall now bound $\mathcal{J}_E$, imposing energy conservation in a more local sense. Proceeding with the MEP, we then introduce the Lagrange multiplier $(1/\chi_E)$ for the constraint $\mathcal{J}_E^2 < C_E(\mathbf{r})$, and a Lagrange multiplier $b(\mathbf{r})$ for the constraint (43). The term $b\delta\dot{\mathcal{E}}$ in (16) is then replaced by $\int b(\mathbf{r})\delta[\mathcal{J}_\mathbf{q} \cdot \nabla\overline{\psi'} - \nabla \cdot \mathcal{J}_E)]\ d^2\mathbf{r} + \int (1/\chi_E(\mathbf{r}))\delta\mathcal{J}_E{}^2/2\ d^2\mathbf{r} = \int [b(\nabla\overline{\psi'}) \cdot \delta\mathcal{J}_\mathbf{q} + (\nabla b + \mathcal{J}_E/\chi_E) \cdot \delta\mathcal{J}_E]\ d^2\mathbf{r}$. The vanishing of the term in $\delta\mathcal{J}$ yields again the eddy fluxes (41, 42) with a local temperature $b(\mathbf{r})$. The vanishing of the term in $\delta\mathcal{J}_E$ gives a "heat conduction" relation $\mathcal{J}_E = -\chi_E\nabla b$. Combining with (43), this yields $\nabla \cdot (\chi_E\nabla b) = -\mathcal{J}_\mathbf{q} \cdot \nabla\psi'$. Expressing $\mathcal{J}_\mathbf{q}$ by (42), we get an elliptic equation determining b:

$$\nabla \cdot (\chi_E\nabla b) - A_E q_2(\nabla\psi')^2 b = -A_E(\nabla\overline{q}) \cdot (\nabla\psi'), \tag{44}$$

with the boundary condition corresponding to the absence of the normal energy flux $\mathbf{n} \cdot \mathcal{J}_E\ |_{\partial\mathcal{D}}$,

$$\mathbf{n} \cdot \nabla b\ |_{\partial\mathcal{D}} = 0. \tag{45}$$

The two-moment approximation (21-22) can be used as before, replacing ψ by ψ'.

The result now depends on two unknown diffusivities χ_E and A_E. Equ. (44) is similar to a Helmholtz equation, so that the solution b is smooth. Assuming a large diffusivity χ_E, b becomes uniform and satisfies the integral condition (18) (with ψ replaced by ψ'). It seems more reasonable to assume that χ_E is proportional to A_E, with a coefficient depending on the local quantities q_2 and b. Then dimensional arguments yield

$$\chi_E \sim A_E/b^2. \tag{46}$$

A similar derivation could be adapted to a multi-level baroclinic model.

8. Conclusions and Discussion

Approaching the complexity of natural systems like oceans requires a hierarchy of deterministic and statistical models at successive scales. A deterministic model predicts the structure of "objects", which are steady solutions, waves, or instability modes. When many of these objects interact in a complex way, their influence on the larger scales must be described in a statistical sense. This procedure is well-established at the molecular level: molecules are objects with a well-defined structure, the solution of the Schrödinger equation. Equilibrium statistical mechanics of molecular motion is then used to predict thermodynamic properties of the fluid, while kinetic models provide transport properties, like diffusivity. The resulting continuous medium equations can be used to determine fluid structures occurring at a macroscopic scale.

Turbulence modeling proceeds with similar ideas but suffers from the difficulty of finding the appropriate elementary objects. Here we have considered fluid particles as elementary objects. They mix differently from molecules because of the conservation of their area and PV. However, the local strain stretches fluid particles so they lose their identity after a typical time inverse of the rate of strain. This process of enstrophy cascade toward small scales is in competition with the spatial mixing of fluid particles described by our statistical approach. Various studies indicate that the statisitical mechanics predictions are indeed quantitatively correct when the flow organization is sufficiently fast in comparison with this straining time. This occurs for instance during vortex merging. This process can be understood as a tendency to increase entropy, and the resulting organization into an axisymmetric structure is quantitatively predicted as a statistical equilibrium. Discrepancies are, however, observed at the periphery of the vortex, where strain dominates vorticity.

Many numerical computations of 2D turbulence are initiated by a random field at small scale. Then organized vortices are formed, and evolve on longer time scales by successive mergings. Then the discrepancies in strain-dominated regions tend to accumulate and the statistical equilibrium prediction of the final state from the initial condition is not good (however it improves greatly when the prediction is made from an intermediate state, taken just before the last merging events, as shown by Brands et al. [31]). When such intermediate vortices exist in the subgrid-scale (which was not the case in our reference oceanic model), the probability distribution of PV levels is non-gaussian, which would change the drift term in (21). Various comparisons, however, indicate that this effect of isolated vortices is not essential on the mean fluxes (but it could be important for more subtle effects controlled by rare events).

Note finally that the present approach is specific to the parameterization of mesoscale eddies. Internal gravity waves also contribute to mixing, and must be described by an appropriate statistical mechanics. Finally, in regions of strong shear, such as boundary layers, fully three-dimensional turbulence prevails, for which empirical models developed for engineering applications can be adapted.

References

1. Holloway, G. (1992) Representing topographic stress for large-scale ocean models, *J. Phys. Oceanogr.* **22**, 1033-1046.
2. Alvarez, A. and Tintoré, J. (1998) Topographic stress: Importance and parameterization, In *Ocean Modeling and Parameterization*, E.P. Chassignet and J. Verron (Eds.), Kluwer Academic Publishers, 327-350.
3. Robert, R. and Sommeria, J. (1991) Statistical equilibrium states for two-dimensional flows, *J. Fluid Mech.* **229**, 291-310.
4. Chavanis, P. and Sommeria, J. (1996) Classification of Self-organized Structures in Two-dimensional Turbulence: The case of a bounded domain, *J. Fluid Mech.* **314**, 267-297.
5. Chavanis, P. and Sommeria, J. (1998) Classification of robust isolated vortices in

two-dimensional turbulence, *J. Fluid Mech.*356, 259-296.
6. Kazantsev, E., Sommeria, J., and Verron, J. (1998) Subgrid-scale eddy parameterization by statistical mechanics in a barotropic ocean model, *J. Phys. Ocean.*, in press.
7. Chavanis, P. (1998) Systematic drift experienced by a point vortex in two-dimensional turbulence, *submitted to Phys. Rev. E.*
8. Onsager, L. (1949) Statistical hydrodynamics, *Nuovo Cimento Suppl.* **6**, 279-287.
9. Montgomery, D. and Joyce, G.(1974) Statistical mechanics of negative temperature states, *Phys. Fluids* **17**, 1139-1145.
10. Kraichnan, R.H. (1967) Inertial ranges in two-dimensional turbulence, *Phys. Fluids* **10**, 1417.
11. Kraichnan, R.H. and Montgomery, D. (1980) Two-dimensional turbulence, *Rep. Prog. Phys.* **43**, 547-617.
12. Salmon, R., Holloway, G., and Hendershott, M. C. (1976) The equilibrium statistical mechanics of simple quasi-geostrophic models, *J. Fluid Mech.* **75**, 691-703.
13. Holloway, G. (1986) Eddies, waves, circulation and mixing: statistical geofluid mechanics, *Ann. Rev. Fluid Mech.* **18**, 91-147.
14. Wang, J. and Vallis, G.K., (1994) Emergence of Fofonoff states in inviscid and viscous ocean models, *J. Mar. Res.* **52**, 83-127.
15. Zou, J. and Holloway, G. (1994) Entropy maximization tendency in topographic turbulence, *J. Fluid Mech.* **263**, 361-374.
16. Robert, R. (1990) Etat d'équilibre statistique pour l'écoulement bidimensionnel d'un fluide parfait, *C.R. Acad. Sci. Paris* **311**, Série I, 575-578.
17. Miller, J. 1990 Statistical mechanics of Euler equations in two dimensions, *Phys. Rev. Lett.* **65**, 17, 2137-2140.
18. Sommeria, J., Nore, C., Dumont, T., and Robert, R. (1991) Théorie statistique de la Tache Rouge de Jupiter, *C. R. Acad. Sci. Paris*312, II, 999-1005.
19. Michel, J. and Robert, R. (1994) Statistical mechanical theory of the Great Red Spot of Jupiter, *J. Stat. Phys.* **77**, 3/4, 645-666.
20. Fofonoff, N.P. (1954) Steady flow in a frictionless homogeneous ocean, *J. Mar. Res.* **13**, 254-262.
21. Le Provost, C. and Verron, J. (1987) Wind-driven ocean circulation transition to barotropic instability, *Dynamics of Atmospheres and Oceans* **11**, 175-201.
22. Ierley, G. and Young, W. (1988) Inertial recirculation in a β-plane corner, *J. Phys. Oceanogr.* **18**, 683-689.
23. Griffa, A. and Salmon, R. (1989) Wind-driven ocean circulation and equilibrium statistical mechanics, *J. Mar. Res.* **47**, 457-492.
24. Cummins, P. (1992) Inertial gyres in decaying and forced geostrophic turbulence, *Journal of Mar. Res.* **50**, 545-566.
25. Bretherton, F.P. and Haidvogel, D.B. (1976) Two-dimensional turbulence above topography, *J. Fluid Mech.* **78**, 1, 129-154.
26. Carnevale, G.F. and Frederiksen, J.S. (1987) Nonlinear stability and statistical mechanics of flow over topography, *J. Fluid Mech.* **175**, 157-181.
27. Juttner, B., Thess, A., and Sommeria J., (1995) On the symmetry of self-organized structures in two-dimensional turbulence, *Phys. Fluids* **7** (9), 2108-2110.
28. Robert, R. and Sommeria, J. (1992) Relaxation Towards a Statistical Equilibrium State in Two-Dimensional Perfect Fluid Dynamics, *Phys. Rev. Lett.* **69 (19)**, 2776-2779.
29. Robert, R. and Rosier, C. (1997) The modelling of small scales in 2D turbulent flows: a statistical mechanics approach, *J. Stat. Phys.* **86**, 481-515.
30. Taylor, G.I. (1921) Diffusion by continuous movements, *Proc. London Math. Soc.* **20**, 196-211.
31. Brands, H., Stulemeyer, J., Pasmanter R.A., and Schep, T. J. (1997) A mean field prediction of the asymptotic state of decaying 2D turbulence, *Phys. Fluids* **9** (10), 2815-2817.

32. Chavanis P. and Sommeria J.,1997: A thermodynamical approach to small scale parameterization in 2D Turbulence *Phys. Rev. Lett.* **28** 3302-3305.

TOPOGRAPHIC STRESS: IMPORTANCE AND PARAMETERIZATION

ALBERTO ALVAREZ
Department of Physics and Center for Complex Systems, National Central University, Chung-li, Taiwan, ROC

AND

JOAQUIN TINTORÉ
Departament de Física, Universitat de les Illes Balears and Instituto Mediterráneo de Estudios Avanzados, Palma de Mallorca, Spain

1. Introduction

1.1. STATEMENT OF THE GENERAL PROBLEM

The ocean circulation problem concerns the motion of a rotating stratified and turbulent fluid on a sphere (the Earth) with complex boundaries introduced by the break up of the continents. This nonlinear nature of ocean dynamics generates a wide variety of interesting physical phenomena mainly related to the existence of strong dynamical links among physical processes occurring at different space and time scales. These links range from space scales of centimeters and time scales that might be counted in minutes or hours, up to global motions with time scales of centuries, that control aspects of the Earth's climate. This range of scale interactions shown by ocean dynamics induces the appearance of collective phenomena that are hardly explained by the individual properties of each ocean process. Understanding ocean dynamics requires not only the study of isolated individual ocean processes, but also the collective result emerging from the combination of these individual processes acting at different space and time scales. Therefore the study of ocean circulation becomes an extremely difficult task because it requires determining the whole set of space and time scales that characterize the behaviour of the ocean system.

Concerning numerical simulation of ocean dynamics, the range of scales that numerical models can handle explicitly is limited by the actual com-

E. P. Chassignet and J. Verron (eds.), Ocean Modeling and Parameterization, 327–350.

puter power, the large scale of the global domain, and the time characteristics of ocean studies. These limitations prevent global models from having sufficient resolution to include explicitly the effects of the whole range of space and time scales in ocean studies. Thus the realism of global ocean circulation models is severely restricted.

Given this problem, what can be done? The answer to this question is that models need to be parameterized to include some of the physics missed due to coarse resolution [14, 27].

1.2. STATEMENT OF A PARTICULAR PROBLEM: NEPTUNE EFFECT

It is well known that much of the variability in energy in the ocean occurs on the mesoscale, for which typical time and space scales are on the order of several months and tens of kilometers [10]. Commonly, limitations on computer resources prevent global models from having sufficient resolution to include explicitly the effects of such mesoscale processes. However mesoscale interactions can exert powerful forces on the large scale flows. This happens, for example, with the interaction between mesoscale eddies and bottom topography. An estimate of the force generated by this interaction is an effective stress of 1 to 10 $dyn\ cm^{-2}$ [17]. This is as large as, or even larger than, the typical wind stresses which are thought to be the dominant driving mechanism for the ocean. The topographic stress in the real ocean might not be so large as the magnitude obtained from theoretical estimations. However, this stress might also not be small enough as to be neglected in the computations of large scale ocean models.

Intuitively, one might suppose that the effect of topographic roughness would be to exert an enhanced drag against large scale overlying currents. However this is not exactly correct. For example, Herring [13] and Holloway [15] considered the problem of considering strongly nonlinear eddy fields, idealized as two-dimensional turbulence overlying random topography. They found that the drag due to topography was not even a "drag". In particular, considering a flow which was at rest in the mean while exhibiting a spectrum of eddy activity, the result is that eddies would latch onto the topography so as to force the flow into mean motion. This new effect due to bottom topography was termed the "Neptune Effect" [17].

Because the Neptune Effect is a direct consequence of the eddy field, its effects on the large scale circulation will strongly depend on the resolution of the model, that is, on our ability to incorporate correctly the small scale processes. If the resolution is not good enough to take into account the eddy scales, then perhaps we will lose an important part of the physics of the system. This could drive the ocean model to an incorrect result.

Let us examine this problem by considering a barotropic quasigeostrophic

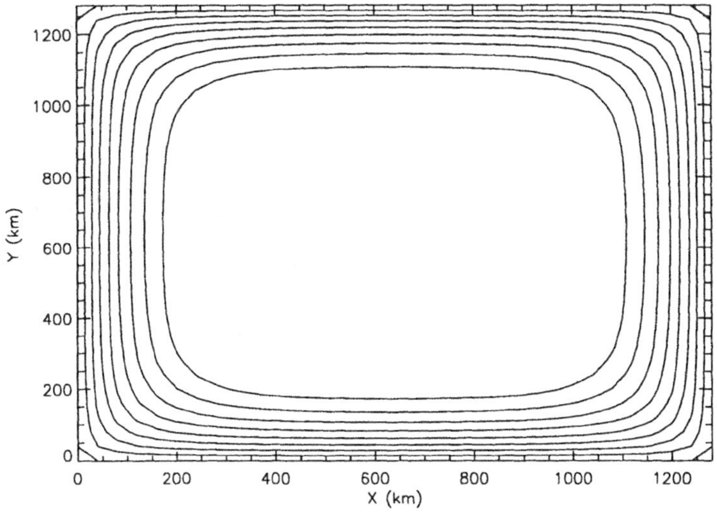

Figure 1. Bottom topography for the numerical experiment. Maximum depth is 5000 m at the center of the basin, minimum depth is 4500 m at the edges. Levels are plotted every 69 m.

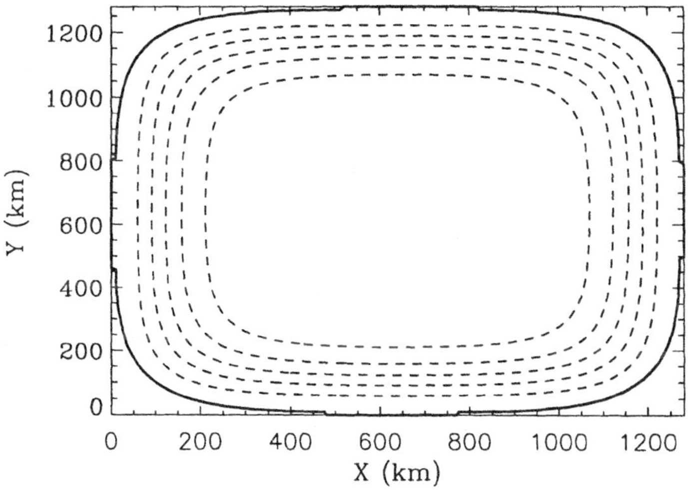

Figure 2. Mean streamfunction for the highest resolution case. Continuous contours denote positive values of the streamfunction, whereas dashed contours denote negative ones. Maximum and minimum values are -23884 and 23884 m^2/s,and levels are plotted every 3981.7 m^2/s.

model of the ocean. We consider this dynamic model because it is complex enough to include the problem that we want to study, i.e. the interaction between eddies and bottom topography, but simple enough to allow us some theoretical understanding and to avoid carrying out very complex numerical experiments. In the quasigeostrophic approximation the dynamics are governed by [28]:

$$\frac{\partial \nabla^2 \psi}{\partial t} + \left[\psi, \nabla^2 \psi + h\right] = 0, \tag{1}$$

$$\psi_{boundary} = 0, \tag{2}$$

where ψ is the streamfunction that provides the horizontal components of the fluid velocity $(u(x,y), v(x,y))$ from

$$u = -\frac{\partial \psi}{\partial y}, \quad v = \frac{\partial \psi}{\partial x}. \tag{3}$$

Here $h = f\Delta H/H_0$, with f the Coriolis parameter, H_0 the mean depth, and $\Delta H(x,y)$ the local deviation from the mean depth. The Poisson bracket or Jacobian is defined as:

$$[A, B] = \frac{\partial A}{\partial x}\frac{\partial B}{\partial y} - \frac{\partial B}{\partial x}\frac{\partial A}{\partial y}. \tag{4}$$

Equation (1) represents the time evolution of the relative vorticity without forcing and dissipation. The proposed experiment consists in solving numerically (1) in an ocean basin (see Figure 1) with different resolutions. To do that, we use the numerical scheme developed in [9] on different grids of 128×128, 64×64 and 32×32 points. The distance between grid points corresponds to 10 km, 20 km, and 40 km respectively, so that the total system size is $L = 1280$ km. The algorithm is based on Arakawa finite differences and the leap-frog method. As initial conditions, we have generated a random eddy field with the same energy for the three resolutions. Figure 2 shows the circulation obtained for the case of highest resolution. As the reader can check, the initial distribution of eddies has generated a large scale flow following isobaths as found in [13, 15, 29]. The same kind of circulation pattern has also been obtained for the coarser resolution cases. However, a detailed analysis of the data reveals important differences among the results obtained with different resolutions. Effectively, although the total energy of our quasigeostrophic ocean is the same for the three cases, the energy of the mean flow decreases with the resolution of the model. A spectral analysis of the velocity field for the three different cases indicates that these energy differences are mainly located in the large scale flow (Figure 3).

We can conclude from this experiment that the influence of the Neptune Effect strongly depends on the resolution of the numerical model. This could

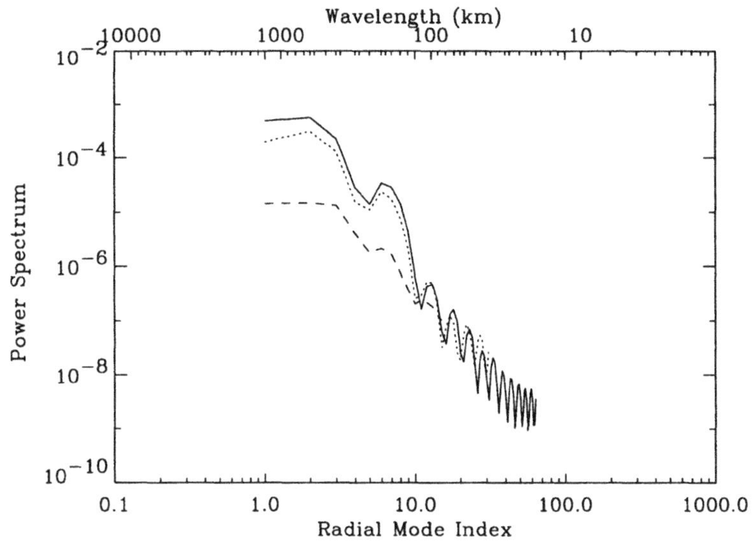

Figure 3. Energy spectrum for the resolutions of 10 (continuous), 20 (dotted) and 40 (dashed) kilometers.

be guessed a priori considering the fact that the Neptune Effect arises as a result of the collective behaviour of the set of eddies. If the eddy field is not adequately resolved, its collective effects can be lost, and an important mechanism driving the ocean circulation may be missed. As a consequence, the results obtained from a coarse resolution model could be wrong. Therefore, a parameterization to correctly include the Neptune Effect in coarse resolutions models is needed. This is the task considered in the next section.

2. Obtaining a Parameterization of the Eddy-topography Interaction

2.1. THE EQUILIBRIUM OCEAN

Ocean dynamics can be studied by means of theoretical schemes. Specific flow mechanisms, say a type of wave propagation or an instability, might be isolated and solved analytically. A harder question is how to obtain the collective result of many such mechanisms acting in a common environment. In these cases, turbulence theory deals with the study of the properties of the collective result on the different processes acting on the fluid system. Such theories yield statistical information on collective behaviour of fluid systems. However, when the underlying mechanisms are more elaborate, the assumptions from turbulence theory can not be applied.

Still another theoretical approach to the study of the ocean may be taken into account. It considers the application of ideas from statistical mechanics as applied to macroscale geophysical flows [16]. Statistical mechanics is useful to obtain parameterizations of the physical processes which cannot yet be included in numerical codes due to their enormous computational cost. In the following subsections we will explore the utility of statistical mechanics to obtain basic information about the large scale ocean flows.

2.1.1. *Statistical description of ocean circulation*

Let us attempt to combine the ideas of statistical mechanics with our knowledge of the dynamics defined by (1). To do this, we consider the dynamics of the Fourier modes of (1):

$$\frac{\partial \zeta_k}{\partial t} = \sum_{k=p+q} A_{kpq} \left(\zeta_p \zeta_q + \zeta_p h_q \right), \tag{5}$$

with $\zeta_k = -k^2 \hat{\psi}_k$, and where the interaction coefficient is:

$$A_{kpq} = (p_x q_y - p_y q_x) p^{-2} \delta_{k,p+q}. \tag{6}$$

The phase-space of our ocean will have as many degrees of freedom as the total number N of modes taken into account in the Fourier expansion, i.e. a state of the ocean will be defined by a set of values $\{\zeta_1, \zeta_2, ..., \zeta_N\}$. When the system is in equilibrium, each point of phase-space $\{\zeta_1, \zeta_2, ..., \zeta_N\}$ will have associated with it a probability of occurrence $\rho\left(\{\zeta_1, \zeta_2, ..., \zeta_N\}\right)$ [hereinafter $\equiv \rho\left(\{\zeta_k\}\right)$].

Several elements should be taken into account in the calculation of the probability distribution $\rho\left(\{\zeta_k\}\right)$. These elements are related to the dynamical properties shown by (1). The basic point is that (1) conserves an infinite set of quantities given by $E = \int dxdy \frac{1}{2} \left(\nabla \psi\right)^2$, which corresponds to the kinetic energy of the system, and by $Q_m = \int dxdy \frac{1}{m} \left(\nabla^2 \psi + h\right)^m$ with m being any positive integer (the magnitude $Q_2 (\equiv Q)$ defines the enstrophy of the system). The existence of these invariants implies that not all the points of phase space are accessible for the ocean. Instead a subset of these will have the chance to be explored by the system. These states that are accessible to the system are completely determined by the initial conditions. Specifically, let us suppose that the initial condition of the ocean fixes an energy $E = E_0$ and an enstrophy value $Q = Q_0$. This means that the initial state is a point of phase space lying on the intersection between the hypersurfaces defined by $E = E_0$ and $Q = Q_0$. The only accessible states will then lie on this intersection between the two hypersurfaces. Considering this fact and assuming ergodicity, then the probability distribution for our ocean is given by (if other invariants are neglected):

$$\rho(\{\zeta_k\}) = C\delta\left(E(\{\zeta_k\}) - E_0\right)\delta\left(Q(\{\zeta_k\}) - Q_0\right), \tag{7}$$

where C is some normalization constant. The probability distribution defined by (7) is known as the microcanonical probability distribution.

The reason that we have only considered the invariants of energy and enstrophy in (7) is because the truncated dynamics (5) only conserves these quadratic invariants [20]. It is beyond the scope of this note to discuss the possible role of the other invariants. Readers interested in the construction of a statistical mechanics retaining all of the invariants are referred to [24].

The microcanonical probability distribution is in general difficult to use for calculations. In order to find a more suitable probability distribution, we will have to relax our physical assumptions. The new assumption takes into account the condition that the isolated ocean is no longer subject to the constraints of constant energy and enstrophy. Instead, mean values of these magnitudes are constrained to be fixed. To compute the new probability distribution $\rho(\{\zeta_k\})$, we will use the fact that entropy is a maximum at equilibrium [19]. Maximization of the entropy functional, keeping as constraints the values of the mean energy and enstrophy, provides the following probability distribution [29]:

$$\rho(\{\zeta_k\}) = \frac{e^{-aE(\{\zeta_k\})-bQ(\{\zeta_k\})}}{\int\prod_k d\zeta_k e^{-aE(\{\zeta_k\})-bQ(\{\zeta_k\})}}, \tag{8}$$

where a and b are the Lagrange multipliers of the maximization problem. The probability distribution defined by (8) is called the canonical probability distribution. The microcanonical and canonical probability distributions are equivalent descriptions of the statistics of the system when the number of modes N is very large (see for example [29]).

The canonical probability distribution (8) can be easily employed to compute the mathematical relations that the fluid satisfies at equilibrium. This relation is defined by a linear dependence of the mean potential vorticity on the mean streamfunction [29]:

$$\left\langle\nabla^2\psi\right\rangle + h = \frac{a}{b}\langle\psi\rangle. \tag{9}$$

Considering the fact that for large scale flows the term in (9) corresponding to the relative vorticity, $\langle\nabla^2\psi\rangle$, is usually small compared to the ambient vorticity term h, the relation (9) can then be simplified as in [17]:

$$h \approx \frac{a}{b}\langle\psi\rangle. \tag{10}$$

That is, we obtain that the mean streamfunction "follows" the isobaths. This is an important feature. Independent of our initial state, the system shows an equilibrium state characterized by the existence of a well-defined large scale circulation that follows the isobaths with anticyclonic circulation (cyclonic) over bumps (depressions) in the Northern Hemisphere.

We have already seen in subsection 1.2 that this statistical tendency has been termed the Neptune Effect. We have also studied its dependence on the resolution of the model. In the next subsection, we will describe a parameterization of this tendency based on the results obtained from the application of statistical mechanics to the large scale ocean summarized in the result (10).

2.1.2. *Neptune Effect parameterization*

Based on the previous results, Holloway [17] suggested implementing the Neptune Effect within numerical ocean models as a tendency of the system towards a defined equilibrium state. This equilibrium state of the ocean would be based on the equilibrium state obtained from the inviscid quasi-geostrophic theory, given by (9). The subgrid scale tendency is expected to occur on account of the difference field between the instantaneous velocity field and the velocity field of the defined equilibrium state of the ocean. As a first choice, Holloway [17] suggested appending a relaxation term towards the equilibrium velocities into the model equations:

$$\frac{\partial u}{\partial t} = model\,[u] - \frac{(u - u^*)}{\tau}, \tag{11}$$

where $u^* = \hat{z} \times \nabla\psi^*$ and $\psi^* = -fLH$ (where f is the Coriolis parameter, H the bottom topography, and L an adjustable parameter which depends on the mean energy and mean enstrophy of the system) being the theoretical equilibrium state. τ is a relaxation time which could have spatial and temporal variation.

In order to obviate the determination of the parameter τ, Holloway [17] suggested another expression for the parameterization derived from (11). The new suggestion consisted of the correction of the explicit representation of the eddy viscosity term, $A\nabla^2 u$. The eddy viscosity term was then replaced by a form centered on u^*, that is, $A\nabla^2\,(u - u^*)$:

$$\frac{\partial u}{\partial t} = model\,[u] + A\nabla^2\,(u - u^*)\,. \tag{12}$$

This is more scale selective than the preceding suggestion (11), and also avoids the need to specify τ. Of course, the eddy coefficient A must still be specified, but that will be required either explicitly or implicitly in any case. Hence this second approach has the further advantage of requiring fewer ad hoc specifications. The new numerical codes including this statistical tendency of the eddy field are denoted hybrid models.

2.2. THE OPEN OCEAN

Although the parameterization developed in the last subsection has been employed successfully in the numerical simulation of specific areas of the ocean (see for example [1], [18], and [31]), a host of objections can be raised to it. First, [17] did not provide a physical basis for the suggestion of making the parameterization linear dependent of the difference between the instantaneous state and the maximum entrophy state. If the dynamics should be changed, this must be carried out in a consistent way. Second, another objection arises because statistical dynamics describes isolated systems, whereas the role of direct forcing and dissipation in ocean circulation are manifestly important. Therefore the next objective in our approach will be to analyse the case of the forced and viscous ocean. The outcome of this study will show that the Neptune Effect parameterization can be rigorously obtained from the physical properties of the open ocean. This point will be the main task of the following subsection.

2.2.1. *The viscous and unforced ocean: minimum enstrophy states*

In order to progress in our research, we will now increase the complexity of (1), considering the role that dissipation plays in a quasigeostrophic fluid with underlying topography. Inclusion of dissipation is a way to force the system to exchange energy with the environment, that is, our quasigeostrophic ocean will no longer be an isolated system. Following the work of Bretherton and Haidvogel [7], let us consider our ocean modeled by:

$$\frac{\partial \nabla^2 \psi}{\partial t} + \left[\psi, \nabla^2 \psi + h\right] = \nu \nabla^2 \left(\nabla^2 \psi\right), \tag{13}$$

where ν is the viscosity parameter and the other terms have already been defined in the previous section. Because no input of energy is considered, it is clear that due to the action of the viscosity term the final state of (13) is motionless. Although the final state achieved by (13) has no particular interest, the transient state prior to reaching the rest state shows a very interesting phenomenon closely related to the physical nature of turbulent two-dimensional flows. It is well known [28] that a particular characteristic of the dynamics described by (1) or (13) is that enstrophy systematically cascades to very small scales at which it is dissipated, while energy remains at large scales where the action of dissipation is weak due to the scale selective nature of the viscosity term. The fact that the energy drag is much slower than the enstrophy drag, induces the generation of quasi-stationary states characterized by having the minimum enstrophy compatible with a given value of energy. This minimal flow pattern can be computed by minimizing the enstrophy, keeping constant energy as a constraint, that is, the flow pattern must verify:

$$\delta\left[Q+\mu E\right]=0, \tag{14}$$

where μ is the Lagrange multiplier of the problems. Equation (14) implies:

$$\nabla^2\psi+h=\mu\psi. \tag{15}$$

The relevance of (15) is that, although it is a transient state, it shows a tendency of the open system to generate large scale currents following isobaths. The question now is whether or not this is the only mechanism in an open ocean to generate such a tendency. In other words, we still have to explore the existence of other physical mechanisms that could generate the same tendency as expressed by (9) or (15).

2.2.2. *The stochastically forced ocean: noise-sustained currents*

Once the case of the viscous ocean has been analysed, the next step consists of again increasing the complexity of the ocean model by considering some type of forcing. Because we are interested in studying the effects of small scales on large scales when bottom topography is present, this forcing must be directly related to the action of small scales.

It is common in physics to employ a noise term as a representation of the scales that are fast and small enough to be included in the dynamics. For example in geophysical problems, the baroclinic nature of ocean circulation can be usefully represented for large scales by randomly forced barotropic models [36, 2]. In these barotropic models, the random forcing formulation is a simple Markovian process that provides a reasonable representation of the baroclinic nonlinear interactions [11, 21]. A different way in which noise can enter the description of fluid systems is by considering the interactions taking place at scales below the resolution of a computer model. Addition of noise is a (crude) way of including some influence from unresolved scales [36].

The influence of small noise on large scale fluid dynamics is, in principle, expected to be small. Several studies [13, 15, 34] have addressed the existence of stationary currents following isobaths in a variety of conditions involving random noise. However, from these studies it is not possible to infer a clear relation between noise and the physical nature of these currents. The reason is that scale selective (i.e., viscosity) dampings were considered so that the generation of directed currents can be explained in terms of minimum enstrophy states [7]. As we have already seen, this physical mechanism involving minimum enstrophy solutions does not require the existence of noise.

Now we try to highlight the role of noise in the appearance of steady currents in such open systems. To do this, we are going to consider topographic turbulence damped by Rayleigh friction (bottom friction) instead of by viscosity. This introduces important physical differences with respect to the case studied in paragraph 2.2.1, because Rayleigh friction provides a scale-unselective way of dissipating energy from the system and thus does not produce transient minimum enstrophy states. In this way, avoiding one of the possible mechanisms for explaining the mean flows that were present in the simulations of [13], [15], and [34], we can see if the combined effect of noise and nonlinear terms is sufficient to generate mean flows when bottom topography is present.

Let us consider then the following quasigeostrophic dynamics:

$$\frac{\partial \nabla^2 \psi}{\partial t} + \left[\psi, \nabla^2 \psi + h\right] = -\epsilon \nabla^2 \psi + F \ , \tag{16}$$

where ϵ^{-1} is the bottom friction decay time, $F(x, y, t)$ is a Gaussian white-noise process with zero mean and correlations $\langle F(x, y, t) F(x', y', t')\rangle = D\delta(x - x')\delta(y - y')\delta(t - t')$, and the other parameters retain the same definition as given in subsection 1.2.

We solve (16) numerically using the scheme previously described in subsection 1.2 on a grid of 128×128 points, that is, keeping the same resolution as in the high resolution case of subsection 1.2. The consistent way of introducing the stochastic term into the leap-frog scheme can be found in [30, 12, 2]. The random number generator employed in the simulation is detailed in [33]. The amplitude of the forcing, $D = 2 \times 10^{-9}\ m^2/s^2$, and the e-folding decay constant, $\epsilon = 3 \times 10^{-7} s^{-1}$ of the damping mechanism have been chosen in order to obtain final velocities of several centimeters per second. The topographic field (Figure 4) is randomly generated from an isotropic spectrum containing, with equal amplitude and random phases, all the Fourier modes corresponding to scales between 80 km and 300 km. The model was run for 5×10^5 time steps (corresponding to 126 years) after a statistically stationary state was reached. The average streamfunction was then computed from the configurations obtained at each time step. The results obtained from this numerical experiment show the generation of a stationary mean flow correlated with bottom topography and with velocity values on the order of 3 cm/s (Figure 5). This numerical result indicates the existence of a physical mechanism, different from the minimum enstrophy tendency studied in the previous paragraph, that is able to organize random inputs of energy in a mean state different from rest, characterized by currents with a clear dependence on bottom topography.

The following explanation for the emergence of these currents has been suggested by [2]: As shown in subsection 1.2, invariance of energy and potential enstrophy has important implications for the inviscid and unforced dynamics. In particular, statistical properties can be correctly computed by assuming that at long time scales the system explores completely, or

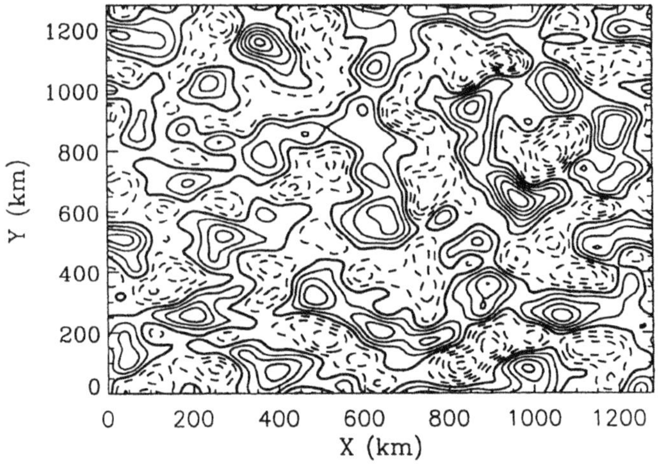

Figure 4. Depth contours of randomly generated bottom topography. Maximum depth is 381.8m and minimum depth -381.8m over an average depth of 5000m. Levels are plotted every 63.6m. Continuous contours are for positive deviations with respect to the mean, whereas dashed contours denote negative ones.

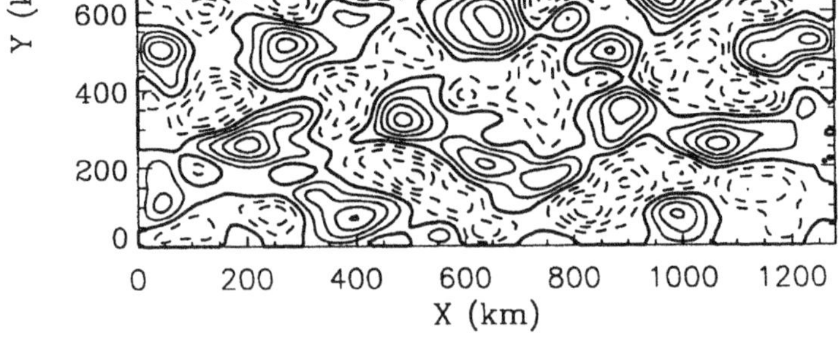

Figure 5. Mean streamfunction computed by time averaging when a stationary state has been achieved. Maximum and minimum values are 1120.2 and -1120.2 m^2/s. Levels are plotted every 186.7 m^2/s. Continuous contours denote positive values of the streamfunction, whereas dashed contours denote negative ones.

at least samples with enough significance, the hypersurface of constant energy and potential enstrophy (for a discussion see [24]). When damping is introduced, the system will be brought to the vicinity of the rest state in phase space. But random forcing will keep some motion present. Since the effect of the nonlinear terms in the unforced and undamped dynamics was to drive the system along the intersection of the constant-energy and constant-potential-enstrophy hypersurfaces, it is then reasonable to expect that. in the presence of forcing and damping, evolution will proceed on average close to the hypersurfaces of mean constant energy and mean potential enstrophy that can be determined by the damping and forcing coefficients. If this hypothesis holds, the most likely probability distribution compatible with such constraints, that is the one obtained from the maximization of the entropy functional subjected to the constraints of constant mean energy and potential enstrophy, should be an approximation to the real stationary probability distribution [19]. The validity of the heuristic considerations above, like the validity of general statistical mechanics arguments in any interesting physical situation, would be hard to establish rigorously. In this specific case of a randomly forced ocean, ergodicity or at least efficient sampling of the constant energy and potential enstrophy hypersurfaces would certainly be more justified than in the unforced case, because of the additional mixing tendency in phase-space provided by random forcing.

Entropy maximization subjected to constraints of constant mean energy and enstrophy leads formally to the same expression as in the inviscid and unforced case, that is, the predicted flow is given by [2]:

$$\nabla^2 \langle\psi\rangle + h = c\langle\psi\rangle\,, \tag{17}$$

where $c = a/b$ with a and b being the Lagrange multipliers of the maximization problem. This linear relationship between the mean potential vorticity and the mean streamfunction implies, considering that for large scale motions the relative vorticity is smaller than the ambient vorticity, that $\langle\psi\rangle$ is proportional to h, that is $\langle\psi\rangle$ describes mean flows following the isobaths. From the numerical simulation we can see that the correlation between topography and mean streamfunction, as predicted by (17), is obvious by comparing Figures 4 and 5. The agreement can be made more quantitative by plotting $\nabla^2 \langle\psi\rangle + h$ versus $\langle\psi\rangle$, both obtained from the numerical simulation. This is done in Figure 6. Equation (17) predicts a linear relationship. Figure 6 shows a clear linear trend but there is also an evident scatter in the data. The linear correlation coefficient is 0.940. Further analysis of the validity of theoretical predictions can be done by fitting the data in Figure 6 to (17) to obtain a numerical value for the constant c. This procedure gives a value of $c = (5.30 \pm 0.02) \times 10^{-9}\ m^{-2}$. The predicted value of c ob-

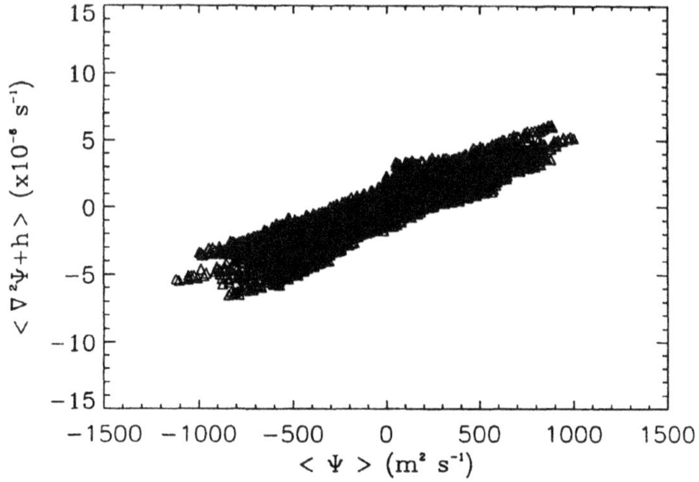

Figure 6. Scatter plot of mean potential vorticity versus mean streamfunction obtained from numerical simulation. A gross linear tendency appears that resembles the linear tendency observed in inviscid calculations. Each symbol is obtained from a different position in the simulation domain.

tained from the theory, by considering the magnitudes of the mean energy and potential enstrophy obtained in the stationary state, is $5.5 \times 10^{-9} m^{-2}$, which represents a deviation of 4% with respect to the value obtained by numerical simulation. Readers can find a more detailed analysis of these noise-sustained currents in [2].

Presently, general ocean circulation models are deterministic. However, based on the phenomena studied in this subsection, it seems reasonable to hypothesize that stochastic models will provide more realistic answers about ocean dynamics and more concrete information about the possible existence and relevance of these noise-sustained currents in the real ocean. However, several warnings with respect to stochastic models must be given. First, the numerical schemes developed for integrating stochastic differential equations are quite limited compared to the numerical schemes for the deterministic case. The correct numerical implementation of noise in an ocean model could become an extremely difficult task. Second, computer speed becomes slower if the required amount of random numbers is large. This would be the situation in a numerical simulation of the ocean. For these reasons we still have to search for a parameterization that can describe the effects of viscosity and random forcing in a deterministic way. Then we will be able to incorporate in the numerical model the physical effects induced by noise and topography while keeping the advantages of working with deterministic models. Therefore the objective in the next subsection will be to find a deterministic description of the phenomena of noise-sustained currents.

2.2.3. *Getting the Neptune Effect parameterization: The coarse grained ocean model*

It seems clear now that a parameterization of the Neptune Effect should be based on the physical properties of a model including noise and viscosity, instead of on the properties of an isolated ocean. Due to these considerations, we will increase the complexity of the previous quasigeostrophic model by considering the joint effect of viscosity and noise. To do this, we consider an ocean model governed by the dynamics:

$$\frac{\partial \nabla^2 \psi}{\partial t} + \lambda \left[\psi, \nabla^2 \psi + h \right] = \nu \nabla^4 \psi + F \; , \tag{18}$$

where ν is the viscosity parameter, $F(x, y, t)$ is any kind of external forcing (noise term), and $h = f\Delta H/H_0$, with f the Coriolis parameter, H_0 the mean depth, and $\Delta H(x, y)$ the local deviation from the mean depth. λ is a bookkeeping parameter introduced to allow perturbative expansions in the interaction term. The physical case corresponds to $\lambda = 1$. As already defined, $[,]$ is the Poisson bracket or Jacobian and ψ is the streamfunction.

Our purpose now is to analytically calculate the closed effective equations of motion for the large scales of the flow, i.e. to obtain a dynamics that is only defined for the large scales while taking into account the effect of the small scales. This effect of the small scales will be described in terms of the parameters and variables of the large scale field. The new dynamics describing only the behaviour of the large scales will be obtaining by coarse-graining the small scales. Due to its complexity, it is beyond the scope of this paper to give a description of the coarse-graining method; however, readers with special interest in this particular subject are referred to [35]. To apply coarse-grainig to (18), we have assumed for the relative vorticity forcing F a Gaussian stochastic process with zero mean and correlations given by $\left\langle \hat{F}_{\mathbf{k}}(\omega) \hat{F}_{\mathbf{k}'}(\omega') \right\rangle = Dk^{-y} \delta(\mathbf{k} + \mathbf{k}') \delta(\omega + \omega')$. $\hat{F}_{\mathbf{k}}(\omega)$ denotes the Fourier transform of $F(x, y, t)$, $\mathbf{k} = (k_x, k_y)$, and $k = |\mathbf{k}|$. The process is then white in time but has power-law correlations in space. This choice gives us enough freedom to model a variety of processes. For example, $y = 0$ corresponds to white-noise in space also.

We use the Fourier components of the streamfunction $\hat{\psi}_{k\omega}$ or equivalently the relative vorticity $\zeta_{k\omega} = -k^2 \hat{\psi}_{k\omega}$. This variable satisfies:

$$\zeta_{k\omega} = G^0_{k\omega} F_{k\omega} + G^0_{k\omega} \lambda \sum_{k=p+q, \omega=\Omega+\Omega'} A_{kpq} \left(\zeta_{p\Omega} \zeta_{q\Omega'} + \zeta_{p\Omega} h_q \right) , \tag{19}$$

where the interaction coefficient is:

$$A_{kpq} = (p_x q_y - p_y q_x) p^{-2} \delta_{k,p+q} \tag{20}$$

and the bare propagator is:

$$G^0_{k\omega} = (-i\omega + \nu k^2)^{-1}. \tag{21}$$

$0 < k < k_0$, with k_0 an upper cut-off. Following the method described above, one can eliminate the modes $\zeta^>_k$ with k in the shell $k_0 e^{-\delta} < k < k_0$, and substitute their expressions into the equations for the remaining low-wavenumber modes $\zeta^<$ with $0 < k < k_0 e^{-\delta}$. To second order in λ, the resulting equation of motion for the modes $\zeta^<$ is [3]:

$$\frac{\partial \nabla^2 \psi^<}{\partial t} + \lambda \left[\psi^<, \nabla^2 \psi^< + h^< \right] = \nu' \nabla^4 (\psi^< - g h^<) + F' , \tag{22}$$

where

$$\nu' = \nu \left(1 - \frac{\lambda^2 S_2 D (2+y) \delta}{32 (2\pi)^2 \nu^3} \right), \tag{23}$$

$$g(\lambda, D, \delta, \nu, y) = \frac{\lambda^2 D S_2 (y+4) \delta}{16 (2\pi)^2 \nu^3}, \tag{24}$$

where S_2 is the area of the unit sphere in two dimensions and $F'(x, y, t)$ is an effective noise which turns out to be also a Gaussian process with mean value and correlations given by:

$$< F'(x, y, t) >= - \frac{\lambda^2 D S_2 (4+y) \delta}{16 (2\pi)^2 \nu^2} \nabla^4 h^<, \tag{25}$$

$$\left\langle \left(\hat{F}'_k(\omega) - \left\langle \hat{F}'_k(\omega) \right\rangle \right) \left(\hat{F}'_{k'}(\omega') - \left\langle \hat{F}'_{k'}(\omega') \right\rangle \right) \right\rangle = D k^{-y} \delta(k + k') \delta(\omega + \omega'). \tag{26}$$

Equations (22)-(26) are the main result in this section. They give the dynamics of long wavelength modes $\psi^<_k$ with $0 < k < k_0 e^{-\delta}$. They are valid for small λ or, when $\lambda \approx 1$, for small width δ of the elimination band. The effects of the eliminated short wavelengths on these large scales are described in the new structure of the viscosity operator and the corrections to the noise term F'.

The action of the dressed viscosity term $\nabla^4(\psi^< - g h^<)$ is no longer to drive large scale motion towards rest, but towards a motion state $\approx g h^<$ characterized by the existence of flow following the isolevels of bottom perturbations $h^<$. This ground state would characterize the structure of the mean pattern. The energy in this ground state is determined by the function $g(\lambda, D, \delta, \nu, y)$, which measures the influence of the different terms (nonlinearity, noise, viscosity). Relation (24) shows that while nonlinearities and noise increase the energy level of the ground state, high values of the viscosity parameter would imply a reduction of the strength of the ground state motion due to the damping effect that viscosity exerts over small scales. The other mechanism that reinforces the existence of pattern formation arises from the fact that the dressed noise has a mean component as a result of the small scale elimination. This mean value given by (25) also favours motions controlled by the underlying topography. The overall effect on the large scale motions appears as a renormalization of the viscosity operator, in such a way that it favors relaxation towards a state correlated with topography instead of towards rest. This last statement recovers, from a more systematic justification, the parameterization previously suggested from a heuristic point of view [17] in the context of large scale ocean models, which we have discussed in subsection 2.1.

3. Importance of the Neptune Effect in Large Scale Ocean Circulation

In the last section we have carried out an extensive study to obtain a parameterization of the Neptune Effect. Starting from the simplest concepts of an unforced quasigeostrophic model of the ocean circulation, we have increased the complexity of the model to build in a coherent way the parameterization of the effects that eddy-topography interaction has on large scale motions. There is still one question to be answered. This question is related to the real role that the Neptune Effect plays in large scale circulations. As previously pointed out, the experimental determination of the relevance of the Neptune Effect seems difficult to carry out. The only possibility of making this test is by considering ocean models more realistic than the quasigeostrophic one. In this case, the circulation obtained from the numerical model with and without parameterization can be compared with the circulation inferred from observations. The agreement or not of the numerical result with observations will give us with an idea about the relevance of the Neptune Effect in a previously determined region of the ocean.

In this section we will consider a set of "realistic" numerical experiments of the large scale circulation of the Western Mediterranean (hereinafter WM). The reasons for choosing the WM can be summarized as follows: First because of the availability of observations to compare with the numerical results, and second because the WM is a sea characterized by a high degree of mesoscale activity that would imply in principle an important control of the Neptune Effect on the large scale circulation of this sea.

Before describing the set of numerical simulations of the large scale circulation of the WM, we will summarize the main features of this circulation in order to compare them a posteriori with the numerical results.

3.1. LARGE SCALE CIRCULATION IN THE WESTERN MEDITERRANEAN

The circulation in the Mediterranean Sea is affected by the surface energy and mass exchange with the atmosphere. Evaporation exceeds total precipitation and runoff by an average rate of 0.76×10^5 m^3/s [6]. The resulting water deficit is compensated for by a two-layer flow in the Strait of Gibraltar. In this strait, Atlantic Water (AW) flows eastward in the upper 200 m and Mediterranean water (MW) flows westward into the Atlantic. AW becomes rapidly more saline in its eastward migration and is then named Modified Atlantic Water (MAW). MW is essentially composed of Mediterranean Deep Water (MDW, formed in winter in the Gulf of Lyon), Levantine Intermediate Water found at approximately 500 m (LIW, formed in the eastern basin) and Surface Mediterranean Water (SMW) found in the upper 300 m, with surface salinities higher that 37.5. SMW can be also considered "old" MAW.

The following descriptive diagram for the large scale surface circulation in the WM has been proposed by [25, 26] (see Figure 7): the circulation begins with the inflow of AW into the Alboran Sea through the Gibraltar Strait. Recent studies [32] have shown that AW usually forms a set of anticyclonic gyres in the Alboran Sea. During their transition in the Alboran Sea, AW becomes saltier, originating the MAW. The outflow of MAW from the Alboran Sea constitutes a well-defined coastal current, the Algerian Current, following the African coast [4]. This eastward current is characterized by the formation, in summer, of large scale anticyclonic eddies (diameter on the order of 100 km) that can be advected northeastwards towards the Balearic Sea. In the Strait of Sicily, part of the MAW flow enters into the eastern basin while the remainder circulates into the northern area of the western Mediterranean. The remaining flow constitutes the Ligurian Current, after the Corsica Channel, and the Northern Current which, flowing southwestwards following the French and Spanish coasts, ends in the Alboran Sea.

3.2. NUMERICAL SIMULATIONS

Several experiments were developed to test the influence of the Neptune Effect in the surface large scale circulation of the WM. To do this, a grid of 98×57 points was created in the horizontal scale, including two reservoirs

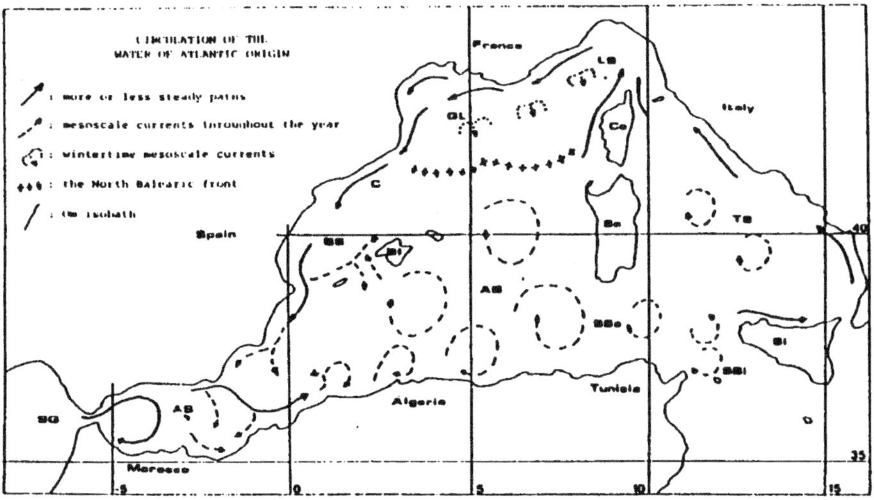

Figure 7. The circulation of the water of Atlantic origin in the Western Mediterranean (Millot, 1987).

to simulate the Atlantic Ocean and the Eastern Mediterranean. In the vertical we fixed 23 layers ranging from 5 to 2760 m. Bottom topography was obtained from [5]. Three islands were included: Corsica-Sardinia, Mallorca, and Ibiza. The model was forced with monthly wind data from [23]. Surface T and S forcing and initial data were taken from [22]. In the two reservoirs there are no wind and surface forcings. In these two places, T and S fields are constantly relaxed to Levitus data with a relaxation time of 60 days. The Strait of Gibraltar has one velocity grid point and a depth of 360 m at its shallowest point.

In this experiment, a test to establish the importance of the climatological atmospheric forcing was first carried out. Second, both climatological and Neptune forcing have been considered. The results are then compared with the circulation pattern suggested by [25, 26]. All the results shown correspond to the month of December at twenty years of simulation, when an almost stationary state is achieved.

3.2.1. *Results*

We proceed first to describe the results obtained in the control case. Figure 8 shows model results after 20 years of simulation without the Neptune Effect at 80 m. In this experiment the transport through the Strait of Gibraltar is 0.84 Sv, very similar to the value of 0.76 Sv obtained from observations [8].

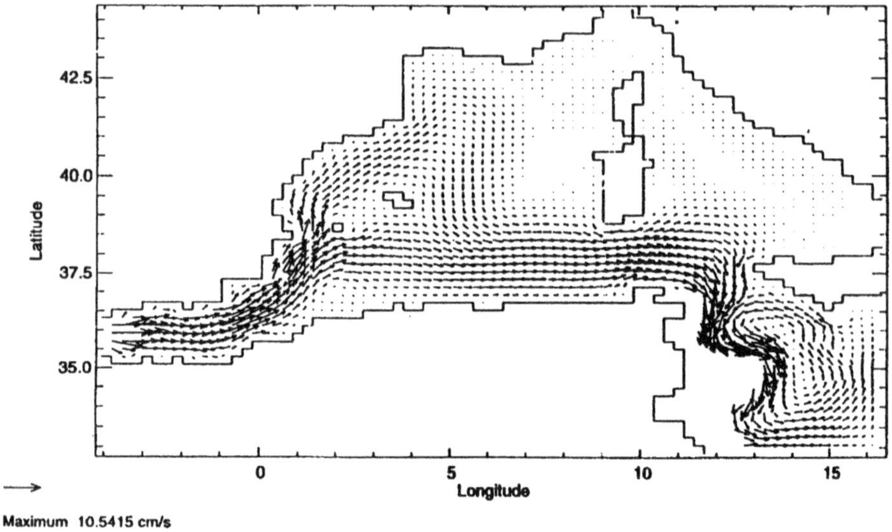

Figure 8. Model circulation at 80 m depth without eddy-topography interaction.

The transport in the Strait of Sicily is 1.14 Sv, in agreement with values obtained by [6] for this strait. With these adequate transports we expect the main effects of the straits to be well represented in the numerical simulations. For the control case where only climatological and buoyancy forcing are considered, the large scale circulation at 80 m is strongly influenced by the straits flows as is evident from Figure 8. The mean feature is the existence of a wide eastward current flowing in the Algerian sub-basin. However, it is interesting to point out that this current, although flowing in the correct sense, does not present the characteristics of the Algerian Current, that is, a narrow eastward flow following the African coast [4]. Also, there is no southwestward flow following the Iberian coast. Another model deficiency is the lack of a significant circulation in the Tyrrhenian/ Liguro sub-basins. The inclusion of the Neptune Effect has important consequences in the numerical results. The most relevant of these consequences is the existence of a clear cyclonic pattern in the large scale circulation in the WM (see Figure 9). In this experiment the transports through the Straits of Gibraltar and Sicily are 0.86 Sv and 1.24 Sv, respectively. The results for the circulation of the upper layer can be summarized in the following way (Figure 9): the AW enters in the WM through the Strait of Gibraltar. Part of this flow follows the African coast in a narrow coastal current that can be identified with the Algerian Current, another part enters in the Algero-Provençal sub-basin, and finally another is deflected towards the Balearic

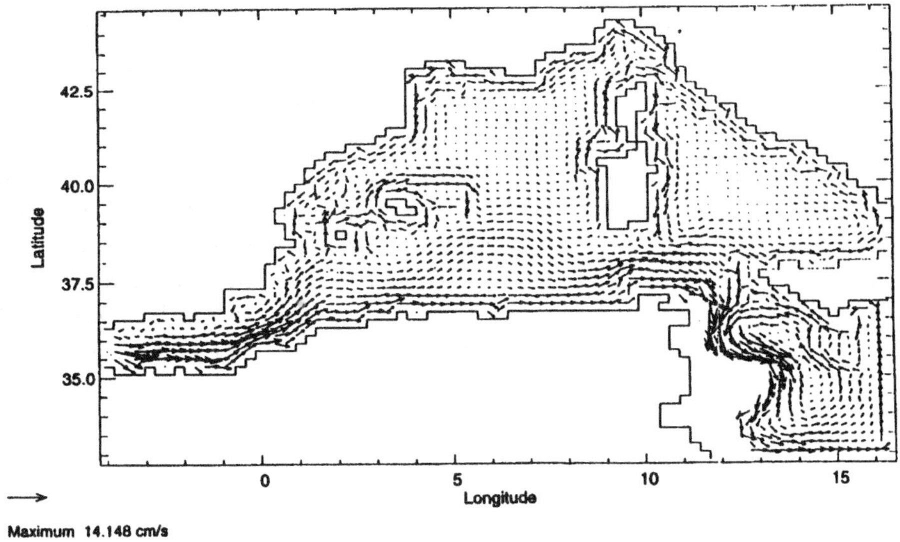

Figure 9. Model circulation at 80 m depth with eddy-topography interaction.

sub-basin, originating the Balearic front. Some of this MAW recirculates together with the outflow from the Strait of Sicily in a well-defined cyclonic pattern following the Italian, French, and Spanish coasts, reaching the Alboran Sea. It is clear from these set of experiments that the inclusion of the Neptune Effect leads to a better agreement between observations and large scale model circulation driven by climatological wind forcing. Particularly remarkable are the results obtained on a sub-basin scale, for example on the width and direction of the Algerian Current and on the circulation in the Balearic sub-basin.

4. Conclusions

One of the actual problems in physical oceanography is related to the influence that bottom topography can exert on the ocean circulation. In this report, efforts have been made to highlight the possible role that bottom topography may play in large scale ocean dynamics.

From the large scale point of view, bottom topography could be one important factor driving ocean circulation. This importance would be due to the influence that interactions between eddies and bottom topography have on large scale flows. Commonly, the mean effect of eddies on large scale circulation was thought to contribute to dissipation terms, and therefore models usually include an eddy viscosity term. However, this assumption is not so clear when bottom topography is present, since dynamical tendencies of an interacting eddy field overlying bottom topography are not to drive the system towards rest but towards a state characterized by mean flows following isobaths. In this sense, bottom topography might be considered as a link between mesoscale and large scale flows, and between the fluctuating component of the flow and the mean one. The first link becomes evident when considering a bottom topography with large scale components, because then the generation of a large scale mean flow will be obtained. The second link is seen from the theoretical and numerical results obtained in section 2 of this lecture.

The dynamical tendency of the eddy-topography interaction might not be faithfully performed in ocean models that cannot adequately resolve eddies, thus presenting a problem of subgrid scale parameterization. We have seen from the analysis of the forced and viscous quasigeostrophic model how this dynamical tendency can be parameterized. This parameterization obtained in a physically consistent way does not contain any parameter that requires an ad hoc specification. In this sense the parameterization is close, providing a measure of the relevance of the Neptune Effect once the energy of the mesoscale activity is known.

The utility in large scale ocean modeling of the theoretical ideas developed in section 2 has been tested using an OGCM in a realistic experiment of the large scale circulation in the WM. Two conclusions can be reached from this set of experiments. First, the similarity of the results obtained from the experiments including the Neptune Effect parameterization to the circulation inferred from observations indicates the ability of the parameterization to capture some of the effects of the eddy-topography interaction missed by the coarse resolution. Second, it is important to point out the relevance that the Neptune Effect seems to have in the circulation of the Western Mediterranean Sea. This relevance could also be expected in other areas of the world oceans.

Acknowledgments

Finantial support was provided by the MAST program MATTER MAS3-CT96-0051 (EU). Alberto Alvarez was also supported by the FPIE fellowship from the Spanish Ministry of Education and from the National Central University. This support is gratefully acknowledged.

References

1. Alvarez A., Tintoré J., Holloway G., Eby M., Beckers J. M. (1994) The effect of the topographic stress on the Western Mediterranean circulation, *J. Geophys. Res.* **99**, 16053-16064.

2. Alvarez A., Hernandez E., Tintoré J. (1997) Noise-sustained currents on quasigeostrophic turbulence over topography, *Physica A* **247**, 312-326.
3. Alvarez A., Hernandez E., Tintoré J. (1998) Noise rectification in quasigeostrophic forced turbulence, submitted to *Phys. Rev. Letters.*
4. Arnone A., Wiesenberg D. A., Saunders K. D. (1991) The origin and characteristics of the Algerian Current, *J. Geophys. Res.* **95**, 1587-1599.
5. Beckers J. M. (1992), M.A. Thesis, Universite de Liege, Belgium, 360 pp.
6. Bethoux J. P. (1979) Budgets of the Mediterranean Sea. Their dependence on the local climate and on the characteristics of the Atlantic Waters, *Oceanol. Acta* **2**, 157-163.
7. Bretherton F. P., Haidvogel D. B. (1976) Two-dimensional turbulence above topography, *J. Fluid Mech.* **78**, 129-154.
8. Bryden H. L., Brady E. C., Pillsbury R. D. (1988) Flow through the Strait of Gibraltar, in J. L. Almazan, T. Kinder, H. Bryden and G. Parrilla (eds.),*Seminario sobre la oceanografia fisica del estrecho de Gibraltar*, Seceg, Madrid.
9. Cummins P. F. (1992) Inertial gyres in decaying and forced geostrophic turbulence, *J. Mar. Res.* **50**, 545-566.
10. Danabasoglu G., McWilliams J. C., Gent P. R. (1994) The role of mesoscale tracer transports in the global ocean circulation, *Science*, **264**, 1123-1126.
11. Edwards S. F. (1964) The statistical dynamics of homogeneous turbulence, *J. Fluid Mech.* **18**, 239-273.
12. Greiner A., Strittmatter W., Honerkamp J. (1988) Numerical integrations of stochastic differential equations, *J. Stat. Phys.* **51**, 95-108.
13. Herring J. R. (1977) On the statistical theory of two-dimensional topographic turbulence, *J. Atmos. Sci.* **34**, 1731-1750.
14. Holland R. W., McWilliams J. C. (1987) Computer modeling in physical oceanography from the global circulation to turbulence, *Physics Today* **40**, 51-67.
15. Holloway G. (1978) A spectral theory of nonlinear barotropic motion above irregular topography, *J. Phys. Oceanogr.* **8**, 414-427.
16. Holloway G. (1986) Eddies, waves, circulation and mixing: statistical geofluid mechanics, *Ann. Rev. Fluid Mech.* **18**, 91-147.
17. Holloway G. (1992) Representing topographic stress for large-scale ocean models, *J. Phys. Oceanogr.* **22**, 1033-1046.
18. Holloway G., Sou T., Eby M. (1995) Dynamics of the circulation of the Japan Sea, *J. Mar. Res.* **53**, 539-569.
19. Katz A. (1967) *Principles of statistical mechanics*, W.H. Freeman & Co., San Francisco; Grandy W. T. (1987) *Foundations of statistical mechanics*, Kluwer Academic Publishers, Dordrecht.
20. Kraichnan R. H., Montgomery D., Rep. Prog. Phys., **43**, 547 (1980).
21. Leith C. E. (1971) Atmospheric predictability and two-dimensional turbulence, *J. Atmos. Sci.* **28**, 145-161.
22. Levitus S. (1982) Climatological atlas of the world ocean, *NOAA Prof. Paper*, US Government Printing Office, Washington, DC, 173 pp.
23. May P. W. (1992) Climatological flux in Western Mediterranean Sea, part 1: wind and wind stresses, *NORDA Rep.* **54**, 56 pp.
24. Miller J., Weichman P. B., Cross M. C. (1992) Statistical mechanics, Euler's equation and Jupiter's Red Spot, *Phys. Rev. A* **45**, 2328-2359.
25. Millot C. (1987) Circulation in the Western Mediterranean Sea, *Oceanol. Acta* **10**, 143-149.
26. Millot C. (1991) Mesoscale and seasonal variabilities of the circulation in the Western Mediterranean, *Dyn. Atmos. Oceans* **15**, 179-214.
27. Neelin D. J., Marotzke J. (1994) Representing ocean eddies in climate models, *Science* **264**, 1099-1100.
28. Pedlosky J. (1987) *Geophysical fluid dynamics*, Springer-Verlag, New York.
29. Salmon R., Holloway G., Hendershot M. C. (1976) The equilibrium statistical me-

chanics of simple quasi-geostrophic models, *J. Fluid Mech.* **75**, 691-703.
30. Sancho J. M., San Miguel M., Katz S., Gunton J. (1982) Analytical and numerical studies of multiplicative noise. *Phys. Rev.* **A 26**, 1589-1609.
31. Sou T., Holloway G., Eby M. (1996) Topographic stress and Caribean Sea circulation, *J. Geophys. Res.* **101**, 16449-16453.
32. Tintoré J., Gomis D., Alonso S., Parrilla G. (1991) Mesoscale dynamics and vertical motion in the Alboran Sea, *J. Phys. Oceanogr.* **21**, 811-823.
33. Toral R., Chakrabarti A. (1988) Generation of gaussian distributed random numbers by using a numerical inversion method, *Computer Phys. Comm.* **74**, 327-334.
34. Treguier A. M. (1989) Topographically generated steady currents in barotropic turbulence, *Geophys. Astrophys. Fluid Dynamics* **47**, 43-68.
35. Yakhot V., Orszag S. A. (1986) Renormalization group analysis of turbulence I. Basic theory, *J. Sci. Comp.* **1**, 1-51.
36. Williams G. P. (1978) Planetary circulations: 1. Barotropic representation of Jovian and terrestrial turbulence, *J. Atmos. Sci.* **35**, 1399-1426.

LARGE-EDDY SIMULATIONS OF THREE-DIMENSIONAL TURBULENT FLOWS: GEOPHYSICAL APPLICATIONS

OLIVIER MÉTAIS
Centre National de la Recherche Scientifique,
Laboratoire des Écoulements Géophysiques et Industriels,
Grenoble, France.

1. Introduction

Direct-numerical simulation of turbulence (DNS) consists of solving explicitly all the scales of motion, from the largest l_I to the Kolmogorov dissipative scale l_D. It is well known from the statistical theory of turbulence that l_I/l_D scales like $R_l^{3/4}$, where R_l is the large-scale Reynolds number $u'l_I/\nu$ based upon the rms velocity fluctuation u'. Therefore, the total number of degrees of freedom necessary to represent the whole span of scales of a three-dimensional turbulent flow is of the order of $R_l^{9/4}$ in three dimensions. With the presently available computers, the DNS is then limited to Reynolds numbers which are several orders of magnitude smaller than those encountered in the ocean, the atmosphere, or most of the industrial facilities. In order to increase the Reynolds number in the simulations, it is necessary to introduce a subgrid-scale model representing the action of scales smaller than Δx, the computational mesh, upon the explicitly resolved scales. This is the basis of the Large-Eddy Simulation (LES) techniques.

2. Large-Eddy Simulation (LES) Formalism

The formalism of the large-eddy simulations for incompressible flows is the following: one considers a spatial filter G of width Δx, which filters out the *subgrid-scales* of wavelength $< \Delta x$. The filtered field is defined as

$$\bar{u}_i(\vec{x},t) = \int u_i(\vec{y},t)G(\vec{x}-\vec{y})d\vec{y} = \int u_i(\vec{x}-\vec{y},t)G(\vec{y})d\vec{y}, \qquad (1)$$

E. P. Chassignet and J. Verron (eds.), Ocean Modeling and Parameterization, 351–372.

and the subgrid-scale field is the departure of the actual flow with respect to the filtered field:

$$u_i = \bar{u}_i + u'_i \ . \tag{2}$$

Because of the commutative properties of the filter and the derivatives, the continuity equation $\partial \bar{u}_j / \partial x_j = 0$ for the filtered field still holds. One considers now the incompressible Navier-Stokes equations within the Boussinesq approximation, in a frame rotating with constant angular velocity Ω about the x_3 axis, and applies the filter G to the equations. One obtains:

$$\frac{\partial \bar{u}_i}{\partial t} + \frac{\partial}{\partial x_j}(\bar{u}_i \bar{u}_j) = -\frac{1}{\rho_0}\frac{\partial \bar{p}}{\partial x_i} + 2\epsilon_{ij3}\Omega \bar{u}_j - g_i \delta_{i3}\frac{\bar{\rho}}{\rho_0} + \frac{\partial}{\partial x_j}\left\{\nu\left(\frac{\partial \bar{u}_i}{\partial x_j} + \frac{\partial \bar{u}_j}{\partial x_i}\right) + T_{ij}\right\} \tag{3}$$

$$\frac{\partial \bar{\rho}}{\partial t} + \frac{\partial}{\partial x_j}(\bar{\rho}\bar{u}_j) = \frac{\partial}{\partial x_j}\left\{\kappa \frac{\partial \bar{\rho}}{\partial x_j} + R_{ij}\right\} \tag{4}$$

ρ is the density, and $\bar{p}$ the filtered pressure. T_{ij} and R_{ij} are the subgrid-scale tensors:

$$\begin{aligned} T_{ij} &= \bar{u}_i \bar{u}_j - \overline{u_i u_j} \\ &= -\left(\overline{u'_i u'_j} + \overline{\bar{u}_i u'_j} + \overline{u'_i \bar{u}_j} + \overline{\bar{u}_i \bar{u}_j} - \bar{u}_i \bar{u}_j\right) \end{aligned} \tag{5}$$

$$\begin{aligned} R_{ij} &= \bar{\rho}\bar{u}_j - \overline{\rho u_j} \\ &= -\left(\overline{\rho' u'_j} + \overline{\bar{\rho} u'_j} + \overline{\rho' \bar{u}_j} + \overline{\bar{\rho}\bar{u}_j} - \bar{\rho}\bar{u}_j\right) \ , \end{aligned} \tag{6}$$

which have to be modelled. Most subgrid-scale models make eddy-viscosity and eddy-diffusivity assumptions (Boussinesq's hypothesis) in order to model the subgrid-scale tensors:

$$T_{ij} = 2\nu_t \ \bar{S}_{ij} + \frac{1}{3}T_{ll} \ \delta_{ij} \ ; \quad R_{ij} = \kappa_t \ \frac{\partial \bar{\rho}}{\partial x_j} \tag{7}$$

where

$$\bar{S}_{ij} = \frac{1}{2}\left(\frac{\partial \bar{u}_i}{\partial x_j} + \frac{\partial \bar{u}_j}{\partial x_i}\right) \tag{8}$$

is the deformation tensor of the filtered field. The generalized LES Boussinesq equations are then written as:

$$\frac{\partial \bar{u}_i}{\partial t} + \frac{\partial}{\partial x_j}(\bar{u}_i \bar{u}_j) = -\frac{1}{\rho_0}\frac{\partial \bar{P}}{\partial x_i} + 2\epsilon_{ij3}\Omega \bar{u}_j - g_i \delta_{i3}\frac{\bar{\rho}}{\rho_0} + \frac{\partial}{\partial x_j}\left\{(\nu + \nu_t)\left(\frac{\partial \bar{u}_i}{\partial x_j} + \frac{\partial \bar{u}_j}{\partial x_i}\right)\right\} \tag{9}$$

$$\frac{\partial \bar{\rho}}{\partial t} + \frac{\partial}{\partial x_j}(\bar{\rho}\bar{u}_j) = \frac{\partial}{\partial x_j}\left\{(\kappa + \kappa_t)\,\frac{\partial \bar{\rho}}{\partial x_j}\right\} \tag{10}$$

where $\bar{P} = \bar{p} + (1/3)\rho_0 T_{ll}$ is a modified pressure.

The question now is to determine the eddy-viscosity $\nu_t(\vec{x}, t)$ and eddy-diffusivity $\kappa_t(\vec{x}, t)$. Most of the existing models, except for the dynamic model discussed below, assume a proportionality between ν_t and κ_t with a constant Prandtl number

$$Pr_t = \nu_t/\kappa_t = \text{const.} \tag{11}$$

2.1. SMAGORINSKY'S MODEL

The most widely used eddy-viscosity model was proposed by Smagorinsky [41]. As a meteorologist, Smagorinsky was simulating a two-layer quasi-geostrophic model, in order to represent large (synoptic) scale atmospheric motions. He introduced an eddy-viscosity which was supposed to model three-dimensional turbulence that has approximately a 3D Kolmogorov $k^{-5/3}$ cascade in the subgrid scales. Smagorinsky's model still proves very popular for several applications, since the pioneering work of Deardorff [10] for the channel flow.

In Smagorinsky's model, a sort of mixing-length assumption is made, in which the eddy-viscosity is assumed to be proportional to the subgrid-scale characteristic length scale Δx, and to a characteristic turbulent velocity based on the second invariant of the filtered-field deformation tensor. The model is

$$\nu_t = (C_S \Delta x)^2 |\bar{S}| \ , \tag{12}$$

where the local strain rate is defined by $|\bar{S}| = \sqrt{2\bar{S}_{ij}\bar{S}_{ij}}$. If one assumes that $k_C = \pi/\Delta x$, the cutoff wavenumber in Fourier space, lies within a $k^{-5/3}$ Kolmogorov cascade $E(k) = C_K \epsilon^{2/3} k^{-5/3}$ (C_K is the Kolmogorov constant), one can analytically determine the constant C_S. It is then found (see e.g. [30]):

$$C_S \approx \frac{1}{\pi}\left(\frac{3C_K}{2}\right)^{-3/4} . \tag{13}$$

The newly developed subgrid-scale models try to tackle the inherent weaknesses attached to the eddy-viscosity and eddy-diffusivity assumption given by (7) and to subgrid-scale models of Smagorinsky's type:

- These assume a one-to-one correlation between the subgrid-scale stress and the large-scale strain rate tensors. This is highly questionable and has never been verified experimentally or numerically (see e.g. [7], [31]).

- Equation (13) yields $C_S \approx 0.18$ for a Kolmogorov constant of 1.4. In fact, most workers prefer $C_S = 0.1$ (which represents a reduction by nearly a factor of 4 of the eddy-viscosity), a value for which Smagorinsky's model behaves reasonably well for free-shear flows and for channel flow [37]. This clearly indicates that C_S is not a universal constant, and that assuming that $k_C = \pi/\Delta x$ lies within a $k^{-5/3}$ Kolmogorov cascade is too strong of a constraint, since even the smallest resolved scales can be far from isotropic.

Some of these weaknesses are addressed by the various subgrid-scale models that have been developed since Smagorinsky's original model. These models are presented in the following sections. The details of the various models can be found in [28], [26].

2.2. SPECTRAL EDDY-VISCOSITY MODELS

We now work in Fourier space, in the context of three-dimensional isotropic turbulence. The filter consists of a sharp cut-off filter clipping all the modes larger than k_C, where $k_C = \pi/\Delta x$ is the cut-off wavenumber. The concept of k-dependent eddy-viscosity was first introduced by Kraichnan [19] for three-dimensional isotropic turbulence, in the following way: if $T_{>k_C}(k,t)$ is the kinetic energy transfer across the cutoff k_C, corresponding to triadic interactions such that $k < k_C, p$ and (or) $q > k_C$ (see [27], for details), one poses

$$\nu_t(k|k_c) = -\frac{T_{>k_C}(k,t)}{2k^2 E(k,t)} , \tag{14}$$

in such a way that the kinetic energy spectrum $E(k,t)$ in the resolved scales $(k \leq k_C)$ satisfies:

$$[\frac{\partial}{\partial t} + 2(\nu + \nu_t(k|k_c))k^2]\ E(k,t) = T_{<k_C}(k,t) , \tag{15}$$

where $T_{<k_C}(k,t)$ is the kinetic energy transfer corresponding to resolved triads such that $k,p,q \leq k_C$. Using the Eddy-Damped Quasi-Normal Markovian (E.D.Q.N.M.) theory, one of the statistical closures of isotropic turbulence (see [27]), and assuming that k_C lies within a Kolmogorov cascade, we find that the eddy-viscosity may be written

$$\nu_t(k,k_C) = 0.441\ C_K{}^{-3/2} \left[\frac{E(k_C)}{k_C}\right]^{1/2} \nu_t^*(\frac{k}{k_C}) , \tag{16}$$

where $E(k_C)$ is the kinetic-energy spectrum at the cutoff k_C, and $\nu_t^*(k/k_C)$ is a non-dimensional eddy-viscosity, constant and equal to 1 for $k/k_C <\approx 0.3$, and rising for higher k up to $k/k_C = 1$ (cusp behaviour, [19]). This cusp is due to the predominance of "local" transfers across k_C: near the cutoff wavenumber, the main non-linear interactions between the resolved and unresolved scales involve the smallest eddies of the former and the largest eddies of the latter. The constants and the form of ν_t^* can be determined using E.D.Q.N.M. theory. It can approximately be expresssed as:

$$\nu_t^* \left(\frac{k}{k_C}\right) = 1. + 34.5 e^{-3.03(k_C/k)} \ . \tag{17}$$

For the LES in Fourier space, the spectral eddy-viscosity (16) is plugged into the Navier-Stokes equation for the Fourier transform $\hat{u}_i(\vec{k},t)$ of the velocity field:

$$[\frac{\partial}{\partial t} + \nu k^2 + \nu_t(k, k_C) k^2] \hat{u}_i(\vec{k},t) = -i k_m P_{ij}(\vec{k}) \int_{\vec{p}+\vec{q}=\vec{k}}^{p,q \leq k_C} \hat{u}_j(\vec{p},t) \hat{u}_m(\vec{q},t) d\vec{p} \ , \tag{18}$$

where $P_{ij}(\vec{k})$ is the projector onto the plane perpendicular to $\vec{k}$.

One of the nice features of this model is that it can be easily extended to the case where k_C is not situated within a $k^{-5/3}$ Kolmogorov cascade. Indeed, as shown by Métais and Lesieur [36], the scaling by $\sqrt{E(k_C)/k_C}$ of the eddy-viscosity still holds when k_C lies within a k^{-m} spectral range as long as $m \leq 3$. Métais and Lesieur [36] indeed proposed, on the basis of the E.D.Q.N.M. theory, that the eddy-viscosity should be

$$\nu_t(k, k_C) = 0.31 \ \frac{5-m}{m+1} \sqrt{3-m} \ C_K{}^{-3/2} \left[\frac{E(k_C)}{k_C}\right]^{1/2} \nu_t^* \left(\frac{k}{k_C}\right) \ . \tag{19}$$

Note that the value 0.441 is recovered for $m = 5/3$. See below for an application of this model to the turbulent channel flow.

2.3. STRUCTURE-FUNCTION MODEL

This model, due to Métais and Lesieur [36], consists of using the spectral eddy-viscosity in physical space, while taking into account the intermittency of turbulence. If the cusp behaviour is discarded, the eddy-viscosity given by equation (16) can be determined locally in physical space, through the local second-order velocity structure function. We then obtain:

$$\nu_t(\vec{x}, \Delta, t) = 0.105 \ C_K^{-3/2} \Delta x \ [\bar{F}_2(\vec{x}, \Delta x, t)]^{1/2} \ , \tag{20}$$

where $\bar{F}_2$ is the structure function of the filtered signal

$$\bar{F}_2(\vec{x}, \Delta x) = \left\langle \|\vec{\bar{u}}(\vec{x},t) - \vec{\bar{u}}(\vec{x}+\vec{r},t)\|^2 \right\rangle_{\|\vec{r}\|=\Delta x} \ . \tag{21}$$

One of the drawbacks of this model is the absence of a cusp near k_C. However, E.D.Q.N.M. data show that the exponential form given in equation (17) can be correctly approximated by a power law of the type:

$$\nu_t^*\left(\frac{k}{k_C}\right)=\left(1.+\nu_{tn}^*\left(\frac{k}{k_C}\right)^{2n}\right) . \tag{22}$$

with $2n \approx 3.7$. Lesieur and Métais [28] have shown that ν_{tn}^* can be determined by considering the energy balance between explicit and subgrid-scale transfers. This yields:

$$\nu_{tn}^* = 0.512\left(\frac{3n}{2}+1\right) . \tag{23}$$

In fact, the E.D.Q.N.M. value of $2n = 3.7$ is not so far from the exponent $2n = 4$ which would be obtained with a Laplacian operator iterated twice. Therefore, Lesieur and Métais [28] proposed a physical-space turbulent dissipative operator based upon the structure-function model and taking into account the "cusp" behaviour:

$$2\frac{\partial}{\partial x_j}\left[\nu_t^{SF}\bar{S}_{ij}\right]+\nu_t^{(2)}(\Delta x)^4\left(\frac{\partial^2}{\partial x_j^2}\right)^3\bar{u}_i \quad , \tag{24}$$

where $\bar{S}_{ij}$ is the deformation tensor of the field $\bar{u}_i$. $\left(\partial^2/\partial x_j^2\right)^3$ designates the Laplacian operator iterated three times. ν_t^{SF} is given by the r.h.s. of equation (20) multiplied by $0.441/(2/3)$, $\nu_t^{(2)} = \nu_t^{SF} \times \nu_{t2}^*$, and ν_{t2}^* is given by equation (23) with $n = 2$. The expression (24) is interesting in the sense that it provides an eddy dissipation combining the structure function model with a hyperviscosity $(\nabla^2)^3\overline{u}_i$. The latter represents in physical space the action of the cusp in Kraichnan's spectral eddy-viscosity. The importance of the presence of a cusp will be demonstrated in Section 3.3.

One of the important features of the model is its ability to easily take account of non-uniform (but orthogonal) grids. This is of crucial importance for geophysical flow simulations, which frequently involve large aspect ratios. Let $\Delta c = (\Delta x_1 \Delta x_2 \Delta x_3)^{1/3}$ be the geometric mean of the meshes in the three spatial directions. One takes into account Kolmogorov's [18] law which states that the second-order velocity structure-function scales like $(\epsilon r)^{2/3}$, where r is the distance between the points. Thus the eddy-viscosity is interpolated by replacing Δx by Δc, with (in the six-point formulation)

$$\begin{aligned} F_2(\vec{x},\Delta c) &= \tfrac{1}{6}\Sigma_{i=1}^3[\|\vec{u}(\vec{x})-\vec{u}(\vec{x}+\Delta x_i\vec{e}_i)\|^2 \\ &\quad +\|\vec{u}(\vec{x})-\vec{u}(\vec{x}-\Delta x_i\vec{e}_i)\|^2]\left(\tfrac{\Delta c}{\Delta x_i}\right)^{2/3} , \end{aligned} \tag{25}$$

where $\vec{e}_i$ is the unit vector in direction x_i.

2.4. HYPER-VISCOSITY

The model given by equation (24) bears some resemblance to hyper-viscosity models which are widely used in the study of geophysical flows because of their simplicity. Indeed, the hyperviscosity consists in replacing the molecular dissipative operator $\nu\nabla^2$ by $(-1)^{\alpha-1}\nu_\alpha(\nabla^2)^\alpha$, where α is a positive integer. As opposed to equation (24), ν_α is here a constant (positive) coefficient which has to be adjusted. This has been widely used in two-dimensional isotropic turbulence (see [4]), with $\alpha = 2$ or $\alpha = 8$, as a way to shift the dissipation to the neighbourhood of k_c. This allows for a reduction of the number of scales strongly affected by viscous effects, and has rendered it possible in the case of two-dimensional turbulence to demonstrate the existence of coherent vortices.

In three-dimensional turbulence, this approach was used by Bartello *et al.* [3] to study the influence of a solid-body rotation, with surprisingly good results.

2.5. SELECTIVE AND FILTERED STRUCTURE-FUNCTION MODELS

As with Smagorinsky's model, however, the Structure-Function (SF) model is too dissipative for transition in a boundary layer for instance, and does not behave well in a channel: in fact, the spectrum $E_{\vec{x}}(k_C)$ is sensitive to the low-frequency oscillations. To overcome the difficulty with transition, two improved versions of the SF model have been developed: the selective structure-function model (SSF), and the filtered structure-function model (FSF). The dynamic model is another way of adapting the eddy-viscosity to the local conditions of the flow (see next section).

The selective structure-function (SSF) model was developed by David [9] (see also [8], [28] for details). The idea is to switch off the eddy-viscosity when the flow is not three-dimensional enough. The three-dimensionalization criterion is the following: one measures the angle between the vorticity at a given grid point and the average vorticity at the six closest neighbouring points (or the four closest points in the four-point formulation). If this angle exceeds 20^0, the most probable value according to simulations of isotropic turbulence at a resolution of $32^3 \sim 64^3$, the eddy-viscosity is turned on. Otherwise, only molecular dissipation acts. In the filtered structure-function (FSF, see [11], [8], [28]), the filtered field $\bar{u}_i$ is submitted to a high-pass filter in order to get rid of low-frequency oscillations which affect the local kinetic-energy spectrum $E(k_C)$.

2.6. SCALE-SIMILARITY AND MIXED MODELS

The lack of correlation between the subgrid-scale stress and the large-scale strain rate tensors has led Bardina *et al.* [2] to propose an alternative subgrid-scale model called the scale similarity model. This is based on a double filtering approach and on the idea that the important interactions between the resolved and unresolved scales involve the smallest eddies of the former and the largest eddies of the latter. They suggest evaluating the subgrid tensor as:

$$T_{ij} = \bar{\bar{u}}_i \bar{\bar{u}}_j - \overline{\bar{u}_i \bar{u}_j} \ . \tag{26}$$

The analysis of DNS and experimental data [2], [31] have shown that the modelled subgrid-scale stress deduced from (26) exhibits a good correlation with the real (measured) stress. However, when implemented in LES calculations, the model hardly dissipates any energy. It is therefore necessary to combine it with an eddy-viscosity type model such as Smagorinsky's model to produce the "mixed" model. Along the lines of the Bardina *et al.* model, new formulations have been proposed to correct this lack of dissipation. Liu *et al.* [31] have proposed the following model:

$$T_{ij} = C_L \left(\tilde{\bar{u}}_i \tilde{\bar{u}}_j - \widetilde{\bar{u}_i \bar{u}_j} \right) \ , \tag{27}$$

where C_L is a dimensionless coefficient. The operator $\tilde{.}$ consists of a second filter of different width [31]. This concept of double filtering can be taken one step further, leading to the dynamic models presented in the next section.

2.7. DYNAMIC MODELS

This class of models is based on Germano's [15] work (see also [16], [28]). A new filter $< . >$ (called the test filter, generally of width $2\Delta x$) is applied to $\bar{u}$, giving a "test-field" $< \bar{u} >$. The evaluation of the energetic transfers between the test field and $\bar{u}$, based on the so-called Germano's identity, permits in principle the recalculation of a time-varying Smagorinsky's constant $C = C_S^2$ which is local in space. However the problem is ill-conditioned, since C is given by a tensorial equation yielding five equations for one unknown. To remove the singularity, Lilly [32] proposed to evaluate C by variational techniques, via a least-square approach. However, this yields many negative values for $C(\vec{x}, t)$, which rapidly lead to computational instability. The cure which is often adopted to deal with this problem consists of averaging over the direction of flow homogeneity. For example, Germano *et al.* [16] and Piomelli [38] obtained very good results by averaging in planes parallel to the walls in their channel flow simulation. Note that the use of Smagorinsky's model as a base for the dynamic procedure is not compulsory, and any of the subgrid-scale models can be a candidate. As an example, Zang *et al.* [42] have successfully applied this procedure to the scale-similarity model proposed by Bardina *et al.* [2]. More recently, El-Hady and Zang [12] have shown that the "dynamic" structure-function model yields better results than the "dynamic" Smagorinsky model in a transitional compressible boundary layer.

2.8. ANISOTROPIC SUBGRID-SCALE MODELS

As stressed above, the subgrid-scale tensor and buoyancy flux given by (7) are assumed to be strictly proportional to the grid-scale strain rate tensor and buoyancy flux, respectively. Abbà *et al.* [1] have recently proposed an anisotropic formulation of (7) using eddy-viscosity and eddy-diffusivity tensors instead of scalar ones :

$$T_{ij} - \frac{1}{3} T_{ll}\ \delta_{ij} = 2 \sum_{r,s} \nu^t_{ijrs} \bar{S}_{rs} - \frac{2}{3} \delta_{ij} \sum_{l,r,s,} \nu^t_{llrs} \bar{S}_{rs}\ . \tag{28}$$

This formulation allows for a better description of the small-scale anisotropy. This model in conjunction with the dynamic procedure previously described has been used successfully in Large-Eddy Simulations of turbulent natural convection [1].

2.9. SUBGRID-SCALE MODELS VALIDATION

Subgrid-scale models may be validated by comparing the numerical results with the predictions of existing theories. This is generally only possible for flow in simple or "idealized" geometries. A good example is the isotropic three-dimensional turbulence which constitutes a good test to check the model ability to reproduce a proper $k^{-5/3}$ Kolmogorov's cascade (see e.g. [36]). Shear flow prototypes like mixing layers, wakes, jets and channel flows are also often used as a validation of LES since they allow for detailed comparisons of the computed statistics and the computed flow structures with those from laboratory experiments or field measurements. Direct Numerical Simulations at the highest Reynolds number allowed by the computer resources may also be performed to test the LES. DNS/LES comparisons are however limited to flows corresponding to Reynolds numbers that are very low in comparison with those generally encountered in geophysical flows.

3. LES of Flows of Geophysical Interest

We next present applications of some of the previously described subgrid-scale models to turbulent flows encountered in geophysical situations. We first consider the simple shear-flow prototype that constitutes the mixing

Run	ε_{2D}	ε_{3D}	Re	$L_x/\delta_i,\ L_y/\delta_i,\ L_z/\delta_i$	N_x, N_y, N_z
DNSQ2D	10^{-3}	10^{-4}	100	$140, 28, 14$	$480, 96, 48$
FSFQ2D	10^{-4}	10^{-5}	-	$112, 28, 14$	$384, 96, 48$

TABLE 1. Recapitulative table of the simulations presented here. (L_x, L_y and L_z denote the domain's dimensions in the streamwise, transverse and spanwise directions, respectively. The corresponding numbers of collocation points N_x, N_y and N_z are such that the meshes are cubic of side $\Delta \approx 0.29\ \delta_i$.)

layer. Next, we show the ability of the LES to correctly reproduce the dynamics of coherent vortices present in wall-bounded shear flows subjected to solid-body rotation. Finally, we apply LES techniques to the computation of atmospheric and oceanic eddies resulting from baroclinic instability.

3.1. SPATIALLY GROWING MIXING LAYER

The layer developing between two parallel flows of different velocities is called the mixing layer. The two streams develop at their interface a Kelvin-Helmholtz type instability which is eventually responsible for the generation of large quasi-two-dimensional vortices, often called the Kelvin-Helmholtz vortices. These geometrically simple flows have been the subject of extensive laboratory studies (see e.g. [6]) as well as numerical studies, because of their importance to understanding mixing processes of momentum and of heat. For instance, this type of instability plays a crucial role in the turbulent mixing at the bottom of the oceanic mixed-layer.

Here we compare a DNS at low Reynolds number ($Re = 100$) with a LES using the filtered structure-function subgrid-scale model of a spatially developing incompressible mixing layer between two streams of velocity U_1 and U_2 ($U_1 > U_2$). Further details can be found in Silvestrini *et al.* [39]. The inflow consists of a hyperbolic-tangent velocity profile

$$\bar{u}(y) = \frac{U_1 + U_2}{2} + \frac{U_1 - U_2}{2} \tanh \frac{2y}{\delta_i} \quad . \tag{29}$$

where δ_i is the vorticity thickness. Two small-amplitude random perturbations of Gaussian p.d.f. are superimposed onto this profile: the first is three-dimensional (i.e., a function of y, z and t); its kinetic energy is denoted $\varepsilon_{3D}U^2$. The second (of energy $\varepsilon_{2D}U^2$) depends only on y and t. The ratio $\varepsilon_{2D}/\varepsilon_{3D}$ is here set to 10, corresponding to a quasi-two-dimensional perturbation. The DNS and the LES are henceforth referred to as DNSQ2D and FSFQ2D respectively. It is important to notice that the low Reynolds

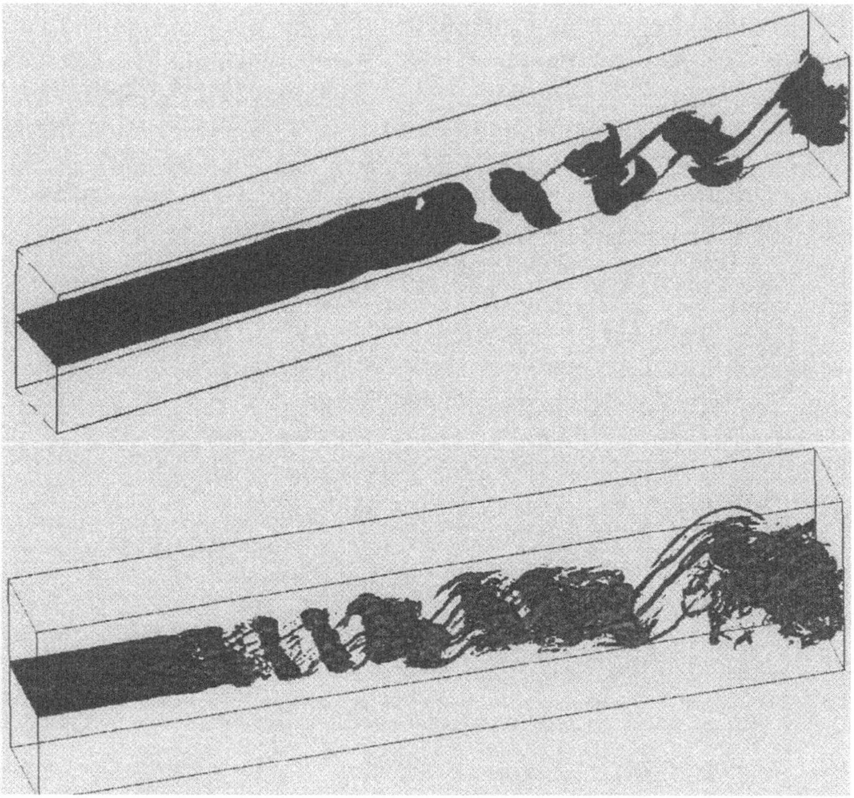

Figure 1. Perspective views of isovorticity surface: top, run DNSQ2D, $||\vec{\omega}|| = \omega_i/3$; bottom, run FSFQ2D, $||\omega|| = 2/3\omega_i$.

number DNS requires more grid points than the LES, which will turn out to be much more turbulent.

Precise numerical methods are compulsory when one wants to precisely describe the three-dimensional vortex dynamics. We solve the complete Navier-Stokes and LES equations for an incompressible fluid in a parallelepipedic domain. Sixth-order compact finite-differences [25] are used in the longitudinal direction x, along with pseudo-spectral methods on yz planes. Periodicity is assumed in the spanwise direction z. Sine/cosine expansions are used in the transverse direction y, enforcing free-slip boundary conditions. Non-reflective outflow boundary conditions are approximated by a multi-dimensional extension of Orlansky's discretization scheme, with limiters on the phase velocity (see [17], for a detailed description of the numerical code).

Figure 1 (top) shows an isosurface of the vorticity modulus. The vorticity sheet undergoes longitudinal oscillations leading to a first roll-up further downstream. Subsequently, the Kelvin-Helmholtz vortices exhibit successive pairings. An important feature consists of thin intense longitudinal hairpin vortices which are stretched between the Kelvin-Helmholtz rollers as in Bernal and Roshko's [5] experiment. The vorticity magnitude within the longitudinal vortices peaks at $2\omega_i$, where $\omega_i = 2U/\delta_i$ is the maximal vorticity magnitude introduced at the inlet. Run LESQ2D (Figure 1, bottom) is visibly much more turbulent than DNSQ2D: maximal vorticity magnitude is $\approx 4\ \omega_i$ for the whole run. Roll-up and pairing events occur much faster than in the DNS. Notice the complexity of the dynamics with a cluster of three fundamental Kelvin-Helmholtz vortices undergoing a first pairing and, at its downstream end, a billow made of 4 fundamental KH vortices whose second pairing is in progress. Experimentally observed trends such as the doubling of the spanwise spacing of the longitudinal vortices at every pairing also seem to be correctly reproduced.

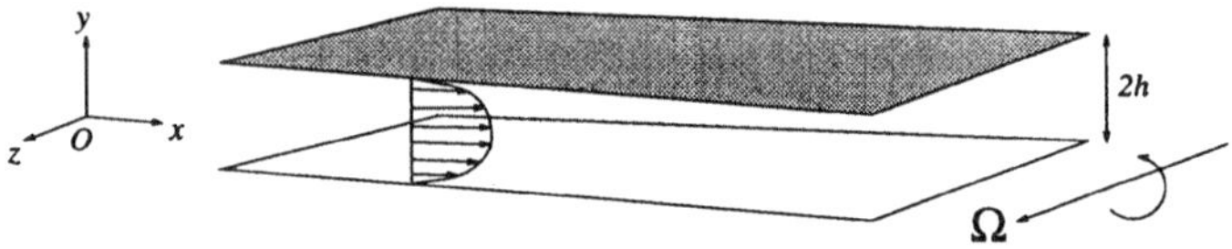

Figure 2. Schematic view of the rotating channel.

3.2. ROTATING CHANNEL FLOW

Turbulent or transitional shear flows in a rotating frame have been extensively studied due to their importance in many geophysical and engineering applications. Within these flows, the local Rossby number, which characterizes the relative importance of inertial and Coriolis forces, can vary significantly. Typical values of the Rossby number $R_o = U/2\Omega D$, where U is a characteristic velocity associated with the eddy and D its diameter, are on the order of 0.05 in mesoscale oceanic eddies and in Jupiter's Great Red Spot, 0.3 for large synoptic-scale atmospheric perturbations, and 2.5 for the atmospheric wake of a small island. Turbulence in rotating fluids finds numerous industrial applications in turbo-machinery; e.g., the turbulent characteristics of the flow in blade passages of radial pumps and compressor impellers determine the efficiency of these devices. Turbulence is also of great importance for the cooling of the fluid inside the blades. Depending upon the magnitude of the radial velocity, the Rossby number within rotating machines can range from values close to unity to very small ones (of the order 0.05).

Laboratory experiments, theoretical works, numerical simulations and atmospheric and oceanic observations show that there are three basic effects associated with rotating bounded- or free-shear flows. (i) If the shear vorticity is parallel and of the same sign as the rotation vector (cyclonic rotation), the flow is made more two-dimensional. (ii) If the two vectors are anti-parallel (anticyclonic rotation), destabilization is observed at moderate rotation rates (high Rossby numbers), while (iii) two-dimensionalization is recovered for fast rotation. It is easy to show that the asymmetry between the cyclonic and anticyclonic cases can only be explained by considering the influence of rotation on the growth of three-dimensional perturbations.

Numerical investigations of planar mixing layers and wakes subjected to solid-body rotation of axis parallel to the basic vorticity have been performed by Lesieur *et al.* [29] and Métais *et al.* [35]. They have shown that, for small Rossby numbers, the rotation renders the flow more two-dimensional both in the cyclonic and anticyclonic cases. At moderate anticyclonic rotation rates, a critical value of the Rossby number was found for which the flow is strongly destabilized. The flow three-dimensionalization, in that case, is due to an intense stretching of the ***absolute*** vortex lines yielding intense longitudinal absolute hairpin vortices. Lamballais *et al.* [21], [22], [23] have performed a detailed analysis of the vorticity field issued from a DNS of the channel flow. They have checked that similar intense absolute vorticity stretching mechanisms take place yielding intense longitudinal hairpin vortices. Our purpose here is to show the ability of the LES to accurately reproduce the detailed vorticity dynamics even in the presence of external forces like solid-body rotation.

We simulate the turbulent channel flow (turbulent flow between two infinite parallel plates). x, y and z are respectively the streamwise direction, the direction normal to the channel walls, and the spanwise direction. The rotation vector is oriented along the spanwise direction $\vec{\Omega} = (0, 0, \Omega)$ (see Figure 2). The vorticity vector associated with the mean velocity profile is parallel to $\vec{\Omega}$ near one channel wall and antiparallel near the opposite wall; we refer to flow near these two particular walls as cyclonic and anticyclonic, respectively. Let U_m be the bulk velocity. The non-dimensional parameters are the Reynolds number $Re = U_m \, 2h/\nu$ and the global Rossby number $Ro_g = 3 \, U_m/2\Omega \, h$. A strong rotation regime corresponds to small values of Ro_g, while infinite Ro_g denotes the case without rotation. In the following, most of the quantities are normalized by U_m and h. However, the superscript $^+$ indicates the conventional boundary-layer scaling by the friction velocity u_τ and the kinematic viscosity ν, and we define the microscopic Reynolds number as $h^+ = u_\tau h/\nu$.

As previously stressed, a detailed analysis of the three-dimensional flow structures requires high-order numerical schemes. To solve the incompressible three-dimensional Navier-Stokes equations we used the same numerical code as for the previously described mixing layer case. It combines pseudo-spectral methods in the periodicity directions (streamwise and spanwise directions) and high-order finite difference schemes in the direction normal to the wall (compact schemes).

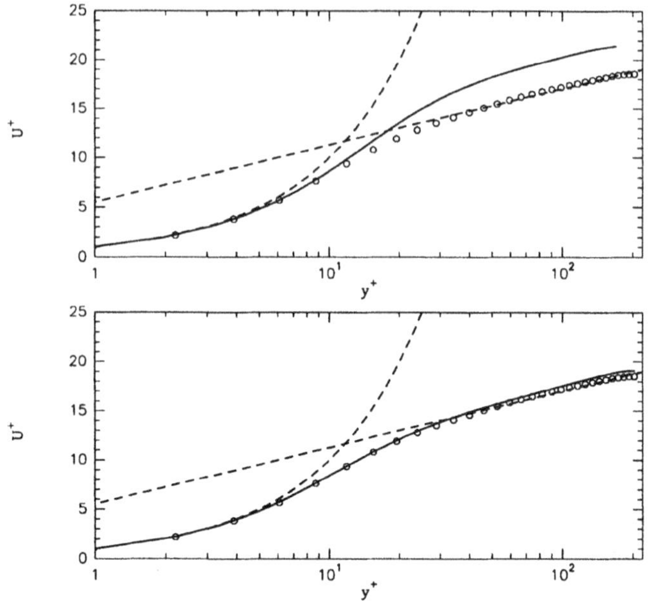

Figure 3. Mean velocity profiles in wall units. Lines : present simulation ($Re = 6666$); symbols : Piomelli [38] ($Re = 6500$). Case A (top) : $m = 5/3$ ($h^+ = 181$) ; Case B (bottom) : dynamic evaluation of $m(y,t)$ ($h^+ = 204$).

3.2.1. *Importance of the subgrid-scale model*

We first present the LES of the non-rotating turbulent channel flow at $Re = 6666$ ($h^+ = 204$). Details can be found in [23], [20]. The size of the computational domain in the streamwise, wall-normal and spanwise directions is $2\pi h \times 2h \times \pi h$, and $64 \times 65 \times 32$ grid points are used. Two distinct LES based on spectral eddy-viscosity are performed: the first one with m maintained at a fixed value of $m = 5/3$ (case A; see equation 16), the second one based upon the dynamic-spectral model (equation 19) with $m(y,t)$ (case B). We have checked that the eddy-viscosity is zero in regions close to the wall, up to $y^+ \approx 10$ where longitudinal velocity fluctuations become very intense, due to the low- and high-speed streaks. Therefore, and since the first grid point is taken close to the wall ($y^+ = 1$), our LES has the interesting property of behaving like a DNS in the vicinity of the wall, which enables us to accurately capture the turbulent events occurring in this region. The correction given by the spectral dynamic model thus provides a correct near-wall behaviour, without any recourse to "hand-tuned" constants. Figure 3 shows the mean velocity profile, compared with the LES of Piomelli [38] using the dynamic model of Germano previously described. The latter is known to agree very well with experiments at these low Reynolds numbers. By comparison with the LES carried out with the classical spectral-cusp model with $m = 5/3$, the case B simulation coincides much better with Piomelli's data.

3.2.2. *Spatial organization of the rotating turbulent channel flow*

Here we compare the flow structures given by LES with a spectral dynamic model at $Re = 14000$ with those given by DNS at $Re = 5000$. We investigate the rotation regime for which anticyclonic destabilization is achieved: this corresponds to rotation rates such that $Ro_g > 1$. The case corresponding to $Ro_g = 6$ is shown. Figure 4 clearly shows that the rotation greatly modfies the vortex organization. Near the cyclonic wall (top wall in the figure), the iso-vorticity surfaces are nearly flat indicating a weak growth of the perturbation. On the anticyclonic side, the increase of the turbulence level is associated with the longitudinal vorticity intensification. The LES is capable of reproducing all the characteristic features of the flow organization. The subgrid-scale model is indeed able to capture cyclonic relaminarization with inactive turbulent motions as well as detailed turbulent flow organization on the anticyclonic side. It is also clear that the LES does reproduce features expected from turbulence at higher Reynolds number, and displays much more vortical activity in the small scales that the DNS.

3.3. BAROCLINIC EDDIES

Our goal here is to study, in the case of a baroclinic jet subjected to a rapid rotation, the nature of the coherent vortices and in particular the asymmetry between cyclonic and anticyclonic eddies. We will quantify this asymmetry as a function of the characteristic parameters: the Rossby and Froude numbers. Finally, LES will be used to study frontal secondary instabilities at high Reynolds numbers. The following results are described in detail by Garnier *et al.* [14], [13].

We consider a domain of spatial extents L_x, L_y and H in, respectively, the zonal x, meridional y and vertical z directions. We investigate the destabilization and the subsequent transition to turbulence of a baroclinic mean state consisting of a zonal flow $U(y, z)$ and a potential temperature (or density) profile $\Theta(y) + \overline{\Theta}(z)$ in thermal wind equilibrium. These are given

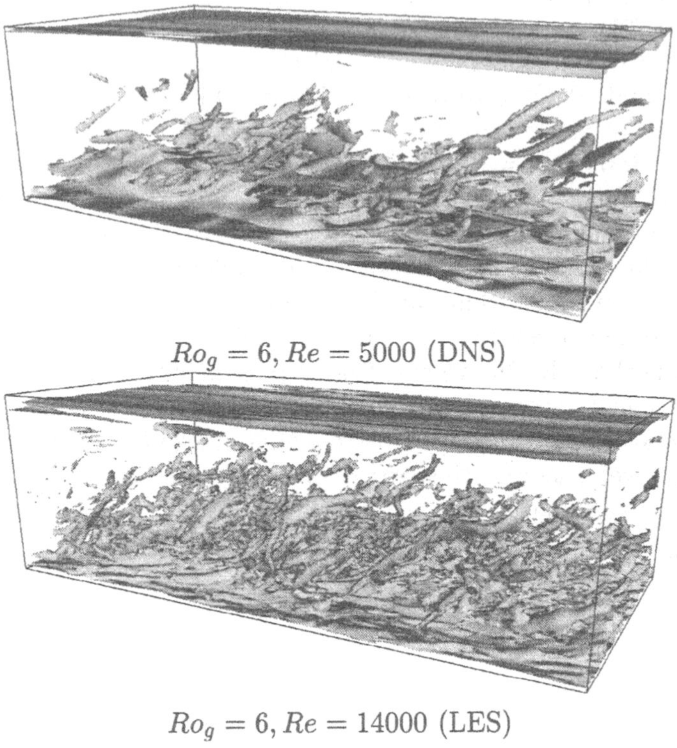

Figure 4. **Isosurfaces of vorticity modulus $\omega = 3$ for $Ro_g = 6$.**

by the following analytical expressions :

$$U(y,z) = \frac{U}{\cosh^2(2y/\delta)}(2z/H - 1)\ ;\ \Theta(y) = -\Delta\Theta \tanh(2y/\delta) \qquad (30)$$

where $\Delta\Theta = \theta_0 f U \delta/(gH)$ is the meridional variation of potential temperature. θ_0 is a reference potential temperature and f is the Coriolis parameter, assumed to be constant. The velocity profile corresponds to two Bickley jets of thickness δ, blowing in opposite directions (a westerly jet with maximum velocity U at the upper lid $z = H$, and an easterly jet with maximum velocity $-U$ at the lower lid $z = 0$). The meridional temperature gradient is concentrated in the central band (along the jet axis) and has a characteristic thickness δ. The fluid is stably stratified, and the spatial variations of the Brunt-Väisälä frequency $N = \sqrt{(g/\theta_0)\, d\overline{\Theta}/dz}$ are neglected. No surface friction or Ekman layer effects have been included in the present simulations. The characteristic nondimensional parameters are: the Rossby number $Ro = U/f\delta$, the Froude number $F = U/NH$, the Burger number $Bu = (NH/f\delta)^2$, the Reynolds number $Re = U\delta/\nu$, and the aspect ratio $\alpha = H/\delta$. Note that instead of the Burger number, one can consider the parameter $Ro/F = \sqrt{Bu} = r_I/\delta$, where $r_I = NH/f$ is the internal Rossby radius of deformation. The parameter Ro/F controls the width of the frontal zone.

Garnier *et al.* [13] have first carried out DNS for low-amplitude perturbations, which enabled them to perform a linear stability analysis for various degrees of baroclinicity of the basic state (measured by the parameter Ro/F). They identified a critical value of Ro/F equal to 1.5, separating a regime of strongly baroclinically unstable modes ($Ro/F < 1.5$) from a regime characterized by weakly unstable modes of barotropic nature ($Ro/F > 1.5$). Here we concentrate on the first regime $Ro/Fr < 1.5$, with $Ro/Fr = 0.5$. The Rossby number is fixed at 0.1. The numerical code is similar to that of the channel flow study of the previous section, except that here compact finite-difference schemes are used in two spatial directions. We show results both from DNS at a Reynolds number of 400 and from LES at $Re = 10,000$. The resolution is $193 \times 64 \times 20$ points both for the DNS and the LES.

3.3.1. *Direct numerical simulations*

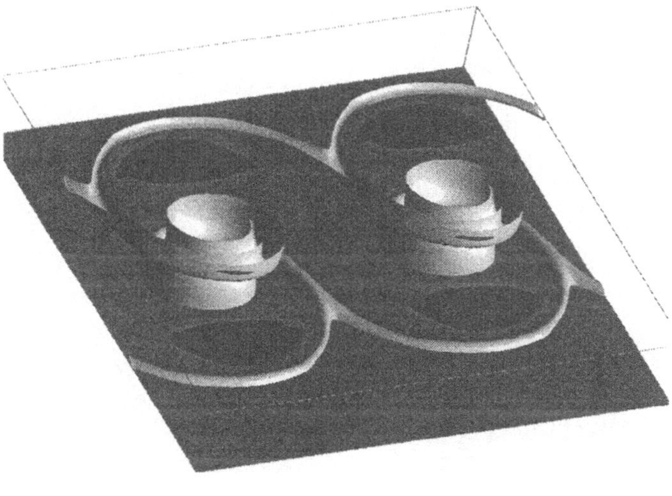

Figure 5. Iso-surfaces of vertical vorticity corresponding to $|\omega_z| = 2\omega^i_{2D}$; light-gray: cyclonic vorticity, black: anticyclonic vorticity. $t = 47r_I/U$ (from [13]).

Figure 5 shows the vorticity structure once the instability has fully developed for $Ro/Fr = 0.5$. We observe the formation of long-lived cyclonic eddies of strong intensity, composed of nearly two-dimensional cores between which develop arms of intense cyclonic vorticity, connected to the cyclonic vorticity columns and located in the vicinity of the horizontal boundaries. The appearance of these intense cyclonic arms leads to the entrainment of the adjacent anticyclonic vorticity and then to its strong three-dimensionalization. The maximum vorticity within the cyclones reaches about 8 times the initial vorticity of the jet U/δ. This corresponds to $\omega_z \approx 0.8f$ in the cyclone core, and to a local Rossby number $\omega_z/f = 0.8$. Anticyclones, located close to the lids, are associated with a weaker intensity ($|\omega_z| \approx 0.6f$). Very intense thin fronts connecting the consecutive cyclones have formed in the vicinity of the lids. The surface front exhibits the typical cusp-like shape, often seen on synoptic weather charts. Following the traditional nomenclature, the cold front corresponds to the front on the western edge of the cusp, while the warm front corresponds to the eastern edge. The northern part of the warm front coincides with the maximum cyclonic vorticity (about 30 times the initial value).

Let us now focus on the preferential cyclonic vorticity amplification mechanisms by considering the various production terms in the vorticity dynamics. Neglecting dissipation effects, the vorticity equation within the Boussinesq approximation may be written as follows:

$$\frac{d\omega_z}{dt} = \vec{\omega}_h \cdot \vec{\nabla}_h w + (\omega_z + f)\frac{\partial w}{\partial z} \tag{31}$$

where h denotes the horizontal direction. Vorticity might only be produced by the horizontal vorticity tilting or by the absolute vertical vorticity stretching. One can estimate the relative importance of these two terms by normalizing (31). For small Ro and for small α, is possible to show that vorticity is mainly produced by the absolute vorticity stretching (see [34]). Our numerical simulations have confirmed that the stretching term is the main production term. It is responsible for the asymmetry between cyclonic and anticyclonic vorticity amplification. Indeed, positive feed-back effects accelerate the cyclonic production, as $(\omega_z + f)$ increases while ω_z grows. Conversely, when anticyclonic vorticity increases (for $-f < \omega_z < 0$), its production term decreases. Note that that the vertical stretching term in (31) induces stronger asymmetry when the initial vorticity of the jet becomes of the order of f (or equivalently when Ro approaches unity). In that case, the absolute anticyclonic vorticity $(\omega_z + f)$ is indeed initially very weak so that anticyclones are hardly amplified. Conversely, for very small values of Ro, the intensity and shape of cyclonic and anticyclonic structures are almost symmetrical (see [13]).

3.3.2. *Large-eddy simulations*

In order to reach Reynolds numbers closer to those encountered in atmospheric or oceanic flows, we now represent the effects of the subgrid-scales on the large scale field by using the subgrid-scale structure function model with a cusp given by equation (24). In Figure 6, the time evolution of the

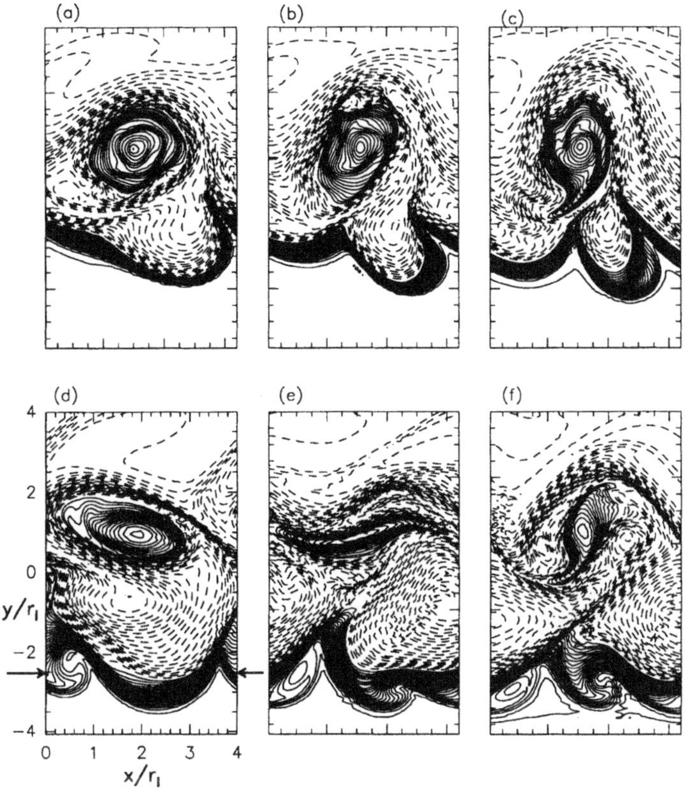

Figure 6. Potential temperature at the lower lid, at $t = 49,\ 51.5,\ 55.,\ 56.5,\ 57.,\ 58.\ r_I/U$ in (a)-(f), for the LES. Contour interval is $0.05\Delta\theta$ (from [13]).

surface potential temperature field is shown after the occlusion. The warm surface cyclone is a long-lived structure that remains at the same location and becomes zonally elongated. We have checked that the frontal regions are much steeper in the LES indicating more energy near the wavenumber cut-off. The steepening of the fronts is associated with the appearance of a secondary instability resulting in a local intensification of the vertical vorticity on the cold front. We have checked that this instability seems to take place in regions where the local values of the Rossby and Froude number $Ro(\vec{x})$ and $Fr(\vec{x})$ satisfy the criterion $Ro(\vec{x})/Fr(\vec{x}) \leq 1.5$. The instability eventually leads to the formation of a first small-scale secondary cyclone (characterized by the temperature spiraling contours) (Figures 6-

c-f). The horizontal wavelength of this secondary cyclone is of the order of r_I (typically 800 km). Concurrently, another secondary cyclone forms (Figures 6-e-f), of scale and intensity comparable to that of the first one. Note that the maximum intensity in the secondary warm fronts reaches 20 times the initial vorticity ($2f$), so that strong ageostrophic effects are present in these regions. It is important to notice that if the structure-function model without cusp is used, excessive accumulation of energy is observed at the smallest scales, eventually leading to numerical divergence. This demonstrates the importance of the cusp-like behaviour and the feasibility of the subgrid-scale model given by (24) for the LES of geophysical flows with quasi-two-dimensional regions and sharp frontal regions.

4. Concluding Remarks

Development of LES techniques is blooming at present. This is attributable to new ideas in subgrid-scale modelling and to tremendous progress in scientific computing. Once confined to very simple flow configurations, the field is evolving to include more and more complex geometries and external forces. Together with DNS, LES is able to perform deterministic predictions (of flows containing coherent vortices, for instance) and to provide statistical information. The latter is very important for assessing and improving statistical models, in particular for turbulent flows submitted to external forces (stratification, rotation, ...) or compressibility effects. There is no doubt that the complexity of problems tackled by LES will continue to increase, and that this has already had a decisive impact on industrial modelling and flow control. As far as geophysical flows are concerned, LES has been extensively used in atmospheric studies related to the boundary layers (see Mason [33], for a review). It has been more recently applied to helping to understand ocean boundary layer processes [see Large [24] in the same volume]. LES results by Skyllingstad *et al.* [40] have been successfully tested against in situ measurements of the upper ocean microstructures. As shown in the present paper, LES techniques undoubtedly allow for a better understanding of the turbulence and the dynamics of vortices over a wide range of scales. The results of Skyllingstad *et al.* [40] have also clearly demonstrated that the LES provides valuable information for building improved oceanic parameterizations of crucial regions such as the mixed layer.

Acknowledgments

Computations were carried out at the IDRIS (Institut du Développement et des Ressources en Informatique Scientifique, Paris). I am indebted to P. Comte, E. Garnier, E. Lamballais, M. Lesieur and J. Silvestrini for useful discussions and for contributions to the work presented here.

References

1. Abbà, A., Cercignani, C., Valdettaro, L., and Zanini, P. (1997) LES of turbulent thermal convection, in J.P. Chollet, P.R. Voke and L. Kleiser (eds.), *Direct and Large Eddy Simulation II*, Kluwer Academic Publishers, pp. 147–156.
2. Bardina, J., Ferziger, J.H., Reynolds, W.C. (1980) Improved subgrid model for large-eddy simulation, *AIAA paper N° 80-1357.*
3. Bartello, P., Métais, O., and Lesieur, M. (1994) Coherent structures in rotating three-dimensional turbulence. *J. Fluid Mech.* **273**, 1–29.
4. Basdevant, C., Sadourny R. (1983) Modélisation des échelles virtuelles dans la simulation numérique des écoulements bidimensionnels. *J. Mec. Theor. et Appl.*, Numéro Spécial, 243–269.
5. Bernal, L.P., and Roshko, A. (1986) Streamwise vortex structure in plane mixing layer, *J. Fluid Mech.* **170**, 499–525.
6. Brown, G., and Roshko, A. (1974) On density effects and large structure in turbulent mixing layers, *J. Fluid Mech.* **64**, 775–816.
7. Clark, R.A., Ferziger, J.H., and Reynolds, W.C. (1979) Evaluation of subgrid-scale models using an accurately simulated turbulent flow, *J. Fluid Mech.* **91**, 1–16.
8. Comte, P., Ducros, F., Silvestrini, J., David, E., Lamballais, E., Métais, O. and Lesieur, M. (1994) Simulation des grandes échelles d'écoulements transitionnels, *AGARD Conference Proceedings 551*, pp. 14.1–14.12.
9. David, E. (1993) *Modélisation des Ecoulements Compressibles et Hypersoniques: une Approche Instationnaire.* PhD thesis. National Polytechnic Institute, Grenoble.
10. Deardorff, J.W. (1970) A numerical study of three-dimensional turbulent channel flow at large Reynolds number, *J. Fluid Mech.* **41**, 453–80.
11. Ducros, F., Comte, P., and Lesieur, M. (1996) Large-eddy simulation of transition to turbulence in a boundary layer developing spatially over a flat plate, *J. Fluid Mech.* **326**, 1–36.
12. El-Hady, N.M., and Zang T.A. (1995) Large-eddy simulation of nonlinear evolution and breakdown to turbulence in high-speed boundary layers, *Theoret. Comput. Fluid Dynamics* **7**, 217–240.
13. Garnier, E., Métais, O., and Lesieur, M. (1998) Synoptic and frontal-cyclone scale instabilities in baroclinic jet flows, *J. Atmos. Sci.*, in press.
14. Garnier, E., Métais, O., and Lesieur, M. (1996) Instabilités primaire et secondaire dans un jet barocline, *C.R. Acad. Sci. Paris* Série II b **323**, 161–168.
15. Germano, M., (1992) Turbulence, the filtering approach, *J. Fluid Mech.* **238**, 325–336.
16. Germano, M., Piomelli, U., Moin, P., and Cabot, W. (1991) A dynamic subgrid-scale eddy-viscosity model, *Phys. Fluids A.* **3**(7), 1760–1765.
17. Gonze, M.A. (1993) *Simulation Numérique des Sillages en Transition à la Turbulence.* PhD thesis. National Polytechnic Institute, Grenoble.
18. Kolmogorov, A.N. (1941) The local structure of turbulence in incompressible viscous fluid for very large Reynolds numbers. *Dokl. Akad. Nauk SSSR* **30**, 301–305.
19. Kraichnan, R.H. (1976) Eddy viscosity in two and three dimensions. *J. Atmos. Sci.* **33**, 1521–1536.
20. Lamballais, E. (1996) *Simulations numériques de la turbulence dans un canal plan tournant*, PhD thesis. National Polytechnic Institute, Grenoble.
21. Lamballais, E., Lesieur, M., Métais, O. (1996) Effects of spanwise rotation on the vorticity stretching in transitional and turbulent channel flow, *Int. J. Heat and Fluid Flow* **17** (3), 324–332.
22. Lamballais, E., Lesieur, M., and Métais, O. (1996) Influence of a solid-body rotation upon coherent vortices in a channel, *C. R. Acad. Sci.* Série II b **323**, 95–101.
23. Lamballais, E., Métais, O., and Lesieur, M. (1996) Influence of a spanwise rotation upon the coherent-structure dynamics in a turbulent channel flow, in J.P. Chollet, P.R. Voke and L. Kleiser (eds.), *Direct and Large Eddy Simulation II*, Kluwer Academic Publishers, pp. 225–236.

24. Large, W., 1998: Modeling and parameterizing ocean planetary boundary layers. In *Ocean Modeling and Parameterization*, E.P. Chassignet and J. Verron (Eds.), Kluwer Academic Publishers, 81-120.
25. Lele, S.K. (1992) Compact finite difference schemes with spectral-like resolution. *J. Comput. Phys.* **103**, 16–42.
26. Lesieur, M. (1997) Recent approaches in large-eddy simulations of turbulence, in O. Métais and J. Ferziger (eds), *New Tools in Turbulence Modelling*, Les Éditions de Physique, Springer-Verlag, pp. 1–28.
27. Lesieur, M. (1997) *Turbulence in Fluids, Third Revised and Enlarged Edition*, Kluwer Academic Publishers, Dordrecht.
28. Lesieur, M., and Métais, O. (1996) New trends in large-eddy simulations of turbulence", *Annu. Rev. Fluid Mech.* **28**, 45–82.
29. Lesieur, M., Yanase, S., and Métais, O. (1991) Stabilizing and destabilizing effects of a solid-body rotation on quasi-two-dimensional shear layers, *Phys. Fluids A* **3**, 403–407.
30. Lilly, D.K. (1987) J.R. Herring and J.C. McWilliams (eds), *Lecture Notes on Turbulence*, World Scientific, pp. 171–218.
31. Liu, S., Meneveau, C., and Katz, J. (1994) On the properties of similarity subgrid-scale models as deduced from measurements in turbulent jet, *J. Fluid Mech.* **275**,83–119.
32. Lilly, D.K. (1992) A proposed modification of the Germano subgrid-scale closure method, *Phys. Fluids A.* **4** (3), 633–635.
33. Mason, P. J. (1994) Large-eddy simulation: a critical review of the technique, *Q.J.R. Meteorol. Soc.* **120**, 1–26.
34. McWilliams, J.C. (1985) A uniformly valid model spanning the regimes of geostrophic and isotropic, stratified turbulence: Balanced turbulence. *J. Atmos. Sci.* **42**, 1773–1774.
35. Métais, O., Flores, C., Yanase, S., Riley, J.J. and Lesieur, M. (1995) Rotating free shear flows Part 2 : Numerical simulations, *J. Fluid Mech.* **293**, 41–80.
36. Métais, O., and Lesieur M. (1992) Spectral large-eddy simulations of isotropic and stably-stratified turbulence. *J. Fluid Mech* **239**, 157–194.
37. Moin, P., and Kim, J. (1982) Numerical investigation of turbulent channel flow, *J. Fluid Mech.* **118**, 341–377.
38. Piomelli, U. (1993) High Reynolds number calculations using the dynamic subgrid-scale stress model, *Phys. Fluids A* **5** (6), 1484–1490.
39. Silvestrini, J., Comte, P., and Lesieur, M. (1995) DNS and LES of spatial incompressible mixing layer. *Proc. of the 10th Symposium on Turbulent Shear Flows.*
40. Skyllingstad, E.D., Smyth, W.D., Moum, J.N., and Wijsesekera, H. (1998) Upper ocean turbulence during a westerly wind burst: a comparison of Large-Eddy Simulation results and microstructure measurements, *submitted to J. Phys. Oceanogr.*.
41. Smagorinsky, J. (1963) General circulation experiments with the primitive equations, *Mon. Weath. Rev.* **91**, 3, 99–164.
42. Zang, Y., Street, R.L., and Koseff J.R. (1993) A dynamic mixed subgrid-scale model and its application to turbulent recirculating flows. *Phys. Fluids A* **5** (12), 3186–3196.

PARAMETER ESTIMATION IN DYNAMICAL MODELS

GEIR EVENSEN

Nansen Environmental and Remote Sensing Center,
Bergen, Norway

DICK P. DEE

General Sciences Corporation,
Laurel, Maryland, U.S.A.

AND

JENS SCHRÖTER

Alfred-Wegener-Institut für Polar- und Meeresforschung,
Bremerhaven, Germany

Abstract

This paper gives a general introduction to the parameter estimation problem for dynamical models. The basic formulation and methodology in a parameter estimation problem will be discussed and some rather simple examples will be presented. It will be shown that even for linear dynamics the parameter estimation problem becomes nonlinear and may become extremely difficult to solve. Also, to have a well-posed problem with a unique solution, care must be taken when a parameter estimation problem is formulated. The discussion leads to the conclusion that it is possible to estimate poorly known parameters in a model, at least for simple dynamical models, but care must be taken to have a consistent solution. The rule is that all parameters which will be estimated should be added in a penalty function as weak constraints measuring their distance from a first guess in some norm. Some previous works in which data assimilation methods have been used to improve estimates of poorly known model parameters or even the model bias will be briefly reviewed.

E. P. Chassignet and J. Verron (eds.), Ocean Modeling and Parameterization, 373–398.

1. Introduction

Data assimilation and inverse methods are normally used for generating estimates of dynamical variables, taking into account both the information about the dynamics from a dynamical model and the information about the true state which is contained in a set of measurements. Such techniques have also been proposed as a tool for parameter estimation in dynamical models, although only a few works in oceanography have so far considered the parameter estimation problem in this context. The idea is that it should be possible to use mathematical tools to formulate inverse problems for parameter estimation, if there is additional information available about the true state of the ocean in the form of measurements. Thus, one may attempt to search for model parameters resulting in a model solution which is close to the observed state.

The parameter estimation problem has been more extensively treated in other fields and many applications have been based on the theory outlined in [10] and [11]. Most relevant are the applications in groundwater flow, [7], [8], and [9].

Currently there only exist a few publications in which such inverse calculations have been carried out for oceanographic applications, and these have considered relatively simple dynamical models. Examples are [22], which estimated the phase speeds in a reduced-gravity model of equatorial waves, using a strong constraint variational formulation, and [24] (with a correction in [25]), which used a modified one-dimensional Ekman layer model in combination with variational optimal control techniques to simultaneously estimate the surface wind drag coefficient and the vertical profile of the eddy viscosity from observed data. This was a rather early work in this field, and there were some errors in the formulation and solution method which made the problem ill-posed. In a paper [14], this problem was revisited and solved using more recent methodologies. In the paper [21], a strong constraint formulation for a quasi-geostrophic model was used to estimate the initial streamfunction for the model, the Rossby radius of deformation, the friction coefficient, and two parameters describing the curl of the wind stress. Two other works, [23] and [26], attempt to determine the "optimal" nudging coefficients using adjoint techniques in a parameter estimation context. Some of the results from these papers will be discussed in Section 5, and have clearly made a very important contribution to our current understanding of the parameter estimation problem. A recent review, [20], discusses several previous parameter estimation applications in atmospheric and oceanic models.

One would expect that these methods would be much more widely used. However, the parameter estimation problem often becomes extremely complicated to solve and is nonlinear even if the dynamical model itself is linear. It is in many cases extremely difficult to find a proper solution. It is also a fact that most of the scientists who are working in data assimilation are still busy solving the general data assimilation problem for realistic Ocean General Circulation Models (OGCM) and have given the parameter estimation problem less attention. However, we still believe the parameter estimation problem should be further explored, and there is still a great deal of challenging work to be done in the years to come. The major interest will probably be parameter estimation in OGCMs, but possibly even more interest will be related to calibration of other types of dynamical models, e.g., marine ecosystem and sediment models.

This paper will discuss in some detail the general parameter estimation problem and how it can be formulated in a consistent manner. A general solution procedure will be presented which can be applied with a large class of dynamical models. Some of the previously published works on parameter estimation will be briefly reviewed and their methodology and results will be discussed.

2. Formulation of a Parameter Estimation Problem

We will now look at the general formulation of the parameter estimation problem and discuss the properties of the equations that will have to be solved.

2.1. A MODEL AND OBSERVATIONS

For purposes of illustration, we will use a a simple nonlinear ordinary differential equation when formulating and discussing the parameter estimation problem. The model for ψ with an initial condition is given as

$$\frac{d\psi}{dt} = f(\alpha, \psi), \tag{1}$$

$$\psi(0) = \psi_0. \tag{2}$$

Here f is a nonlinear model operator and it is assumed that α is a poorly known parameter or constant in the model. The system as it is has one unique solution for a given value of α. Thus if the value of α is to be improved, we need additional information. This can be given by introducing a set of observations of the model variable taken at various locations in time (and space if the space dimensions are included). The vector of observations can be written as

$$\boldsymbol{d} = \boldsymbol{\mathcal{M}}[\psi], \tag{3}$$

where the linear operator $\boldsymbol{\mathcal{M}}$ relates the observations to the model variable.

As an example of the linear measurement operator $\mathcal{M}[\psi]$, which relates the observations $\boldsymbol{d}$ to the state $\psi(t)$, a direct measurement would have a measurement functional of the form

$$\mathcal{M}_i[\psi] = \int_0^T \psi(t)\delta(t - T_i)\,dt = \psi(T_i), \tag{4}$$

with T_i as the measurement location in time, and the subscript i denoting a component of the measurement functional which is a vector with dimension m equal to the number of measurements.

In general the problem becomes over-determined since it will not be possible to chose one value for α that gives a model trajectory that fits the observations exactly; thus, no solution exists when the observations are introduced as additional constraints. This is resolved by allowing the observations to contain errors with some prescribed statistical properties. The measurement equation then becomes

$$\boldsymbol{d} = \mathcal{M}[\psi^t] + \epsilon, \tag{5}$$

where ψ^t is the unknown true state and ϵ is a vector of measurement errors with mean $\overline{\epsilon} = 0$ and error covariance $\overline{\epsilon\epsilon^{\mathrm{T}}} = w^{-1}$.

The overbar denotes some kind of averaging operator. In stochastic estimation theory this is the *expectation operator*, defined in terms of the set of all possible realizations of the stochastic-dynamic system whose state is to be estimated [19]. The precise definition of this operator requires the joint probability distribution of forecast and observation errors. In a more practical setting, it might represent an *ensemble average*, defined in terms of certain computational realizations of a deterministic model of the system (e.g., a forecast model) [18]. In many operational data assimilation systems, the ensemble average is simply replaced by a *time average* – that is, forecast and observation error means and covariances are computed by averaging in time the errors associated with a single realization of the system [1].

There are now two essential items that should be noted:

1. There will *always* be some information available about realistic or physically acceptable values of the poorly known parameter α. Such information could be that the value of the parameter should stay within some physically acceptable limits. This information should always be included in the inverse formulation through a first guess value and its uncertainty.
2. Unless such a first guess estimate is included in the formulation, it will in many cases become practically impossible to determine how many observations are needed to find a unique solution for the parameters. Alternatively, given a particular set of observations, it cannot be ensured that there exists a unique solution. This is primarily due to the

fact that the information in the observations may not be independent. There may then be several values for the parameters leading to solutions which interpolate the data exactly.

Thus, the first guess value of the parameter should as a rule always be included in the inverse formulation. As will be seen, this ensures a unique solution even without any observations.

An equation for the parameter can now be defined as

$$\alpha = \alpha_0 + \tilde{\alpha}, \tag{6}$$

where α_0 is now the first or best guess value and $\tilde{\alpha}$ is the error in α_0 assumed to have statistics defined by $\overline{\tilde{\alpha}} = 0$ and $\overline{\tilde{\alpha}^2} = Q_{\tilde{\alpha}\tilde{\alpha}}$. The inverse of the covariance is defined as $W_{\tilde{\alpha}\tilde{\alpha}} = Q_{\tilde{\alpha}\tilde{\alpha}}^{-1}$.

We will now look at some different formulations of the inverse problem.

2.2. STRONG CONSTRAINT FORMULATION: ALTERNATIVE ONE

The parameter estimation problem can now be formulated by seeking a value of α which results in the model solution which is closest to the observations. Since we have introduced statistical assumptions involving only the two first moments (mean and covariance) it is natural to use a least squares estimator and we define the penalty function

$$\begin{aligned} \mathcal{J}_1[\alpha] &= \tilde{\alpha} W_{\tilde{\alpha}\tilde{\alpha}} \tilde{\alpha} + \boldsymbol{\epsilon}^{\mathrm{T}} \boldsymbol{w} \boldsymbol{\epsilon} \\ &= (\alpha - \alpha_0) W_{\tilde{\alpha}\tilde{\alpha}} (\alpha - \alpha_0) + (\boldsymbol{d} - \boldsymbol{\mathcal{M}}[\psi])^{\mathrm{T}} \boldsymbol{w} (\boldsymbol{d} - \boldsymbol{\mathcal{M}}[\psi]). \end{aligned} \tag{7}$$

Thus we attempt to search for a value of α, not too far from α_0, which gives the model solution closest to the observations. We then have to minimize the penalty function (7) with respect to α, under the constraint that the initial condition and the model equation are satisfied exactly. This is a so-called constrained minimization problem, which can most easily be solved by creating a Lagrangian functional in which the model and the initial conditions are added as strong constraints using Lagrangian multipliers, μ and λ,

$$\mathcal{L}_1[\psi, \alpha, \mu, \lambda] = \mathcal{J}_1 + \mu(\psi - \psi_0) + \int_0^T \lambda \left(\frac{d\psi}{dt} - f(\alpha, \psi) \right) dt. \tag{8}$$

This functional can now be minimized using variational calculus. Clearly, the variation with respect to μ and λ just gives us back the forward model (1) with its initial condition (2) which must be satisfied exactly. The variation with respect to ψ results in an equation for λ with a final condition at $t = T$. This is the so-called adjoint equation which can be integrated backward in time forced by the data misfits. It is given in Section 2.6 as (22) with final condition (23), at which point the Euler–Lagrange equations will be discussed in some more detail. The variation with respect to α results in an expression for the gradient of the Lagrangian with respect to α which can be used in a descent algorithm (see equation 24 in Section 2.6).

2.3. STRONG CONSTRAINT FORMULATION: ALTERNATIVE TWO

The formulation above does not take into account that the initial condition may contain some errors. An error in the initial condition will lead to a situation in which the estimated value for α also attempts to correct for this, and a poor estimate for α may be found.

This problem can be easily resolved by allowing for errors in the initial condition of the form

$$\psi(0) = \psi_0 + a, \tag{9}$$

where the error a is assumed to satisfy the statistical assumptions $\overline{a} = 0$ and $\overline{a^2} = Q_{aa}$. Again we define the weight as the inverse of the variance $W_{aa} = Q_{aa}^{-1}$. The first guess initial condition can then be added as a weak constraint and the penalty function becomes

$$\begin{aligned} \mathcal{J}_2[\psi(0), \alpha] &= \tilde{\alpha} W_{\tilde{\alpha}\tilde{\alpha}} \tilde{\alpha} + a W_{aa} a + \boldsymbol{\epsilon}^{\mathrm{T}} \boldsymbol{w} \boldsymbol{\epsilon} \\ &= (\alpha - \alpha_0) W_{\tilde{\alpha}\tilde{\alpha}} (\alpha - \alpha_0) + (\psi - \psi_0) W_{aa} (\psi - \psi_0) \\ &+ (\boldsymbol{d} - \boldsymbol{\mathcal{M}}[\psi])^{\mathrm{T}} \boldsymbol{w} (\boldsymbol{d} - \boldsymbol{\mathcal{M}}[\psi]). \end{aligned} \tag{10}$$

Thus, a solution is searched for which is close to the observations $\boldsymbol{d}$, the initial condition ψ_0, and the first guess value α_0, while the model equation is assumed to be satisfied exactly. Again a Lagrangian is constructed to include the forward model equation

$$\mathcal{L}_2[\psi, \alpha, \lambda] = \mathcal{J}_2 + \int_0^T \lambda \left(\frac{d\psi}{dt} - f(\alpha, \psi) \right) dt. \tag{11}$$

The Euler-Lagrange equations will be similar to those for the Lagrangian (8), but now containing an additional term accounting for the errors in the initial condition. See Section 2.6 for a discussion of the Euler-Lagrange equations.

2.4. WEAK CONSTRAINT GENERAL INVERSE FORMULATION

Another problem with the formulations above is that they assume the model equation is perfect except for the poorly known α. Thus, it has not been taken into account that the model may be corrupted by other errors, such as those related to insufficient resolution, neglected physics, linearizations, and other approximations that are commonly adapted in most dynamical models to make the equations more tractable. Thus, if there are model errors of significant magnitude compared to the real world that are observed by the measurements, a similar situation arises as for the previous case, in which the estimate for the parameter α attempts to correct for other errors than those related to the particular parameterization.

There is also another issue, as discussed in detail in [16], which becomes important for nonlinear models. If the model is describing unstable or chaotic dynamics such as the ocean and atmosphere, the strong constraint formulation will have a particular problem estimating the correct initial conditions. This is caused by the strong sensitivity of the model solution and also by the penalty function with respect to small changes in the initial conditions. By including the model as a weak constraint, this problem is completely eliminated, and we obtain a better-posed problem to solve [16]. The reason is that the inverse estimate can deviate from the exact model trajectory and forget long past and future information (relative to the predictability time of the dynamics). Thus, after a few predictability times, the estimate will not be greatly influenced by the inital condition, if there are also sufficient observations to assimilate.

Thus, we should account for general model errors in the formulation. This can easily be done by allowing the model equation to contain errors. To sum up, we now have the following system of equations defining the problem:

$$\frac{d\psi}{dt} = f(\alpha, \psi) + q, \tag{12}$$

$$\psi(0) = \psi_0 + a, \tag{13}$$

$$\alpha = \alpha_0 + \tilde{\alpha}, \tag{14}$$

$$\boldsymbol{d} = \boldsymbol{\mathcal{M}}[\psi^t] + \epsilon. \tag{15}$$

The only new variable is $q(t)$, which represents the model errors other than those related to the value of α. Again we make a hypothesis about the model errors; $\overline{q(t)} = 0$ and $\overline{q(t_1)q(t_2)} = Q_{qq}(t_1, t_2)$. Note also that we have neglected all cross correlations between the different error terms, which in most cases is a valid assumption (the errors are normally generated by independent sources). The inverse of $Q_{qq}(t_1, t_2)$ is defined as $W_{qq}(t_1, t_2)$ through the relation

$$\int_0^T W_{qq}(t_1, t_2) Q_{qq}(t_2, t_3)\, dt_2 = \delta(t_1 - t_3). \tag{16}$$

We can now formulate the general weak constraint inverse problem for the system (12–15) in terms of a penalty function

$$
\begin{aligned}
\mathcal{J}[\psi,\alpha] &= \tilde{\alpha} W_{\tilde{\alpha}\tilde{\alpha}} \tilde{\alpha} + \int_0^T \int_0^T q(t_1) W_{qq}(t_1,t_2) q(t_2)\, dt_1\, dt_2 + a W_{aa} a + \boldsymbol{\epsilon}^{\mathrm{T}} \boldsymbol{w} \boldsymbol{\epsilon} \\
&= (\alpha - \alpha_0) W_{\tilde{\alpha}\tilde{\alpha}} (\alpha - \alpha_0) \\
&+ \int_0^T \int_0^T \left\{ \left(\frac{d\psi(t_1)}{dt_1} - f(\alpha, \psi(t_1)) \right) \right. \\
&\qquad \left. W_{qq}(t_1,t_2) \left(\frac{d\psi(t_2)}{dt_2} - f(\alpha, \psi(t_2)) \right) \right\} dt_1\, dt_2 \\
&+ (\psi - \psi_0) W_{aa} (\psi - \psi_0) + (\boldsymbol{d} - \boldsymbol{\mathcal{M}}[\psi])^{\mathrm{T}} \boldsymbol{w} (\boldsymbol{d} - \boldsymbol{\mathcal{M}}[\psi]).
\end{aligned}
\tag{17}
$$

It should now be noted that if the statistical hypothesis about normally distributed errors is correct, then the minimum of $\mathcal{J}$ (which is the optimal variance minimizing estimate) will be the maximum likelihood estimate for ψ and α. Thus, using weights which are the inverses of the error covariances is justifiable.

By minimizing this variational functional, an estimate is searched for which minimizes the distance from the observations, the initial condition, and the first guess value for α, and at the same time "almost" satisfies the dynamical model equation.

2.5. EULER–LAGRANGE EQUATIONS

It is now instructive to present the Euler-Lagrange equations for the penalty function $\mathcal{J}$, which is derived by taking the variation of $\mathcal{J}$ with respect to ψ and α. Thus, ψ and α define an extremum of the penalty function if

$$
\mathcal{J}[\psi + \delta\psi, \alpha] - \mathcal{J}[\psi, \alpha] = \mathcal{O}(\delta\psi^2), \tag{18}
$$

$$
\mathcal{J}[\psi, \alpha + \delta\alpha] - \mathcal{J}[\psi, \alpha] = \mathcal{O}(\delta\alpha^2), \tag{19}
$$

when $\delta\psi$ and $\delta\alpha$ approach zero.

One then obtains

$$
\frac{d\psi}{dt} = f(\alpha,\psi) + \int_0^T Q_{qq}(t,t_1)\lambda(t_1)dt_1, \tag{20}
$$

$$
\psi(0) = \psi_0 + Q_{aa}\lambda(0), \tag{21}
$$

$$
\frac{d\lambda}{dt} = -f_\psi(\alpha,\psi)\lambda - \boldsymbol{\mathcal{M}}^{\mathrm{T}}[\delta]\boldsymbol{w}(\boldsymbol{d} - \boldsymbol{\mathcal{M}}[\psi]), \tag{22}
$$

$$
\lambda(T) = 0, \tag{23}
$$

$$\alpha = \alpha_0 + Q_{\tilde{\alpha}\tilde{\alpha}} \int_0^T f_\alpha(\alpha, \psi)\lambda dt. \tag{24}$$

Since the model is nonlinear, a Taylor expansion is used for $f(\alpha, \psi)$, and the first derivatives of f with respect to α and ψ are introduced as f_α and f_ψ. Note that λ is defined through the relation (20). From this general set of equations, we obtain the Euler–Lagrange equations for the Lagrangian $\mathcal{L}_2$ by setting $Q_{qq} = 0$ and for $\mathcal{L}_1$ by setting $Q_{qq} = Q_{aa} = 0$.

The expression $\mathcal{M}[\delta]$ is a short notation for a vector of elements

$$\mathcal{M}_i[\delta(t - t_2)] = \int_0^T \delta(t - t_2)\delta(t_2 - T_i)\, dt_2, \tag{25}$$

which are equal to 1 when the model time variable is equal to T_i and zero elsewhere.

The system of Euler–Lagrange equations defines the extrema of $\mathcal{J}$. The system consists of the original forward model forced by a term that is proportional to the adjoint variable λ in (20). The magnitude of this term is defined by the model error covariance; thus large model errors give a large contribution through the forcing term. The forward model is integrated from an initial condition which also contains a similar correction term proportional to the adjoint variable. The equation for λ can be integrated backward in time from a final condition, while forced by delta functions scaled by the residual between the measurement and forward model estimate at each measurement location. Thus, the forward model needs knowledge of the adjoint variable to be integrated, and the backward model needs the forward variable at measurement locations. We therefore have a coupled boundary value problem in time, in which the forward and backward models must be solved simultaneously.

2.6. MODEL BIAS

A final but not less important remark is related to a bias in the model. It is normally easier to remove an eventual bias in observations, e.g., by collecting control observations using other instruments, than to remove the bias in the model. If initial conditions are estimated from observations, they may also be unbiased. Thus the important problem will be to account for model bias which we have assumed is zero in the formulation above. Clearly, if there is a significant bias in the model, this will corrupt both the inverse problem for the model solution and the parameter estimation problem. We may end up with a totally unphysical estimate for α which attempts to correct for a model bias caused by a completely different physical approximation.

Estimation of model bias may be considered as a special case of the parameter estimation problem. If the model (12) is rewritten as

$$\frac{d\psi}{dt} = f(\alpha, \psi) + \beta + q, \tag{26}$$

β can now represent the unknown model bias which is constant in time. Thus, there is no approximation related to setting $\overline{q(t)} = 0$. An equation is needed for β,

$$\beta = \beta_0 + \tilde{\beta}. \tag{27}$$

It is assumed that β_0 is a first guess value for the bias which would typically be zero in many applications. The error $\tilde{\beta}$ in the first guess for the bias is assumed to be normally distributed with $\overline{\tilde{\beta}} = 0$ and $\overline{\tilde{\beta}^2} = Q_{\tilde{\beta}\tilde{\beta}}$. The inverse of the covariance is defined as $W_{\tilde{\beta}\tilde{\beta}} = Q^{-1}_{\tilde{\beta}\tilde{\beta}}$. A weak constraint variational formulation for this problem is

$$\begin{aligned}
\mathcal{J}[\psi, \alpha] &= \tilde{\alpha} W_{\tilde{\alpha}\tilde{\alpha}} \tilde{\alpha} + \tilde{\beta} W_{\tilde{\beta}\tilde{\beta}} \tilde{\beta} \\
&\quad + \int_0^T \int_0^T q(t_1) W_{qq}(t_1, t_2) q(t_2)\, dt_1 dt_2 + a W_{aa} a + \boldsymbol{\epsilon}^{\mathrm{T}} \boldsymbol{w} \boldsymbol{\epsilon} \\
&= (\alpha - \alpha_0) W_{\tilde{\alpha}\tilde{\alpha}} (\alpha - \alpha_0) + (\beta - \beta_0) W_{\tilde{\beta}\tilde{\beta}} (\beta - \beta_0) \\
&\quad + \int_0^T \int_0^T \left\{ \left(\frac{d\psi(t_1)}{dt_1} - f(\alpha, \psi(t_1)) - \beta \right) \right. \\
&\qquad\qquad \left. W_{qq}(t_1, t_2) \left(\frac{d\psi(t_2)}{dt_2} - f(\alpha, \psi(t_2)) - \beta \right) \right\} dt_1 dt_2 \\
&\quad + (\psi - \psi_0) W_{aa} (\psi - \psi_0) + (\boldsymbol{d} - \boldsymbol{\mathcal{M}}[\psi])^{\mathrm{T}} \boldsymbol{w} (\boldsymbol{d} - \boldsymbol{\mathcal{M}}[\psi]).
\end{aligned} \tag{28}$$

The Euler–Lagrange equations for this problem become

$$\frac{d\psi}{dt} = f(\alpha, \psi) + \beta + \int_0^T Q_{qq}(t, t_1) \lambda(t_1) dt_1, \tag{29}$$

$$\psi(0) = \psi_0 + Q_{aa} \lambda(0), \tag{30}$$

$$\frac{d\lambda}{dt} = -f_\psi(\alpha, \psi)\lambda - \boldsymbol{\mathcal{M}}[\delta] \boldsymbol{w} (\boldsymbol{d} - \boldsymbol{\mathcal{M}}[\psi]), \tag{31}$$

$$\lambda(T) = 0, \tag{32}$$

$$\alpha = \alpha_0 + Q_{\tilde{\alpha}\tilde{\alpha}} \int_0^T f_\alpha(\alpha, \psi) \lambda dt, \tag{33}$$

$$\beta = \beta_0 + Q_{\tilde{\beta}\tilde{\beta}} \int_0^T \lambda dt. \tag{34}$$

Thus, we are now estimating both the value for α and the bias β in the model. In Section 4 an alternative method will be illustrated in which the model bias is corrected using a sequential data assimilation algorithm.

3. A solution Method for the Parameter Estimation Problem.

At this stage, it should be pointed out that the inverse problem may become highly nonlinear, not only because of the nonlinearities in the model, but also by the introduction of the parameter as an additional variable in the system. It may in many cases be impossible to find a solution using traditional methods involving the Euler–Lagrange equations. However, a methodology will now be presented that has proven successful with linear and weakly nonlinear models.

3.1. ITERATION OF PARAMETERS

The system of Euler–Lagrange equations is strongly coupled through the parameters and the dependent variable ψ, and it is also nonlinear. To simplify the system it is possible to define an iteration for the unknown parameters. Note that the equations for α and β, (33) and (34), are the gradient of the penalty function with respect to the parameters. By defining the iterations for the parameters as

$$\alpha^{k+1} = \alpha^k - \gamma_\alpha \left(\alpha^k - \alpha_0 - Q_{\tilde{\alpha}\tilde{\alpha}} \int_0^T f_\alpha(\alpha^k, \psi^k) \lambda^k \, dt \right), \tag{35}$$

$$\beta^{k+1} = \beta^k - \gamma_\beta \left(\beta^k - \beta_0 - Q_{\tilde{\beta}\tilde{\beta}} \int_0^T \lambda^k \, dt \right), \tag{36}$$

the problem is now reduced to solving a sequence of inverse problems for ψ and λ for each iterate k of the parameters α and β. The two factors γ_α and γ_β are just the step lengths in the gradient descent iterations. The new estimates for ψ and λ are then used to iterate the parameters through (35) and (36). Thus, if these iterations converge, the problem is now reduced to solving the following equations for each iteration (iteration superscripts have been dropped):

$$\frac{d\psi}{dt} = f(\alpha, \psi) + \beta + \int_0^T Q_{qq}(t, t_1) \lambda(t_1) dt_1, \tag{37}$$

$$\psi(0) = \psi_0 + Q_{aa} \lambda(0), \tag{38}$$

$$\frac{d\lambda}{dt} = -f_\psi(\alpha, \psi)\lambda - \boldsymbol{\mathcal{M}}^{\mathrm{T}}[\delta] \boldsymbol{w} (\boldsymbol{d} - \boldsymbol{\mathcal{M}}[\psi]) \tag{39}$$

$$\lambda(T) = 0. \tag{40}$$

These are the Euler–Lagrange equations for the standard inverse or data assimilation problem without any parameters to estimate.

3.2. SOLUTION OF THE "ALTERNATIVE ONE" PROBLEM

The problem defined by $\mathcal{L}_1$ can now be solved easily since both Q_{aa} and Q_{qq} are zero. This removes the coupling between the forward and backward equations in (37–40), and given an estimate for the parameters one can always find ψ from one forward integration and λ from a following backward integration. It is suggested that this approach be used to test whether the iterations of the parameters converge before one of the more sophisticated methods discussed below is implemented.

3.3. SOLUTION OF THE "ALTERNATIVE TWO" PROBLEM

The assumption that the model is perfect, $Q_{qq} = 0$, apparently simplifies the Euler–Lagrange equations (37–40). A data assimilation or inverse method that is very commonly used for this special case is the so-called "adjoint method". This method iterates the initial condition in the same way as has been done for the parameters above, using the fact that the equation (38) is just the gradient of $\mathcal{L}_2$ with respect to the initial conditions. We can define the iteration

$$\psi^{k+1} = \psi^k - \gamma_\psi(\psi^k - \psi_0 - Q_{aa}\lambda^k), \tag{41}$$

where again γ_ψ is a step length in the gradient descent method.

Thus, it is now possible to solve the parameter estimation problem by iterating the equations (35), (36), and (41). Each iteration provides an estimate for the two parameters and the initial condition, which makes it possible to solve for the current iterate of ψ and λ needed to update the parameters and the initial condition again. Note that various strategies should be tried, such as using several iterations for the initial condition, to achieve convergence for ψ and λ before updating the parameters again. This may be especially important for high dimensional problems.

Note that this method searches for the solution in a space with dimension equal to the number of state variables in the model. For many realistic ocean models this dimension becomes extremely large, $\mathcal{O}(10^{5-8})$.

3.4. REPRESENTER SOLUTION FOR A LINEAR INVERSE PROBLEM

The representer method is an elegant methodology that can solve the linear inverse problem without any iterations and in which the solution is searched for in a space with dimension equal to the number of measurements.

Assume now that the model dynamics are linear. The cases with weakly nonlinear and strongly nonlinear dynamics will be treated next. The model operator $f(\alpha, \psi)$ is now, for illustrational purposes, written as $f\alpha\psi$, where f is now a scalar for this particular model and $\alpha \simeq 1$ can be interpreted as a poorly known correction to f. For this linear model we now get the Euler–Lagrange equations of the form (where the parameters are now assumed given)

$$\frac{d\psi}{dt} = f\alpha\psi + \beta + \int_0^T Q_{qq}(t,t_1)\lambda(t_1)dt_1, \tag{42}$$

$$\psi(0) = \psi_0 + Q_{aa}\lambda(0), \tag{43}$$

$$\frac{d\lambda}{dt} = -f\alpha\lambda - \boldsymbol{\mathcal{M}}[\delta]^{\mathrm{T}}\boldsymbol{w}(\boldsymbol{d} - \boldsymbol{\mathcal{M}}[\psi]), \tag{44}$$

$$\lambda(T) = 0. \tag{45}$$

Start by assuming that a solution can be written as

$$\psi(t) = \psi_F(t) + \boldsymbol{b}^{\mathrm{T}}\boldsymbol{r}(t), \tag{46}$$

$$\lambda(t) = \lambda_F(t) + \boldsymbol{b}^{\mathrm{T}}\boldsymbol{s}(t), \tag{47}$$

where the dimensions of the vectors are all equal to the number of measurements, m. Assuming this form for the solution is equivalent to saying that the minimizing solution is a first guess model solution plus a linear combination of time-dependent influence functions or representers $\boldsymbol{r}(t)$, one for each measurement. For a comprehensive discussion of this method, see [2].

Inserting (46) and (47) into the Euler–Lagrange equations (42–45) and choosing first guesses ψ_F and λ_F that satisfy unforced exact equations

$$\frac{d\psi_F}{dt} = f\alpha\psi_F + \beta + \int_0^T Q_{qq}\lambda_F, \tag{48}$$

$$\psi_F(0) = \psi_0, \tag{49}$$

$$\frac{d\lambda_F}{dt} = -f\alpha\lambda_F, \tag{50}$$

$$\lambda_F(T) = 0, \tag{51}$$

gives us the following system of equations for the vector of representers $\boldsymbol{r}$ and corresponding adjoints $\boldsymbol{s}$:

$$\boldsymbol{b}^{\mathrm{T}}\left(\frac{d\boldsymbol{r}}{dt} - f\alpha\boldsymbol{r} - \int_0^T Q_{qq}(t,t_1)\boldsymbol{s}(t_1)dt_1\right) = 0, \tag{52}$$

$$\boldsymbol{b}^{\mathrm{T}}\left(\boldsymbol{r}(0) - Q_{aa}\boldsymbol{s}\right) = 0, \tag{53}$$

$$\boldsymbol{b}^{\mathrm{T}}\left(\frac{d\boldsymbol{s}}{dt} + f\alpha\boldsymbol{s}\right) + \boldsymbol{\mathcal{M}}^{\mathrm{T}}[\delta]\boldsymbol{w}\left(\boldsymbol{d} - \boldsymbol{\mathcal{M}}[\psi_F + \boldsymbol{b}^{\mathrm{T}}\boldsymbol{r}]\right) = 0, \tag{54}$$

$$\boldsymbol{b}^{\mathrm{T}}\boldsymbol{s}(T) = 0. \tag{55}$$

By defining the so far undetermined vector of coefficients $\boldsymbol{b}$ to be given by

$$\boldsymbol{b} = \boldsymbol{w}\left(\boldsymbol{d} - \boldsymbol{\mathcal{M}}[\psi_F + \boldsymbol{b}^{\mathrm{T}}\boldsymbol{r}]\right), \tag{56}$$

equation (54) now becomes

$$\boldsymbol{b}^{\mathrm{T}}\left(\frac{d\boldsymbol{s}}{dt} + \boldsymbol{s} + \boldsymbol{\mathcal{M}}[\delta]\right) = 0, \tag{57}$$

which now decouples the forward and backward Euler–Lagrange equations.

Using the fact that $\boldsymbol{b}$ in general is nonzero, we now have the following set of equations in addition to (48–51): first equations (52) and (53) become

$$\frac{d\boldsymbol{r}}{dt} = f\alpha\boldsymbol{r} + \int_0^T Q_{qq}\boldsymbol{s}\,dt, \tag{58}$$

$$\boldsymbol{r}(0) = Q_{aa}\boldsymbol{s}, \tag{59}$$

for the representers, and from equations (57) and (55) we obtain

$$\frac{d\boldsymbol{s}}{dt} = -f\alpha\boldsymbol{s} - \boldsymbol{\mathcal{M}}[\delta], \tag{60}$$

$$\boldsymbol{s}(T) = 0, \tag{61}$$

for the "adjoints" of the representers.

From equation (56), one easily obtains the linear system

$$(\boldsymbol{\mathcal{M}}^{\mathrm{T}}[\boldsymbol{r}] + \boldsymbol{w}^{-1})\boldsymbol{b} = \boldsymbol{d} - \boldsymbol{\mathcal{M}}[\psi_F], \tag{62}$$

which can be solved as soon as the representer matrix $\boldsymbol{\mathcal{M}}^{\mathrm{T}}[\boldsymbol{r}]$ has been generated.

The equations (60) and (61) for $\boldsymbol{s}$ can now be solved as a sequence of final value problems, since they are decoupled from the forward equations (58) and (59) for the representers. As soon as $\boldsymbol{s}$ is found, the representers can be solved for. This provides the information needed for solving the system (62) for $\boldsymbol{b}$, and the final estimate is found by solving the Euler–Lagrange equation of the form

$$\frac{d\psi}{dt} = f\alpha\psi + \beta + \int_0^T Q_{qq}\lambda\,dt, \tag{63}$$

$$\psi(0) = \psi_0 + Q_{aa}\lambda(0), \tag{64}$$

$$\frac{d\lambda}{dt} = -f\alpha\lambda - \boldsymbol{\mathcal{M}}^{\mathrm{T}}[\delta]\boldsymbol{b}, \tag{65}$$

$$\lambda(T) = 0. \tag{66}$$

The numerical load is $2m+3$ model integrations, with m the number of measurements, but note that only 2 states need to be stored in space and time. If the solution is constructed directly from (46), all the representers need to be stored.

In a parameter estimation problem for a linear model the procedure is as follows:

1. Start with $\alpha^0 = \alpha_0$ and $\beta^0 = \beta_0$.
2. Solve the current iterate of the Euler-Lagrange equations (42–45) for ψ and λ using the representer method.
3. Use ψ and λ in the iterations (35) and (36) to find new estimates for the parameters.
4. Go back to step 2 and repeat until convergence.

The methodology outlined in this Section was used to estimate a vertical diffusion parameter and the wind-drag coefficient in a one-dimensional Ekman model in [14].

Note that this approach can be used also for the "alternative two" problem, and in the Appendix a recently developed approach for speeding up the representer method is presented. This new approach completely avoids the calculation of the representers. The low numerical cost for this algorithm actually makes the representer method a very attractive approach for solving inverse problems.

3.5. REPRESENTER SOLUTION FOR A WEAKLY NONLINEAR INVERSE PROBLEM

The representer method cannot be used directly to solve nonlinear inverse problems. However, if one can define a convergent sequence of linear iterates of the nonlinear model, the inverse problem for each linear iterate can be solved using the representer method. As an example consider the equation

$$\frac{\partial u}{\partial t} + u\frac{\partial u}{\partial x} = \cdots. \tag{67}$$

If the solution of this equation can be found from the iteration

$$\frac{\partial u_k}{\partial t} + u_{k-1}\frac{\partial u_k}{\partial x} = \cdots, \tag{68}$$

then one can also define a convergent sequence of linear inverse problems that can be solved exactly using representer expansions. In the papers [6],

[4], and [3], this approach was used. The major conclusion from these papers is that the representer method can be used with realistic atmospheric prediction models with an affordable numerical cost, and that the iteration on the Euler–Lagrange equation does converge quickly for the particular models used. The representer method can be characterized as the optimal solution method for linear and weakly nonlinear inverse problems, both in terms of the problem minimized (full weak constraint problem) and in terms of numerical cost.

3.6. DISCUSSION

Note that the equation for $\boldsymbol{b}$, (62), is similar to the one solved in the analysis scheme in the standard Kalman filter [15]. Actually the only difference is that here we have a covariance matrix with corresponding representers or influence functions in space and time, while in the Kalman filter the time dimension is excluded. Thus, the solution is sought in a similar form. In the Kalman filter, the representer or influence functions are defined as the measurements of the error covariance matrix ($\boldsymbol{MP}$), with $\boldsymbol{M}$ the measurement matrix and $\boldsymbol{P}$ the error covariance matrix. A similar interpretation can be used here if we define the error covariance $C(t_1, t_2)$ for the first guess solution, which results in an expression for the representers

$$\boldsymbol{r} = \mathcal{M}[C]. \tag{69}$$

It can be shown that the representers are the measurements of the reproducing kernel of the inner product defined by the penalty function, and the reproducing kernel is also the error covariance of the first guess solution [2].

It can also be shown that if we seek a solution of the form

$$\psi(t) = \psi_F(t) + \boldsymbol{b}^{\mathrm{T}}\boldsymbol{r}(t) + g(t), \tag{70}$$

where $g(t)$ is an arbitrary function orthogonal to the space spanned by the representers, then the minimizing solution must satisfy $g(t) \equiv 0$ [2].

This also shows that the solution is sought in the m-dimensional space spanned by the representers. Thus, we have reduced the infinite dimensional problem defined by the penalty function to an m-dimensional problem.

3.7. STRONGLY NONLINEAR INVERSE PROBLEMS

In highly nonlinear cases, one may obtain a solution using a brute force direct substitution method such as the gradient descent method used for the Lorenz model in [16]. If the penalty function $\mathcal{J}$ is discretized in time (and space if the spatial dimensions are included), it is relatively easy to derive an analytical expression for the gradient of $\mathcal{J}$ with respect to the full model state ψ and the parameter α, and a minimum can be found using a descent algorithm. It may of course be a local minimum if the penalty function becomes nonlinear enough, and in that case one may get a step further using statistical methods such as Hybrid Monte Carlo and annealing [17].

4. Model Bias Correction in a Sequential Data Assimilation System

Here we briefly review a different approach, recently developed, for dealing with model bias [13]. Those authors considered the effect of model bias on state estimates produced by a sequential data assimilation algorithm, and showed how such an algorithm can be modified to properly account for this effect. Their approach was motivated by the fact that the majority of current operational data assimilation systems used for weather prediction and climate studies involve sequential estimation algorithms. Typically in atmospheric four-dimensional data assimilation, a short-term (e.g., 6h) forecast is used as a background estimate for the true atmospheric state, and this background estimate is then combined with new observations to produce a so-called analysis. The analysis is subsequently used to initialize the forecast model, and the process is repeated. These assimilation systems are extremely large and complex, involving more than 10^6 state variables, and very little is known about the error characteristics of the forecast model. In this application, it is therefore of great practical importance that any algorithm for bias correction is not highly sensitive to assumptions about the nature of the bias, and that its implementation does not require a complete redesign of the assimilation system.

The main idea behind the sequential approach is that one can estimate the time-mean forecast error directly from the observations, assuming that the observations themselves are unbiased. This estimate can then be used to correct the forecast prior to producing the analysis. Thus, rather than estimating a set of model bias parameters, the algorithm estimates and corrects for the effect of the model bias upon the short-term forecast. As it turns out, sequential estimation of the time-mean forecast error can be accomplished by means of existing components of a sequential data assimilation system. It is this feature in particular that makes this algorithm interesting from a practical point of view.

4.1. NOTATION

Let the n-vector $\boldsymbol{\psi}_k^f$ denote a model forecast valid for time t_k, and $\boldsymbol{\psi}_k^t$ the unknown true state at that time. The *forecast error* is then simply

$$\boldsymbol{e}_k^f \equiv \boldsymbol{\psi}_k^f - \boldsymbol{\psi}_k^t. \tag{71}$$

For a m_k-vector $\boldsymbol{d}_k$ of measurements generated by a particular instrument at time t_k, the *observation error* is

$$\boldsymbol{\epsilon}_k \equiv \boldsymbol{d}_k - \boldsymbol{M}_k \boldsymbol{\psi}_k^t. \tag{72}$$

The linear m_k-vector function $\boldsymbol{M}_k$ is the *discrete forward observation operator*, mapping model variables to the data type associated with the instrument.

We use the following notation for the forecast error means and covariances:

$$\mathbf{b}_k^f \equiv \overline{\boldsymbol{e}_k^f}, \qquad \boldsymbol{P}_k^f \equiv \overline{(\boldsymbol{e}_k^f - \mathbf{b}_k^f)(\boldsymbol{e}_k^f - \mathbf{b}_k^f)^{\mathrm{T}}}. \tag{73}$$

We use the roman letter $\mathbf{b}$ here to distinguish forecast bias from model bias parameters $\boldsymbol{\beta}$; generally $\mathbf{b}_k^f = \mathbf{b}_k^f(\boldsymbol{\beta})$.

We assume that observations are unbiased, and

$$\boldsymbol{R}_k \equiv \overline{\boldsymbol{\epsilon}_k \boldsymbol{\epsilon}_k^{\mathrm{T}}}. \tag{74}$$

We do not consider possible covariances between forecast and observation errors here.

We may use the observation operator in (72) to compare forecasts with observations. The *observed-minus-forecast residuals* defined by

$$\boldsymbol{v}_k = \boldsymbol{d}_k - \boldsymbol{M}_k \boldsymbol{\psi}_k^f \tag{75}$$

contain information about the actual observation and forecast errors. This is clearly expressed by

$$\boldsymbol{v}_k = \boldsymbol{\epsilon}_k - \boldsymbol{M}_k \boldsymbol{e}_k^f, \tag{76}$$

which is obtained from (71) and (72).

The residual means and covariances are easily obtained from (76):

$$\overline{\boldsymbol{v}_k} \approx -\boldsymbol{M}_k \mathbf{b}_k^f, \tag{77}$$

$$\overline{(\boldsymbol{v}_k - \overline{\boldsymbol{v}_k})(\boldsymbol{v}_k - \overline{\boldsymbol{v}_k})^{\mathrm{T}}} \approx \boldsymbol{R}_k + \boldsymbol{M}_k \boldsymbol{P}_k^f \boldsymbol{M}_k^{\mathrm{T}}. \tag{78}$$

4.2. FORECAST BIAS ESTIMATION

We can rewrite (76) as

$$\boldsymbol{v}_k = -\boldsymbol{M}_k \mathbf{b}_k^f + \boldsymbol{\eta}_k, \tag{79}$$

where $\boldsymbol{\eta}_k$ is a random vector with first and second moments

$$\overline{\boldsymbol{\eta}_k} \approx 0, \tag{80}$$

$$\overline{\boldsymbol{\eta}_k \boldsymbol{\eta}_k^{\mathrm{T}}} \approx \boldsymbol{R}_k + \boldsymbol{M}_k \boldsymbol{P}_k^f \boldsymbol{M}_k^{\mathrm{T}}. \tag{81}$$

This follows from (76) and (78) by noting that $\boldsymbol{\eta}_k = \boldsymbol{v}_k - \overline{\boldsymbol{v}_k}$.

Equation (79) can be regarded as a *measurement model* for the forecast bias $\mathbf{b}_k^f$. It expresses the relationship between the observations, the forecast, and the forecast bias. For each fixed time t_k, the dimension m_k of $\boldsymbol{v}_k$ is generally less than the dimension n of $\mathbf{b}_k^f$, so that (79) cannot be inverted without additional information.

This is a familiar problem: incomplete observations of an unknown quantity must be supplemented with information from a *state model* in order to compensate for the lack of data. When the unknown is the state of a natural system (e.g., the Earth's atmosphere), some kind of predictive evolution model may be used for this purpose (e.g., a general circulation model). In case the unknown quantity is forecast bias, however, it is not immediately obvious what form the state model should take.

Generally the model should express any known or assumed spatial, temporal, and multivariate relationships satisfied by the bias field. For example, suppose that we wish to estimate the time-mean forecast error, averaged over a time period exceeding synoptic time scales. A reasonable state model for $\mathbf{b}_k^f$ is then simply *persistence*, i.e.,

$$\mathbf{b}_k^f = \mathbf{b}_{k-1}^f. \tag{82}$$

From (79), (80), and (82), we easily obtain a recursive estimate for the forecast bias $\mathbf{b}_k^f$:

$$\widehat{\mathbf{b}}_k^f = \widehat{\mathbf{b}}_{k-1}^f - \boldsymbol{L}_k \left[\boldsymbol{v}_k + \boldsymbol{M}_k \widehat{\mathbf{b}}_{k-1}^f \right] \tag{83}$$

where $\boldsymbol{L}_k$ is – for now – an arbitrary matrix. When initialized with an *a priori* estimate $\widehat{\mathbf{b}}_0^f$, equation (83) is a sequential algorithm for estimating the forecast bias associated with a data assimilation system. The algorithm uses the observed-minus-forecast residuals $\boldsymbol{\epsilon}_k$ produced by the system.

We can show [13] that the optimal choice of $\boldsymbol{L}_k$ is

$$\boldsymbol{L}_k = \boldsymbol{P}^b_{k-1}\boldsymbol{M}^{\mathrm{T}}_k \left[\boldsymbol{M}_k\boldsymbol{P}^f_k\boldsymbol{M}^{\mathrm{T}}_k + \boldsymbol{R}_k\right]^{-1} \tag{84}$$

with $\boldsymbol{P}^b_k$ the error covariance of the bias estimate $\widehat{\mathbf{b}}^f_k$:

$$\boldsymbol{P}^b_k \equiv \overline{(\widehat{\mathbf{b}}^f_k - \mathbf{b}^f_k)(\widehat{\mathbf{b}}^f_k - \mathbf{b}^f_k)^{\mathrm{T}}}. \tag{85}$$

It is possible to derive a recursion for $\boldsymbol{P}^b_k$ as well; combined with (84) this yields the Kalman filter for the system (79, 82). It is more practical, however, to specify the covariance matrix $\boldsymbol{P}^b_k$ based on simplifying assumptions. This is the usual procedure for obtaining the forecast and observation error covariances $\boldsymbol{P}^f_k$ and $\boldsymbol{R}_k$ for operational data assimilation systems.

An example of a class of bias estimation error covariance specifications is simply

$$\boldsymbol{P}^b_k = \gamma\boldsymbol{P}^f_k, \tag{86}$$

where γ is a scalar. Assuming that the forecast error covariances themselves are modeled in terms of a set of parameters $\boldsymbol{\alpha} \equiv (\alpha_1, \ldots, \alpha_N)$ [12], a more general class of specifications is

$$\boldsymbol{P}^b_k(\boldsymbol{\alpha}) = \gamma\boldsymbol{P}^f_k(\boldsymbol{\alpha}). \tag{87}$$

The parameter vector $\boldsymbol{\alpha}$ may include, for example, variance parameters, spatial decorrelation length scales, etc.

4.3. ON-LINE BIAS CORRECTION

We can apply the algorithm (83)–(84) to estimate forecast bias associated with any existing data assimilation system, using stored observed-minus-forecast residuals. Moreover, if the system involves sequential analysis updates, forecast bias can be estimated on the fly. We can then use the estimates to correct the biased forecasts on-line and to produce unbiased analyses. The complete algorithm is:

forecast: (88)

$$\psi^f_k = \boldsymbol{F}_k\widetilde{\psi}^a_{k-1} \tag{89}$$

$$\widetilde{\psi}^f_k = \psi^f_k - \widehat{\mathbf{b}}^f_{k-1} \tag{90}$$

bias update: (91)

$$\widehat{\mathbf{b}}^f_k = \widehat{\mathbf{b}}^f_{k-1} - \boldsymbol{L}_k\left[\boldsymbol{d}_k - \boldsymbol{M}_k\widetilde{\psi}^f_k\right] \tag{92}$$

$$\boldsymbol{L}_k = \boldsymbol{P}_k^b \boldsymbol{M}_k^{\mathrm{T}} \left[\boldsymbol{M}_k \boldsymbol{P}_k^f \boldsymbol{M}_k^{\mathrm{T}} + \boldsymbol{R}_k\right]^{-1} \tag{93}$$

$$\widetilde{\widetilde{\psi}}_k^f = \psi_k^f - \widehat{\mathbf{b}}_k^f \tag{94}$$

analysis: (95)

$$\widetilde{\psi}_k^a = \widetilde{\widetilde{\psi}}_k^f + \boldsymbol{K}_k \left[d_k - \boldsymbol{M}_k \widetilde{\widetilde{\psi}}_k^f\right] \tag{96}$$

$$\boldsymbol{K}_k = (\boldsymbol{P}_k^f - \boldsymbol{P}_k^b)\boldsymbol{M}_k^{\mathrm{T}} \left[\boldsymbol{M}_k(\boldsymbol{P}_k^f - \boldsymbol{P}_k^b)\boldsymbol{M}_k^{\mathrm{T}} + \boldsymbol{R}_k\right]^{-1}. \tag{97}$$

The operator $\boldsymbol{F}_k$ represents the forecast model; $\widetilde{\psi}_k^f$ and $\widetilde{\widetilde{\psi}}_k^f$ are the *a priori* and *a posteriori* bias-corrected forecasts, respectively, and $\widetilde{\psi}_k^a$ is the unbiased analysis. Note that for $\boldsymbol{P}_k^b = 0$ and with $\widehat{\mathbf{b}}_0^f \equiv 0$, the algorithm reduces to the familiar sequential analysis update equations.

5. Some Previous Parameter Estimation Works

The paper [21] pointed out several important issues regarding the formulation of the parameter estimation inverse problem. They used a quasi-geostrophic ocean model for the Gulf Stream extension and assimilated sea surface heights from Geosat radar altimeter data. A strong constraint formulation was used, in which the penalty function included the data misfit and a Laplacian smoothing term plus a quasi-geostrophic model as strong constraints, and it was minimized using the adjoint method. The control variables were the initial streamfunction for the model, the Rossby radius of deformation, the friction coefficient, and two parameters describing the wind-stress curl.

From the discussion in the previous sections this appears to be an "ill-posed" problem, since none of the first guesses are penalized. Without the smoothness constraint the problem was not well-posed, and it was possible to fit the observations closely and to obtain a penalty function smaller than the noise level. This also led to an estimate for the initial streamfunction which was very noisy and physically unacceptable. Furthermore, it could not be ensured that this was the only initial field fitting the observations. Thus, the problem was likely to be underdetermined.

By adding a Laplacian smoothness constraint on the initial conditions the problem becomes a "semi-norm". Thus the situation improves considerably, and a smooth estimate for the initial condition is found. However, it can still not be proved that this solution is unique, not even for linear dynamics. For example, assume that no observations are available. In this case there are infinitely many solutions. If a single observation is included, there will still be infinitely many solutions. One could continue adding observations, but it is impossible to know how many are needed to close the problem. One could be led to believe that if the number of observations is larger than the number of control variables, this would be sufficient. However, in general this is not the case. The reason is that the observations will normally not be independent. For example, two observations may be located close together (say within one Rossby deformation length). They will then provide some dependent information.

A very important result from this study was that when using the smoothness constraint it was impossible to estimate the diffusion parameter. The problem is that with a smooth initial condition the parameter attempts to become negative, which leads to an exponential growth of the noise that remains in the model solution. Clearly, the reason for this is that the noisier solution will have a better chance to get close to the noisy observations than a smooth model solution.

What has been learned from this work is that it is important to penalize the deviation from the first guesses of all control variables. This will ensure that the estimated amplitudes stay within a physically acceptable range. Furthermore, a regularization constraint is needed for all control variables. Here a smoothness constraint was used for the initial conditions. Alternatively such a constraint can be incorporated by using nondiagonal weights in the penalty function.

In the studies [24] and [25], some similar problems were encountered. In [24], a penalty function was defined in which the first guesses of the diffusion coefficient and the wind drag coefficient were penalized in addition to the residual between the model results and the observations. However, there was no penalty on the first-guess initial conditions. Without such a penalty, every choice of initial conditions can be used, and there may exist many initial conditions resulting in a solution that interpolates the data and gives a penalty function equal to zero as shown in [5]. [25] included such a term although the resulting "gradient" equation for the initial condition was incorrect. With a correct equation for the initial condition, the only missing issue would be the regularization of the vertical diffusion parameter. This is needed to ensure that a smooth estimate is found for the diffusion parameter.

The paper [14] reformulated this inverse problem using a weak constraint formulation with proper penalties on the first guesses and regularizations of all control parameters. The Euler–Lagrange equations were derived and solved by the methodology outlined in the previous sections using iterations on the parameters and the representer method.

There was a substantial difference in the values for the diffusion coefficient and the wind drag obtained in the two studies. It should be noted that [24] *replaced* the first-guess values of the parameters with the *current estimate* in each iteration of the parameters. These should of course be kept constant. Clearly, [24] solved a different inverse problem in each iteration and did not have any real penalty of the first guesses at all. Actually, it is not clear from their figures that the iterations did converge.

6. Summary

A general introduction has been given to the formulation and solution of the parameter estimation problem for dynamical models. A methodology has been presented in which the dynamical model with its initial condition and poorly known parameters are all treated as weak constraints in a variational inverse formulation. The variational formulation penalizes deviations from the exact model equation and distances from the first guesses for the control variables. The minimum of the variational functional is an estimate which almost satisfies the model equations, and at the same time is "close" to the observations and the first guesses of the initial condition and parameters. From this general weak constraint formulation it is easy to obtain the Euler–Lagrange equations for suboptimal or less general strong constraint formulations. The basic method proposed here for estimating poorly known parameters defines a gradient descent iteration for the parameters, while the remaining inverse problem (with the parameters assumed given) is solved using the representer method [2]. The method should be applicable for many problems as long as they are not too strongly nonlinear. For an extensive discussion of an application of the method, see [14], in which the vertical diffusion and the wind-drag were estimated using the method described here.

Appendix: Speeding Up the Representer Method

With a large number of observations, the standard formulation of the representer method becomes very expensive to compute. However there are ways to get around this problem as discussed in [3]. First make the following observations:

1. Only $\boldsymbol{b}$ is needed to compute the estimate from (63–66).
2. If a conjugate gradient method is used to solve (62), then the representer matrix $\boldsymbol{\mathcal{M}}^{\mathrm{T}}[\boldsymbol{r}]$ itself is not needed; only the representer matrix times a vector must be computed in each iteration.

Point 2 above can be illustrated by the following: first rewrite the linear system (62) for the coefficients $\boldsymbol{b}$ as

$$\boldsymbol{A}\boldsymbol{b} = \boldsymbol{h}, \tag{98}$$

with $\boldsymbol{A} = \boldsymbol{\mathcal{M}}^{\mathrm{T}}[\boldsymbol{r}] + \boldsymbol{w}^{-1}$ and $\boldsymbol{h} = \boldsymbol{d} - \boldsymbol{\mathcal{M}}[\psi_F]$. Then we define the quadratic functional

$$\mathcal{J}_{\boldsymbol{b}}[\boldsymbol{b}] = \frac{1}{2}\boldsymbol{b}^{\mathrm{T}}\boldsymbol{A}\boldsymbol{b} - \boldsymbol{b}^{\mathrm{T}}\boldsymbol{h}, \tag{99}$$

where the gradient with respect to $\boldsymbol{b}$ becomes

$$\nabla_{\boldsymbol{b}}\mathcal{J}_{\boldsymbol{b}} = \boldsymbol{A}\boldsymbol{b} - \boldsymbol{h}. \tag{100}$$

Thus, the minimum of $\mathcal{J}_{\boldsymbol{b}}$ is also where the gradient $\boldsymbol{A}\boldsymbol{b} - \boldsymbol{h} = 0$. A gradient descent iteration can now be defined as

$$\boldsymbol{b}^{k+1} = \boldsymbol{b}^{k} - \gamma_{\boldsymbol{b}}\left(\boldsymbol{A}\boldsymbol{b}^{k} - \boldsymbol{h}\right), \tag{101}$$

where $\gamma_{\boldsymbol{b}}$ is just a factor used in the gradient descent algorithm. Thus, the system (98) can be solved using the iteration given by (101), in which only matrix-vector multiplications need to be calculated. In practical settings a more sophisticated iteration should be used such as conjugate gradient methods.

Starting by measuring (46), we get

$$\boldsymbol{\mathcal{M}}[\psi(t)] = \boldsymbol{\mathcal{M}}[\psi_F(t)] + \boldsymbol{\mathcal{M}}^{\mathrm{T}}[\boldsymbol{r}(t)]\boldsymbol{b}. \tag{102}$$

If we assume a problem with the first guess solution $\psi_F \equiv 0$ (obtained by setting the initial condition equal to zero), we obtain

$$\boldsymbol{\mathcal{M}}[\psi(t)] = \boldsymbol{\mathcal{M}}^{\mathrm{T}}[\boldsymbol{r}(t)]\boldsymbol{b}, \tag{103}$$

and thus, measuring the minimizing solution is equivalent to computing the representer matrix times $\boldsymbol{b}$. Now, define an arbitrary estimate $\tilde{\psi}$, which is obtained from the Euler–Lagrange equations with $\psi_0 = 0$, $\beta = 0$, and by replacing $\boldsymbol{b}$ with $\boldsymbol{b}^k$; i.e., $\tilde{\psi}$ is the solution of the system

$$\frac{d\tilde{\psi}}{dt} - \tilde{\psi} = Q_{qq}\lambda, \tag{104}$$

$$\tilde{\psi}(0) = Q_{aa}\lambda(0), \tag{105}$$

$$\frac{d\lambda}{dt} + \lambda = -\boldsymbol{\mathcal{M}}^{\mathrm{T}}[\delta]\boldsymbol{b}^k, \tag{106}$$

$$\lambda(T) = 0, \tag{107}$$

obtained from (63–66). Then, measuring $\tilde{\psi}$ will result in a vector equal to $\boldsymbol{\mathcal{M}}^{\mathrm{T}}[\boldsymbol{r}(t)]\boldsymbol{b}^k$. Thus, in each iteration of the conjugate gradient algorithm, the system above is solved to calculate $\tilde{\psi}$ from a given $\boldsymbol{b}^k$. Then $\tilde{\psi}$ is measured to generate $\boldsymbol{\mathcal{M}}^{\mathrm{T}}[\boldsymbol{r}(t)]\boldsymbol{b}^k$, and hence $(\boldsymbol{\mathcal{M}}^{\mathrm{T}}[\boldsymbol{r}(t)] + \boldsymbol{w}^{-1})\boldsymbol{b}^k$, which is needed in the algorithm.

The numerical cost for finding the minimizing solution is now equal to $2j + 3$ model integrations, where j is the required number of iterations in the conjugate gradient algorithm to solve the system of linear equations (62). Note that the "real" dimension of this system is equal to the number of independent degrees of freedom in the observations, which in many cases is order of magnitudes less than the state variables to be estimated. Preconditioning can be used to speed up the conjugate gradient algorithm, i.e., if an approximation to the matrix of the linear system is available this will normally lead to a significant reduction in the number of iterations. In [3], an approach was used in which an approximation to the representer matrix was calculated on a coarse grid from $2m$ integrations, which can be performed relatively quickly.

Acknowledgement

This work has been supported by the EC–MAST–III BASYS project (MAS3–CT96–0058).

References

1. P. Bartello and H. L. Mitchell. A continuous three-dimensional model of short-range forecast error covariances. *Tellus, Ser. A*, 44:217–235, 1992.
2. A. F. Bennett. *Inverse Methods in Physical Oceanography*. Cambridge University Press, 1992.
3. A. F. Bennett, B. S. Chua, and L. M. Leslie. Generalized inversion of a global numerical weather prediction model. *Meteorol. Atmos. Phys.*, 60:165–178, 1996.
4. A. F. Bennett, L. M. Leslie, C. R. Hagelberg, and P. E. Powers. Tropical cyclone prediction using a barotropic model initialized by a generalized inverse method. *Mon. Weather Rev.*, 121:1714–1729, 1993.
5. A. F. Bennett and R. N. Miller. Weighting initial conditions in variational assimilation schemes. *Mon. Weather Rev.*, 119:1098–1102, 1990.
6. A. F. Bennett and M. Thorburn. The generalized inverse of a nonlinear quasi-geostrophic ocean circulation model. *J. Phys. Oceanogr.*, 22:213–230, 1992.
7. J. Carrera and S. Neuman. Estimation of aquifer parameters under transient and steady state conditions: 1. maximum likelihood method incorporating prior information. *Water Resour. Res.*, 22:199–210, 1986.
8. J. Carrera and S. Neuman. Estimation of aquifer parameters under transient and steady state conditions: 2. uniqueness, stability, and solution algorithms. *Water Resour. Res.*, 22:211–227, 1986.
9. J. Carrera and S. Neuman. Estimation of aquifer parameters under transient and steady state conditions: 3. application to synthetic and field data. *Water Resour. Res.*, 22:228–242, 1986.
10. G. Chavent. Identification of functional parameters in partial differential equations. In R. E. Goodson and M. Polis, editors, *Identification of Parameters in Distributed Systems*, pages 31–48. American Society of Mechanical Engineers, New York, 1974.
11. G. Chavent. Identification of distributed parameter system: About the output least square method, its implementation, and identifiability. In R. Isermann, editor, *Identification and System Parameter Estimation, Proceedings of the Fifth IFAC Symposium, Darmstadt, Federal Republic of Germany, 24–28 September 1979*, volume 1,

pages 85–97, 1979.

12. D. P. Dee. On-line estimation of error covariance parameters for atmospheric data assimilation. *Mon. Weather Rev.*, 123:1128–1145, 1995.
13. D. P. Dee and A. M. da Silva. Data assimilation in the presence of forecast bias. *Q. J. R. Meteorol. Soc.*, 1998. In press.
14. M. Eknes and G. Evensen. Parameter estimation solving a weak constraint variational formulation for an Ekman model. *J. Geophys. Res.*, 102(C6):12,479–12,491, 1997.
15. G. Evensen. Using the extended Kalman filter with a multilayer quasi-geostrophic ocean model. *J. Geophys. Res.*, 97(C11):17,905–17,924, 1992.
16. G. Evensen. Advanced data assimilation for strongly nonlinear dynamics. *Mon. Weather Rev.*, 125:1342–1354, 1997.
17. G. Evensen and N. Fario. A weak constraint variational inverse for the Lorenz equations using substitution methods. *J. Meteor. Soc. Japan*, 75(1B):229–243, 1997.
18. P. L. Houtekamer, L. Lefaivre, J. Derome, H. Ritchie, and H. L. Mitchell. A system simulation approach to ensembles prediction. *Mon. Weather Rev.*, 124:1225–1242, 1996.
19. A. H. Jazwinski. *Stochastic Processes and Filtering Theory*. Academic, San Diego, Calif., 1970.
20. I. M. Navon. Practical and theoretical aspects of adjoint parameter estimation and identifiability in meteorology and oceanography. *Dyn. Atmos. Oceans*, 27:55–79, 1997.
21. J. Schröter, U. Seiler, and M. Wenzel. Variational assimilation of Geosat data into an eddy-resolving model of the Gulf stream extension area. *J. Phys. Oceanogr.*, 23:925–953, 1993.
22. O. M. Smedstad and J. J. O'Brien. Variational data assimilation and parameter estimation in an equatorial Pacific Ocean model. *Prog. Oceanogr.*, 26:179–241, 1991.
23. D. R. Stauffer and J.-W. Bao. Optimal determination of nudging coefficients using the adjoint equations. *Tellus, Ser. A*, 45:358–369, 1993.
24. L. Yu and J. J. O'Brien. Variational estimation of the wind stress drag coefficient and the oceanic eddy viscosity profile. *J. Phys. Oceanogr.*, 21:709–719, 1991.
25. L. Yu and J. J. O'Brien. On the initial condition in parameter estimation. *J. Phys. Oceanogr.*, 22:1361–1364, 1992.
26. X. Zou, I. M. Navon, and F. X. L. Dimet. An optimal nudging data assimilation scheme using parameter estimation. *Q. J. R. Meteorol. Soc.*, 118:1163–1168, 1992.

ON THE LARGE-SCALE MODELING OF SEA ICE AND SEA ICE–OCEAN INTERACTIONS

THIERRY FICHEFET AND HUGUES GOOSSE
Institut d'Astronomie et de Géophysique Georges Lemaître,
Université Catholique de Louvain, Louvain-la-Neuve, Belgium

AND

MIGUEL A. MORALES MAQUEDA
Department of Mathematics,
Keele University, Keele, Staffordshire, United Kingdom

1. Introduction

Sea ice is ice that forms in the ocean as a result of the freezing of seawater. Unlike ice in calm freshwater, such as that found in quiescent lakes and slow current rivers, sea ice forms not only at the surface but also underwater. This is due to the fact that, for typical ocean salinities, the seawater density increases monotonically with decreasing temperature, and therefore surface cooling generates continued convective overturning until the entire oceanic mixed layer reaches the freezing temperature. At this stage, additional heat losses produce supercooled water and ice growth commences. Upon generation of the ice crystals, most of the salt content of seawater is expelled out of the ice lattice. A consequence of this salt rejection process is that, during sea ice formation, the salinity and hence the density of the surrounding waters increase. Convective mixing induced by the expulsion of salt carries relatively warm pycnocline water up into the mixed layer and may retard further freezing or even bring it to a halt. As freezing progresses, minute ice particles collect at the top few millimeters of the sea. This is known as frazil ice. If there is wind, swell, or wave action, frazil ice agglomerates into pan-shaped pieces, termed ice pancakes. Through welding and rafting (sliding of one piece of ice up over another), pancake ice amalgamates into larger blocks of ice, called ice floes, which are the basic constituents of the sea ice cover, or ice pack. Once the ice cover is formed, it will go on thickening mainly via accretion of congelation ice at the ice-water interface. As congelation ice forms, there is a tendency to trap brine in small pockets

E. P. Chassignet and J. Verron (eds.), Ocean Modeling and Parameterization, 399–422.

within the ice, the amount of salt retained increasing with the freezing rate. The presence of brine pockets considerably alters the mechanical strength, heat capacity, thermal conductivity, and electromagnetic properties of the ice. In the course of ice aging, the brine will slowly leak out through the coarse structure of sea ice. By the time it is one year old, sea ice normally has become relatively pure. A detailed account of the growth, structure, and properties of sea ice can be found in Weeks and Ackley (1986).

The fluxes of latent heat required to form or melt an ice cover of, say, 1 m thickness are of the same order of magnitude as typical interannual fluctuations of the surface shortwave and longwave radiative fluxes. Sea ice is therefore very sensitive to seasonal and interannual variations in the atmospheric and oceanic fields. At present, the extent of Arctic and sub-Arctic sea ice oscillates between September and March from $\sim 9 \times 10^6$ km^2 to $\sim 16 \times 10^6$ km^2. There is a marked longitudinal asymmetry in the distribution of the ice cover relating to the land-sea disposition. In the Southern Ocean, the winter ice edge stretches circumglobally around Antarctica. The seasonal range of ice extent is much greater than in the Northern Hemisphere: from $\sim 4 \times 10^6$ km^2 in late February to $\sim 19 \times 10^6$ km^2 in September. Interannual variability in total ice extent and in the timing of ice extent extremes is very large both in the Northern and Southern Hemispheres (Gloersen *et al.*, 1992). It must be realized that not all of the area encircled by the ice edge is actually covered by ice. Winds and currents keep the ice in a virtually ceaseless motion that pulls the ice apart and creates open water areas within the ice cover, known as leads. The abundance of leads is measured by the ice concentration, namely the fraction of ice-covered ocean in a given oceanic area. In the central Arctic, the ice concentration is about 0.97 in winter and declines to 0.80 or less in summer. In the stormy Southern Ocean, divergent motions and large oceanic heat fluxes keep the ice concentration below 0.90 all year long.

Regarding sea ice motion, three large-scale features characterize the ice drift in the Northern Hemisphere (Fig. 1a): the Beaufort Gyre, an anticyclonic gyre located in the Canadian Basin of the Arctic; the Transpolar Drift, which carries ice from the eastern Arctic through the North Pole to Fram Strait; and the East Greenland Drift, which moves southward with the East Greenland Current. In the Antarctic, two basin-scale motion features can be distinguished (Fig. 1b): the Antarctic Circumpolar Drift, a cyclonic circulation that transports ice from west to east all around Antarctica; and the East Wind Drift, an easterly coastal flow that also circumscribes the continent. The large-scale patterns of ice thickness are intimately associated with those ice motions. In the Arctic Ocean, there is a regular increase in winter ice thickness from the Eurasian sector toward the Canadian Archipelago and the north coast of Greenland, a pattern resulting

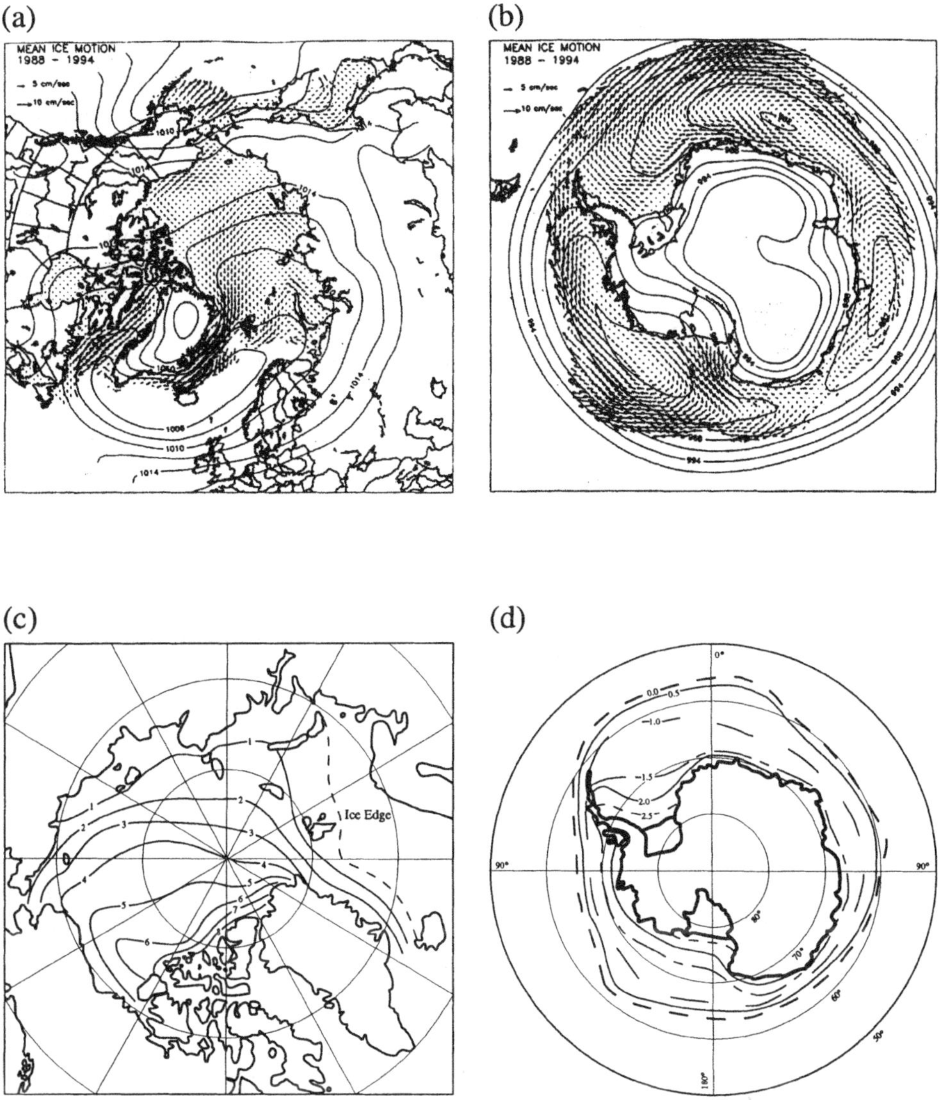

Figure 1. Mean sea ice motions in the Northern (a) and Southern (b) Hemispheres for 1988-94 (after Emery *et al.*, 1997). Sea level pressure isobars (in hPa) are shown as continuous lines. (c) Winter mean ice drafts (i.e., depths of underwater ice; in m) in the Northern Hemisphere (after Bourke and Garrett, 1987). (d) September mean ice thicknesses (in m) in the Southern Hemisphere (after Budd, 1991).

from advection by the Transpolar Drift (Fig. 1c). The average ice thickness is 3-4 m (Bourke and Garrett, 1987). In the Antarctic, the thickness of the September ice pack gradually decreases from 1.5-2.5 m near the coastline to about 0.5 m or less close to the ice edge, and the average ice thickness does not exceed 1 m (Budd, 1991).

Sea ice is a key constituent of the climate system as it profoundly modifies the surface momentum, heat, and mass exchanges at high latitudes. Due to its high albedo and low thermal conductivity, sea ice alters the radiative and turbulent components of the surface heat balance, cutting the absorption of shortwave radiation by as much as 80% in the case of snow-covered ice and reducing the turbulent heat fluxes by 1 to 2 orders of magnitude. In addition, sea ice intercepts most of the snow falling during winter, thus preventing it from immediately contributing to the ocean freshwater balance. Finally, a thick and compact ice pack also hinders the free exchange of momentum between atmosphere and ocean. Another facet of the influence of sea ice on polar climate is related to the fact that it alters the seasonal cycle of the atmospheric and oceanic fields. In the first place, the release and absorption of latent heat that respectively accompany the growth and decay phases of the ice cover tend to delay the seasonal surface temperature extremes. In the second place, the influx of salt into the ocean during ice formation and the corresponding freshwater input during ice melting tend to alter the density structure of the upper ocean and hence the ocean thermohaline circulation. The coupling of the ice drift to the aforementioned processes introduces new important phenomena. On the one hand, ice motion is responsible for the formation within the ice cover of open water areas through which air-sea exchanges are much enhanced compared to ice-covered regions. On the other hand, ice motion combines with ice production and destruction to create a net annual poleward transport of heat in the atmosphere and of salt in the ocean. A detailed discussion of the role of sea ice in shaping the climate of polar regions can be found in The Polar Group (1980).

It is believed that a number of positive and negative feedbacks linking sea ice to variations in surface albedo, cloud cover, atmospheric water vapor, and ocean thermohaline circulation play an important part in determining the Earth's climate and its sensitivity to external and internal perturbations (Dickinson, 1985). Indeed, atmosphere-ocean general circulation model (AOGCM) simulations of the climate transient response to increasing concentrations of greenhouse gases predict that the largest surface warmings, 3-7°C, will occur over the Arctic Ocean and sub-Arctic seas (e.g., Murphy and Mitchell, 1995). However, it is unfortunate that for technical and historical reasons, global climate models have so far included rather crude representations of both sea ice thermodynamics and dynamics, especially the latter. To what extent is the insufficient treatment of sea ice processes responsible for the accentuated sensitivity of these models over polar regions? Elucidating this point is a major challenge in current climatic change modeling.

The sea ice properties and processes that need to be taken into account in a comprehensive sea ice model can be divided into three categories: (1) thermodynamic processes, (2) dynamic processes, and (3) processes that couple ice thermodynamics and dynamics. The thermodynamic component of the model determines the freezing/melting rate of ice given the atmospheric and oceanic heat and mass fluxes. Ice thermodynamics can be subdivided into two components, one associated with vertical processes and one dealing with lateral processes. The former aims at calculating the accretion/ablation rates at the top and bottom boundaries of the ice, whereas the latter determines the changes in ice concentration that result from ice-ocean heat fluxes in leads. The dynamic component of the model consists of a momentum balance equation, which determines the ice drift as a function of the atmospheric and oceanic dynamical forcings, and an ice rheology, which prescribes the dependence of internal ice stresses on ice deformation and ice thickness characteristics. The thermodynamic-dynamic coupling is done via (1) an advection scheme, which works out the transport of volumetric ice variables, such as ice mass and sensible and latent heat contents, and (2) the ice thickness distribution formulation. The ice thickness distribution describes the evolution of the different thickness categories into which ice breaks as a result of dynamical processes.

Below we present an overview of current sea ice and sea ice–ocean interaction modeling efforts. Section 2 reviews the state-of-the-art parameterizations of the aforementioned components of large-scale sea ice models. Section 3 discusses the numerical treatment of sea ice–ocean interactions. In both cases, we examine the relative importance of the different sea ice and sea ice–ocean exchange processes for realistically simulating the behavior of the sea ice–ocean coupled system.

2. Modeling the Sea Ice Processes

2.1. SEA ICE THERMODYNAMICS

Ice formation, accretion, and ablation are all ultimately determined by the heat and mass fluxes at the surface of the ocean and at the air-ice and ice-water interfaces. The surface energy budget includes shortwave and longwave radiative forcings, sensible and latent heat exchanges, and conductive heat flux at the top ice boundary. At the ice bottom and lateral walls, the heat balance is established between the heat conduction within the ice and the oceanic sensible heat flux.

The small amount of brine (~15 ppt at most) that remains trapped in the ice after freezing has a large impact on the effective heat capacity of sea ice. This is because temperature variations inside ice cause brine pockets to either melt or freeze. The latent heat fluxes involved in these phase changes tend to retard ice heating in summer and ice cooling in autumn. In addition, brine has the important effect of decreasing heat conduction. However, the major factor in controlling heat conduction through ice is the accumulation of snowfall. Snow has a heat conductivity which is 5-6 times smaller than that of ice, and therefore the snow cover acts as a thermal shield between air and ice. This is not the only important effect of snow. Snow also modifies the surface albedo, reducing the amount of absorbed solar radiation by 30-40% with respect to that of snow-free ice. Another snow-related process is the formation of snow ice, i.e., the transformation of snow into ice due to flooding of the ice floes. When the load of snow depresses the snow-ice boundary underwater, seawater infiltrates the submerged snow and freezes, forming a snow ice cap and releasing latent heat and salt. The contribution of snow ice to the total mass balance of the Antarctic ice cover can be as large as 15-37% in the Ross Sea (Jeffries and Adolphs, 1997). Finally, one more physical process manifests itself during the decay phase of the ice cover: melt ponding. Melt ponds are small bodies of meltwater retained at the surface of the ice. These features substantially decrease the surface albedo, thus contributing to the acceleration of the thawing process (Ebert and Curry, 1993).

The role of leads in a thermodynamic ice cover is two-fold. In summer, they act in a similar way to melt ponds, considerably increasing, by their low albedo, the amount of solar radiation available for ice melting (Maykut and McPhee, 1995). In fall and winter, leads act as ice factories, since turbulent heat losses to the atmosphere are up to 2 orders of magnitude greater over open water than over ice.

2.1.1. *Vertical growth and decay of sea ice*

Energy transfer within snow and ice is determined from the vertical heat diffusion equation:

$$\rho c_p \frac{\partial T}{\partial t} = k \frac{\partial^2 T}{\partial z^2} + \kappa i_0 \ (1 - \alpha_{su}) F_{sw} \exp(-\kappa \ z), \tag{1}$$

where ρ, c_p, and k are respectively the density, specific heat, and thermal conductivity of the material (snow or ice), T is temperature, t is time, z is the downward vertical coordinate, i_0 is the fraction of net shortwave radiation that penetrates the snow or bare ice, α_{su} is the albedo, F_{sw} is the incoming shortwave radiation, and κ is the bulk extinction coefficient. Boundary conditions for (1) are obtained from the heat budget equations at the ice top and bottom. At the upper surface, the energy balance is given by

$$\begin{aligned} B_{su}(T_{su}) &= (1 - i_0)(1 - \alpha_{su}) F_{sw} + \varepsilon_{su} F_{lw} - \varepsilon_{su} \sigma T_{su}^4 + F_h + F_{le} - F_{cs} \\ &= 0, \end{aligned} \tag{2}$$

where F_{lw} is the incoming longwave radiation, ε_{su} is the emissivity, σ is the Stefan-Boltzmann constant, F_h and F_{le} are the turbulent fluxes of sensible heat and latent heat, respectively, and F_{cs} is the conductive heat flux from below the surface. Downward fluxes are taken to be positive. If the balance in (2) requires the surface temperature T_{su} to be above the melting point, T_{su} is held at that point and the excess of energy leads to snow or ice melting:

$$\left(\frac{\partial h_s}{\partial t}\right)_{su,abl} = -\frac{B_{su}\left(T_{su}\right)}{L_s}, \text{ if } h_s > 0; \; \left(\frac{\partial h_i}{\partial t}\right)_{su,abl} = -\frac{B_{su}\left(T_{su}\right)}{L_i}, \text{ if } h_s = 0, \tag{3}$$

where the subscript *abl* refers to ablation, h_s and h_i are the snow and ice thicknesses, respectively, and L_s and L_i are the volumetric latent heats of fusion of snow and ice, respectively. At the base of the ice, any imbalance between the conductive heat flux at the ice bottom F_{cb} and the heat flux from the ocean to the ice F_w is compensated by accretion (*acc*) or ablation of ice:

$$\left(\frac{\partial h_i}{\partial t}\right)_{b,acc-abl} = \frac{F_{cb} - F_w}{L_i}. \tag{4}$$

The above four equations are the backbone of all thermodynamic sea ice models. Models can nevertheless differ from one another regarding their treatment of processes such as the time evolution of snow and ice albedos, penetration of solar radiation into the ice, latent heat fluxes within ice associated with brine pocket phase changes, and snow ice formation. The model of Maykut and Untersteiner (1971; hereafter referred to as MU71) stands as one of the most elaborate thermodynamic models developed so far. It uses a small vertical increment (0.1 m), which allows for an accurate resolution of (1). The surface albedo is prescribed as a monthly-varying parameter, and the summer albedo does not take into account the presence of melt ponds over melting ice. The net shortwave radiation entirely contributes to surface melting if a snow cover exists, while 17% of this radiation penetrates the ice if there is no snow. In order to simulate the effects of brine within the ice, the volumetric specific heat and thermal conductivity of the ice are made dependent on both temperature and salinity, the vertical distribution of the latter being specified from documented equilibrium profiles. The model does not include a parameterization of snow ice formation.

In the last few years, improved variants of the MU71 model have been presented. Ebert and Curry (1993) introduced a parameterization of the surface albedo that takes into account the dependence of albedo on cloud coverage and on the state of the surface (dry or melting snow, bare ice with/without meltwater ponds). They questioned the value of the fraction of shortwave radiation absorbed at the surface in the MU71 model and suggested that solar absorption should roughly double that of MU71. They also included in their model a parameterization of meltwater ponds. Eicken *et al.* (1995) incorporated a semi-empirical salinity model in the MU71 model as well as a parameterization of snow ice formation.

Semtner (1976; hereafter referred to as S76) proposed a number of simplifications of the MU71 model in order to make it more suitable for long-term integrations. The vertical resolution is drastically reduced: in all, the model consists of three layers of variable thickness, one for snow and two for ice (3-layer model). The treatment of brine pockets significantly differs from that in MU71. During snow-free periods, the penetrating solar radiation absorbed by the ice is stored in a heat reservoir, which represents internal melting (the second term on the right-hand side of (1) is therefore removed). The reservoir has a maximum storage capacity of, say, 30-50% of the heat required to melt all the ice. Once the reservoir is replete, the energy excess goes toward enhancing surface melting. At the onset of the cold season, energy is extracted from the reservoir to prevent the temperature of the upper layer of the ice from falling below the freezing point, thereby simulating the refreezing of brine pockets. The ice volumetric specific heat and thermal conductivity are kept constant; therefore no vertical profile of ice salinity is required. S76 introduced an even simpler model in which the storage of sensible and latent heat is ignored (0-layer model). By adequately tuning the surface albedo and the thermal conductivities of snow and ice, the 0-layer model was made capable of producing an annual mean ice thickness fairly close to that of the 3-layer model. In contrast, the amplitude and phase of the seasonal cycle of ice thickness were poorly represented. Due to their simplicity and their ability to incorporate the most relevant thermodynamic processes, the 3- and 0-layer models of S76 have been, and are being, widely used.

The importance of the above-mentioned processes for an adequate representation of the seasonal cycle of the ice cover can be appreciated in a series of sensitivity experiments with the 3-layer thermodynamic-dynamic sea ice model of Fichefet and Morales Maqueda (1997; hereafter referred to as FMM97). Fig. 2a shows the seasonal variations of the Northern Hemisphere ice volume for a control run with the full model and a sensitivity run with a 0-layer version of the model. Neglecting the sensible heat capacity of ice and the latent heat storage in brine pockets leads to an appreciable reduction in total ice volume. This is basically due to the absence in the 0-layer model of the moderating effect of brine pockets on summer surface melting. This is made evident in Fig. 2b, where the volume cycle from the control run is displayed against that from a run in which the sensible heat storage contribution is retained but not the contribution from latent heat. Without the "brine damping" mechanism, all of the absorbed shortwave

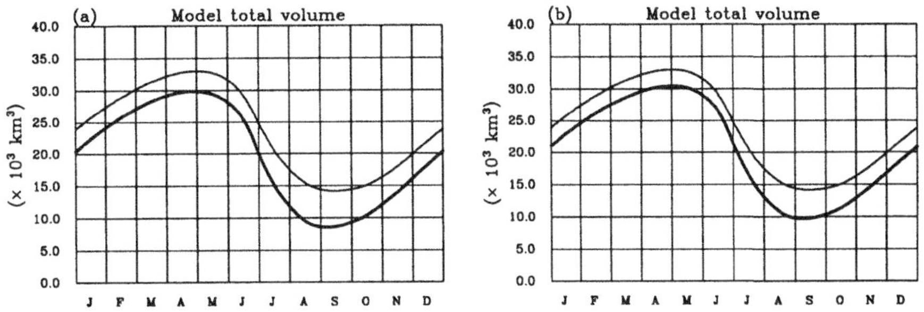

Figure 2. (a) Simulated seasonal cycles of total ice volume in the Northern Hemisphere from a 0-layer model (thick curve) and a 3-layer model (thin curve). (b) Same as Fig. 2a, except for a model with sensible heat capacity but no latent heat storage (thick curve) and a 3-layer model (thin curve).

energy is directed to surface ice ablation, which produces a noticeable decrease in the equilibrium ice volume. Owing to the fact that the Antarctic ice pack is perennially snow covered, little shortwave radiation enters the ice. Consequently, the modeled Southern Ocean ice cover is rather insensitive to brine pocket effects.

The role of the snow cover in determining the thermodynamic growth and decay of sea ice has also been investigated by FMM97 in two sensitivity experiments. In the first, snow accumulation on top of the ice was set to zero. The main effect of ignoring the snow cover is that the winter ice volume increases. In fact, the absence of snow induces stronger cooling of the ice surface during the cold season, which stimulates heat conduction and hence basal accretion. Since snow melts away fairly rapidly in the Northern Hemisphere at the beginning of the melting season, the Arctic summer ice is almost insensitive to the neglect of the snow cover. By contrast, the volume of the summer ice cover in the Southern Ocean is reduced due to (1) the increase in surface albedo and (2) the absence of snow ice formation. In a second experiment, snow was incorporated back into the simulation, but snow ice formation was not permitted. While this has no influence on the snow/sea ice cover in the Northern Hemisphere, where snow ice formation is very rare, the annual mean volume of the Southern Hemisphere ice was reduced by $\sim$10%.

2.1.2. *Lateral growth and decay of sea ice*

The parameterization of open water within the pack requires the introduction of a new variable, the ice concentration A. Time variations in A depend on the heat budget of the lead area B_l. The computation of B_l involves the balance between the net surface heat flux and the vertical oceanic heat flux. If the model allows for a horizontal temperature gradient between the under-ice water and the water in leads, horizontal mixing will also contribute to B_l (e.g., Parkinson and Washington, 1979).

If $B_l < 0$, ice of thickness h_0 is accumulated onto the side of the existing ice:

$$\left(\frac{\partial A}{\partial t}\right)_{acc} = -\frac{(1-A)\,B_l}{L_i h_0}. \tag{5}$$

In general, h_0 depends on the time step δt. If δt is on the order of 1 day, a reasonable value for h_0 is ~0.5 m. In many large-scale sea ice models, a minimum lead fraction is prescribed to simulate the opening of leads owing to unresolved dynamical effects (e.g., Parkinson and Washington, 1979; Darby and Willmott, 1993). However, a better way of dealing with this effect is to take into consideration the contribution of shearing deformation to the formation of open water (Stern and Rothrock, 1995).

In early sea ice models (Parkinson and Washington, 1979), changes in A during the melting period were considered to be due only to ablation at the ice lateral walls. Another approach postulates that heat gains in leads contribute entirely to basal melting. If it is assumed that the ice thickness over the fractional area A is uniformly distributed between 0 and $2\bar{h}_i$ (where $\bar{h}_i$ is the mean ice thickness) and that ablation rates are independent of the ice thickness, bottom melting leads to an attending reduction in A given by

$$\left(\frac{\partial A}{\partial t}\right)_{abl} = -\frac{A}{2\bar{h}_i}\,\Gamma\left(-\frac{\partial \bar{h}_i}{\partial t}\right), \tag{6}$$

where Γ is the Heaviside step function. This parameterization, or variants thereof, have been used in numerous studies (e.g., Darby and Willmott, 1993; FMM97). Note that this formulation of the thermodynamic opening of leads as an effect of ice basal melting lends itself to straightforward incorporation in multi-level models (i.e., models in which more than one ice thickness category are considered in each grid point; see Section 2.3).

Fig. 3 illustrates the crucial role played by leads in shaping the seasonal cycle of the ice cover. It shows the seasonal evolutions of the total ice volume in the Northern and Southern Hemispheres in a simulation with the model of FMM97, in which the ice concentration was prescribed to be unity. A remarkable decrease in ice volume results from the omission of leads. This is due to the dramatic slowdown of the ice growth during fall and winter as a consequence of the reduction in turbulent and radiative longwave heat losses at the surface.

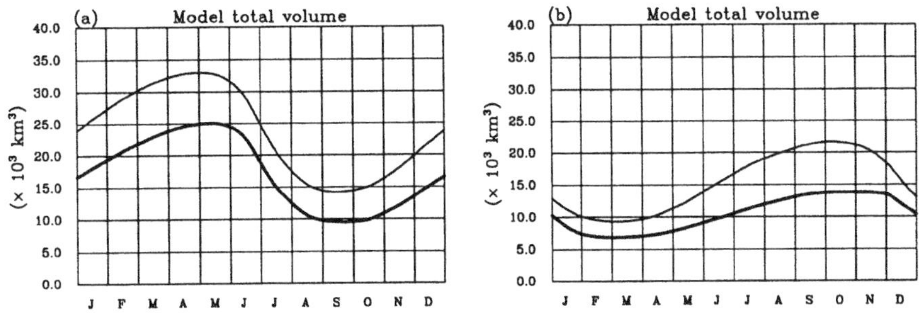

Figure 3. (a) Seasonal cycles of total ice volume in the Northern Hemisphere for a simulation without leads (thick curve) and a full-model simulation (thin curve). (b) Same as Fig. 3a, except for the Southern Hemisphere.

2.2. SEA ICE DYNAMICS

At large scale, sea ice can be considered as a two-dimensional continuum moving in response to atmospheric and oceanic dynamical forcings. The ice momentum balance is established between the forces generated by the winds, the oceanic currents, and the tilt of the sea surface, on one hand, and the ice inertia, Coriolis force, and internal ice stresses, on the other hand. The internal ice forces arise from the interaction between ice floes in relative motion, and are ultimately determined by the mechanical behavior of individual floes. Ice-ice interactions act as integrators of the external air and water forcings, and are a decisive factor in determining the ice drift characteristics. Furthermore, they are also responsible for the creation of material failures in the ice which are at the origin of the opening of leads within the ice cover (Tremblay and Mysak, 1997).

2.2.1. *Ice momentum balance*

The sea ice momentum balance can be written as

$$m\left[\frac{\partial \mathbf{u}_i}{\partial t} + (\mathbf{u}_i \cdot \boldsymbol{\nabla})\,\mathbf{u}_i\right] = -mf\mathbf{k} \times \mathbf{u}_i + \boldsymbol{\tau}_{ai} + \boldsymbol{\tau}_{wi} - mg\boldsymbol{\nabla}\xi + \mathbf{F}, \qquad (7)$$

where m is the snow-plus-ice mass per unit area, $\mathbf{u}_i$ is the sea ice velocity, f is the Coriolis parameter, $\mathbf{k}$ is the upward unit normal, $\boldsymbol{\tau}_{ai}$ and $\boldsymbol{\tau}_{wi}$ are the forces per unit area due to air and water drags respectively, g is the acceleration due to gravity, ξ is the sea surface dynamic height, and $\mathbf{F}$ is the force per unit area due to variations in internal ice stress.

Scale analysis indicates that ice inertial forces are very weak compared to other forces acting on the pack and can therefore be neglected (unless one is interested in the study of inertial oscillations). According to Steele *et al.* (1997), the dominant terms of the ice momentum balance in the Arctic are the air and water stresses and the ice interaction term. The Coriolis and surface tilt forces are much weaker, although the latter is non-negligible on times scales longer than 1 year. Due to the largely divergent ice flow in the Antarctic, the ice interaction term is probably less important there.

2.2.2. *Ice interaction and ice rheology*

The internal ice force is expressed in terms of the divergence of the two-dimensional stress tensor, $\mathbf{F} = \nabla \cdot \boldsymbol{\sigma}$. For a two-dimensional isotropic continuum, the most general constitutive law is given by

$$\sigma = 2\eta\dot{\epsilon} + \left[(\zeta - \eta)\, T(\dot{\epsilon}) - \frac{P}{2}\right] \mathbf{I}, \tag{8}$$

where η and ζ are shear and bulk viscosities respectively, $\dot{\epsilon}$ is the strain rate tensor, $T(\dot{\epsilon})$ is the trace of $\dot{\epsilon}$, P is the ice strength, and $\mathbf{I}$ is the two-dimensional unity tensor. The variables ζ, η, and P all depend on the two invariants of the tensor $\dot{\epsilon}$. The particular formulation of such functional dependence characterizes the ice rheology.

The nonlinear viscous-plastic rheology of Hibler (1979) is the most commonly used so far. In this rheology, viscous creep occurs when the stress state lies inside the region bounded by an elliptical yield curve in principal stress space, while when the stress state is located on the yield curve, plastic flow takes place. Under divergent ice motion, the viscous-plastic rheology does not allow tensile stresses to occur (the ice is said to move in "free drift"). In contrast, the ice resists both convergent and shear deformations. A widely used parameterization of the maximum compressive stress is $P = P^* A\bar{h}_i \exp[-C\ (1 - A)]$, where P^* and C are empirical parameters. A simplification of the viscous-plastic rheology in which shear stresses are neglected was proposed by Flato and Hibler (1992). This rheology is termed the cavitating fluid rheology. The cavitating fluid formulation yields reasonable results if shearing is weak, but it can lead to a large overestimation of the ice velocity and to anomalous ice motion fields when shear stresses are large (FMM97; Steele *et al.*, 1997). This is illustrated in Figs. 4 and 5, which show the differences in annual mean ice velocity and ice thickness in both hemispheres between experiments employing the cavitating fluid and viscous-plastic rheologies. Anomalous patterns of ice circulation and ice thickness are generated by the cavitating fluid rheology in vast regions of the western Arctic and the western Southern Ocean. The simulated annual mean ice velocity and ice thickness fields in the viscous-plastic rheology experiment are also plotted. Note that these modeled fields favorably compare with the observed fields (Fig. 1).

Recently, Tremblay and Mysak (1997) have introduced a dynamic model based on a granular material rheology. This is an appealing parameteriza-

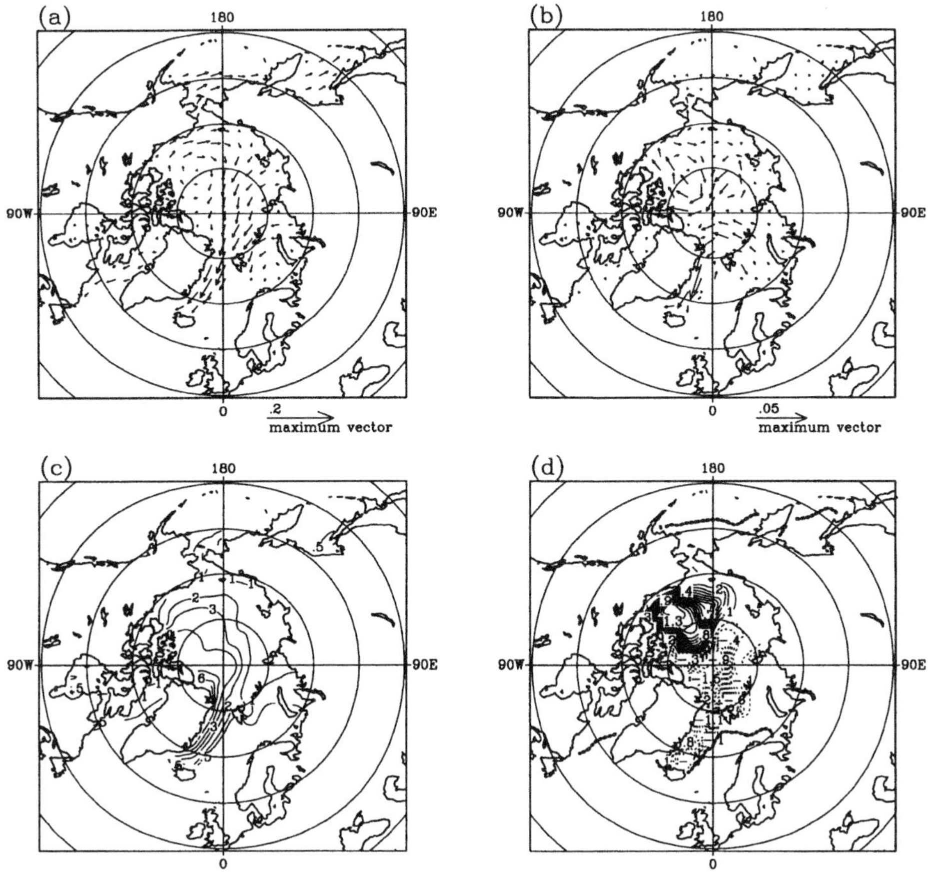

Figure 4. (a) Simulated Northern Hemisphere annual mean ice velocities in an experiment using the viscous-plastic rheology. Scale vector is 0.2 m s^{-1}. (b) Differences in Northern Hemisphere annual mean ice velocity between an experiment using the cavitating fluid rheology and the experiment shown in (a). Scale vector is 0.05 m s^{-1}. Plots (c) and (d) are the same as (a) and (b), respectively, except for the annual mean ice thickness. Contour intervals are 0.5 and 0.1 m, respectively. The thick outer line in (d) is the simulated annual mean ice edge.

tion because it explicitly accounts for the fact that sea ice is a collection of individual ice floes. In particular, the opening of leads as a result of shear deformation is naturally embedded in the formulation.

2.3. THERMODYNAMICS-DYNAMICS COUPLING, ICE THICKNESS DISTRIBUTION

Ice thermodynamics and dynamics mutually modify each other. The ice accretion/ablation rate depends on ice thickness, with open water and thin

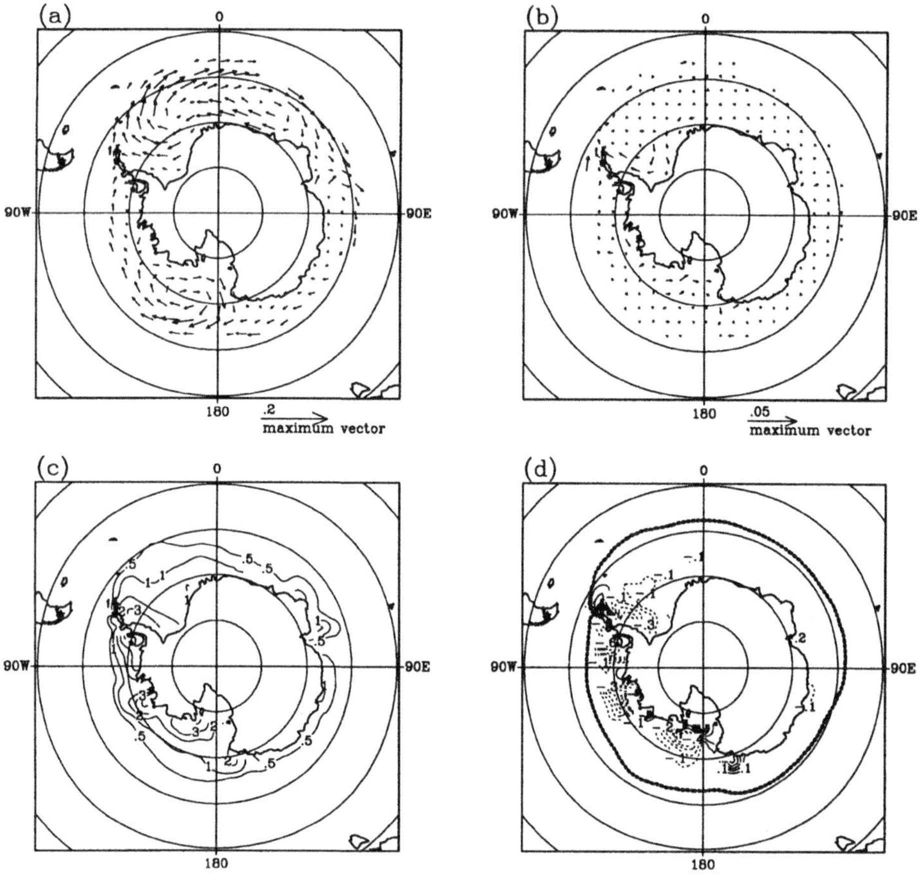

Figure 5. As in Figure 4, except for the Southern Hemisphere.

ice regions experiencing faster ice growth in winter and more intense melting in summer than thick ice regions. Thus, at large scale, the location of regions of ice growth and decay is intimately related to the geographical distribution of ice thickness, which depends in turn on ice advection patterns. At small scale as well, dynamical opening and closing of leads and build-up of pressure ridges have similar effects in altering the thermodynamic evolution of ice. Conversely, ice motion characteristics are influenced by those of ice thickness, which largely control the amount of stress that ice can transmit.

The coupling between thermodynamics and dynamics is encapsulated in a system of equations that express the conservation of volumetric ice variables such as snow/ice volume, of sensible heat content per unit area, and of brine latent heat content per unit area. If Ψ represents any of these quantities, the generic law is

$$\frac{\partial \Psi}{\partial t} = -\nabla \cdot (\mathbf{u}_i \Psi) + S_\Psi, \tag{9}$$

where S_Ψ is the rate of change of Ψ due to thermodynamic processes.

A problem that obviously arises here is that the continuity equation for ice volume does not allow, by itself, the determination of the mean ice thickness and concentration, let alone a distribution of ice thicknesses. Some hypotheses must therefore be adopted regarding the way in which ice thickness and concentration are affected by advection, in order to derive these quantities from the continuity equation. To achieve this goal, most sea ice models to date have postulated a 2-level ice thickness distribution, in which the ice cover at a given point is described by two variables, its mean thickness $\bar{h}_i$ and its concentration A. The evolutions of $A\bar{h}_i$ and A are then determined by the conservation law (9) with the appropriate thermodynamic terms (as discussed in Section 2.1) and with the constraint $0 \leq A \leq 1$.

A more complete approach to this problem consists of defining a thickness distribution G. The thickness distribution is a function of the ice thickness h_i, such that $\int_{h_0}^{h_1} G \, dh_i$ determines the fractional area covered by ice of thickness $h_0 \leq h_i \leq h_1$ (Thorndike *et al.*, 1975). Note that $\bar{h}_i = \int_\epsilon^\infty h_i G \, dh_i$ and $A = \int_\epsilon^\infty G \, dh_i$, where ϵ is an arbitrarily small positive number. The continuity equation for the ice volume can now be written as follows:

$$\frac{\partial G}{\partial t} = -\nabla \cdot (\mathbf{u}_i G) - \frac{\partial (sG)}{\partial h_i} + \Phi, \tag{10}$$

where s is the thermodynamic growth/melt rate for ice of thickness h_i and Φ is a redistribution function. The redistribution function depends on h_i and on G itself. It describes changes in G due to deformation events (lead opening or closing and ice rafting and ridging), thus redistributing ice from one thickness category to another. Although there is flexibility in the parameterization of Φ, it must satisfy a number of physical constraints, namely conservation of area (since $\int_0^\infty G \, dh_i = 1$), volume, and energy. Ice models that adopt this approach are called multi-level models (e.g., Hibler, 1980). Although computationally expensive, multi-level models allow for a greatly improved representation of the ice cover compared to 2-level models. For this reason, as computing power increases, they are expected to supersede 2-level models.

3. Modeling Sea Ice–Ocean Interactions

Exchanges of momentum, heat, and mass between sea ice and ocean take place both at the ice lateral walls and at the ice base. Lateral heat and mass exchanges have been succinctly examined in Section 2.1 and shown to be very important. In the remainder of this section, only vertical exchanges taking place at the ice bottom will be considered. Nevertheless, it is interesting to note that lateral ice-ocean momentum transfer, although neglected in most models, can be very significant as well. Steele *et al.* (1989) have performed an extensive study of the role of floe size and ice concentration on effective ice-ocean stresses. They have demonstrated that neglecting the form drag of the leading edge of the ice can induce a strong overestimation of the ice velocity when the ice concentration is low ($A < 0.3$) and the floe size is small.

An additional comment regards the treatment of volume changes due to ice growth and melt in ice-ocean models. In the majority of models, the vertical extent of the topmost oceanic grid box is not affected by ice; in other words, ice is dealt with as if it were resting on top of the ocean. Although taking into account the actual volume changes caused by the presence of sea ice is a conceptually more satisfactory approach, it requires tackling a number of delicate technical problems and probably does not represent a major practical advantage.

3.1. EXCHANGE OF MOMENTUM

The ice-ocean stress is usually defined by a classical quadratic drag law (McPhee, 1980; Shirasawa and Ingram, 1991) which takes the form

$$\boldsymbol{\tau_{iw}} = \rho_w c_{iw} \left|\boldsymbol{u_i} - \boldsymbol{u_w}\right| (\boldsymbol{u_i} - \boldsymbol{u_w}) , \tag{11}$$

where ρ_w is the seawater density, c_{iw} is the drag coefficient, and $\boldsymbol{u_w}$ is the surface oceanic current. Models in which the uppermost ocean velocity vector is determined at a depth sufficiently close to the surface (maximum 2-5 m deep) can utilize this parameterization (e.g., Häkkinen and Mellor, 1992; Goosse, 1997). It must be stressed, however, that c_{iw} depends on the precise depth at which the oceanic current is given, and its value must be deduced from observations made well below the deepest ice protrusions (ice keels or hummocks) so that their effect on the total drag force is appropriately accounted for (McPhee *et al.*, 1987; Shirasawa and Ingram, 1991).

Most ocean models have too coarse a vertical resolution for the above parameterization to be applicable. Instead, the formulation

$$\boldsymbol{\tau_{iw}} = \rho_w c_g \left|\boldsymbol{u_i} - \boldsymbol{u_g}\right| \left[\cos\theta\,(\boldsymbol{u_i} - \boldsymbol{u_g}) + \sin\theta\, \boldsymbol{k} \times (\boldsymbol{u_i} - \boldsymbol{u_g})\right] \tag{12}$$

is used (Hibler and Bryan, 1987). Here c_g is the geostrophic drag coefficient, $\boldsymbol{u_g}$ is the geostrophic ocean current at the surface, and θ is a turning angle that takes into account the cross-isobaric veering of the Ekman boundary layer. It is assumed that ocean currents at a depth of 50-100 m give a reasonable approximation of u_g. For simplicity, the drag coefficient and turning angle are generally taken as constant, with typical values of $c_g = 5.5 \times 10^{-3}$ and $\theta = 20°$ (McPhee, 1980). However, it has been demonstrated that c_g and θ strongly depend on the vertical structure of the water column as well as on the buoyancy forcing (McPhee, 1983; Mellor *et al.*, 1986).

3.2. EXCHANGE OF HEAT

Ice-ocean heat exchange is ultimately achieved through a very thin viscous sublayer, on the order of 1 millimeter thick, in which molecular diffusion is the dominant mechanism for heat transfer. Since it is not possible to resolve this sublayer in a three-dimensional model, its effect must be parameterized. Various approaches have been suggested (McPhee *et al.*, 1987). McPhee (1992) has proposed a formulation in which the ice-ocean heat flux is expressed as a simple function of the ocean temperature and the turbulent mixing:

$$F_w = \rho_w c_{pw} c_h u_{*iw} \left(T_{ml} - T_{fw} \right), \tag{13}$$

where c_{pw} is the specific heat of seawater, c_h ($= 6 \times 10^{-3}$) is the exchange coefficient, u_{*iw} is the friction velocity, T_{ml} is the mixed layer temperature, and T_{fw} is the freezing point of seawater, a function of ocean salinity. (For a surface salinity of 34 ppt, T_{fw} has a value of $-1.8°$C. T_{fw} decreases by ~0.056°C for an increase of 1 ppt in salinity.) The value of c_h seems to be valid over a wide range of conditions. This unsophisticated parameterization yields results that are nearly as good as those from more complex ones.

Neglecting the moderating effect of the viscous sublayer on the net heat transfer leads to much higher exchange coefficients and thus to heat exchanges that, for the same mixed layer temperature and salinity, can be 10 to 100 times stronger than those observed (McPhee *et al.*, 1987). A similar problem exists in models postulating the "ice bath paradigm", in which the assumption is made that the under-ice mixed layer is always at the freezing point (Hibler and Bryan, 1987; FMM97). In these models, the heat flux at the base of the ice is specified so as to instantaneously compensate for any heat gain in the mixed layer in order to maintain its temperature at the freezing point.

However, both diffusive parameterizations such as that in (13) and "ice bath" parameterizations are bound to yield similar equilibrium heat fluxes. Rather than a strong discrepancy in F_w between these two cases, there will be a relatively small difference in the oceanic temperature below the ice. In contrast, transient behaviors can be significantly affected. The sensitivity of ice growth and melting processes to the parameterization of the basal heat flux has recently been analyzed with a one-dimensional model by Holland *et al.* (1997).

3.3. EXCHANGE OF MASS

As mentioned above, the salt content in sea ice tends to decrease as the ice becomes older (from up to 15 ppt for newly formed ice to 4 ppt or less for multiyear ice). Simulating the evolution of sea ice salinity is not an easy problem, and most models simply stipulate a constant sea ice salinity ($\sim$ 5 ppt). The largest errors introduced by this assumption will occur shortly after ice has formed, when sea ice salinity undergoes large temporal variations.

Ocean models generally deal with freshwater surface fluxes due to precipitation/evaporation, ice growth/melt, and river runoff in terms of an equivalent salt flux. Mass exchanges between sea ice and ocean are thus expressed in terms of a salt flux given by

$$F_{salt} = S_w \left(\frac{\partial m_s}{\partial t} \right)_{abl} + (S_w - S_i) \left(\frac{\partial m_i}{\partial t} \right)_{acc-abl}, \tag{14}$$

where S_w and S_i are the salinities of seawater and sea ice respectively, m_s is the mass of snow resting on top of sea ice per unit area, and m_i is the mass of ice per unit area. The first term on the right-hand side of (14) corresponds to the freshwater flux into the ocean associated with snow melting. The second term relates to salinity fluxes occurring during ice growth and decay. Another term must be added if snow ice is formed (FMM97).

3.4. IMPORTANCE OF SEA ICE–OCEAN INTERACTIONS IN GLOBAL SEA ICE–OCEAN MODELS

Sea ice–ocean momentum, heat, and mass exchanges have significant repercussions on both the mean state and variability of the global ocean. In order to assess the impact of these fluxes on the large-scale ocean circulation, Goosse (1997) recently carried out three sensitivity experiments with a sea ice–ocean general circulation model. These three experiments were specifically designed to separately appraise the influence of momentum, heat, and mass exchanges, respectively, on the behavior of the ice-ocean system. Each experiment consisted of suppressing or modifying a particular component of the coupling between sea ice and ocean. The results of the experiment were then compared with those of a reference run in which the interaction was included.

In the first experiment, mechanical interactions between the ice cover and the ocean were eliminated by applying climatological wind stresses at the ocean surface, irrespective of the presence or absence of ice. Local modifications of the surface stress under ice can be rather large in the Bellingshausen, Amundsen, and western Weddell Seas, and in the central Arctic,

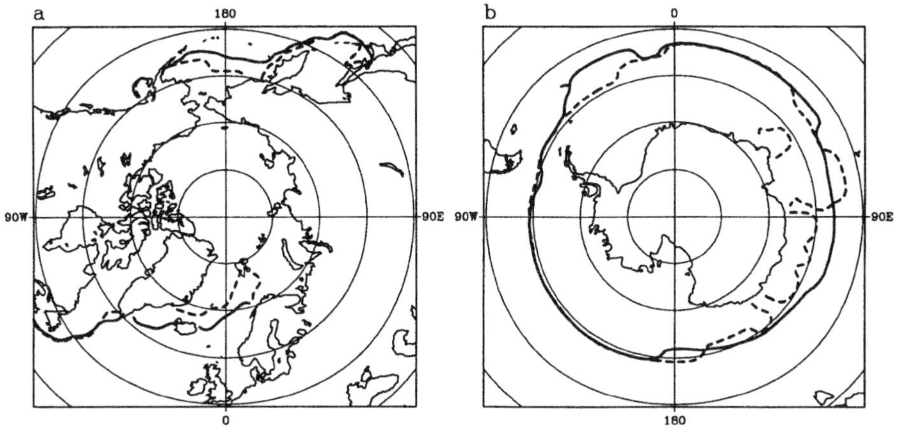

Figure 6. Sensitivity of a coupled sea ice–ocean model to a 5-fold decrease in the ice-ocean heat exchange coefficient, c_h. Locations of the Northern and Southern Hemisphere ice edges at the moment of maximum ice extent (a and b, respectively) for a control experiment with $c_h = 6 \times 10^{-3}$ (dashed lines) and a sensitivity experiment with $c_h = 1.2 \times 10^{-3}$ (solid lines).

where the presence of a thick and compact ice cover tends to attenuate the air-ocean momentum transfer in the control experiment. However, the modification of the surface stress by ice seems to have no major influence on the general circulation of the World Ocean, and, in particular, it does not significantly affect the location of the regions of convective overturning nor the deep water formation rates.

The second experiment aimed at assessing the sensitivity of the ice-ocean system to a decrease in the oceanic sensible heat flux. The few observations available show that the seasonal cycle of F_w is very pronounced in both hemispheres. In summer, F_w is normally very weak. In winter, it increases to at most ~2 W m^{-2} in the central Arctic (Wettlaufer, 1991), although it can be much larger at locations where the ice meets warm waters, as for example the Greenland Sea (Hibler and Bryan, 1987). In the Southern Ocean, F_w attains typical winter values of 10-30 W m^{-2} (Gordon and Huber, 1990; Martinson, 1993). Goosse (1997) reduced c_h in (13) by a factor of 5. The weaker heat exchange leads to a noticeable increase in ice extent during winter in both hemispheres (Fig. 6). The expansion of the ice cover is particularly strong in regions such as the Norwegian Sea, where heat transported by the ocean prevents ice formation in the reference case. In the sensitivity experiment, this heat is not made available at the surface because of the small exchange coefficient. This reduction in surface heat losses under sea ice in the Norwegian Sea produces a warming of the upper ocean that in turn stabilizes the water column and weakens

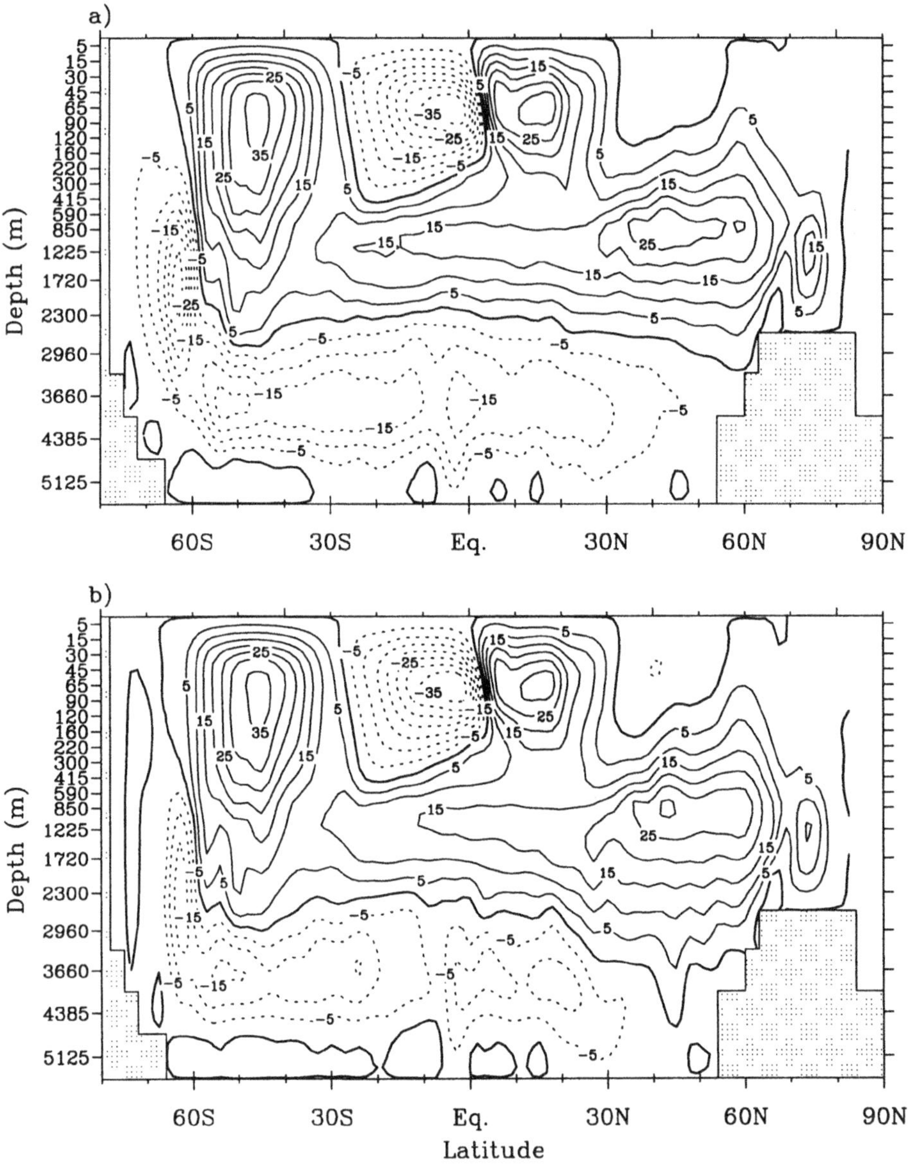

Figure 7. Zonally integrated meridional overturning streamfunction for (a) the reference run and (b) the sensitivity experiment in which salt/freshwater exchanges are neglected. Contour interval is 5 Sv.

convection. Paradoxically, in some regions of the ice pack, such as the Kara and Ross Seas, the oceanic warming is large enough to actually dominate the dampening effect due to the smallness of the heat exchange coefficient. This results in larger oceanic heat fluxes and thinner ice than in the control experiment. Moreover, the oceanic warming is not confined to polar latitudes. The weakened convective activity in the experiment yields a decrease in the production and export of North Atlantic Deep Water and Antarctic Bottom Water, which ultimately yields an increase in the global mean temperature of the ocean of 1.1°C.

A third experiment consisted of neglecting the surface salinity and freshwater fluxes that take place during ice growth and melting. In practice, this amounts to postulating that sea ice has the same salinity as the surrounding seawater. Such an assumption has a remarkable impact on the simulated distribution of salinity in the polar oceans. Surface salinity tends to decrease (up to 5 ppt) in regions of net ice production and to increase in regions of net ice melting. These changes in salinity are accompanied by concomitant changes in seawater density that affect the thermohaline circulation. For instance, over the Antarctic continental shelf, where ice production can be as large as 3 m yr^{-1} at some places, the removal of the salt rejection process from the simulation leads to a complete shutdown of convection there, and, as illustrated in Fig. 7, the export of Antarctic Bottom Water from the Southern Ocean to other basins is accordingly much reduced (see also Stössel *et al.*, 1997).

4. Summary and Conclusions

Due to the profound modifications that sea ice induces in the momentum, heat, and mass transfers between atmosphere and ocean, sea ice and sea ice–ocean interactions have a significant influence on the mean state and variability of the polar and global climates. Indeed, sea ice exerts a tight control on the air temperature patterns and evaporation/precipitation rates at high latitudes. In addition, sea ice is a crucial factor in determining the Earth's surface and planetary albedos and in driving the circumglobal oceanic thermohaline circulation. On these grounds, the inclusion of appropriate representations of sea ice and sea ice–ocean interactions in global climate models is imperative. In this paper, we have offered an overview of the most salient modeling efforts that have been devoted to the numerical treatment of sea ice and sea ice–ocean exchanges during the last 30 years.

The sea ice modeling investigations carried out thus far have shown that a realistic parameterization of sea ice can only be accomplished if both ice thermodynamic and dynamic processes are adequately taken into account. Among the thermodynamic processes that must be included in a sea ice model, the following must be emphasized: (1) the strong dependence of the snow and ice albedos on surface conditions; (2) the thermal inertia of the snow-ice system and, in particular, the buffering effect associated with volume changes in internal brine pockets; (3) the distribution of sea ice in different thickness categories within a given oceanic area, which greatly affects the ice growth/melt rate; (4) the presence of a snow cover on top of sea ice and the transformation of snow into ice due to flooding of the snow; and (5) the existence of a variable fraction of open water within the pack. Regarding the dynamic processes, it has been shown that sea ice transport is notably sensitive to the choice of ice rheology. In this respect, there is ample opportunity for further research into improving the mathematical treatment of internal ice forces. Finally, there is increasing awareness among the sea ice modeling community about the role played by the dynamic processes in the continuous modification of the state of the ice cover, especially via the opening and closing of leads and the formation of ice ridges. The development of large-scale parameterizations of these processes is one of the major challenges to be addressed in the immediate future.

The importance of sea ice–ocean interactions in global sea ice–ocean models has also been discussed. The heat and mass fluxes at the ice-water interface seem to have a significant influence both regionally, in tailoring the state of the water column and, in particular, the strength of diffusive and convective vertical mixings, and globally, in shaping the large-scale ocean circulation. Yet little is known about the way in which modifications of these fluxes can affect the behavior of the global ocean and its response to natural and anthropogenic perturbations of the climatic system. Unfortunately, it is only recently that these issues have begun to be investigated, and their final clarification will only be possible with the use of fully coupled atmosphere-sea ice-ocean general circulation models.

Acknowledgments

We are grateful to J. Ballabrera, A. Beckman, V. Bhushan, J. Staneva, and A.J. Willmott for their valuable comments on an earlier version of the paper. We heartily thank Mr. A. Looms, who kindly helped us in producing the figures. T.F. and H.G. are Research Associate and Research Assistant at the National Fund for Scientific Research (Belgium), respectively. M.A.M.M. gratefully acknowledges support from a research grant awarded via the 4th Round of the Antarctic Special Topic Awards administered by the British Antarctic Survey. Part of the results presented in this paper were obtained within the scope of the Global Change and Sustainable Development Programme (Belgian State, Prime Minister's Services, Federal Office for Scientific, Technical, and Cultural Affairs, contract CG/DD/09A), the Concerted Research Action 092/97–154 (French Community of Belgium, Department of Education, Research, and Formation), and the Environment and Climate Programme (European Commission, contract ENV4–CT95–0102). All of this support is gratefully acknowledged.

References

Bourke, R.H. and Garrett, R.P. (1987) Sea ice thickness distribution in the Arctic Ocean, *Cold Reg. Sci. Technol.* **13**, 259–280.

Budd, W.F. (1991) Antarctica and global change, *Clim. Change* **18**, 272–299.

Darby, M.S. and Willmott, A.J. (1993) A simple time-dependent coupled ice-ocean model with application to the Greenland-Norwegian Sea, *Tellus* **45**, 221–246.

Dickinson, R.E. (1985) Climate Sensitivity, in S. Manabe (ed.), *Issues in Atmospheric and Oceanic Modelling, Part A: Climate Dynamics, Adv. Geophys., 28*, Academic Press, Orlando, pp. 99–129.

Ebert, E.E. and Curry, J.A. (1993) An intermediate one-dimensional thermodynamic sea ice model for investigating ice-atmosphere interactions, *J. Geophys. Res.* **98**, 10,085–10,109.

Eicken, H., Fischer, H. and Lemke, P. (1995) Effects of the snow cover on Antarctic sea ice and potential modulation of its response to climate change, *Ann. Glaciol.* **21**, 369-376.

Emery, W.J., Fowler, C.W. and Maslanik, J.A. (1997) Satellite-derived maps of Arctic and Antarctic sea ice motion: 1988 to 1994, *Geophys. Res. Let.* **24**, 897–900.

Fichefet, T. and Morales Maqueda, M.A. (1997) Sensitivity of a global sea ice model to the treatment of ice thermodynamics and dynamics, *J. Geophys. Res.* **102**, 12,609-12,646.

Flato, G.M. and Hibler III, W.D. (1992) Modeling pack ice as a cavitating fluid, *J. Phys. Oceanogr.* **22**, 626–651.

Gloersen, P., Campbell, W.J., Cavalieri, D.J., Comiso, J.C., Parkinson, C.L. and Zwally, H.J. (1992) *Arctic and Antarctic Sea Ice, 1978–1987: Satellite Passive-Microwave Observations and Analysis*, NASA Spec. Publ. SP-511.

Goosse, H. (1997) *Modelling the large-scale behaviour of the coupled ocean-sea ice system*, Ph. D. Thesis, Université Catholique de Louvain, Louvain-la-Neuve.

Gordon, A.L. and Huber, B.A. (1990) Southern Ocean winter mixed layer, *J. Geophys. Res.* **95**, 11,655–11,672.

Häkkinen, S. and Mellor, G.L. (1992) Modeling the seasonal variability of a coupled Arctic ice-ocean system, *J. Geophys. Res.* **97**, 20,285–20,304.

Hibler III, W.D. (1979) A dynamic thermodynamic sea ice model, *J. Phys. Oceanogr.* **9**, 815–846.

Hibler III, W.D. (1980) Modeling a variable thickness sea ice cover, *Mon. Weather Rev.* **108**, 1943–1973.

Hibler III, W.D. and Bryan, K. (1987) A diagnostic ice-ocean model, *J. Phys. Oceanogr.* **17**, 987–1015.

Holland, M.M., Curry, J.A. and Schramm, J.L. (1997) Modeling the thermodynamics of a sea ice thickness distribution. 2. Sea ice/ocean interactions, *J. Geophys. Res.* **102**, 23,093–23,107.

Jeffries, M.O. and Adolphs U. (1997) Early winter ice and snow thickness distribution, ice structure and development of the western Ross Sea pack ice between the ice edge and the Ross Ice Shelf, *Antarct. Sci.* **9**, 188–200.

Martinson, D.G. (1993) Ocean heat and seasonal sea ice thickness in the Southern Ocean, in W.R. Peltier (ed.), *Ice in the Climate System, NATO ASI Series, I12*, Springer-Verlag, New York, pp. 597–609.

Maykut, G.A. and McPhee, M.G. (1995) Solar heating of the Arctic mixed layer, *J. Geophys. Res.* **100**, 24,691–24,703.

Maykut, G.A. and Untersteiner, N. (1971) Some results from a time-dependent thermodynamic model of sea ice, *J. Geophys. Res.* **76**, 1550–1575.

McPhee, M.G. (1980) An analysis of pack ice drift in summer, in R.S. Pritchard (ed.), *Sea Ice Processes and Models*, University of Washington Press, Seattle, pp. 62–75.

McPhee, M.G. (1983) Turbulent heat and momentum transfer in the oceanic boundary layer under melting pack ice, *J. Geophys. Res.* **88**, 2827–2835.

McPhee, M.G. (1992) Turbulent heat flux in the upper ocean under sea ice, *J. Geophys. Res.* **97**, 5365–5379.
McPhee, M.G., Maykut, G.A. and Morison, J.H. (1987) Dynamics and thermodynamics of the ice/upper ocean system in the marginal ice zone of the Greenland Sea, *J. Geophys. Res.* **92**, 7017–7031.
Mellor, G.L., McPhee, M.G. and Steele, M. (1986) Ice-seawater turbulent boundary layer interaction with melting or freezing, *J. Phys. Oceanogr.* **16**, 1829–1846.
Murphy, J.M. and Mitchell, J.F.B. (1995) Transient response of the Hadley Centre coupled ocean-atmosphere model to increasing carbon dioxide, Part II: Spatial and temporal structure of response, *J. Clim.* **8**, 57–80.
Parkinson, C.L. and Washington, W.M. (1979) A large-scale numerical model of sea ice, *J. Geophys. Res.* **84**, 311-337.
The Polar Group (1980) Polar atmosphere-ice-ocean processes: A review of polar problems in climate research, *Rev. Geophys. Space Phys.* **18**, 525–543.
Semtner, A.J., Jr. (1976) A model for the thermodynamic growth of sea ice in numerical investigations of climate, *J. Phys. Oceanogr.* **6**, 379–389.
Shirasawa, K. and Ingram R.G. (1991) Characteristics of the turbulent oceanic boundary layer under sea ice. Part I: A review of the ice-ocean boundary layer, *J. Mar. Syst.* **2**, 153–160.
Steele, M., Morison, J.H. and Untersteiner, N. (1989) The partition of air-ice-ocean momentum exchange as a function of ice concentration, floe size and draft, *J. Geophys. Res.* **94**, 12,739–12,750.
Steele, M., Zhang, J., Rothrock, D. and Stern, H. (1997) The force balance of sea ice in a numerical model of the Arctic Ocean, *J. Geophys. Res.* **102**, 21,061–21,079.
Stern, H.L. and Rothrock, D.A. (1995) Open water production in Arctic sea ice: Satellite measurements and model parameterizations, *J. Geophys. Res.* **100**, 20,601–20,612.
Stössel, A., Kim, S.J. and Drijfout S.S. (1997) The impact of Southern Ocean sea ice in a global ocean model, *J. Phys. Oceanogr.*, submitted.
Thorndike, A.S., Rothrock, D.A. Maykut, G.A. and Colony, R. (1975) The thickness distribution of sea ice, *J. Geophys. Res.* **80**, 4501–4513.
Tremblay, L.-B. and Mysak, L.A. (1997) Modeling sea ice as a granular material, including the dilatancy effect, *J. Phys. Oceanogr.* **27**, 2342–2360.
Weeks, W.F. and Ackley, S.F. (1986) The growth, structure and properties of sea ice, in N. Untersteiner (ed.), *The Geophysics of Sea Ice, NATO ASI Series, B146*, Plenum Press, New York, pp. 9–164.
Wettlaufer, J.S. (1991) Heat flux at the ice-ocean interface, *J. Geophys. Res.* **96**, 7215-7236.

OCEAN MODELING IN ISOPYCNIC COORDINATES

RAINER BLECK
Department of Meteorology and Physical Oceanography,
Rosenstiel School of Marine and Atmospheric Science,
University of Miami, Miami, FL, U.S.A.

1. Introduction

Electronic computers were made available to civilian researchers in the late 1940s, shortly after World War II. Numerical Weather Prediction – at that time the repetitive solution of a 2-dimensional elliptic partial differential equation describing the evolution of the height of a mid-tropospheric isobaric surface – was among the first mathematical problems tackled with that new technology.

A decade or so later, when computers were fast enough to allow adding a third (vertical) dimension to NWP models, mathematical complexities created by variable terrain height led to the development of a special vertical coordinate $\sigma = p/p_{surface}$ which removed mountain-induced "bumps" from the computational domain [14] . The σ coordinate quickly became the coordinate of choice in NWP models, but the community eventually realized that the search for the perfect vertical model coordinate was far from over. One can say that it is still going on today.

Accuracy and stability problems created by the vertical advection terms in the prognostic equations formulated in z, p, or σ coordinates led to early experimentation with models formulated so as to minimize the role of these terms [21]. While isentropic (θ) coordinate models achieve this goal at least under adiabatic flow conditions, such models never became popular because of concerns about numerical stability at the lower boundary whose location in θ space is irregular even in the absence of mountains and varies with time. Advances in *hybrid* coordinate technology have laid these problems to rest today, allowing isentropic NWP models — specifically, mixed σ-θ coordinate models — finally to achieve operational status.

Modeling the oceanic circulation has lagged behind NWP for reasons having to do with insufficient economic incentives, lack of observations, and

E. P. Chassignet and J. Verron (eds.), Ocean Modeling and Parameterization, 423–448.

the combination of long time scales with the small size of oceanic eddies which makes ocean modeling a more resource-intensive task. Thus, only a few large research institutions were able to commit the resources needed for developing ocean models. Perhaps inevitably, the ocean model "gene pool" remained sparse over the years, and instances of model "inbreeding" were common. Only today, as more and more academic researchers gain access to large computers, do we see the diversity in ocean modeling reminiscent of the NWP culture of the 1960s and '70s.

Models using density or, more precisely, *potential* density as vertical coordinate are one example of the growing ocean model diversity. Development of such models has been driven by the community's increasing awareness of numerical errors inherent in z and σ coordinate models — errors having to do with the physically inappropriate (though mathematically correct) split of transport and diffusion processes into vertical and horizontal components.

Isopycnic (constant potential density) models suppress such errors to a large extent by, essentially, switching the role of density and depth as dependent and independent variables. Not only does this reduce inconsistencies between vertical and horizontal transport terms, but it allows one to hide truncation errors associated with the remaining horizontal transport terms behind the "smoke screen" of isopycnic mixing.

The purpose of this article is to review the numerical tools needed for solving equations in which depth is a dependent variable and potential density an independent variable, and to present results illustrating the unique diagnostic capabilities of such models.

2. Dynamic Equations

Primitive-equation models using potential density as vertical coordinate [13] [6] typically contain 3 or 4 prognostic equations – one equation for the horizontal velocity vector, a mass continuity or layer thickness tendency equation, and either one or two conservation equations for buoyancy-related variables (density, salt, temperature). Since not all coordinate layers in an isopycnic model are necessarily constant-density layers (prime example: the surface mixed layer discussed below), the equations will be written here for a nondescript, generalized vertical coordinate s [3]. Density ρ and its inverse, specific volume α, represent *in-situ* conditions, but it will be shown later that the equations remain valid if these variables are corrected for compressibility effects, i.e., if they are interpreted as *potential* density and its inverse.

The equations, in the order mentioned above, are

$$\frac{\partial \mathbf{v}}{\partial t_s} + \nabla_s \frac{\mathbf{v}^2}{2} + (\zeta + f)\mathbf{k} \times \mathbf{v} + \left(\dot{s}\frac{\partial p}{\partial s}\right)\frac{\partial \mathbf{v}}{\partial p} + \nabla_\alpha M$$

$$= -g\frac{\partial \boldsymbol{\tau}}{\partial p} + \left(\frac{\partial p}{\partial s}\right)^{-1} \nabla_s \cdot \left(\nu \frac{\partial p}{\partial s} \nabla_s \mathbf{v}\right) \tag{1}$$

$$\frac{\partial}{\partial t_s}\left(\frac{\partial p}{\partial s}\right) + \nabla_s \cdot \left(\mathbf{v}\frac{\partial p}{\partial s}\right) + \frac{\partial}{\partial s}\left(\dot{s}\frac{\partial p}{\partial s}\right) = 0 \tag{2}$$

$$\frac{\partial}{\partial t_s}\left(\frac{\partial p}{\partial s}T\right) + \nabla_s \cdot \left(\mathbf{v}\frac{\partial p}{\partial s}T\right) + \frac{\partial}{\partial s}\left(\dot{s}\frac{\partial p}{\partial s}T\right) = \nabla_s \cdot \left(\nu\frac{\partial p}{\partial s}\nabla_s T\right) + \mathcal{H}_T \tag{3}$$

Here, $\mathbf{v} = (u, v)$ is the horizontal velocity vector, p is pressure, T represents any one of the model's thermodynamic variables (temperature, salinity, buoyancy), $\zeta = \partial v/\partial x_s - \partial u/\partial y_s$ is the relative vorticity, $M \equiv \phi + p\alpha$ is the Montgomery potential, $\phi \equiv gz$ is the geopotential, f is the Coriolis parameter, $\mathbf{k}$ is the vertical unit vector, ν is the *isopycnic* eddy viscosity, and τ is the wind- and/or bottom-drag induced shear stress vector. $\mathcal{H}_T$ represents the sum of the diabatic source terms acting on T. Subscripts indicate the variable held constant during partial differentiation. Distances in x, y direction (as well as their time derivatives $\dot{x} \equiv u$ and $\dot{y} \equiv v$) are measured in the projection onto a horizontal plane, a convention which eliminates metric terms related to the slope of the s surface.

Isopycnic models are typically formulated as *layer* models, meaning that a number of prognostic variables are not carried on coordinate surfaces but represent conditions averaged over individual layers; they must therefore be advanced in time using layer-averaged equations. Averaging over a layer bounded by two surfaces s_{top}, s_{bot} entails multiplying the equation in question by $\partial p/\partial s$, integration over the interval $\Delta s = s_{bot} - s_{top}$, and division by $\Delta p = p(s_{bot}) - p(s_{top})$. The ocean surface, the bottom of the mixed layer, the sea floor, and all isopycnic layer interfaces are considered s surfaces here.

The layer-averaged continuity equation (2) becomes a prognostic equation for the layer weight per unit area or pressure thickness Δp:

$$\frac{\partial}{\partial t_s}\Delta p + \nabla_s \cdot (\mathbf{v}\Delta p) + \left(\dot{s}\frac{\partial p}{\partial s}\right)_{bot} - \left(\dot{s}\frac{\partial p}{\partial s}\right)_{top} = 0. \tag{4}$$

The expression $(\dot{s}\partial p/\partial s)$ represents the vertical mass flux – taken to be positive if in $+p$ or downward direction – across an s surface.

Layer-averaging the momentum equation (1) changes the shear stress term (first term on the right) into

$$\frac{g}{\Delta p}(\boldsymbol{\tau}_{top} - \boldsymbol{\tau}_{bot})$$

while the lateral momentum mixing term integrates to

$$(\Delta p)^{-1}\nabla_s \cdot (\nu \Delta p \nabla_s \mathbf{v}). \tag{5}$$

All other terms in (1) retain their formal appearance during the averaging operation.

Wind-induced stress is assumed to be zero at the bottom of the uppermost coodinate layer, which therefore must not be allowed to become shallower than the Ekman layer. Bottom stress is partitioned among coordinate layers in accordance with the assumption that it linearly decreases from $-\rho c_D|\mathbf{v}|\mathbf{v}$ at the sea floor to zero over a prescribed depth interval of order 10 m. $\mathbf{v}$ in this formula is the velocity vector averaged over those 10 m.

The layer-integrated form of (3) is

$$\frac{\partial}{\partial t}T\Delta p + \nabla_s \cdot (\mathbf{v}T\Delta p) + \left(\dot{s}\frac{\partial p}{\partial s}T\right)_{bot} - \left(\dot{s}\frac{\partial p}{\partial s}T\right)_{top} = \nabla_s \cdot (\nu \Delta p \nabla_s T) + \mathcal{H}_T. \tag{6}$$

The above prognostic equations are complemented by diagnostic equations including the hydrostatic equation

$$\frac{\partial M}{\partial \alpha} = p, \tag{7}$$

an equation of state linking T, S, p to α, and an equation governing cabbeling in models where both T and S are treated as prognostic thermodynamic variables. The interlayer mass flux $(\dot{s}\partial p/\partial s)$ must also be be prescribed. At the bottom of the surface mixed layer, this flux is the result of mixed-layer *entrainment* and *detrainment* processes and can be quite large, while at interior interfaces it reflects *diapycnal* mixing processes which typically are weak.

The quickest way to demonstrate the connection between diapycnal mixing and interlayer mass exchange is to start from the diffusion equation in z coordinates,

$$\left(\frac{\partial \theta}{\partial t}\right)_z = \frac{\partial F}{\partial z},$$

where F is the diffusive flux of the variable θ (for example, $-K\partial\theta/\partial z$), and to use the identity $(\partial a/\partial b)_c(\partial b/\partial c)_a(\partial c/\partial a)_b = -1$ to exchange dependent and independent variables. The final result of these manipulations is

$$\left(\frac{\partial z}{\partial t}\right)_\theta = -\frac{\partial F}{\partial \theta}.$$

The derivative $(\partial z/\partial t)_\theta$ in this equation expresses the rate at which a divergent or convergent flux F causes the surface $\theta = const.$ to migrate vertically through the fluid, i.e., the rate at which mass is transferred from one side of the θ surface to the other.

Diapycnal mixing can easily be suppressed in isopycnic models, both physically and numerically, by setting $\dot{s}\partial p/\partial s = 0$. Coordinate layers remain materially insulated from each other in this case.

3. Numerical Issues

Selecting a set of differential equations is an essential first step in building a circulation model, but before the equations can be solved on a computer, they must be converted into a set of *algebraic* equations. This conversion process is the essence of "numerical modeling", a field that has seen remarkable growth since the first failed attempts at NWP early in this century [18].

Certain tools needed for numerically solving fluid dynamics equations in (x, y, ρ) space were not available in the 1960s when ocean models were first conceived; it was therefore appropriate to design those first-generation circulation models around a geometric vertical coordinate.

In this section, the main tools employed in solving the primitive equations in (x, y, ρ) space will be discussed.

3.1. LAYER OUTCROPPING

An isopycnic model may be viewed as a stack of shallow-water models in which the individual 2-dimensional submodels interact primarily through hydrostatically transmitted pressure forces. Since the top-to-bottom density range in a water column varies with geographic location, not all isopycnic coordinate layers can be expected to exist everywhere in the computational domain. Hence, provisions must be made for layers in the model to "vanish", i.e., to reach zero thickness. This is what is referred to as the layer outcropping problem. A closely related problem is that of coordinate layers intersecting the sea floor.

Methods for numerically solving the layer thickness equation (4) in a way that allows the limit $\Delta p = 0$ to be reached – but not overshot – became available in the late 1960s [27]. These methods belong to a by now large and diverse group of transport schemes designed to preserve the positive-definiteness of the transported variable [22]. Typically, they are 2-step schemes where the variable in question (layer thickness Δp in our case) is first advected by a simple *forward-upstream* scheme that is guaranteed to maintain positive-definiteness of the advected field. The positive-upstream step is followed by an *antidiffusive* step, so named because it is designed to counteract the diffusive effects of the first step without jeopardizing the positive-definiteness and, optionally, the monotonicity of the transported field. (Preserving monotonicity is akin to not permitting new maxima or minima to appear.)

It is of interest to note that Oberhuber [13] solves the outcropping problem in his OPYC model by keeping track of the lateral extent of each coordinate layer. In the case of the Miami Isopycnic Coordinate Ocean Model (MICOM), the decision to use one of the general-purpose transport schemes was made because of the ease with which they can be vectorized and parallelized.

3.2. TOPOGRAPHIC INTERACTIONS

Swift meridional currents in the ocean are generally found near basin boundaries where sidewall drag helps close the potential vorticity budget. The credibility of an ocean model therefore depends strongly on its ability to correctly portray the interaction between boundary currents and topography. One important issue is the computation of the horizontal pressure gradient normal to the boundary.

The method used in MICOM for evaluating the pressure gradient near topographic obstacles dates back to [4] and is outlined more fully in [5]. The guiding design principle is to assure that a resting ocean with horizontal layer interfaces remains at rest. This is achieved by switching to noncentered differencing if one of the two M values in the centered finite-difference formula for $\partial M/\partial x$ or $\partial M/\partial y$ in (1) is associated with a massless grid cell on the sea floor. (If both M values are in massless grid cells, there is no need to compute a pressure force.)

It is fair to say that the above method has withstood the test of time and that steep topography is a "non-problem" in MICOM.

3.3. ISOPYCNIC TRANSPORT EQUATIONS

The equation $dT/dt = 0$ expressing conservation of a variable T in x, y, z space can be stated either in *flux* form,

$$\frac{\partial \rho T}{\partial t} = -\frac{\partial u \rho T}{\partial x} - \frac{\partial v \rho T}{\partial y} - \frac{\partial w \rho T}{\partial z}, \tag{8}$$

or in *advective* form,

$$\frac{\partial T}{\partial t} = -u\frac{\partial T}{\partial x} - v\frac{\partial T}{\partial y} - w\frac{\partial T}{\partial z}. \tag{9}$$

The two forms do not *a priori* lead to the same result if solved numerically. However, the near-constancy of ρ in the ocean blurs the distinction between them to such an extent that it is possible to advance T in z coordinate models using the simple advective form (9) without jeopardizing global conservation of T.

The conservation equation analogous to (8) governing transport of T in isopycnic layer models is

$$\frac{\partial}{\partial t}(T\Delta p) = -\frac{\partial}{\partial x}(uT\Delta p) - \frac{\partial}{\partial y}(vT\Delta p) \tag{10}$$

where we have omitted the diapycnal transport component for simplicity's sake. In this case, Δp cannot be treated as quasi-constant the way ρ can be treated in the z coordinate version of the same conservation law. Rather, Δp must be allowed to vary significantly from one location and one time step to the next. This precludes the use of a simple advective equation analogous to (9) in isopycnic models. Furthermore, after $T\Delta p$ has been advanced over one time step and is being divided by Δp to obtain the new T value, provisions must be in place for the case where Δp has dropped to zero during that time step.

Experience has shown that the strong temporal and spatial variability of the Δp field requires use of positive-definite, and preferably monotonicity-preserving, transport schemes not only for Δp but for thermodynamic variables as well. Unfortunately, the extra computer time needed in isopycnic models to advect thermodynamic variables, tracers, etc., by an elaborate 2-step scheme cancels the potential speed advantage that isopycnic models have over z coordinate models due to the fact that they can resolve baroclinic structures (fronts etc.) with considerably fewer grid levels than z coordinate models. In fact, the computational burden associated with solving (10) makes it advisable to treat only one thermodynamic variable – salinity – as prognostic variable in isopycnic layers while inferring the other – potential temperature – from the salinity and potential density. While explicit transport and diffusion of two thermodynamic variables in ρ coordinates seems redundant, it does permit modeling of cabbeling effects and helps guarantee global conservation of both T and S in the model.

3.4. BUOYANCY FORCING

Isopycnic models were initially developed for simulating the wind-driven circulation. While mechanically driven isopycnic models seem to be devoid of a thermodynamic component, the conservation equation for potential density is, in fact, included in the set of model equations in the form of a statement prohibiting mass exchange between coordinate layers. The absence of a transport/diffusion component across layer interfaces under adiabatic flow conditions was already mentioned as the main reason for developing isentropic NWP models and isopycnic ocean models.

Conceptual problems arise in isopycnic models if the forces driving the circulation include buoyancy fluxes, not only because buoyancy is an independent variable in these models, but also because discretization of the governing equations in the vertical implies that only a finite number of discrete density values are permitted in the model.

The fact that buoyancy is an independent variable requires that another variable must be found that can properly respond to extraneously imposed buoyancy changes. This variable is the isopycnic layer thickness. Specifically, buoyancy gains at the surface must be translated in isopycnic models into mass transfer from heavier to lighter layers; in case of surface buoyancy loss the direction of mass transfer is in the other direction.

Difficulties created by the discrete density structure can be addressed in isopycnic models by allowing density in the uppermost model layer (the surface mixed layer) to vary continuously in x, y, and t. Adding this extra nonisopycnic layer clearly does not remove the physical incompatibility between the continously varying sea surface density field and the discrete density structure in the model interior. Rather, the "clash" between the two representations is only shifted from the sea surface to the bottom of the nonisopycnic layer where vertical mass exchange must be allowed to take place in a way that mimics the physics of mixed-layer entrainment and detrainment.

While there are no numerical difficulties in entraining water from constant-density layers into a nonisopycnic layer, the opposite process of mixed-layer detrainment clearly is problematic because water cannot simply be transferred out of the mixed layer unless its density matches that of an interior isopycnic layer.

A sensible way to deal with the detrainment problem is to divide the uppermost model layer conceptually into a "physical" and a "fossil" mixed layer (the fossil mixed layer consisting of water that at one time was, but no longer is, in the physical mixed layer and is now awaiting transfer to an interior isopycnic layer). During detrainment, which occurs when the ocean is warmed or freshened from above, the fossil mixed layer grows at the expense of the physical mixed layer.

The fossil mixed layer is now conceptually divided into two sublayers, and buoyancy is exchanged between the two sublayers until the upper sublayer assumes the buoyancy of the physical mixed layer while the lower sublayer assumes the buoyancy of an interior coordinate layer. (The unknown in this problem is the depth of the interface separating the two sublayers.) This vertical redistribution of buoyancy allows one to combine the upper sublayer with the physical mixed layer into a single, vertically homogeneous coordinate layer, while the water from the lower sublayer can be detrained into the targeted isopycnic layer. Thus, during mixed-layer detrainment episodes the uppermost model layer is usually deeper than the physical mixed layer.

3.5. COMPRESSIBILITY OF SEA WATER

Eddy-induced mixing in a stratified fluid preferentially takes place along surfaces on which laterally displaced fluid particles experience no restoring buoyancy force. These surfaces are referred to as *neutral* [15] [11]. The ocean is a near-incompressible, weakly stratified medium. If it were totally incompressible, constant-density and neutral surfaces would coincide. With compressibility being nonzero, neutral surfaces in the real ocean coincide with *potential* density surfaces referenced to a nearby pressure level.

Ocean models whose coordinate surfaces are neutral can be expected to simulate eddy-induced mixing far more accurately than models operating on a rigid lattice of grid points. (This has to do with the component represenation of the mixing process mentioned in the Introduction.) While isopycnic models come close to this ideal, especially if "isopycnic" is understood to mean constant *potential* density, the physical properties of sea water are such that perfect alignment of coordinate surfaces with neutral surfaces is difficult to achieve near water mass boundaries at depths differing from the chosen reference pressure. This issue will be discussed further in the section on presently unsolved problems.

Accurate simulation of mixing processes is only one concern in climate-oriented ocean modeling; another one is accurate representation of circulations driven by temperature and salinity contrasts imposed on the ocean by the atmosphere. The term in the dynamic equations controlling the onset and maintenance of buoyancy-driven circulations is the curl of the 3-dimensional pressure acceleration, $\nabla \times \alpha \nabla p = (\nabla \alpha) \times (\nabla p)$; this term spawns the familiar solenoidal term $\oint \alpha dp$ in Bjerknes' Circulation Theorem. We see from it that only *isobaric* density contrasts are capable of driving circulations; in other words, density contrasts resulting from the weight of the overlying fluid alone are dynamically passive.

The practice of defining coordinate layers in isopycnic models in terms of *potential* density provides a convenient mechanism for eliminating dynamically passive density variations from the dynamic equations. This point deserves a closer look. As long as the adiabatic compressibility coefficient $\kappa = \rho^{-1}(\partial\rho/\partial p)_{adiab}$ is constant, or at most a function of pressure, it can be shown that

$$\frac{\nabla_p \alpha}{\alpha} = \frac{\nabla_p \alpha_{pot}}{\alpha_{pot}} \tag{11}$$

where α_{pot} is the inverse potential density and ∇_p is the 2-dimensional gradient operator on an isobaric surface. With α differing from α_{pot} by no more than 2% in the world ocean, we conclude that

$$\nabla_p \alpha \sim \nabla_p \alpha_{pot}.$$

Integration of this equation over p (a step equivalent to integrating the thermal wind equation) yields

$$\nabla_p \phi \sim \nabla_p \phi^* \tag{12}$$

where ϕ^* is a geopotential-like function obtained by substituting α_{pot} for α on the right-hand side of the hydrostatic equation $\partial\phi/\partial p = -\alpha$. Since $\nabla_p \phi$ equals the horizontal pressure gradient, we conclude that circulation-inducing pressure forces resulting from spatial variations in the α field can be computed, with at most a 2% error, from the knowledge of the α_{pot} field alone. This obviates the need to carry *in situ* density as a variable in isopycnic models.

Unfortunately, sea water compressibility is not only a function of pressure, but also of temperature: water near the freezing point is more compressible than water a few degrees warmer. Since water density near the freezing point is mainly a function of salinity, we can also state: given two low-temperature water parcels of equal density but different salt content, the less saline parcel is more compressible.

This so-called thermobaric effect invalidates the assumptions on which (11) and subsequent conclusions are based, and there are in fact situations where thermobaricity drives circulations that cannot be captured by the standard set of isopycnic model equations. A solution recently proposed by [19], and successfully tested by [24], consists of splitting the compressibility coefficient κ into a p- and a T-dependent part and using only the former in determining the potential density of a water parcel. The new variable, named α^* to distinguish it from the genuine α_{pot}, allows us to replace (11) by

$$\frac{\nabla_p \alpha}{\alpha} = \frac{\nabla_p \alpha^*}{\alpha^*}$$

and retain (12) with the stipulation that ϕ^* is obtained by integrating a hydrostatic equation of the form $\partial\phi^*/\partial p = -\alpha^*$.

The variable α^* – the inverse of what in [19] is called "virtual" potential density ρ^* – does not appear to obey any conservation law. In particular, surfaces of constant ρ^* are not material, even under adiabatic conditions. The strategy followed in MICOM is therefore to retain the variable ρ_{pot}, which *is* material under adiabatic conditions, as vertical coordinate, and to treat ρ^* as an (x, y, ρ_{pot}, t)-dependent diagnostic variable.

3.6. DEEP CONVECTION

Isopycnic coordinate models turn into single-layer shallow-water models in regions where the water column is homogeneous. Loss of stratification can either be caused by a divergent surface flow field (which brings dense subsurface water to the surface) or by buoyancy loss to the atmosphere. In the real ocean, surface buoyancy loss triggers convection which hydrostatic models, the model class to which MICOM belongs, are unable to simulate correctly. Parameterization of buoyant convection is therefore an important issue in such models.

Compared to the level of sophistication reached by convective parameterization schemes in modern NWP models, the so-called *convective adjustment* scheme used in today's hydrostatic ocean models is embarassingly simple. In isopycnic models, convective adjustment is implemented by checking in each water column whether an isopycnic layer of nonzero thickness exists whose density is less than that of the surface mixed layer. If this is the case, the mass from the layer in question is transferred to the mixed layer and the thickness of the donor layer is set to zero.

4. Solved and Presently Unsolved Problems

Isopycnic ocean models of the type represented by MICOM and Oberhuber's OPYC model have been in service for several years now, and their general utility is no longer being questioned. The question that remains in many people's mind, however, is whether these models are in an overall sense superior to the first-generation fixed-grid models. Some of the issues that in the past have delayed model acceptance, and some that still do because they either have not yet been tackled or are actually unsolvable, will be discussed in this section.

4.1. LAYER OUTCROPPING

The layer outcropping problem, although it was laid to rest 10 or 15 years ago, still is regarded by many as the number one problem in isopycnic modeling. "Laying to rest" in this case does not mean that layer outcropping as such has been eliminated as a physical or numerical phenomenon. It means that numerical methods have been developed – for example, the massless layer approach [4] [25] – which are sufficiently robust to eliminate layer outcropping as a hindrance to further advances in isopycnic modeling. One aspect still in need of attention is the numerical treatment of the wedge of unbounded potential vorticity along outcrop lines caused by vanishing layer thickness [1].

4.2. SURFACE PROCESSES IN A DISCRETIZED DENSITY ENVIRONMENT

The discrete density structure in isopycnic models and its impact on mixed layer physics parameterization is a more pressing concern. The treatment of sea surface density ρ_{surf} most consistent with the discrete layer paradigm would be to set ρ_{surf} at every point x, y to the value associated with the uppermost mass-containing (outcropping) coordinate layer. When buoyancy forcing was introduced into MICOM, the disadvantages of having to formulate air-sea interaction processes on the basis of a piecewise constant ρ_{surf} field were considered severe, and the strategy of allowing the uppermost coordinate layer to have characteristics different from those of the remaining layers – namely, to be nonisopycnic – was regarded as the lesser evil.

As outlined earlier, buoyancy fluxes across the air-sea interface in an isopycnic model must be translated into mass fluxes – specifically, mass fluxes through the bottom of the mixed layer. It was also discussed that in the present MICOM implementation the numerical mismatch between the nonisopycnic surface mixed layer and the isopycnic model layers can lead to temporary differences between the physical mixed-layer depth and the thickness of the uppermost model layer. If maintaining the identity between mixed-layer depth and uppermost layer thickness is considered important in a particular model application, then this mismatch, caused by the presence of "fossil" mixed-layer water in the top coordinate layer, becomes a model error. However, it is a true truncation error in the sense that it can be reduced to any degree desired by increasing vertical model resolution, i.e., by adding more layers. Nevertheless, not all modelers accustomed to truncation errors in z space find it easy to tolerate this particular shortcoming.

As a first step toward improving mixed-layer physics in MICOM, it is planned to split the uppermost model layer consisting of the surface mixed layer and the underlying "fossil" mixed layer into two separate layers. To discourage unlimited buildup of the so-defined fossil layer, the detrainment mechanism described earlier for the mixed layer should in this case be applied to the fossil layer.

The above procedure does not represent a radical departure from the presently used Kraus-Turner [8] mixed-layer parameterization scheme. More elaborate schemes are available today and are expected to eventually find their way into isopycnic models.

4.3. THERMOBARICITY

Another area of concern is the treatment of the equation of state, and in particular the traditional practice of regarding the ocean in isopycnic models as stratified but incompressible. The view that isopycnic models treat the ocean as incompressible is actually not quite correct. As pointed out earlier, density changes caused by the weight of the overlying fluid are dynamically passive. The resulting dynamic equivalence of *in situ* and potential density makes it appear as though the model equations are founded on an incompressibility assumption when, in fact, they are not.

As discussed earlier, a satisfactory method now exists for incorporating thermobaric effects into the modeling of buoyancy-driven circulations. However, thermobaricity has other consequences, some of which possibly belong to the class of principally unsolvable problems in isopycnic modeling. The consequences referred to have to do with the multivalued nature of $\rho_{pot}(z)$, that is, with coordinate folding.

Outside convective regions, the ocean is stably stratified, but to verify this one must measure the static stability in terms of potential density referenced to the "local mean" pressure. The variable ρ_{pot} defining the layer structure in isopycnic models, on the other hand, must be based on a "global" choice of reference pressure. The work of Pingree [15], Reid [17], and others contains numerous examples of how spatial salinity variations combined with a nonlocal reference pressure can lead to grossly erroneous estimates of the ρ_{pot} difference between pairs of water parcels, including the appearance of static instability when, in fact, the column is stably stratified. Note the emphasis here on *salinity* variations, which seemingly contradicts the notion that adiabatic compressiblity κ depends only on T and p. To resolve the contradiction, one must realize that in an isopycnic reference frame T variations are tightly coupled to S variations.

There is no single reference pressure value that would *not* create coordinate folding at least somewhere in the world ocean. The best one can do therefore is to find a reference pressure that minimizes instances of multivaluedness in $\rho_{pot}(z)$ in the model initial conditions, and to smear out the vertical ρ_{pot} distribution in the remaining overturning regions. The neutrally stratified patches of water hereby created will, of course, not be intersected by coordinate surfaces and thus will suffer from poor vertical resolution unless one chooses to fill them with coordinate surfaces defined in a different manner (hybrid coordinates).

Short of causing coordinate overturning, an inappropriately chosen reference pressure will also ruin the alignment of coordinate surfaces with neutral surfaces in regions marked by pronounced salinity variations. The adverse impact of this on the accuracy of mixing simulations could in principle be lessened by borrowing a tool originally designed for z coordinate models, namely, the rotation of the main axes of the mixing tensor [16].

4.4. CONVECTION

The convective adjustment scheme described earlier is attractive in its simplicity but has the disadvantage of causing temporal discontinuities in layer thickness in the model. In the long run, convective adjustment schemes in isopycnic models should therefore be replaced by schemes that parameterize the sinking of dense plumes from the mixed layer to the interior layers. In such schemes, the interior layers would gradually gain mass at the expense of the mixed layer, causing layer interfaces to rise. This slow ascent of interfaces is analogous to the large-scale sinking motion observed in convectively active regions in the atmosphere; proper simulation of this sinking is considered an important part of atmospheric convective parameterization schemes. The other beneficial aspect of such a scheme, to repeat a point just made, is that the layer thickness field would evolve smoothly in time. Numerical solutions that are not smooth in space and time cannot be relied upon to converge, i.e., to approach the solutions of the governing differential equations with decreasing mesh size.

5. Sample Results

Isopycnic models by now have a long history as process models, i.e., as models focused on a particular fluid-dynamical phenomenon or hypothesis. One reason for this is the simple shallow-water form of the potential vorticity conservation law, which allows this law to be carried over into the finite-difference realm [20] [2] and enables isopycnic layer models to rigorously conserve potential enstrophy in regions of nonzero layer depth.

The claim that isopycnic models are capable of serving as full-fledged general circulation models, on the other hand, is relatively recent. The results chosen for inclusion in this article address this aspect of isopycnic modeling, with emphasis on emerging methods for quantifying the thermohaline circulation.

Oberhuber's OPYC model [13] was designed from the start as a global model. In MICOM, the step from basin-scale to near-global modeling was only taken approximately 3 years ago, and the first truly global version – still with only rudimentary treatment of sea ice – was completed at the Southampton Oceanographic Centre (U.K.) only in 1997. Coupled ocean-atmosphere modeling involving the Community Climate Model (CCM3) developed at the National Center for Atmopheric Research and the ice-free, near-global MICOM version began in earnest in the fall of 1997. A propotype global model with a state-of-the-art sea ice component is presently undergoing testing in Miami and at NASA's Goddard Institute for Space Studies.

5.1. OCEAN-ONLY SIMULATIONS ON EDDY- AND NONEDDY-RESOLVING MESHES

Results shown in this section are taken from MICOM simulations carried out on a near-global domain extending from 65°N to 69°S. In keeping with the tradition of forcing MICOM with *atmospheric* rather than oceanic state variables, mean monthly fields of atmospheric temperature, humidity, wind stress, and total wind speed were derived from the Comprehensive Ocean-Atmosphere Data Set [26]; net radiation from the Oberhuber Atlas [12]; and precipitation from the microwave sounding unit [23].

Due to the lack of ship observations in the Southern Ocean, COADS stress and wind data were blended with model output climatology from the European Centre for Medium Range Forecasting. Specifically, a linear transition zone was defined in the latitude band 30°S – 50°S. North of that zone, COADS data were used exclusively; south of that zone, ECMWF data were used exclusively. To match the generally weaker COADS winds, ECMWF winds, including the stress fields, were reduced by 25%.

Initial conditions for all experiments described here were derived from Levitus [10] climatology.

One disadvantage of formulating the model forcing functions in a physically realistic manner is that the model's freshwater budget is not automatically closed: precipitation is an imposed flux field while evaporation depends on sea surface temperature and hence is model-state dependent. Test runs revealed a net freshwater loss of approximately 1 Sv. To counter that loss, 0.65 Sv of freshwater were added to the precipitation field at 14 major river outflow points; 0.27 Sv of Antarctic ice melt were distributed uniformly along the southern boundary; and 0.08 Sv of Arctic river outflow were distributed uniformly along northern model boundary between Labrador and Norway.

The ocean is bounded by solid walls at the domain edges (65°N and 69°S). Bottom water production in the Norwegian and Weddell Seas, both of which lie outside the model domain, is emulated by specifying downward diapycnal fluxes in select locations next to the lateral boundaries. Prescribed downwelling rates are 6 Sv at the northern boundary and 3 Sv at the southern boundary.

The primary goals of the present suite of experiments are to explore (a) the sensitivity of the thermohaline circulation to thermobaricity and the choice of reference pressure used in defining potential density, and (b) the sensitivity of the poleward heat flux to the horizontal mesh size.

The following experiments are underway at the time of this writing. All of them use 16 layers in the vertical. (See Figs. 2 and 3 for the actual potential density values.)

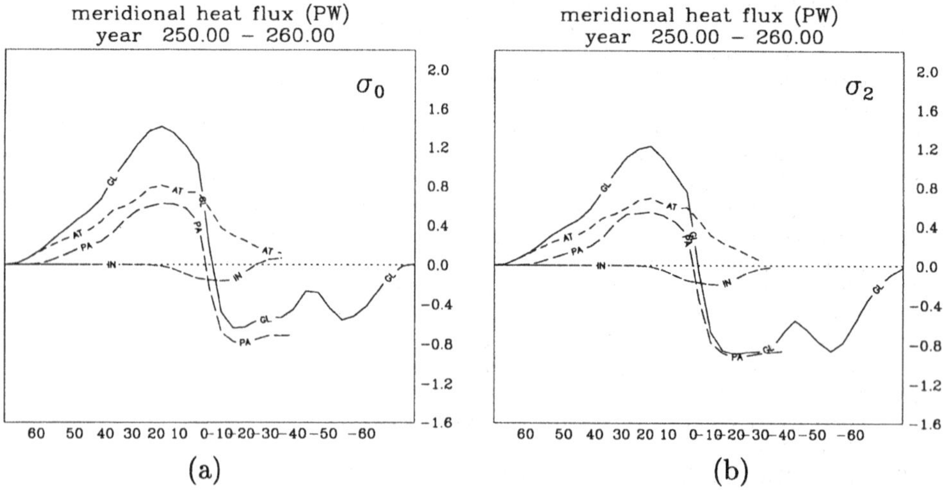

Figure 1. Meridional heat flux (PW) averaged over model years 250–260 in three individual basins and the global ocean from 1.4° model runs. (a) σ_0 coordinate; (b) σ_2 coordinate.

1. Horizontal mesh size $0.225° \times 0.225° \cos(\varphi)$
   (a) Vertical coordinate: σ_0 (potential density referenced to sea level pressure, no thermobaric effects)
   (b) Vertical coordinate: σ_2^* (potential density referenced to 2000 m depth, thermobaric effects included)
2. Horizontal mesh size $1.4° \times 1.4° \cos(\varphi)$
   (a) Vertical coordinate: σ_0 (potential density referenced to sea level pressure, no thermobaric effects)
   (b) Vertical coordinate: σ_2 (potential density referenced to 2000 m depth, no thermobaric effects)
   (c) Vertical coordinate: σ_2^* (potential density referenced to 2000 m depth, thermobaric effects included)

The experiments listed above are being carried out on SGI Origin 2000 computers at Los Alamos National Laboratory, a facility operated by the U.S. Department of Energy. The eddy-resolving (0.225°) experiments involve 20 million grid points and thus progress rather slowly; on 64 processors, the model only runs 90 times faster than the modeled system itself. Thus, a decadal simulation on a dedicated 64-processor machine takes more than one month to complete, which does not leave much room for experimentation. The coarse-mesh (1.4°) experiments run 3600 times faster than

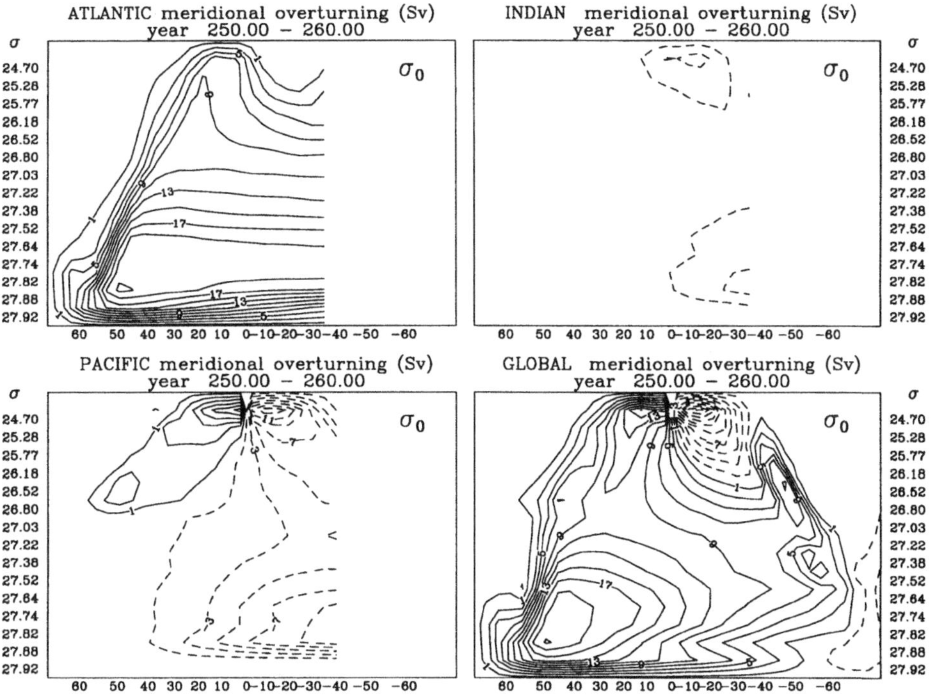

Figure 2. Meridional overturning streamfunction in latitude-density space averaged over model years 250–260 in three individual basins and the global ocean from the $1.4°/\sigma_0$ coordinate model. Contour interval: 2 Sv.

real time on 16 processors; this translates into 10 model years per day, an acceptable speed for century-scale experiments.

Fig. 1 shows the sensitivity of the meridional heat flux to the choice of reference pressure (surface versus 2000 m) used in defining the vertical model coordinate. The biggest differences between the two simulations are seen in the South Pacific and the Southern Ocean where the σ_2 model transports considerably more heat poleward than the σ_0 model.

Figs. 2 and 3 show the meridional overturning motion that accomplishes this heat transport. (Note that horizontal, wind-forced gyre motion becomes part of the "vertical" overturning circulation in density space whenever the pole- and equatorward current branches experience heat loss/gain.) The Atlantic circulation in both figures has the familiar interhemispheric component driven by deep water production near Greenland while the circulation in the Pacific is seen to be dominated by wind-forced equatorial upwelling. The major difference between Fig. 2 and 3 is seen, once again, in the Southern Ocean where deep water production in the σ_2 model drives a thermally direct 13 Sv overturning cell extending into the Pacific. The ther-

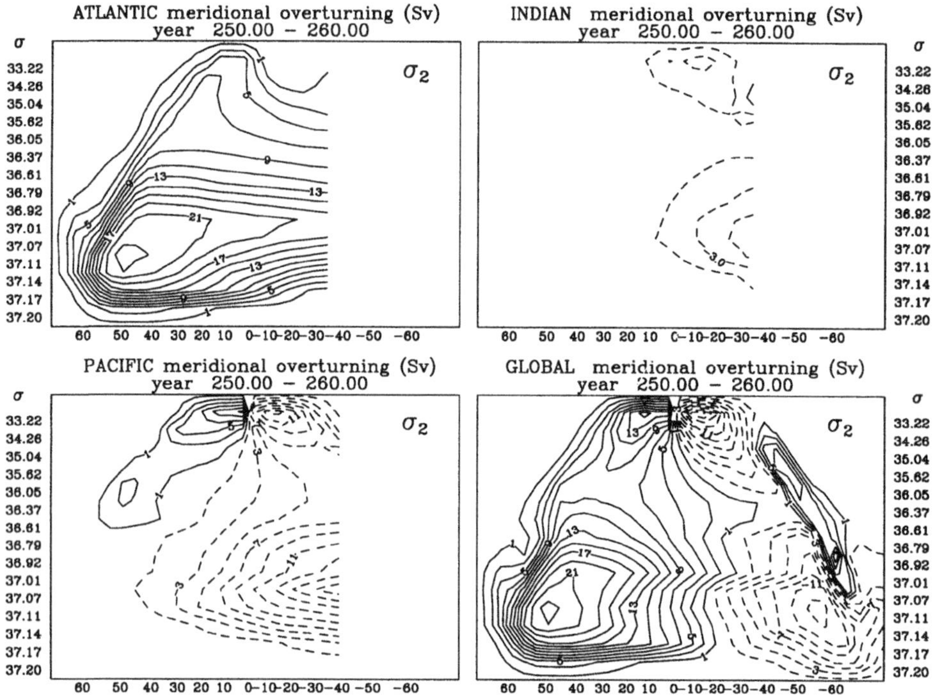

Figure 3. As in Fig. 2, but from the σ_2 coordinate model.

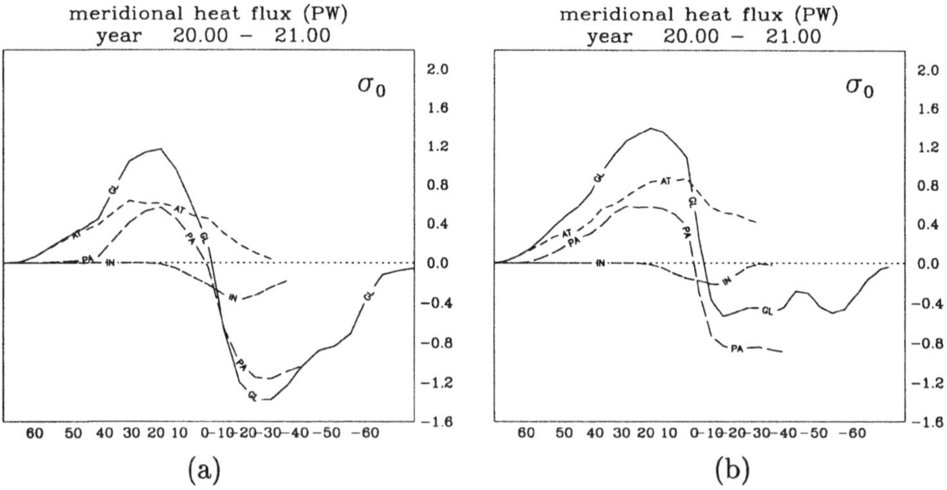

Figure 4. Meridional heat flux (PW) averaged over model year 21 in three individual basins and the global ocean. Vertical coordinate: σ_0. (a) 0.225° model; (b) 1.4° model.

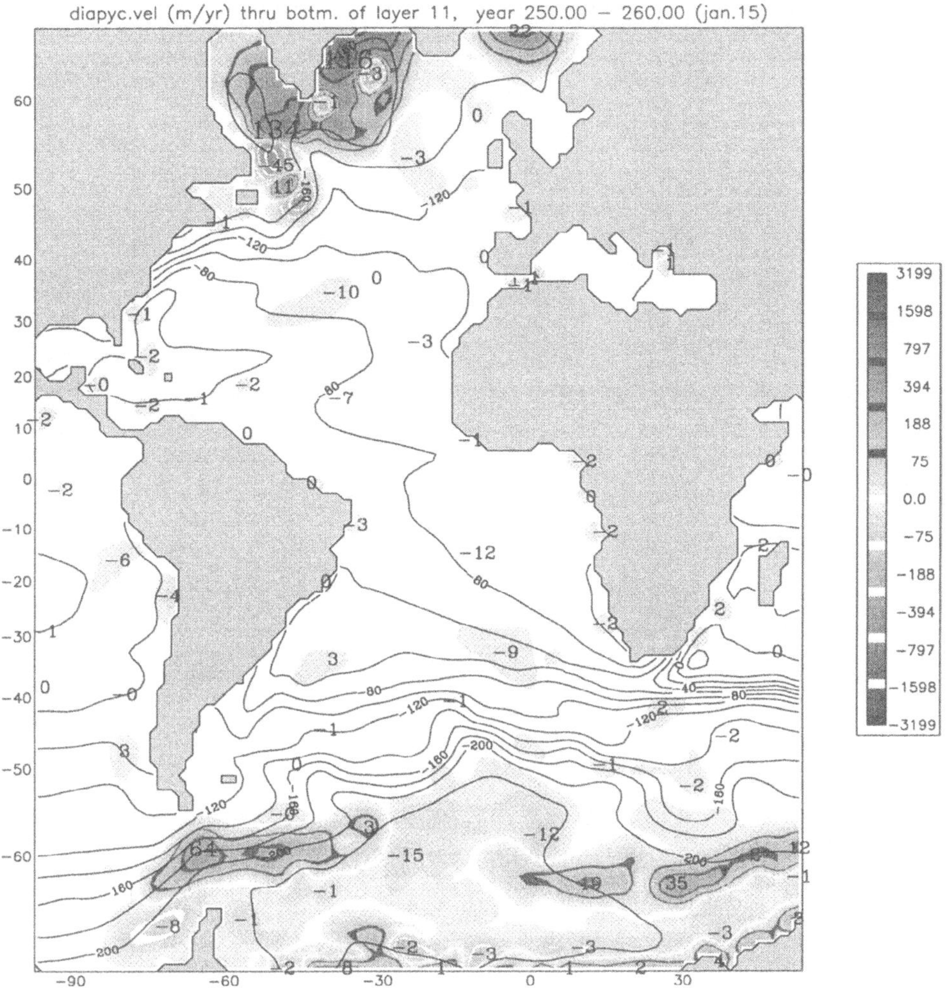

Figure 5. Grey shading: diapycnic flux (m/year) through bottom of model layer 11 (downward motion positive). Numbers: regionally condensed diapycnic fluxes (0.1 Sv). Isolines: sea surface height.

mally indirect near-surface cell between 40°S and 65°S is the Deacon cell created by equatorward Ekman transport across the wind-driven Antarctic Circumpolar Current.

At the time of this writing, the eddy-resolving $0.225°/\sigma_0$ model run has passed the 21-year mark and is therefore not even close to reaching an equilibrium state. One result that seems to emerge is that the meridional heat transport is fairy insensitive to the presence of eddies except in the southern ocean. This is illustrated in Fig. 4 where heat flux curves for model

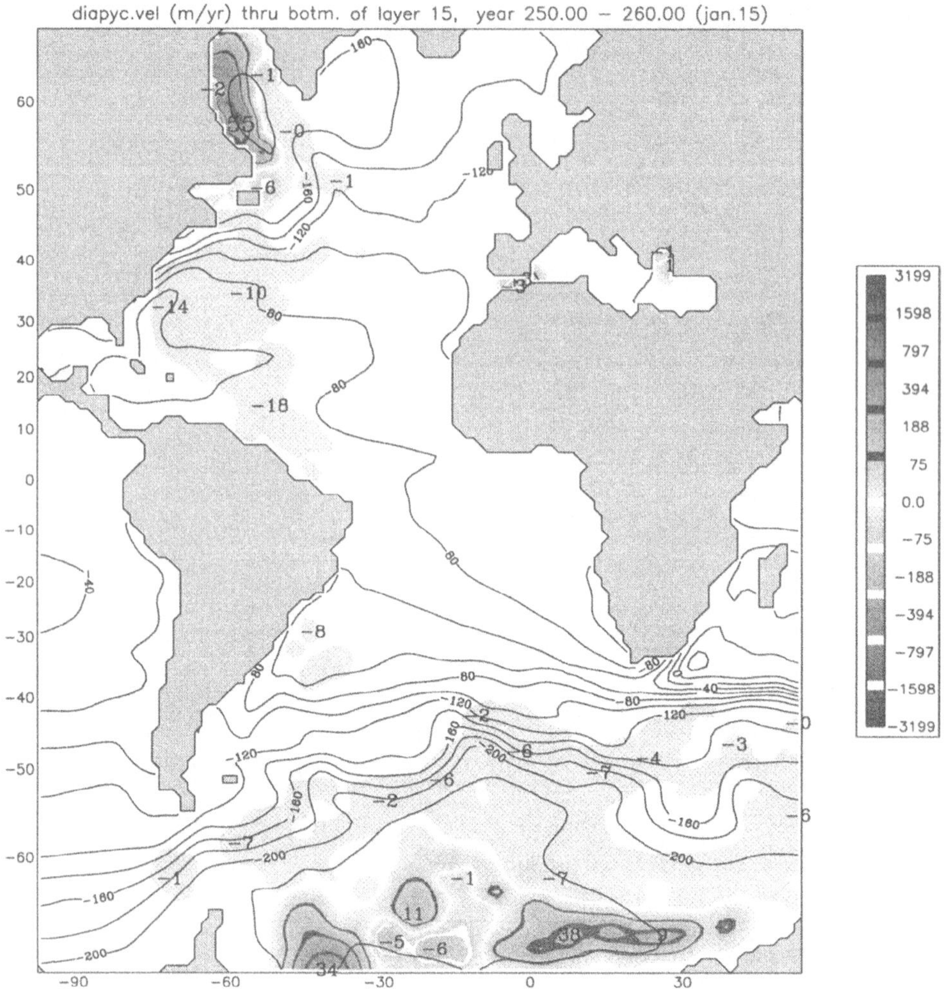

Figure 6. As in Fig. 5, but for bottom of layer 15.

year 12 from the $0.225°/\sigma_0$ and $1.4°/\sigma_0$ experiments are plotted side by side. The heightened sensitivity of the heat flux in the southern ocean to grid size is consistent with the notion that heat transport in basins bounded by meridional barriers is mainly accomplished by mean meridional motion, which coarse-mesh models are able to simulate fairly well, while heat flux across a prediminantly zonal current is accomplished by eddies and hence can only be simulated correctly by an eddy-resolving model. (Note that atmospheric models are *always* eddy-resolving for this reason.)

One of the challenges in ocean modeling is the reduction of the vast

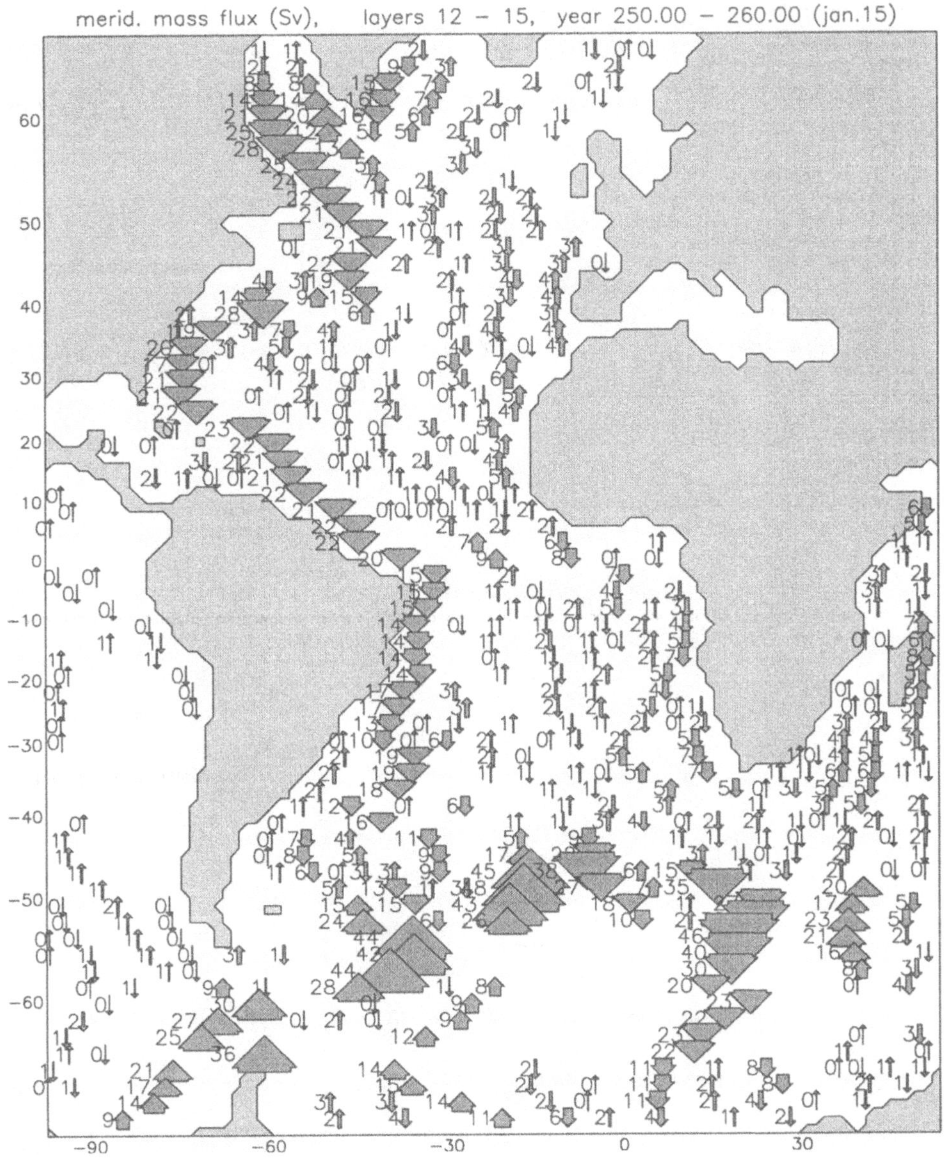

Figure 7. Laterally condensed meridional mass fluxes in layers 12–15 of the $1.4°/\sigma_2$ coordinate model.

amount of computer-generated data for display and diagnostic purposes. The material presented so far gives a highly condensed view of the part played by the ocean in the climate machine. Adding detail to this picture

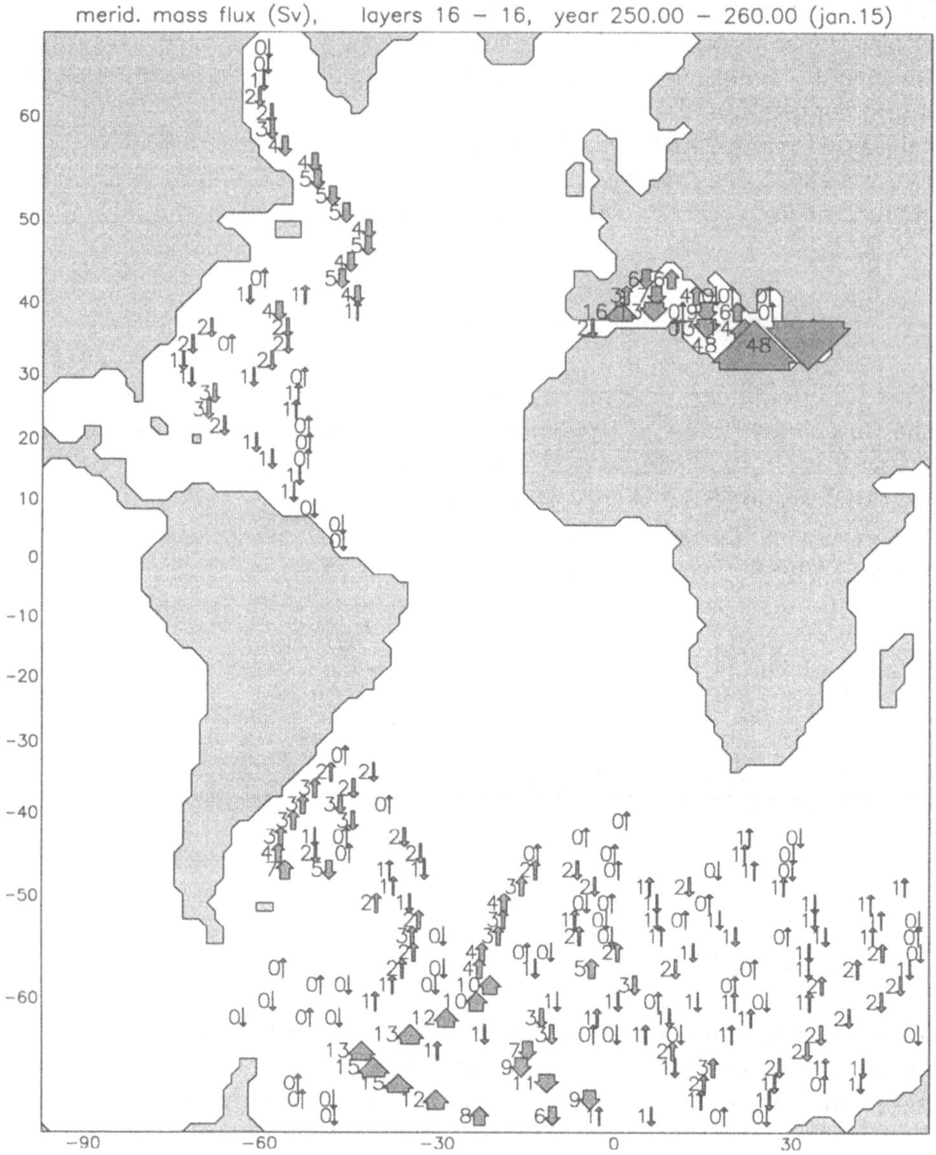

Figure 8. As in Fig. 7, but for layer 16.

without losing the ability to foster understanding and convey quantitative information is a major challenge. The remaining figures illustrate techniques currently being developed in Miami for quantitatively revealing the regional characteristics of the 3-dimensional thermohaline-driven circulation.

Figure 9. Zonal vertical sections at 10°S through the upper 500 m of 2° model ocean. Shaded contours: salinity; unshaded contours: layer interfaces. Top: uncoupled simulation; bottom: coupled simulation.

As long as layer interfaces are defined as constant potential-density surfaces (the reference pressure does not matter here), the 3-dimensional mass flux in an isopycnic model naturally breaks down into isopycnic and diapycnic flux components. Fig. 5 shows the diapycnic flux though the bottom of layer 11 in the Atlantic from the $1.4°/\sigma_2$ simulation. This field, if diagnosed on all interfaces and zonally integrated across the basin, yields the stream functions shown in Fig. 3.

Fig. 5 displays the diapycnic flux field in two ways: as a contoured vertical velocity field (dimensioned length/time), and as a set of spatially distributed discrete flux amounts (dimensioned volume/time). The method used for extracting the latter from the former is analogous to the process of determining the volume of mountains from the terrain height field plotted on a topographic chart. The "mountains" in this case can have positive and negative heights corresponding to subduction/upwelling areas; these must be processed separately. A user-specified parameter regulates the extent to which neighboring mountains (of the same sign) can be lumped together.

Fig. 5 reveals, among other things, that the downward mass flux in the Davis Strait and Denmark Strait combined is 13.4 + 11.6 = 25.0 Sv. Downwelling in the Southern Ocean appears to take place underneath and on the poleward side of the ACC.

The diapycnal flux through the lowest model interface is shown in Fig. 6. We see that 5.5 Sv of bottom water are being produced in the Labrador Current region while net bottom water production along the Atlantic segment of Antarctica is 7.3 Sv (9.2 Sv down, 1.9 Sv up).

What pathways are taken by the currents in the near-bottom layers that are fed by these downwelling fluxes? Figs. 7 and 8 provide a partial answer. The arrows in these figures are constructed by first locating on each parallel the points where the meridional mass flux changes sign, and then summing up the flux values over each interval so defined. Arrows are plotted at flux-weighted positions. Their size gives a rough idea of the magnitude of the mass flux; the exact transport amount is written next to each arrow.

As indicated in Fig. 8 by the decreasing size of the arrows associated with the Labrador Current, layer-16 water produced in the Labrador Sea is absorbed into layers above through diapycnal mixing before reaching the equator. Antarctic bottom water moves north along the South American shelf but returns south before reaching the Rio Grande Rise at 30°S.

The 48 Sv gyre seen in the eastern Mediterranean is a degeneracy resulting from having a basin filled with water whose density lies outside the range of the presently used 15 isopycnic layers.

According to Fig. 7, the North Atlantic Deep Water, whose sources are depicted in Fig. 5, forms a southward-flowing boundary current that can be traced well past the Rio Grande Rise. There, it merges with the deep part of the meandering ACC.

Space limitations do not allow us to show the meridional mass fluxes in layers 1 – 11, as well as a corresponding set of zonal mass fluxes. Sun [24] has taken the next step of distilling the type of information shown in Figs. 5 – 8 into "plumbing diagrams" that show how currents in various basins and in various density classes connect to form the model's representation of the global conveyor belt [7].

5.2. COUPLED COARSE-MESH SIMULATIONS

A near-global MICOM version configured like experiment 2c except for a 2° mesh size has recently been coupled to a T21 version of NCAR's CCM3, and a 50-year simulation starting from Levitus [10] climatology has been completed. None of the measures commonly taken to keep the modeled climate close to the observed one, such as flux corrections or separate spinup phases for the oceanic and atmospheric submodels, are employed at present.

The coupled model quickly settles into a quasi-steady climate state which reproduces most of the basic features of the observed climate. The biggest discrepancy between model climate and reality is found in the subtropical eastern Pacific, where annual sea surface temperature and rainfall patterns in the model are more symmetric with respect to the equator than in the real system. In other words, an ITCZ (intertropical convergence zone) not found in nature develops south of the equator in austral summer, similar to the one north of the equator in boreal summer. The southern ITCZ appears to be caused by anomalously warm surface waters south of the equator.

The reason for this phenomenon, which is also found in coupled simulations employing other models, has not been found yet. It appears to be linked to an El Niño-like, but perpetual, decrease in the strength of the Pacific trade winds. Since surface wind speed enters the formula employed in circulation models for calculating air-sea heat exchange rates, lower wind speeds will tend to lower the rate at which the ocean vents solar heat to the atmosphere. Increasing oceanic temperatures will in turn decrease the mixed-layer depth, accentuating the heat accumulation further.

Fig. 9 shows the effect of this process on the stratification of the upper water column in the south-equatorial Pacific. In an ocean-only simulation (Fig. 9, top) driven by atmospheric data obtained from an independently performed atmosphere-only simulation, the isopycnals in the upper 300 m of the Pacific are steeply inclined in a way that suggests upwelling in the east and wind-driven accumulation of warm water in the west. In the coupled experiment (Fig. 9, bottom), the mixed-layer depth indicated by the uppermost interface is significantly shallower, and there is much less evidence of eastern upwelling and western warm water accumulation.

A further discussion of the factors controlling the strength of the trade wind circulation is beyond the scope of this article.

References

1. Arakawa, A., and Y. G. Hsu, 1990: Energy conserving and potential-enstrophy dissipating schemes for the shallow water equations. *Mon. Wea. Rev., 118*, 1960–1969.
2. —, and V. R. Lamb, 1981: A potential enstrophy- and energy-conserving scheme for the shallow water equations, *Mon. Weather Rev., 109*, 18–36.

3. Bleck, R., 1978: Finite difference equations in generalized vertical coordinates. Part I: Total energy conservation. *Contrib. Atm. Phys.*, *51*, 360–372.
4. —, 1984: An isentropic coordinate model suitable for lee cyclogenesis simulation. *Riv. Meteorol. Aeronaut.*, *44*, 189–194.
5. —, and L. Smith, 1990: A wind-driven isopycnic coordinate model of the north and equatorial Atlantic Ocean. 1. Model development and supporting experiments. *J. Geophys. Res.*, *95C*, 3273–3285.
6. —, C. Rooth, D. Hu, and L. Smith, 1992: Salinity-driven thermocline transients in a wind- and thermohaline-forced isopycnic coordinate model of the North Atlantic. *J. Phys. Oceanogr.*, *22*, 1486–1505.
7. Broecker, W. S., 1991: The great ocean conveyor. *Oceanography, 4*, 79 – 89.
8. Kraus, E.B., and J.S. Turner, 1967: A one-dimensional model of the seasonal thermocline: II. The general theory and its consequences. *Tellus, 19*, 98 – 106.
9. Levitus, S., 1982: *Climatological Atlas of the World Ocean.* NOAA Professional Paper 13, 173 pp.
10. Levitus, S., 1994: Revised version of [9].
11. McDougall, T., 1987: Neutral Surfaces. *J. Phys. Oceanogr.*, *17*, 1950 – 1964.
12. Oberhuber, J.M., 1988: An atlas based on the 'COADS' data set: the budgets of heat, buoyancy, and turbulent kinetic energy at the surface of the global ocean. *Max-Planck-Institut für Meteorologie*, Hamburg, 202 pp. (ISSN 0937-1060).
13. Oberhuber, J. M., 1993: Simulation of the Atlantic circulation with a coupled sea ice-mixed layer-isopycnal general circulation model. Part I: model description. *J. Phys. Oceanogr.*, *23*, 808 – 829.
14. Phillips, N.A., 1957: A coordinate system having some special advantages for numerical forecasting. *J. Meteor.*, *14*, 184 – 185.
15. Pingree, R. D., 1972: Mixing in the deep stratified ocean. *Deep-Sea Res.*, *19*, 549 – 561.
16. Redi, M. H., 1982: Oceanic isopycnal mixing by coordinate rotation. *J. Phys. Oceanogr.*, *12*, 1154 – 1158.
17. Reid, J. L., and R. J. Lynn, 1971: On the influence of the Norwegian-Greenland and Weddell seas upon the bottom waters of the Indian and Pacific oceans. *Deep-Sea Res.*, *18*, 1063 – 1088.
18. Richardson, L. F., 1922: *Weather prediction by numerical process.* Cambridge Univ. Press, reprinted Dover 1965, 236 pp.
19. Rooth, C. G., S. Sun, R. Bleck, and E. P. Chassignet, 1998: Note on the inclusion of themobaricity in numerical ocean models framed in isopycnic coordinates. *J. Phys. Oceanogr.*, subm.
20. Sadourny, R., The dynamics of finite-difference models of the shallow-water equations, *J. Atmos. Sci.*, *32*, 680–689, 1975.
21. Shuman, F. G., 1960: Numerical experiments with the primitive equations. *Proc. Intern. Symp. Num. Wea. Pred.*, Tokyo, 85 – 107.
22. Smolarkiewicz, P. K., and W. W. Grabowski, 1990: The multidimensional positive definite advection transport algorithm: Nonoscillatory option. *J. Comput. Phys.*, *86*, 355-375.
23. Spencer, R. W., 1993: Global Oceanic Precipitation from the MSU during 1979-91 and Comparisons to other Climatologies. *J. Climate, 6*, 1301 – 1326.
24. Sun, S., 1997: Compressibility effects in the Miami Isopycnic Coordinate Ocean Model. *Ph.D. Diss.*, University of Miami, 138pp.
25. —, R. Bleck, and E. Chassignet, 1993: Layer outcropping in numerical models of stratified flows. *J. Phys. Oceanogr.*, *23*, 1877–1884.
26. Woodruff, S.D., R.J. Slutz, R.L. Jenne, and P.M. Steurer, 1987: A comprehensive ocean–atmosphere data set. *Bull. Amer. Meteor. Soc.*, *68,* 1239 – 1250.
27. Zalesak, S., 1979: Fully multidimensional flux-corrected transport algorithms for fluids. *J. Comput. Phys.*, *31*, 335–362.

INDEX

E. P. Chassignet and J. Verron (eds.), Ocean Modeling and Parameterization, 449–451.

Batch number: 10799853

Printed by Printforce, the Netherlands